D0907150

The Physics of
Micro/Nano-Fabrication

MICRODEVICES
Physics and Fabrication Technologies

Series Editors: Ivor Brodie and Julius J. Muray†

SRI International
Menlo Park, California

ELECTRON AND ION OPTICS
Miklos Szilagyi

GaAs DEVICES AND CIRCUITS
Michael Shur

ORIENTED CRYSTALLIZATION ON AMORPHOUS SUBSTRATES
E. I. Givargizov

THE PHYSICS OF MICRO/NANO-FABRICATION
Ivor Brodie and Julius J. Muray

PHYSICS OF SUBMICRON DEVICES
David K. Ferry and Robert O. Grondin

THE PHYSICS OF SUBMICRON LITHOGRAPHY
Kamil A. Valiev

SEMICONDUCTOR LITHOGRAPHY
Principles, Practices, and Materials
Wayne M. Moreau

SEMICONDUCTOR PHYSICAL ELECTRONICS
Sheng S. Li

† *Deceased.*

A Continuation Order Plan is available for this series. A continuation order will bring delivery of each new volume immediately upon publication. Volumes are billed only upon actual shipment. For further information please contact the publisher.

The Physics of Micro/Nano-Fabrication

Ivor Brodie and
Julius J. Muray[†]

SRI International
Menlo Park, California

Plenum Press • New York and London

Library of Congress Cataloging-in-Publication Data

Brodie, Ivor.
   The physics of micro/nano-fabrication / Ivor Brodie and Julius J.
Muray.
      p.   cm. -- (Microdevices)
   Includes bibliographical references and index.
   ISBN 0-306-44146-2
   1. Microelectronics. 2. Thin-film circuits.   I. Muray, Julius J.
II. Series.
TK7874.B725 1992
621.381--dc20   92-28941
                                               CIP

ISBN 0-306-44146-2

© 1992 Plenum Press, New York
A Division of Plenum Publishing Corporation
233 Spring Street, New York, N.Y. 10013

Printed in the United States of America

Dedicated to the memory of

Julius J. Muray
1931–1991

Preface

In assembling the material for *The Physics of Microfabrication* a decade ago, it became clear that the field of building ultrasmall devices was in a state of rapid change. People not only were striving to build microdevices in fields other than microelectronic circuits, such as micro-optical, micromechanical, and microchemical devices, but, perhaps more important, also were striving toward achieving atomic tolerances (less than one nanometer). At the time we felt that if we wrote a book that was restricted to the physical basis of microfabrication, it should be able to cope with *evolutionary* changes in microfabrication processes and thereby not date too quickly. The intervening years, however, (perhaps not unexpectedly) brought with them important *revolutionary* developments in methods for making and viewing ultrasmall devices, reaching well into the nanometer scale—hence the small change in the title of the present volume. Outstanding among these new technologies was the scanning tunneling microscope (STM) and variants thereof, which have effectively brought atomic resolution within our grasp, the expanded use of plasmas for micro/nano-fabrication, and the marriage of organomolecular engineering with more conventional planar technologies through Langmuir–Blodgett films, not only as resists or insulators but as membranes for the control of chemicals that may penetrate or stimulate the membrane in the same manner as the bilipid layer that forms the biological cell wall.

The technologies that we currently use are still largely based on removing material from a larger block, much as a sculptor carves a statue. However, we may now look forward to significant advances in molecular engineering technologies[1, 2]* that will enable us to construct our devices atom by atom,[3] much as a builder constructs a house brick by brick or as nature evolves living cells. It may be argued that molecular engineering has already been employed for simple applications, such as monomolecular film devices[4] and superlattice fabrication.[5]

In living cells, nature has given us extraordinarily diverse and intricate devices operating on the molecular level that we may attempt to emulate. The physics of how these devices function must be better understood before this can be done. Gaining this understanding will require the use of new microinstrumentation

*References cited in the Preface are included with those for Chapter 1.

vii

techniques[6–10] and a more sophisticated physical interpretation[11] of the measurements so obtained.

The Physical Electronics Laboratory of SRI International (formerly Stanford Research Institute) has been pioneering the development of microdevices for over thirty years. In 1961, a landmark paper on electron-beam lithography and its associated technologies was published by K. R. Shoulders (then at SRI), which set the stage for subsequent advancements in this field.[12] He had the foresight to believe that the building of submicron-sized devices to nanometer tolerances was actually within the range of human capabilities using electron-beam lithographic methods.

As with our earlier volume, the intent of the book is to give the uninitiated a basic physical understanding of the tools, technologies, and physical models needed to build, analyze, and understand the mechanism of microdevices, to see what physical limitations may be expected, and to put numbers to the magnitudes involved. Throughout the text we have emphasized microelectronic devices based on planar silicon technology to illustrate how the techniques of microfabrication may be applied. This class of devices was chosen because it has had such a major impact on our everyday life and economy, and provided the impetus and resources for the evolution of microfabrication technologies. Our intention, however, is to encourage workers in related fields to embark on building micro/nano-structures applicable to their needs, and convince them that this task may not be as difficult as they have imagined.

We have provided references that we hope will be useful to the reader who wishes to follow up on a particular subject or who is interested in the historical aspects. As a general rule, we have used SI units in the equations, unless some other units are specified. We have found that workers in a given specialty tend to develop their own mix of units that are used in common parlance. We have used their units, at least for educational purposes, as appropriate.

The field covered by this volume is so broad that we recognize that it cannot be completely up-to-date; and we may have unintentionally failed to properly reference works or papers of significance, for which we apologize in advance.

We wish to acknowledge the assistance provided by past and present members of the Physical Electronics Laboratory scientific staff. We appreciate the support of SRI International management in allowing us to use SRI facilities and materials for the preparation of this book. Special thanks are due to Professor Peter J. Hesketh of the University of Illinois in Chicago for providing the questions given at the end of each chapter, and Dr. J. Macaulay for a critical reading of the entire manuscript of the volume.

We also wish to express our appreciation to our administrative assistant, Joyce Garbutt, for carrying out the arduous task of typing the manuscript, and our publications group for editing, revisions, and illustrations. We thank Plenum Press for commissioning this work. And finally, we thank our wives and families for their support of our efforts.

This volume differs from *The Physics of Microfabrication* in that we have taken the opportunity to add a chapter on plasma physics, rearrange some of the material for a more logical approach, and update the text to include the important advances toward nanofabrication that have taken place since the earlier manuscript was prepared in 1981.

After the manuscript for this book was completed, my co-author, Dr. Julius J. Muray, died unexpectedly in March 1991. May this book stand as a memorial to his scholarship and his broad vision of the world of ultrasmall devices and their fabrication.

I. Brodie

Menlo Park, California

Contents

CHAPTER 3. Plasmas: Physics and Chemistry

CHAPTER 5. Pattern Generation

CHAPTER 6. Microcharacterization

CHAPTER 7. Limits to Nanofabrication

1

Preliminary Survey

1.1. INTRODUCTION

This chapter is intended to enable those readers with no previous experience in the subject matter to gain a brief overview of the evolution of microfabrication technologies from the needs of the electronic microcircuit industry, which was born with the invention of the transistor in late 1947.[13–16]*

Germanium[17,18] was the first semiconducting substrate material used in the early evolutionary stages of device development. Developments in germanium technology (single-crystal technology,[19] controlled doping,[20] purification of semiconductor crystals,[21] and alloying[22]) made possible a number of devices, including the point-contact germanium transistor and the grown-junction and alloy-germanium transistors.

However, germanium was soon replaced by silicon, which has the unique property that in an oxidizing atmosphere a thin, strongly bonded, impervious layer of amorphous silica (SiO_2) is formed. The silica film is important because it can be used both as a protective coating over silicon surfaces (or encapsulation layer) and as a dielectric insulation between circuit functions.

Gallium arsenide (GaAs) has emerged as an important semiconducting material primarily because of its high electron mobility, but also because it is a direct band gap material. Progress toward its more general use in microelectronics has been inhibited by technological difficulties in growing large, defect-free, single crystals and in forming insulating layers. At present, microelectronics remains dominated by silicon, except for those classes of devices where the special properties of GaAs are necessary.

The key concept that enabled the next evolutionary steps to take place was that of planar processing, or planar technology. The idea here was to fabricate patterned layers, one on top of the other, made of materials with different electrical properties. Together the multidecked sandwich of patterned layers is made to form various circuit elements such as transistors, capacitors, and rectifiers, and these are finally

*References 1–12 are cited in the Preface.

1

interconnected by a patterned conducting overlayer to form an integrated circuit (IC).

The layers with different electrical properties can be formed by modifying the substrate, for example, by doping or oxidizing it, or by depositing a layer from an external source, for example, by evaporation or sputtering. Patterning is usually accomplished by the process of lithography. In photolithography the pattern in the form of a photographic transparency is projected onto a surface that has been precoated with a photoresist layer. Photoresist materials have two properties: First, when exposed to light, their solubility in one class of solvents is changed, so that after immersion in such a solvent the projected pattern is replicated in the surface; second, the undissolved regions of the resist (sometimes after a "hardening process") are completely unaffected by ("resist") a second class of solvents, which are able to etch or modify the underlying material. In a "positive" resist the exposed areas become impervious and in a "negative" resist it is the unexposed areas that become impervious.

If these two properties cannot be found in the same material, it may be necessary to pattern an intermediate layer that does have the required resist properties to a particular substrate etching or modifying process. For example, dopants that are pervious or interact with an organic resist are often impervious to silicon dioxide. Thus, an SiO_2 mask is first patterned on the surface and the dopant is subsequently diffused into the silicon substrate through the open areas, forming patterns of doped silicon in the substrate. Some basic steps of the planar process are illustrated in Fig. 1.1. Complete microelectronic devices are made by repetitive application of such processes to build up the layered structures required.

Planar processing also contributes to the economic volume production of integrated circuits. This is because a substantial number of circuits can be fabricated in parallel on a large-area substrate (or "Wafer"). After fabrication, the circuits (dice) are separated by cutting (or "dicing") the wafer. External leads are then connected to each circuit and "packaged" for insertion into circuit boards. Wafers in the range of 100 to 200 cm (4 to 8 inch) in diameter are conventionally used, with dice sizes in the range of 3–10 mm.

1.2. MICROELECTRONIC DEVICES

1.2.1. Background

The physics of semiconductor device operation and the design of integrated circuits have been well treated in the literature.[23,24] The purpose of this section is simply to describe the more common devices at a level necessary for an understanding of the problems encountered in their fabrication.

Semiconducting solids have important physical properties that are used in the operation of electronic devices. A perfect impurity-free semiconductor at low temperature has a filled set of energy states (the valence band) separated from the nearest unfilled states (the conduction band) by an energy gap. Given this circumstance, the material is actually an insulator, since all of the states in the valence band are filled. Thus, the valence electrons are not free to move in response to an applied electric

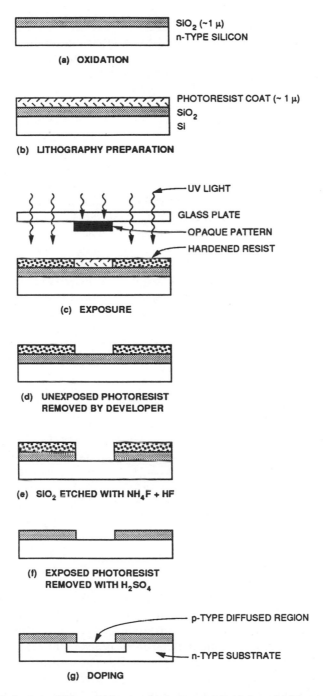

FIGURE 1.1. Basic steps of lithographic microfabrication. (a) Oxidation; (b) lithography preparation; (c) exposure; (d) unexposed photoresist removed by developer; (e) SiO_2 etched with $NH_4F + HF$; (f) exposed photoresist removed with H_2SO_4; (g) doping.

field, and there are no electrons in the conduction band. However, if an electron is excited from the valence to the conduction band, both the electron in the conduction band (a negative carrier) and the resultant hole in the valence band (which behaves like a positive carrier) are free to respond to an electric field and conduct current. Semiconductors differ from insulators in three ways. They have moderate to high electron and hole mobilities, their carrier concentrations can be modified in a controlled fashion by the addition of selected impurities (donors or acceptors), and it is possible to make nonrectifying (ohmic) contact to them with metals.

Semiconductors are classified as n type or p type, depending on whether their conductivity is due to a majority of negative-charge carriers (electrons) or positive-charge carriers (holes). The interface (or junction) between n- and p-type materials is rectifying, since in one direction an applied electric field will drive both carriers toward the junction region where they can recombine, thus allowing current to flow; whereas in the reverse direction the field will drive the majority carriers away from the junction until the internal field balances the applied field, thus limiting the current flow to that due to recombination of the minority carriers. By controlling the purity of a semiconductor material, the number density of majority charge carriers within it can be varied from the very small (intrinsic) to very large. In silicon, number densities greater than $10^{18}/cm^3$ are designated n^+ or p^+. Even in lightly doped silicon the ratio of majority to minority charge carriers is greater than 10^8.

A key device in microelectronic applications is the transistor. This is essentially a device for controlling the current through a circuit by supplying a relatively small current or voltage to an auxiliary electrode.

1.2.2. Bipolar Transistors

The bipolar transistor (Fig. 1.2) consists of a sandwich of two back-to-back pn junctions that share a thin common region called the base (B). The outer regions, which are of the same conductivity type, are called the emitter (E) and the collector (C). A voltage is applied between E and C so as to drive the emitter charge carriers toward the collector. If no current is allowed to flow in the E–B circuit, a retarding field builds up at the B–E junction that prevents the majority charge carriers from E penetrating into the B region. However, if a voltage is applied to B such as to allow current to flow in the E–B circuit, the retarding field is diminished and a large number of majority charge carriers can flow from E into the B region. Most of these charges are injected into the B region, where they are minority carriers, are able to diffuse to C before they recombine with the majority carriers flowing into B. Thus, a small flow of charge carriers from B to E gives rise to a large flow of (oppositely charged) carriers from E to C. The term *bipolar* derives from the fact that the operation of the device depends on the flow of both signs of charge carrier.

The cross section of an *npn* bipolar transistor is shown in Fig. 1.2a. The p-type base is formed by diffusing boron into n-type silicon. Since boron is trivalent, only three electrons are available to bond to its four adjacent silicon atoms. This local site tends to bind an electron from the valence band, leaving it with a hole or vacancy, thus making the material p-type.

The emitter is formed by diffusing atoms with five valence electrons (phosphorus) into the p-type base. Since only four of the phosphorus electrons are needed

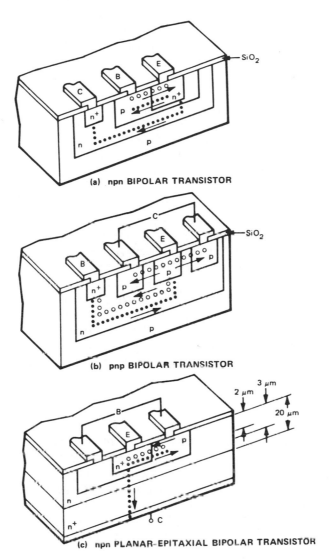

FIGURE 1.2. Bipolar transistors. (a) *npn* bipolar transistor; (b) *pnp* bipolar transistor; (c) *npn* planar epitaxial bipolar transistor. ●, electrons; ○, holes.

to bond to the four surrounding silicon atoms, one electron is easily freed to conduct electric current (*n*-type semiconductor).

For simplicity in manufacturing, a *pnp* transistor (Fig. 1.2b) is generally constructed according to a plan in which the E–C current is lateral rather than vertical. The principles of operation are the same as for an *npn* transistor, except that all of the polarities are reversed. Conections to the circuit elements of both *npn* and *pnp* transistors are made through aluminum conductors deposited over an insulating layer of SiO_2.

A significant development in bipolar technology came with the ability to form a thin layer of high-resistivity silicon over low-resistivity silicon by epitaxial deposition (see Section 1.3.2). The resulting transistor structure has substantially lower collector series resistance and much less stored charge than do conventional diffused devices. The high-resistivity layer significantly increases the switching speed and high-frequency gain. Figure 1.2c shows the cross section of the epitaxial diffused bipolar transistor.

1.2.3. MOS Devices

In 1960 the first metal–oxide–semiconductor (MOS) transistor using silicon was reported.[25] The principle of the MOS transistor was actually conceived in 1930,[26] but because of materials difficulties, they could not be made economically until the early 1960s.[27]

The MOS field-effect transistor[28] (MOSFET, Fig. 1.3) consists of two small highly conductive regions of the same charge carrier type, called the source (S) and the drain (D), which are embedded in a substrate of opposite charge carrier type. The region between S and D (the channel) is covered with a thin layer of insulating material (oxide) that supports a metal electrode called the gate (G). If a voltage is applied between S and D in a direction to drive the charge carriers from S to D, the substrate takes the potential of S, but current cannot flow, as the substrate to D junction is rectifying. In the enhancement mode of operation, a voltage is applied between G and S, thus generating an electric field in the channel region. Depending on the direction of the field, charge carriers of either sign may be forced into the thin channel region near the surface. If these charge carriers are of the same sign as those of S and D, current can "channel" between them. Since the oxide is thin, large fields can be generated in the channel region, with relatively small G voltages for switching current between S and D.

The detailed mechanism for channel formation is complex, but in principle may be understood by first considering the case where S and D are p-type and the channel region is n-type. If the G voltage is negative, the electrons just below G are forced away from the channel region, leaving behind a net positive space charge (a depletion region). The electron energy states are bent upward in this positive space charge region. When the valence band at the semiconductor–insulator interface is raised near the energy of the bulk conduction band edge, electrons can be thermally excited from the valence band at the interface to the bulk conduction band. This leaves a p-type channel (called an inversion layer) at the interface that allows a hole current to flow between the p-type source and drain. In the depletion mode of operation, the thin channel region connecting S and D is made of material of the same charge carrier type as that of S and D. Application of the appropriate electric field depletes the channel of charge carriers, thus disconnecting S from D.

A cross section of an n-channel or enhancement-mode MOSFET is shown in Fig. 1.3a, b. Its starting material is p-type silicon, into which two n^+ "pockets" have been diffused to form the source and drain regions of the transistor. The aluminum gate is separated from the silicon by a thin layer of SiO_2, usually between 90 and 120 nm in thickness. It is from this multilayer Al–SiO_2–Si that the term *MOS* originated.

In MOSFET devices, the thickness and dielectric properties of the SiO_2 film greatly influence the basic characteristics of the transistor. The lack of consistent quality control over these films and the wide resultant distribution of the threshold

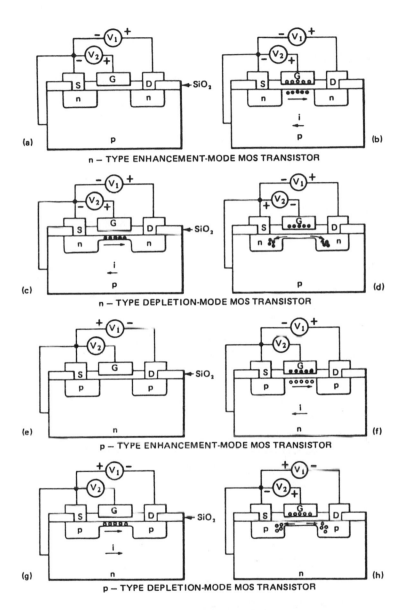

FIGURE 1.3. Field effect transistors. When no voltage is applied to an *n*-type enhancement-mode MOS transistor (a), the drain to substrate *pn* junction is reverse biased and prevents electrons from flowing from the source to the drain. Applying a negative voltage to the gate electrode (b) induces a "channel" of *p*-type substrate below it. Current can flow from the source to the drain via this channel. In an *n*-type depletion-mode MOS transistor (c), there is a continuous channel of *n*-type silicon so that electrons flow from the source to the drain. The transistor normally conducts. When a negative voltage is applied to the gate electrode (d), electrons are expelled from the channel and the transistor no longer conducts. By doping islands of *p*-type material in an *n*-type substrate, the corresponding *p*-type MOS transistors can be made (e, f, g, h).

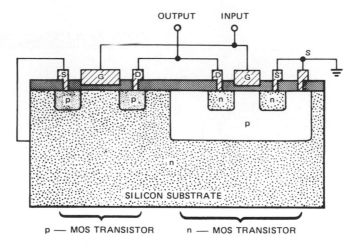

FIGURE 1.4.   Complementary MOS integrated circuits.

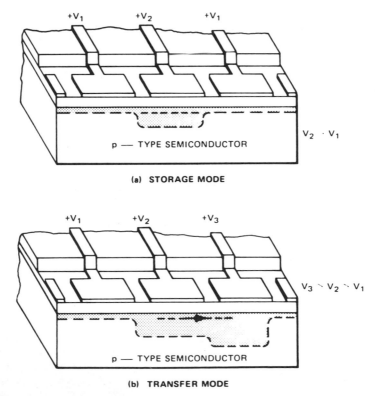

FIGURE 1.5.   Transfer mechanism in a charge-coupled device (CCD). In the storage mode (a), charge is held under the center gate. Application of $V_3 > V_2$ on the right-hand gate causes transfer to the right, as shown in (b), the transfer mode.

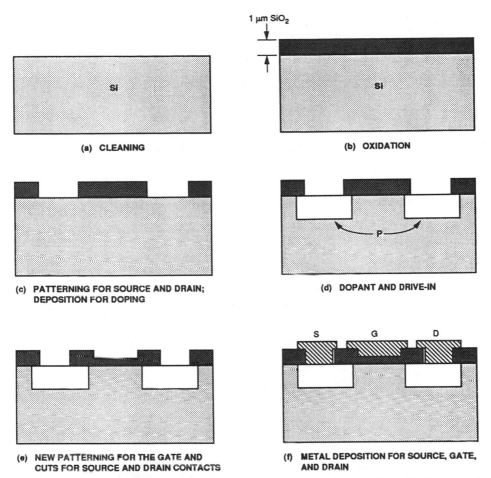

FIGURE 1.6. Process sequence for the fabrication of a p-channel MOS transistor. (a) Cleaning; (b) oxidation; (c) patterning for source and drain; deposition for doping; (d) dopant and drive-in; (e) new patterning for the gate and cuts for source and drain contacts; (f) metal deposition for source, gate, and drain.

voltages are the primary reasons why MOS technology was not employed until the 1960s.

The depletion-mode n-MOSFET is shown in Fig. 1.3c, d. By forming pockets of p-type material in an n-type substrate, the corresponding p-MOS devices can be constructed, as in Fig. 1.3e, h.

A complementary MOS device (CMOS) (Fig. 1.4) contains both n- and p-MOS transistors on a single chip of silicon, forming a balanced circuit. Due to their low-power dissipation, complementary ICs are more popular. However, new configurations are always being explored for large-scale integration (LSI) applications.

MOS technology can be used to make high-quality, precisely controlled capacitors. Charge-coupled devices (CCDs) consist simply of closely spaced arrays of MOS capacitors,[29] as shown in Fig. 1.5. However, the capacitors are built close enough to one another so that the free charge stored in the inversion layer can be transferred to the channel region of the adjacent device. The exchange of charge is controlled by the voltages applied to the metal gates of the MOS capacitors. CCDs have made significant improvements to memory circuits, signal-processing circuits, and TV cameras.

In Fig. 1.6 the process sequence for the fabrication of a p-channel MOS transistor is shown. As a first step, the n-type silicon wafer is cleaned to remove contamination and then oxidized to a thickness of approximately 1 μm. A pattern is then cut through the SiO$_2$ by means of lithography and p-type dopant is deposited only through the "window" in the oxide film. This dopant is then diffused into the silicon

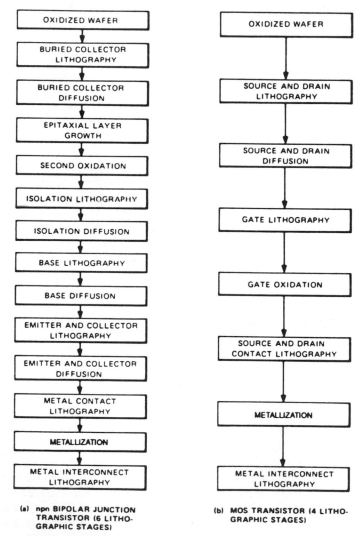

(a) npn BIPOLAR JUNCTION
TRANSISTOR (6 LITHO-
GRAPHIC STAGES)

(b) MOS TRANSISTOR (4 LITHO-
GRAPHIC STAGES)

FIGURE 1.7. Comparison of steps required for fabrication of an *npn* bipolar junction transistor and a MOS transistor. (a) *npn* bipolar junction transistor (six lithographic stages); (b) MOS transistor (four lithographic stages).

and the wafer is again oxidized. After another oxide patterning, a thin (100 nm) gate oxide is grown on the uncovered silicon surface. The oxide mask is then removed in the areas where contact is to be made between the p^+ silicon and aluminum metal conductors. Finally, aluminum is evaporated onto the wafer and patterned to form contacts to the source, gate, and drain.

Because MOS devices do not have to be isolated from each other, MOS transistors can be packed more densely than bipolar transistors, which require isolation. Furthermore, the manufacture of MOS ICs requires fewer steps than do the bipolar ICs (see Fig. 1.7). For these reasons, MOS technology has come to dominate the manufacture of large-scale ICs (10^4–10^8 transistors per chip).

1.3. PLANAR PROCESSING

1.3.1. Background

Planar technology requires being able to create patterned thin layers of materials with various electrical properties on a supporting substrate, to form a connected set of devices. In this section we introduce the layering technologies of epitaxy, oxidation, vacuum evaporation, sputtering, and surface doping, the pattern generation techniques of lithography, and the selective layer removal techniques of etching.[30]

The fabrication steps of a planar p–n diode are shown in Fig. 1.8. Using an n^+ silicon wafer as the starting substrate, a thin layer of n silicon is grown by the epitaxial process. The dopant atoms are added by depositing them on the silicon surface and then diffusing them into those regions of the silicon that are not protected by a layer of SiO_2. The precise delineation or patterning of the SiO_2 film is accomplished lithographically. Since the dopant atoms are introduced from the surface and typical diffusions are on the order of a few microns or less, the active region of the diodes is within a few microns of the surface of the silicon wafer. The remainder of the wafer thickness (typically 560 μm) serves simply to support the surface structure.

An important aspect of planar technology is that each step of the fabrication process is applied to the entire wafer at the same time. This simultaneous fabrication of several semiconductor devices in the same piece of semiconductor produces an IC in each silicon chip and many ICs (several hundred) per wafer.

1.3.2. Layering Technologies

1.3.2.1. Epitaxy. Epitaxy (from the Greek word meaning "arranged upon") describes the growth technique for depositing atoms upon a crystalline substrate so that the lattice structure of the newly grown film duplicates that of the substrate. The key reason for using this growth technique is the ability to grow an extremely pure layer while retaining control of the dopant level. The dopant in the film can be n- or p-type and is independent of the substrate doping. The three types of epitaxial growth processes currently employed are vapor-phase (VPE), liquid-phase (LPE), and molecular beam epitaxy (MBE).

A vapor-phase epitaxial growth system is shown in Fig. 1.9a. Hydrogen gas containing a controlled concentration of $SiCl_4$ is fed into a reactor containing silicon

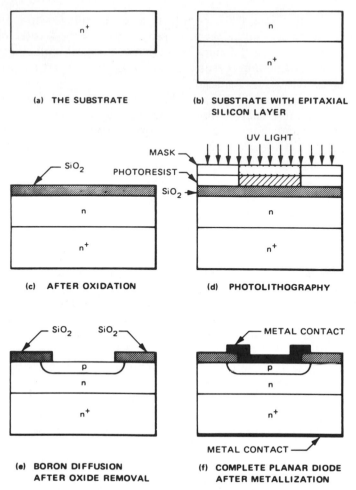

FIGURE 1.8. Fabrication steps of a planar p–n diode. (a) The substrate; (b) substrate with epitaxial silicon layer; (c) after oxidation; (d) photolithography; (e) boron diffusion after oxide removal; (f) complete planar diode after metallization.

wafers in a graphite susceptor. The graphite is inductively heated by a radio-frequency (rf) coil to a high temperature (>1000°C). The high temperature is necessary to permit the deposited atoms to find their proper position in the lattice in order to maintain a single-crystal film. The basic reaction is given by

$$SiCl_4 + 2H_2 \leftrightarrow Si \text{ (solid)} 4HCl$$

This reaction is reversible. The forward reaction produces an epitaxial film on the silicon. The reverse reaction removes or etches the substrate.

The growth rate of the film as a function of the concentration of $SiCl_4$ in the gas is shown in Fig. 1.9b. Notice that the growth rate reaches a maximum and decreases as $SiCl_4$ concentration is increased. This effect is caused by the competing

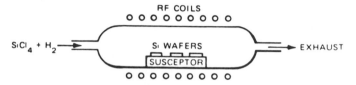

(a) SCHEMATIC OF A VAPOR-PHASE EPITAXIAL GROWTH SYSTEM

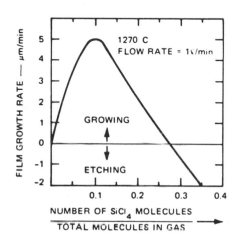

(b) GROWTH RATE OF EPITAXIAL FILM AS A FUNCTION OF PERCENTAGE CONCENTRATION OF SiCl₄ IN GAS

FIGURE 1.9.   Vapor-phase epitaxy (VPE). (a) Schematic of a vapor-phase epitaxial growth system; (b) growth rate of epitaxial film as a function of percentage concentration of $SiCl_4$ in gas.

chemical reaction

$$SiCl_4 + Si(solid) \rightarrow 2SiCl_2$$

Thus, etching of silicon occurs at high concentrations of $SiCl_4$.

Impurity atoms are introduced in the gas stream to grow a doped epitaxial layer. Phosphine (PH_3) is used for n-type doping and diborane (B_2H_3) is used for p-type doping.

For depositing multilayers of different materials on the same substrate, LPE is used. Figure 1.10 shows an LPE apparatus for the epitaxial growth of four different layers. In operation, the sliding solution holder is moved to bring the substrate in contact with the solute. With this method, junctions of different materials (Ge–Si, GaAs–GaP), i.e., heterojunctions, can be fabricated where the layer thicknesses are less than 1 μm.

MBE achieves crystal growth in an ultrahigh vacuum (UHV) environment through the reaction of multiple molecular beams with a heated single-crystal substrate. This process is illustrated in Fig. 1.11, which shows the essential elements for MBE of doped ($Al_xGa_{1-x}As$). Each furnace contains a crucible, which in turn

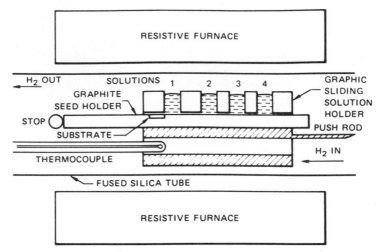

FIGURE 1.10.   Schematic of liquid-phase epitaxial reactor.

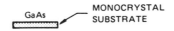

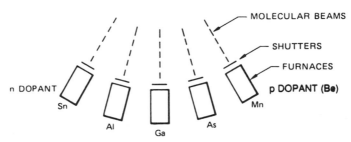

FIGURE 1.11.   Schematic of molecular-beam epitaxial furnaces.

contains one of the constituent elements of the desired film. The temperature of each furnace is chosen so that the vapor pressures of the materials are sufficiently high for free evaporation generation of thermal-energy molecular "beams." The furnaces are arranged so that the central portion of the beam flux distribution from each furnace intersects the substrate. By choosing appropriate furnace and substrate temperatures, epitaxial films of the desired chemical composition can be obtained. Additional control over the growth process is achieved by individual shutters interposed between each furnace and the substrate. Operation of these shutters permits abrupt cessation or initiation of any given beam flux to the substrate.

One of the distinguishing characteristics of MBE is the low growth rate: approximately 1 μm/h or, equivalently, 1 monolayer/s. The molecular beam flux at the substrate can therefore be readily modulated in monolayer quantities, with shutter

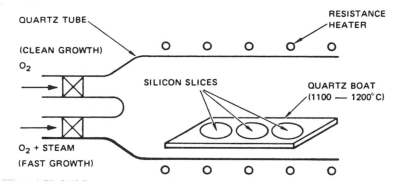

FIGURE 1.12.   Thermal oxidation of SiO_2 layer. From Ref. 30.

operation times below 1 s. MBE brings to microfabrication almost two orders of magnitude improvement in structural resolution in the direction of growth over techniques of LPE and VPE.

MBE has been used to prepare films and layer structures for a variety of GaAs and $Al_xGa_{1-x}As$ devices. These include varactor diodes having highly controlled hyperabrupt capacitance–voltage characteristics, IMPATT diodes, microwave mixer diodes, Schottky-barrier field effect transistors (FETS), injection lasers, optical waveguides, and integrated optical structures. The potential of MBE for future solid-state electronics is greatest for microwave and optical solid-state devices and circuits in which submicrometer layer structures are essential. The inherent adaptability of the process to planar technology and integration also offers significant design opportunities.

Possible longer-term implications of MBE for solid-state electronics are related to its capability of growing extended layer sequences with alternating composition, such as GaAs and AlAs. Such superlattice structures with periodicities of 50 to 100 Å show negative resistance characteristics attributed to resonant tunneling.

1.3.2.2. Oxidation.   A silicon dioxide layer is usually formed on the wafer by the chemical combination of silicon atoms in the semiconductor with oxygen that is allowed to flow over the silicon wafer surfaces while the wafer is heated to a high temperature (900–1200°C) in a resistance-heated furnace, as illustrated in Fig. 1.12.[30] The oxidizing ambient can be either dry or wet oxygen. The chemical reaction takes place according to one of the following equations:

Dry oxidation: $Si(solid) + O_2 \rightarrow SiO_2(solid)$ (1.1)

Steam oxidation: $Si(solid) + 2H_2O \rightarrow SiO_2(solid) + 2H_2$ (1.2)

Oxygen must diffuse through the oxide layer to react with the underlying silicon, thus increasing the oxide thickness from beneath.

1.3.2.3. Doping.   Fabrication of circuit elements requires a method for selectively making *n*- or *p*-type layers by doping the silicon substrate. Until the early 1970s, this task was accomplished by diffusion techniques[30] in a furnace of 950–1280°C, as

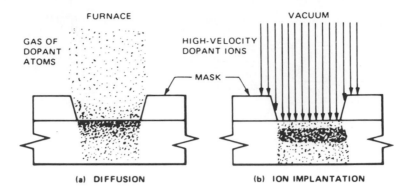

FIGURE 1.13. Comparison of diffusion and ion-implantation process for selective introduction of dopants into the silicon lattice. (a) Diffusion; (b) ion implantation.

Process Parameter	Diffusion	Ion Implantation
Process Control Dopant Uniformity and Reproducibility	· 5% on Wafer · 15% overall	· 1% overall
Contamination Danger	High	Inherently Low
Delineation Techniques	Refractory Ins. Refractory Metals Poly Silicon	Refractory and Non- Refractory Materials: Insulators, Metals, Polymers (Beam Writing)
Environment	Furnace	Vacuum

shown in Fig. 1.13a. The diffusion process involves two steps. First, the dopant atoms are placed on or near the surface of the wafer by deposition from the gas phase or by coating the wafer with a layer containing the desired dopant impurity. This is followed by drive-in diffusion, which moves the dopant atoms farther into the wafer. The shape of the resulting dopant distribution is determined primarily by the manner in which the dopant is placed near the surface, while the diffusion depth depends primarily on the temperature and time of the drive-in diffusion.

Since the early 1970s, selective introduction of dopants into a silicon lattice has also been performed at room temperature by ion implantation,[31] as shown in Fig. 1.13b, and more recently by neutron transformation.[32] These methods lead to better control than that obtained by diffusion.

Most impurity atoms in silicon are situated substitutionally in the lattice sites. These impurities can be relocated whenever empty lattice sites or vacancies exist next to them. At a high temperature (e.g., ~1000°C) many silicon atoms have moved out of their lattice sites and a high density of vacancies exists. If a concentration gradient exists, the impurity atoms move by way of the vacancies, as shown in Fig. 1.14a, and solid-state diffusion takes place. When the crystal is cooled after the diffusion, the vacancies disappear and the impurity atoms that occupy lattice sites are fixed.

Impurity atoms that occupy the voids between atoms are known as interstitial impurities. These impurities move through the crystal lattice by jumping from one

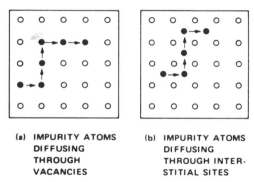

(a) IMPURITY ATOMS
DIFFUSING
THROUGH
VACANCIES

(b) IMPURITY ATOMS
DIFFUSING
THROUGH INTER-
STITIAL SITES

FIGURE 1.14.   Diffusion of impurities through the crystal lattice. (a) Impurity atoms diffusing through vacancies; (b) impurity atoms diffusing through interstitial sites.

interstitial void (or site) to another. At a high temperature, the spacing between atoms is wider, so that impurities can diffuse via interstitial sites, as seen in Fig. 1.14b. When the crystal is cooled, interstitial atoms may return to substitutional sites and become electronically active. Substitution is the diffusion mechanism of boron, phosphorus, and most impurities used in silicon. An important exception is gold, which diffuses primarily by the interstitial mechanism.

In ion implantation, the dopant atoms are ionized and then accelerated to high energies (30–350 keV) by an electric field. A mass-separating magnet eliminates unwanted ion species. After passing through the deflection and focusing control, the ion beam is aimed at the semiconductor target so that the high-energy ions can penetrate the semiconductor surface. The energetic ions lose their energy through collisions with the target nuclei and electrons and finally come to rest. The total distance that an ion travels before coming to rest is called its range (R); the projection of this distance in the direction of incidence is called the projected range (R_p).

The doping profile is usually characterized by the projected range and its standard deviation (ΔR_p). Typically, implantation depths can be varied between 0.01 and 1 μm as the impinging ion energy varies from 10 keV to 1 MeV.[33]

As the accelerated ions collide with silicon in the lattice, a significant amount of damage is imparted to the crystal lattice. Fortunately, most of this damage can be eliminated by annealing the wafers at 700–1000°C, which allows the displaced atoms to move back to their equilibrium lattice positions, thereby restoring the single-crystal structure. Although some diffusion of the dopant takes place during annealing, the temperatures are usually lower and the times shorter than for typical diffusion cycles.

Recently, postdoping high-intensity surface irradiation has been used to eliminate crystal damage of ion implantation. Rapid surface heating causes a thin region to melt and recrystallize from the substrate outwards, forming a high-quality crystal layer and fully annealing the implanted-impurity profile. For slower annealing, the Si may be heated to a temperature below the melting point. Here the damaged implanted regions regrow via the mechanism of solid-phase epitaxy toward the surface, without observable impurity diffusion.

The primary advantage of ion implantation is the process control it offers. By measuring the ion current and implant time, the exact number of dopant ions incorporated in a crystal can be determined. Also, wide variations in impurity concentrations can be achieved, since the amount of dopant introduced in independent of solubility considerations. Furthermore, some control of the distribution of impurity species prior to high-temperature diffusion is possible by regulation of the accelerating voltage.

Another process utilized for doping crystals is the neutron transmutation of some of the silicon atoms to phosphorus by placing the silicon in a nuclear reactor and passing a stream of thermal neutrons ($E_n \sim 0.025$ eV) through it.[31] This process is called neutron transmutation doping (NTD). The consecutive thermal neutron reactions during NTD of Si can be summarized as

$$^{30}\text{Si} + \text{n(thermal)} \rightarrow {}^{31}\text{Si} \tag{1.3}$$

$$^{31}\text{Si} \xrightarrow{t_{1/2}\,=\,2.6\ \text{h}} {}^{31}\text{P} + \beta^- \tag{1.4}$$

$$^{31}\text{P} + \text{n(thermal)} \rightarrow {}^{32}\text{P} \tag{1.5}$$

$$^{32}\text{P} \xrightarrow{t_{1/2}\,=\,14.3\ \text{d}} {}^{32}\text{S} + \beta^- \tag{1.6}$$

From these reactions it is clear that the NTD process is complicated by the fact that ^{31}P formed by the primary transmutation of ^{30}Si can also capture a neutron to give rise to a beta emitter ^{32}P. While the beta decay of ^{31}Si is very short-lived and exhibits no residual radioactivity, that of ^{32}P is moderately long-lived and can result in measurable amounts of radioactivity. The amount of ^{32}P present depends primarily, of course, on the amounts of ^{31}P produced and, to some degree, on neutron flux and the amount of phosphorus originally in the silicon.

Because of the long free paths of thermal neutrons in solids, doping silicon by NTD processing can, of course, only form an unpatterned n-type phosphorus-doped substrate, but it has the advantage of providing an extremely uniform phosphorus concentration in the silicon. Some radiation damage exists after processing in the nuclear reactor, and therefore an anneal is required to restore the lattice and resistivity.

Neutron transmutation-doped silicon has primarily been used in high-power applications, but recent interest has been shown in using it for ICs and semiconductor optical sensors.

1.3.2.4. Cathodic Sputtering. In this thin-film deposition process,[34,35] the material to be deposited is used as a cathode in a system in which a glow discharge is established in an inert gas (Ar or Xe) at a pressure of 10^{-1}–10^{-2} Torr and a voltage of several kilovolts. The substrate to which the film is to be deposited is placed on the anode. The positive ions of the gas created by the discharge are accelerated toward the cathode and arrive with the energy gained in the cathode fall region (see

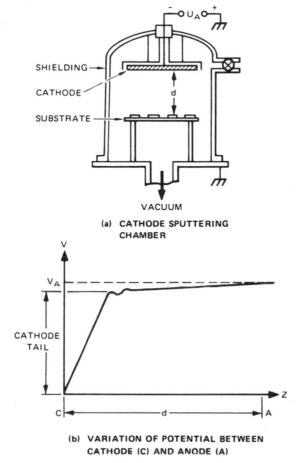

FIGURE 1.15.   Cathode sputtering thin-film deposition process. (a) Cathode sputtering chamber; (b) veriation of potential between cathode (C) and anode (A).

Fig. 1.15). Under the ion bombardment, material is sputtered from the cathode, mostly in the form of neutral atoms, but partly in the form of ions. The liberated components condense on surrounding areas, including the substrates on the anode.

The amount of the material sputtered, Q, in a unit of time under constant conditions is inversely proportional to the gas pressure p and anode–cathode distance d:

$$Q = \frac{kVi}{pd} \tag{1.7}$$

where k is the constant of proportionality, V is the working voltage, and i is the discharge current.

1.3.2.5. Vacuum Evaporation.   Vacuum evaporation[34,35] is the most widely used method for the preparation of thin films. The method is comparatively simple, but it can, under proper experimental conditions, provide films of extreme purity and known structure.

The process of film formation by evaporation proceeds by several physical stages:

- Heating the material to be deposited to a temperature where it is evaporating or sublimating at a fast rate
- Condensation of these particles on a lower temperature substrate
- Rearrangement or modification of the condensed particles for bonding on the surface of the substrate

The rate of film growth depends on the rate of arrival of evaporant particles and the probability of a given particle being retained on the surface. The latter is measured by a parameter called the "sticking" coefficient. The most used substances for thin-film formation are elements or simple compounds whose vapor pressures range from 1 to 10×10^{-3} Torr in the temperature interval 600–1200°C. The pressure and composition of background gases are important if the film properties are sensitive to impurity atoms.

The number of particles with molecular weight M leaving a source at temperature T per unit time per square centimeter can be determined using elementary kinetic theory (Section 3.7.2) and is given by

$$N_e = \frac{P_e}{(2\pi MkT)^{1/2}} = 3.513 \times 10^{22} \frac{P_e}{(MT)^{1/2}} \tag{1.8}$$

where P_e is the vapor pressure in Torr.

Since the substrate is colder than the source, the liberated particles travel in space, with their thermal velocities along a straight line, until collision with another particle. To ensure a straight path for them between the source and substrate, the residual particle concentration in the space must be low, i.e., the space must be sufficiently evacuated. The fraction of particles scattered by collisions with atoms of residual gas is proportional to $[1 - \exp(-d/\lambda)]$, where d is the source–substrate distance and λ is the mean free path of the particles.

The earliest sources were refractory metals such as tungsten, which could be heated by passage of a current. Later, boats of various materials such as molybdenum, graphite, and boron nitride were developed which could contain high-vapor-pressure materials. In addition to direct electrical heating, inductive methods are also in use. By using rf induction and large-diameter inert crucibles, evaporation of a number of different metals in a single evaporator cycle is possible. The major difficulties with this method are contamination of the deposit by the source and source failure.

To circumvent these problems, a focused electron beam can be used to heat a portion of the material while the crucible containing the material is water-cooled.

1.3.2.6. Selective Layer Removal (Etching). After a resist pattern has been delineated by the lithographic process, the film beneath the resist is usually delineated by an etching process.[36] In this section we briefly review the wet chemical and dry etching (ion etching, plasma etching, and reactive plasma etching) processes.

TABLE 1.1. Thin Film Materials and Liquid Etchants Used during Integrated Circuit Manufacture

Film	Etchant	Temperature of etch bath (°C)
SiO_2	HF	20–25
Si_3N_4	H_3PO_4	160–180
Al	$H_3PO_4/HNO_3/CH_2H_3O_2$	40–50
Crystalline/polycrystalline Si	HNO_3HF	20–25

Table 1.1 lists the typical wet chemical etchants used for thin-film materials in the microelectronic industry. Wet chemical or solution etching presents several problems. Photoresists often lose their adhesion to underlying films when exposed to hot acids. Also, as etching proceeds downward, it also proceeds laterally, thereby undercutting the photoresist film and broadening lines. Finally, with the current trend toward submicron geometries, wet chemical etching becomes increasingly difficult because the surface tension of the solutions tends to cause the liquid to bridge the space between two strips of photoresist, thereby preventing etching of the underlying film.

Anisotropic etching of a single-crystal material results from the different etch rates to chemical solvents of its crystallographic planes. The diamond cubic structure of silicon and the Miller indices of two of its planes are illustrated in Fig. 1.16. The $\langle 111 \rangle$ plane of silicon (Fig. 1.16c) is more closely packed than is the $\langle 100 \rangle$ plane (Fig. 1.16b) and etches at a much lower rate. This concept has been used to fabricate a variety of active and passive three-dimensional structures[37] and surface devices.

Figure 1.16d shows the cross section of an anisotropically etched square pyramidal hole in $\langle 100 \rangle$ silicon. The dimensions of the hole are given by the expression

$$W_0 = W_{Si} - \sqrt{2t_{Si}} \tag{1.9}$$

where W_0 is the side of the square apex, W_{Si} is the side of the square base hole in the wafer surface, and t_{Si} is the etched depth. This technique is based on the fact that the etch rate of the $\langle 100 \rangle$ plane is much higher than that for the $\langle 111 \rangle$ plane.

A membrane-type circular orifice (Fig. 1.17) can be fabricated by the anisotropic etching process in combination with a process that takes advantage of the etch resistance of heavily doped p^+ silicon in P-ED, an anisotropic etching solution.[38] At an impurity concentration $N_A \sim 10^{19}/cm^3$, the etch rate of Si in P-ED drops sharply and practically reaches zero at $N_A \geq 7 \times 10^{19}/cm^3$. When a silicon wafer with a heavily doped p^+ surface layer is etched in P-ED, the undoped Si is removed and a p^+ membrane is left, with a thickness equal to the depth of the surface layer and a concentration $N_A \geq 10^{19}/cm^3$. This property has been used to fabricate various device structures incorporating membranes ranging in thickness between 1 and 10 μm.

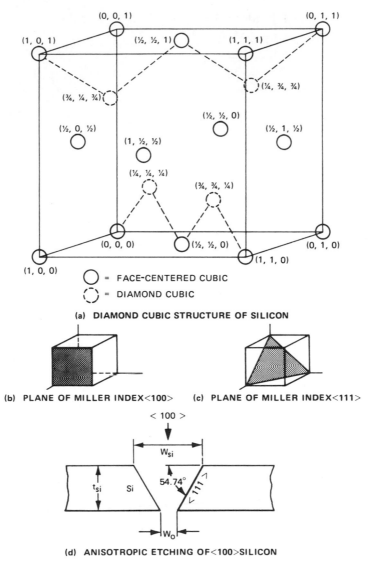

(a) **DIAMOND CUBIC STRUCTURE OF SILICON**

(b) **PLANE OF MILLER INDEX<100>**   (c) **PLANE OF MILLER INDEX<111>**

(d) **ANISOTROPIC ETCHING OF<100>SILICON**

FIGURE 1.16.   Structural characteristics relevant to anisotropic etching of silicon. (a) Diamond cubic structure of silicon; (b) plane of Miller index $\langle 100 \rangle$; (c) plane of Miller index $\langle 111 \rangle$; (d) anisotropic etching of $\langle 100 \rangle$ silicon.

Because of these difficulties in delineating high-resolution patterns by wet etching, dry etching processes using ionized gases have been developed,[39,40] as discussed below.

In ion or sputter etching the surface is eroded by bombardment with energetic ions. In one form of ion etching the substrates are laid directly on the cathode of a discharge tube. It is called ion-beam milling if the ion beam is generated in a plasma source separated from the substrates being milled.

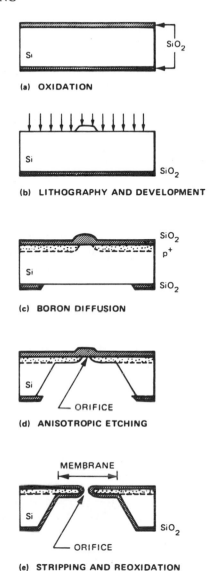

(a) OXIDATION

(b) LITHOGRAPHY AND DEVELOPMENT

(c) BORON DIFFUSION

(d) ANISOTROPIC ETCHING

(e) STRIPPING AND REOXIDATION

FIGURE 1.17. Fabrication of a membrane nozzle. (a) Oxidation; (b) lithography and development; (c) boron diffusion; (d) anisotropic etching; (e) stripping and reoxidation.

The key parameter in ion etching is the sputtering yield, which is defined as the number of sputtered target atoms per incident ion. The sputtering yield depends on the material being etched, the bombarding atom species, its energy, the angle of incidence, and in some instances the composition of the background gas. The sputtering yield bears a relation to the binding energy of the target atoms. It is low for elements such as carbon, silicon, and titanium, and high for gold and platinum. The sputtering yield increases with increasing energy and usually saturates above a few kiloelectron volts, since the penetration of an ion into a target increases without a corresponding increase in the energy lost close to the surface. For this reason, ion etching is usually done with energies below a few kilovolts. The etch rate

$$\delta(\theta) = 9.6 \times 10^{25} \frac{S(\theta)}{n} \cos \theta \left| \frac{\text{Å/min}}{\text{mA/cm}^2} \right| \tag{1.10}$$

where $S(\theta)$ is the sputtering yield, n is the atomic density of the target material in atoms/cm^3, and the cos θ term accounts for the reduced current density at an angle θ off the normal.

Since the sputtering yield is related to the binding energy of the atoms in the material being etched, it is possible to vary its value by introducing reactive gases. These gases can react with the etched surface and vary the binding energy and consequently the etch rate. For example, oxygen will adsorb on fresh surfaces of materials like titanium and silicon during ion etching, form oxides, and reduce the etch rate.[40] On the other hand, activated chlorine- or fluorine-containing species react with the above materials to form loosely bound or even volatile compounds, and thus increase the etch rate.[41]

A system for ion-beam milling is shown in Fig. 1.18. It contains a gun, a neutralizer, and a substrate stage.[42] The gun generates ions in a confined plasma discharge and accelerates them in the form of a beam toward the sample. The neutralizer, usually a hot filament, emits a flux of electrons to keep the sample

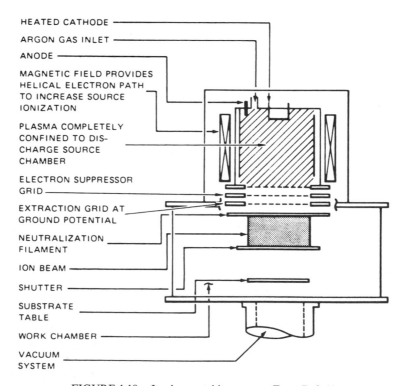

FIGURE 1.18.   Ion-beam etching system. From Ref. 41.

TABLE 1.2.   Etch Rates with Argon Ions,
Normal Beam Incidence[a]

Target material	Å/min
Silicon	360
Gallium arsenide	2600
SiO_2 (thermal)	420
SiO_2 (evaporated)	380
KTFR photoresist	390
AZ 1350 photoresist	600
PMM electron resist	840
Silver	2000
Gold	1600
Platinum	1200
Copper	1000
Palladium	900
Nickel	540
Aluminum	440
Zirconium	320
Niobium	300
Iron	320
Ferrous oxide	660
Molybdenum	400
Titanium	100
Chromium	200
Alumina	130

[a] 1 keV energy, j (current density) = 1.0 mA/cm^2.

neutral. Noble gases are usually used in the ion-milling machines because they exhibit higher sputtering yields relative to other atoms and also avoid chemical reactions. Table 1.2 gives the etch rates for argon-ion beams at normal beam incidence.

Etching of a sample can also be obtained if it is exposed to a gas plasma that contains highly reactive molecular and ionic species so that etching occurs even where there are no high-voltage ions impinging on the samples. This is generally referred to as plasma etching or reactive ion etching (RIE). In plasma etching, undercutting can be controlled to some degree by optimizing the mean free path of the etchant molecules by pressure variation, and the adhesion remains good.

An important consideration with respect to using plasma etching techniques involves the selectivity of an etchant for one film material compared to another. Whereas liquid etchants can be highly selective, such good fortune does not extend to plasma etching. Typical plasma reaction chambers are illustrated in Fig. 1.19a, b. They usually consist of a fused-silica tube in which the samples can be stacked vertically or horizontally. A rf potential is applied between the electrodes in the vacuum station. The rf field ($\approx$ 15-MHz frequency) causes the free electrons present in the plasma to oscillate, resulting in collisions with gas atoms, a certain fraction of which result in ionization of the gas atoms and sustain the plasma. The pressure required to sustain the plasma is from 2 to 20 Torr. The resultant plasma contains chemically active species that etch the sample material. The etching is not a sputtering phenomenon, but a chemical reaction that converts the etched material to a volatile compound. For example, oxygen plasma is used to remove photoresist material by oxidizing the hydrocarbon-based photoresists to volatile products.

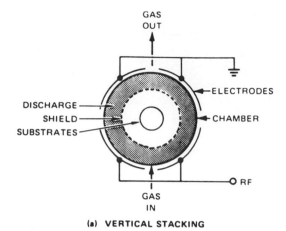

FIGURE 1.19.   Plasma chambers. (a) Vertical stacking; (b) horizontal stacking.

The addition of a chemically active gas to the ion plasma can have the effect of changing the etching rates through the chemical interaction between the substrate and the added gas. The addition of oxygen[42] has the effect of reducing the sputtering yield of metals that oxidize readily (Cr and Al), but it has relatively little effect on inert metals (Au and Pt). The use of a halocarbon (CCl_2F_2 and CCl_2FCClF_2) plasma increases the sputter etch rate of various materials (Si, SiO_2, Al, and photoresist).

We see, therefore, that plasma etching and reactive plasma etching are processes in which the material removal is mainly caused by a chemical reaction that results in a volatile compound. Because of the chemical nature of reactive etching, it can be more efficient than ion etching and also allow a higher degree of control over the relative etch rate of different materials. However, undercutting and edge profiles are more difficult to control than in ion etching. Although ion etching has a very high resolution capability, it is inefficient and offers little flexibility in preferential etching of different materials.

1.3.3. Pattern Generation

1.3.3.1. Background.   Figure 1.20 shows the types of lithographies used for microstructure fabrication.[43] Photolithography (Fig. 1.20a) is the most important technology in the microelectronics industry; it is used routinely down to 2 to 3-μm linewidths. Electron-beam lithography (Fig. 1.20b) is currently used for mask

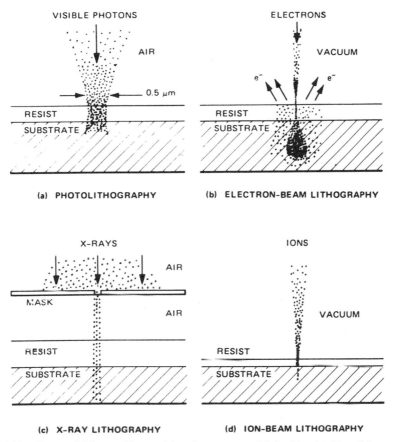

FIGURE 1.20.   Types of lithographies used in microstructure fabrication. (a) Photolithography; (b) electron-beam lithography; (c) X-ray lithography; (d) ion-beam lithography.

making. Electron backscattering limits its practical minimum linewidth to 0.3 μm for high-density microstructures. X-ray lithography (Fig. 1.20c) is usable down to near a few hundred angstroms, but requires a complex absorber mask or a thin-film support structure. Ion-beam lithography (Fig. 1.20d) offers patterned doping capability and a very high resolution ($\sim$100 Å).

The different technologies for lithography encompass two major areas: mask making and image transfer from mask to wafer. As will be discussed later, electron-beam lithography is capable of writing directly on the wafer without use of a mask. Figure 1.21 shows the different combinations currently applied for mask making and image transfer.

1.3.3.2. Contact Photolithography.   In photolithography it is first necessary to produce a mask or transparency of the pattern required.

Mask making begins with a large-scale layout called the artwork. Once the designer has completed a circuit design, the locations of all of the circuit components on the surface of the chip must be determined. Although the chip dimensions are from 0.5 to 5 mm on a side, the artwork must be made many times the actual chip

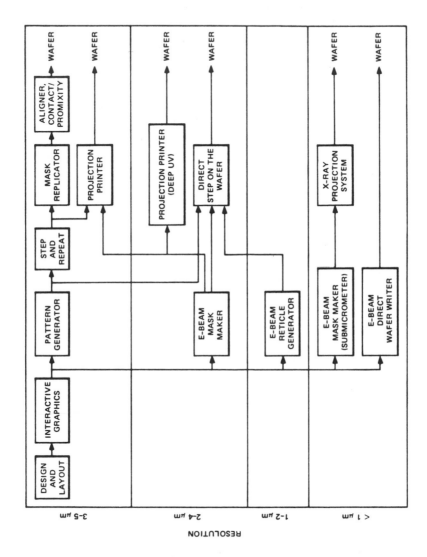

FIGURE 1.21.   Different combinations of electron-beam lithography for mask making and image transfer.

size to avoid large tolerance errors and to be of a reasonable size for human operators to handle. The layout is usually made in the form of a drawing showing the position of the windows that are required for a particular step of the fabrication sequence. Six or more layout drawings are required for a typical circuit. For complex circuits, the layout process can be performed by use of computer-aided graphics, and the computer can generate the drawing.

Next, the artwork is photographed by a large camera. Typically the original artwork will be as much as 500 times the size of the final circuit chip. Consequently,

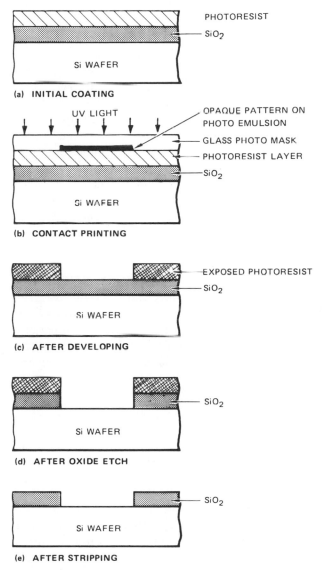

FIGURE 1.22. Steps in the photoresist process. (a) Initial coating; (b) contact printing; (c) after developing; (d) after oxide etch; (e) after stripping.

for a chip 2.5 mm on a side, the artwork may be 125 cm on a side, and a large camera is indeed required. Successive photographs are taken to reduce the artwork first to 100 times, then to 10 times, and finally to exact size on a master plate. The master plate is used in a precision step-and-repeat printer to produce multiple sequential images of the layout on a high-resolution photographic plate. This plate is used as the mask in the photoresist operation to transfer the layout pattern to the wafer surface.

To illustrate the photoresist procedure, we consider the case where small openings are to be made in the silicon dioxide layer covering a silicon wafer. Photoresist is coated on the oxide layer, as shown in Fig. 1.22a, placed in contact with the glass mask containing the pattern of the oxide to be removed, and exposed to light (Fig. 1.22b). During the development process, the unexposed coating is dissolved, leaving an opening in the coating (Fig. 1.22c). The photoresist coating that remains is chemically resistant to the buffered acid solution used to etch through the oxide layer, which produces an opening in the oxide (Fig. 1.22d). The remaining photoresist coating is then removed from the water, and the wafer is ready for the next step.

This method of generating patterns on semiconductor wafers is called contact printing. Contact with the photoresist wears the mask and eventually causes defects. Spacing the mask away from the substrate prevents contact and the defects that result from contact, but causes other problems. The larger spacing increases diffraction of the transmitted light, thus reducing resolution and blurring the individual photoresist features. The degree to which this occurs depends on the actual mask-to-wafer spacing, which may vary across the wafer. Wafer flatness variations (which are especially troublesome at small spacings) and diffraction effects (which are troublesome at large spacings) limit proximity printing with visible light to features no smaller than 7 μm.

1.3.3.3. Projection Photolithography. In projection photolithography an image of the photomask is projected directly onto the photoresist-covered wafer by means of a high-resolution lens between the mask and the wafer. In this system the mask life is potentially limited only by handling damage. In one class of projection printers the entire wafer is illuminated in a single exposure, with a pattern originating in a mask that is usually the same size as the wafer (5–10 cm in diameter). One-to-one projection printers are commercially available that have an image resolution in the range of 2–3 μm and an overall registration accuracy of about 0.3–0.6 μm.[44,45]

In the other class of projection printers, only part of the wafer is exposed to a pattern originating in a mask (or reticule) that is five or ten times larger than the projected image. The wafer is then stepped to a new position, where another part of the wafer can be exposed. By stepping and repeating, the entire wafer is covered with the reticule pattern. Step-and-repeat systems using 9 to 81 steps are also commercially available, and they typically have a resolution in the range of 1–2 μm and an overall registration accuracy of about 0.25–0.5 μm.[46–50]

1.3.3.4. Electron-Beam Lithography. The smallest features that can be formed by the conventional photolithographic process are ultimately limited by the wavelength of light. Current technology can routinely reproduce elements a few micrometers across, and it appears possible to reduce the smallest features to about 1 μm. Electron beams and X rays, however, have wavelengths measured in nanometers and even smaller units, and they are thus capable of producing extremely fine features.

Electron beams are attractive for lithography for additional reasons other than their short wavelengths:

- Electrons can be imaged to form either a pattern or a small point ≤ 100 Å, as opposed to 5000 Å for light.
- Electron beams can be deflected and modulated with speed and precision by electrostatic or magnetic fields.
- The energy and dose delivered to the resist-coated wafer can be controlled precisely.

Electrons are used either by scanning the beam to generate patterns directly from computer programs or by electron imaging through special masks. Electrons from a source can be formed into a pencil-like beam that can be deflected over an electron resist-coated substrate and modulated to draw a desired pattern. The beam can be imaged to a submicron spot, with sufficient current to expose the resist in less than 10^{-7} s. Since up to 10^{10} spots are typically required for a wafer having 0.5×0.5 cm^2 chips and 0.25 μm^2 picture elements (pixels), this extremely high speed is important.

Accurately positioning 10^{10} spots requires a precision of about 1 part in 10^6 along each coordinate axis. This places severe demands on the electron deflection system. Difficulties arise from inaccuracies in digital-to-analog conversion nonlinearities in the electron deflection system, uncertainties in the surface position of the substrate, wafer bow, and electrostatic and magnetostatic disturbances. All of these factors limit the practical range of deflection to about 1 mm. Since this is less than the width of the chip, a mechanical table is used to move the substrate to allow the full field to be exposed. Ensuring that the relative position of the beam and table remain within the required accuracy has led to two divergent approaches. In each, the electron column is used as a scanning electron microscope to locate registration features on the substrate for alignment purposes. In one approach, alignment on separate registration marks is accomplished for each table position. In the other, following electron-beam registration, laser interferometers are used to precisely measure the table movement. Further elctron-beam registration is used occasionally on a single mark to guard against beam drift.

Two scanning systems are used: raster scan and vector control. The raster scan systems methodically cover the entire area of the pattern to be generated, turning the electron beam on or off as demanded by the requirements of the pattern being recorded. The vector-control systems deflect the beam to follow paths specified by the needs of the pattern. A block diagram of an electrom-beam generator developed for microcircuit fabrication is shown in Fig. 1.23. The electron optics associated with electron-beam pattern-generation equipment are similar to those used in electron microscopy. An electron source, normally a heated cathode, supplies free electrons by thermionic emission. These free electrons are subsequently accelerated by electrostatic fields and focused by electromagnetic fields. They are controlled and deflected by combinations of electromagnetic and electrostatic fields to achieve the final result of tracing out a well-defined pattern.

Electrons can also be imaged to form a complete pattern at one time. An electron image-projection system (ELIPS) has been described[51,52] in which a photocathode is deposited on the patterned surface of an optical mask. Ultraviolet (UV) light

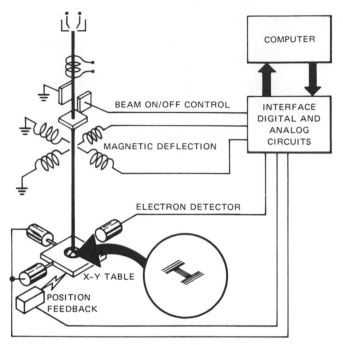

FIGURE 1.23.   Schematic illustration of an electron-beam lithography system.

illuminates the photocathode layer through the substrate in the transparent regions of the mask, causing a patterned emission of electrons from the photocathode. These electrons are imaged by uniform coaxial electrostatic and magnetic fields onto the facing resist-coated substrate. The system is capable of submicron resolution over the full wafer area.

In the past few years, considerable work has been done to apply electron lithography to the fabrication of microelectronic devices. Most of this work has used the scanning electron beam because of its ability to create high-resolution patterns (linewidth ≤ 5000 Å), its programmability, its large depth of focus ($\sim 10 \mu$m), and its capability of providing focus and registration via scanning electron microscopy.

There are two distinct uses of scanning electron-beam lithography in microelectronic device fabrication: using the scanning beam to expose directly a resist on a device substrate, or using the scanning electron beam to create a mask whose pattern can then be transferred onto device substrates by optical means.

1.3.3.5. X-ray Lithography.   The X-ray lithography technique is illustrated in Fig. 1.24. The mask consists of an X-ray-transparent membrane that supports a thin patterned film made of a material that strongly absorbs X rays. It is placed over a substrate coated with a radiation-sensitive resist. A distant "point" source of X rays produced by a focused electron beam illuminates the mask, thus projecting the shadow of the X-ray absorber onto the polymer film. This is the only feasible exposure scheme, since efficient X-ray lenses and mirrors for collimation cannot be made. The inset of Fig. 1.24 illustrates the penumbral shadowing δ, which results from the finite size d of any practical X-ray source. In any given exposure situation, the magnitude of δ can be made as small as required by a proper choice of the parameters s, d, and D.

Figures 1.25 and 1.26 show the processes used to fabricate micrometer surface relief and doping structures for simple and layered substrates after lithography.[53]

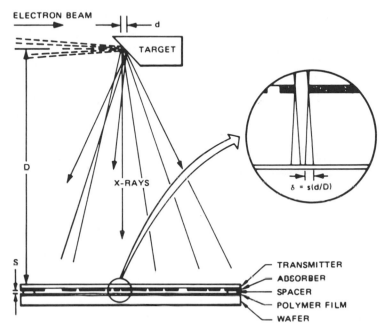

FIGURE 1.24.   X-ray lithography technique.

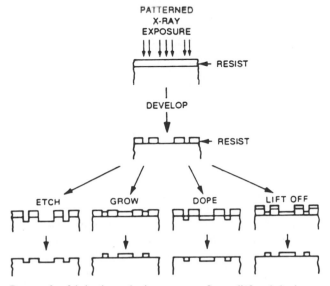

FIGURE 1.25.   Process for fabricating submicrometer surface relief and doping structures for simple substrates.

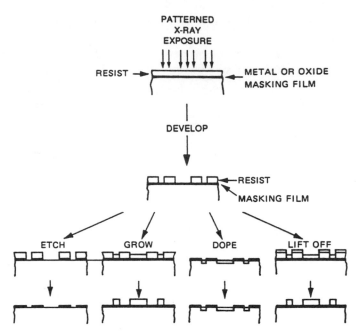

FIGURE 1.26.   Process for fabricating submicrometer surface relief and doping structures for layered substrates.

Generally, following exposure, a development step removes either the exposed regions (positive resist) or the unexposed regions (negative resist), thereby leaving a resist pattern in relief on the substrate surface. Following the creation of the relief structure in the resist, the substrate is processed in one of the following ways:

- By etching a relief structure in it
- By growing material
- By doping
- By depositing material through the open spaces in the resist pattern (lift-off process)

Since the X-ray wavelength used is on the order of 10 Å, diffraction effects are generally negligible, and proximity masking may be used. A partial list of the lithographic tools and associated performance criteria is presented in Table 1.3.

1.3.4. Process Sequence in Silicon Planar Technology

We have now introduced the more important processes that are used in silicon planar technology. For high-volume manufacturing it is desirable to start with a large substrate area so that as many individual circuits (dice or dies) can be made as possible. This places heavy demands on the accuracy of the masks, the uniformity with which the films are deposited, and the precision of pattern overlaying. Nevertheless, wafer sizes up to 8 inches in diameter are used.

TABLE 1.3. A Partial List of Lithography Tools

System	Wavelength or acceleration potential	Minimum linewidth (μm)	Registration method and accuracy (μm)	Throughout (wafers/h)[a]	References
Optical					
Near-contact printers	−4000 Å	2	Automatic or manual ±0.5	50–100	45
Deep UV contact printers	−2500 Å	0.5		5–10	46, 47
1:1 projection printers	−4000 Å	2–3	Manual ±0.8 Manual	40–60	48
Reduction projection systems	−4000 Å	1.25	Manual	10–50	49–51
X-ray systems	4–10 Å	0.2	0.1 for lab models	1–10	52, 53
Scanning-electron beam systems					
EBES (Bell Labs)	15 kV	0.2	Automatic ±0.2	2	54
VS (IBM)	25 kV	0.1	Automatic ±0.1	2–5	41
ELI (IBM)	25 kV	2.5	Automatic ±0.25	20	41
EBM (TI) printers	15 kV	0.25	Automatic ±0.25	2–5	41

[a] Throughout numbers are for 52-mm wafers and include handling and pumpdown time for electron-beam exposure tools.

In Figs. 1.27 and 1.28 we show schematically the sequence of process steps used for large-scale integration (LSI) and, more generally, for semiconductor device manufacture.

After the individual circuit elements have been fabricated in the silicon wafer, they must be interconnected to form an IC. This is accomplished by the process of metallization, in which a metal film ($\sim$0.5–2 μm) is deposited on the wafer and subsequently patterned so that individual transistors, diodes, capacitors, and resistors are interconnected appropriately.

Currently, aluminum or aluminum alloys are used for metallization because they satisfy most of the requirements (good adhesion, low resistivity, and low cost) for metallizing materials. Table 1.4 is a summary of the more common metals used on these films for metallization.

Most of the shortcomings of aluminum metallization arise from the ever-decreasing size of IC elements. Smaller and faster devices require shallower p–n junctions. These junctions cannot withstand heat treatments ($<$500°C) performed after metallization because of the interdiffusion of Si and Al.[54]

Electromigration[55] caused by momentum transfer from the flowing electrons to the stationary metal atoms presents another problem for aluminum metallization. Because of this phenomenon, voids and hillocks are produced along the length of

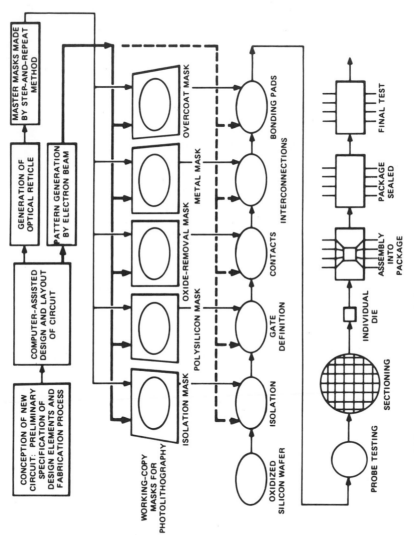

FIGURE 1.27. Process steps in large-scale integration (LSI) applications.

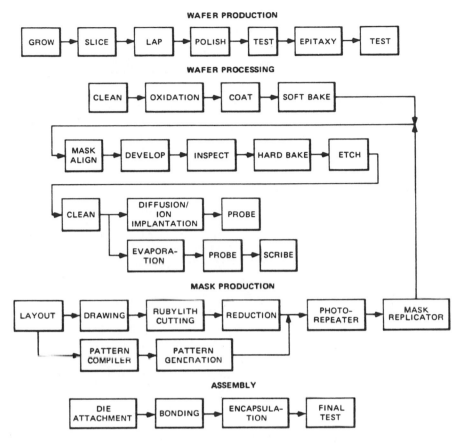

FIGURE 1.28.   General semiconductor production process.

TABLE 1.4.   Properties of Metals Employed for Metalization

Metal	Limiting current density $(A/cm^2 \times 10^5)$	Resistivity $(\times 10^{-6}\Omega \cdot cm)$	Remarks
Ag	4.0	1.59	Poor adhesion
Cu	4.0	1.67	Poor adhesion; corrosion
Au	7.0	2.35	Silicon eutectic at 370°C, poor adhesion
Al	0.5	2.65	Silicon eutectic at 577°C; electromigration
Al + Cu	2.0		
Mg	—	4.45	Extremely reactive
Rh	—	4.51	Poor adhesion
Ir	—	5.3	Poor adhesion
W	20.0	5.6	Difficult etching
Mo	10.0	5.7	Corrosion susceptibility
Pt	—	9.8	Poor adhesion
Ti	—	0.55	

TABLE   1.5.   Thin-Film Materials Used in Microelectronics (Classified by Area of Application)

Interconnections/ terminals	Insulation/ passivation	Encapsulation	Resistive	Capacitive	Semiconducting	
Al	$SiO_2 \cdot P_2O_5$	SiO_2	Ta	SIO_2	Si	InAs
Al alloys (Cu, Si)	SiO_2	$SiO_2 \cdot P_2O_2$	TaN_2	Ta_2O_5	Se	InSb
Cu	Si_3N_4	Al_2O_3	$Cr \cdot SiO$	HfO_2	Te	PbS
Mo–Au	Al_2O_3	Parylene	NiCr	ZrO_2	SiC	PbTe
Ti–Ag	BN	$PbO \cdot B_2O_3 \cdot SiO_2$	SnO_2	$PbTiO_3$	GaAs	CdS
Cr–Ag	SiO		Kanthal		Gap	CdSe
Pt–Au					AlN	ZnSe
Cr–CuAu						
Pb–Sn						

the conductor, which eventually leads to loss of continuity and circuit failure. High-current-density application ($>1 \times 10^5 \, A/cm^2$) requires the use of other metals (e.g., gold) that are not prone to electromigration. With the decrease in feature sizes, high current densities are becoming the rule, rather than the exception.

Although at first sight gold appears to be an attractive material for metallization because of its high conductivity, low corrosion, and electromigration resistance, it cannot be used generally. This is because the adhesion of gold to silicon dioxide is poor and the use of gold establishes a temperature limit (380°C) for subsequent processing because it forms a eutectic silicon. It is, however, used in conjunction with other metal films to improve contact adhesion.

At the present time, refractory metal silicides are being considered as inter-connect materials because they can be highly conducting and yet withstand the temperatures and reagents encountered during processing.[56] Tantalum, molyb-denum, and tungsten silicides are presently favored, although they have resistivities at room temperatures 10 to 30 times that of aluminum.

Corrosion resistance and scratch protection of ICs are achieved by a final "passivation" coating. Materials for passivation and other processes[57] are shown in Table 1.5.

1.4. MICROANALYSIS

1.4.1. Background

In order to be sure that a given microdevice has been constructed according to its design and is functioning in the way that was intended, it is of course necessary to have instruments and tools to measure and test it on the scale to which it has been built. Optical microscopes are insufficient for this purpose since their resolution is limited to a few thousand angstroms in the best of circumstances; hence, electron-beam methods have taken over for microscopic viewing, analysis, and test. Electron beams can be focused into spots as small as 5 Å diameter and manipulated to provide microscopic viewing, chemical analysis, and electrical probing to a precision in the range 10 Å to 100 Å. It is probably fair to say that the contemporaneous development of the scanning electron microscope[8] along with planar processing has been a crucial factor in enabling the scale of integration and circuit design rules to be improved at the incredibly fast rate we have seen. In this section we give a brief introduction to these topics, which are treated more comprehensively in Chapter 6.

1.4.2. Scanning Electron Microscope

The electron microscope was originally developed to provide an electron analogue to the optical transmission microscope. However, in this form sample preparation was extremely time-consuming since either the sample had to be very thin to be sufficiently transparent to electrons, or the surface had to be replicated for viewing in the form of an electron-transparent thin film. The scanning electron microscope (SEM), however, is based on building an image by the now familiar television process of raster scanning in which a tightly focused beam traverses the object to be viewed in a series of closely spaced parallel lines. When the electron beam strikes the surface, various particles may be emitted depending on the energy of the primary beam, namely,

- Secondary electrons
- Photons in the visible range (cathodoluminescence)
- X-ray photons

The secondary electrons are the easiest to detect since they may be subsequently accelerated to high energies and their presence amplified by using scintillation/photo-multiplier techniques (Fig. 1.29). This enables a small number of secondary electrons to be rapidly and quantitatively detected. The secondary emission generated by the primary beam will vary over the surface of the object due to both its composition

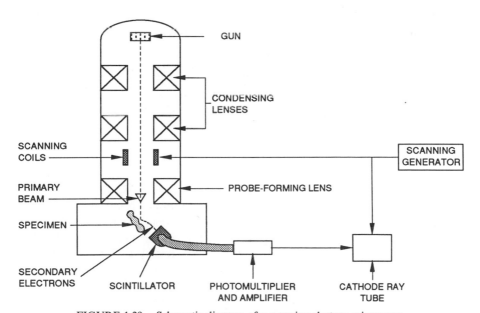

FIGURE 1.29.   Schematic diagram of a scanning electron microscope.

and topology. Thus, every picture element (pixel) on the surface is characterized by the number of secondary electrons it generates that reach the detector. A cathode-ray tube (CRT) display is scanned in synchronism with the beam scanning the object and the intensity of the CRT scan is modulated to reflect the number of secondary electrons emerging from the corresponding pixel of the primary beam scanning the object. In this way a complete picture of the object is displayed and can be recorded on photographic film with a single slow scan. Lower-resolution images are viewed at normal television rates for finding regions of interest and adjusting the magnification. The magnification is given by the ratio of the amplitude of the scan line of the primary beam to the scan line of the CRT beam. Since electron beams have an extremely large depth of focus, the size of the rastered spot does not change significantly with distance, thus it images three-dimensional objects as easily as flat objects. This factor alone gives the SEM a great advantage in microscopy. Another advantage is the ease with which the SEM can vary its magnification and "zoom" in on a region of interest. This is simply accomplished by varying the length of the primary beam scan line, which is controlled by coils that magnetically deflect the primary beam. The resolution is determined by the size of the spot and the scan time Δt for which secondaries are collected from a pixel. Thus, it is necessary to have sufficient current in the beam to enable the relatively short dwell time of the beam on a pixel to generate a detectable number of secondary electrons. For a beam current I (amperes, A), the number of primary electrons incident on a pixel

$$n_i = \frac{I \Delta t}{e} \tag{1.11}$$

and the number collected in the analyzer

$$n_c = \frac{\delta(V) I \Delta t}{e} \cdot f \tag{1.12}$$

where $\delta(V)$ is the secondary electron emission coefficient for an electron (of energy eV) and f is the fraction of secondaries that are detected. For objects of uniform $\delta(V)$, the surface is made viewable because of the variation of f over the surface due to the topography. Typically, f may vary between 0 and 0.3, and $\delta(V)$ between 10^{-2} and 1.

For a good range of contrast, n_c should, as a minimum, vary between 0 and 10^4 in steps of about 100 electrons.

The current that can be focused into a small spot is limited by the current density available from the cathode of the gun providing the primary beam, the beam optics, and the energy spread in the beam. In practice at 20–40 keV, a maximum of about 10^{-11} A can be delivered into an area of 10^4 Å^2 (i.e., a resolution of 100 Å). If 10^{-3} s is allowed for detection, an area of 10^6 pixels can be covered in a reasonable period of time (10^3 s $\cong$ 17 min).

Insertion of these typical values in Eq. (1.12) shows that n_c is just in its minimum range for useful assignment of a pixel. A smaller pixel size would require a cathode for the primary beam gun with substantially higher emission density. The most useful thermionic cathode material is lanthanum hexaboride, which is capable of up to 20 Å/cm² resulting in resolution in the 30 Å range.

The number of cathodoluminescent photons emitted from the surface on bombardment by the primary beam is usually substantially less than the secondary electrons, except for a few efficient materials, and hence this means of detecting is not often used. The generation of X-ray photons is also quite small and the fraction collected even smaller. But the energies of the X-ray photons are characteristic of the materials being bombarded and are therefore useful in analyzing the surface. However, because of the small numbers, the resolution is very poor compared with that obtained with secondary electrons. Typically, impurity elements in the few percent range can be detected with a resolution of a few microns.

1.5. MICRODEVICES

The technologies evolved for ICs have been found useful for fabricating microdevices for many applications, a few of which are discussed below. Other applications may be found in the literature (e.g., Refs. 58–63).

1.5.1. Optical Components

Integrated optics are characterized by microdevices that combine optical and electrical components using thin-film structures on a single substrate. Such integrated devices (waveguides, couplers, lasers, and modulators) are smaller in size, lower in cost, and more complex than the currently used bulk optical components.

The most promising material in integrated optics is the GaAs–GaAlAs heterostructure,[64] because by varying the alloy composition both the index of refraction and the band gap can be changed. This material also has high electro-optic and acousto-optic coefficients,[65] making it usable for modulation and switching. Heterostructures are also used in injection lasers, in which the stimulated emissions are produced by the recombination of carriers injected across a p–n junction.

Most fabrication techniques including photon- and electron-beam lithographies, molecular beam epitaxy, ion milling, epitaxy, and ion implantation[66] are the same as used for Si. This means that fabrication of the optical and electrical circuits can be performed simultaneously on the same chip.

The optical waveguide is the basic integrated optical component. The light wave is usually guided in the middle layer of a three-layer structure of three different material compositions (see Fig. 1.30). The necessary condition for waveguiding in these layers is that the middle layer have a higher refractive index than those of the outer layers so that the wave is *totally reflected at the boundary*. These structures can be manufactured easily using molecular beam epitaxy.

In integrated optics it is necessary for light to be coupled from one waveguide to another. For this purpose, three types of couplers are utilized: prism, grating, and taper. Figure 1.31 shows the structure of an output coupler based on a bulk prism and a prism–output coupler integrated into a heterostructure laser.[67]

In the integrated prism coupler (Fig. 1.31a), a portion of the light wave propagating in the waveguide extends into the region of the prism if the air gap is small

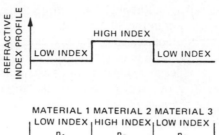

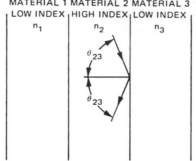

FIGURE 1.30.   Heterostructure optical waveguide.

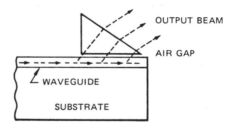

(a) BULK-COMPONENT

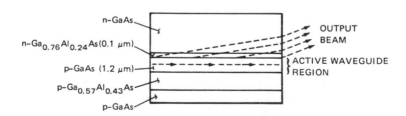

(b) PRISM COMPONENTS INTEGRATED
      INTO A DOUBLE HETEROSTRUCTURE
      DIODE LASER

FIGURE 1.31.   Prism output couplers. (a) Bulk component; (b) prism components integrated into a double heterostructure diode laser.

enough (typically 100 Å). If the refractive index of the prism n_p is higher than the equivalent refractive index of the guided mode $n_{eq} < n_2$, then light will "tunnel" through the air into the prism. Furthermore, if $n_p > n_{eq}$, that wave radiates in the prism at an angle to the original propagation direction and is therefore effectively coupled out. The light may be coupled into the waveguide by launching a beam in the opposite direction into a prism.

In the prism–output coupled diode laser (Fig. 1.31b) the thin (0.1 μm) n-Ga$_{0.76}$Al$_{0.24}$As heterostructure layer replaces the air gap. This region has a lower refractive index ($\sim$3.4) than that of either the n-GaAs substrate ($n \sim 3.6$), which corresponds to the prism, or the p-GaAs active waveguide region ($n \sim 3.6$). Thus, again because of the thin n-Ga$_{0.76}$Al$_{0.24}$As layer, the evanescent tail of the guided mode overlaps the high-index substrate region, and the light is coupled out at an angle given by

$$\sin \theta = (n_s^2 - n_{eq}^2)^{1/2}/n_s \qquad (1.13)$$

where, in terms of the wavelength in the guide, $n_{eq} = \lambda_0\lambda_g$. The prism-coupled output beam illuminates the facet over a large area.

Schematic diagrams of a bulk grating coupler and an integrated-output grating coupler are shown in Fig. 1.32. Here the light propagating in the waveguide interacts with the grating and is diffracted out. For output coupling from GaAs/GaAlAs

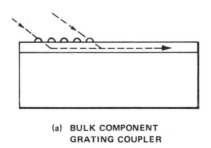

**(a) BULK COMPONENT
GRATING COUPLER**

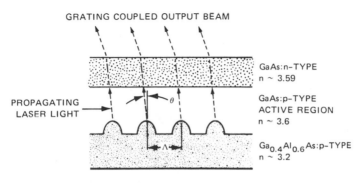

(b) INTEGRATED OUTPUT

FIGURE 1.32. Grating output couplers. (a) Bulk component grating coupler; (b) integrated output.

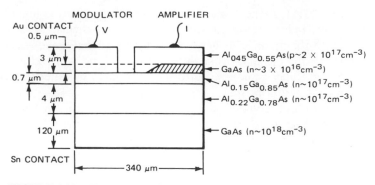

FIGURE 1.33.   Taper coupler integrated into a double heterostructure laser.

lasers, which emit light with a free-space wavelength of ~8500 Å and have an equivalent refractive index n_{eq} of ~3.6, a grating period of ~2400 Å is required if the light is to be diffracted out at right angles to its direction of propagation. Gratings with these short periods are fabricated by ion-beam machining,[68] optical lithography,[69] or holography.[70]

Taper couplers are used to couple a guided mode from one optical waveguide to another. In a taper coupler, one optical waveguide becomes progressively thinner, forcing the wave into a parallel guide. Figure 1.33 shows the structure of a taper coupler integrated into a double heterostructure laser. The output waveguide is coupled into a light modulator.

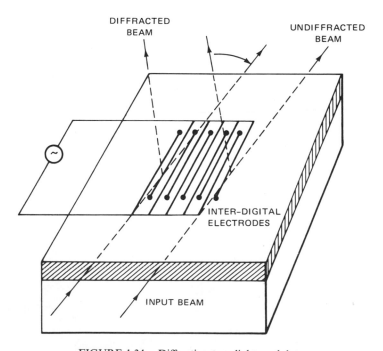

FIGURE 1.34.   Diffraction-type light modulator.

Optical modulators are electrically controlled devices that vary the amplitude, phase, or frequency of light. p–n junctions can be used as modulators because applying a reverse voltage to the junction sweeps out the free carriers and increases the refractive index, thus causing a phase shift. This phase modulation can be converted to amplitude modulation by the use of an external polarizer.

Power propagating in a planar waveguide can be diffracted by a periodic variation in the refractive index (Fig. 1.34). Here the periodic index change is induced by a voltage applied to the electrodes.

1.5.2. X-ray Optical Components

Considerable effort is currently being expended developing X-ray optical components such as waveguide, mirrors, lenses, and zone plates. These components are needed for X-ray instruments for use in microscopy, materials analysis, and microstructure fabrication. The techniques currently used for the fabrication of these X-ray optical components include electron-optical demagnification,[71] sputter deposition of multilayers,[72] and molecular beam epitaxy. Table 1.6 summarizes the present and future fabrication for these components.

1.5.3. Superconducting Junction Devices

A pair of superconductors separated by a thin insulting barrier forms a "Josephson junction." A Josephson junction can be viewed as a weak spot in a superconducting circuit at which a localized difference in quantum phase can be established between two superconducting regions.[73] This phase difference can be monitored by observing the "supercurrent" through the junction. This supercurrent responds to externally applied electric and magnetic fields, and it is this response that leads to a variety of interesting and useful Josephson devices.[74]

Josephson devices operate at liquid-helium temperatures (1–4 K) and are used to measure low-frequency voltages and magnetic fields. They are also used as detectors, mixers, thermometers, and computer elements. Josephson junctions can be constructed in several ways, using microfabrication techniques.[75] The most important planar structures in which microfabrication techniques can be used are shown in Fig. 1.35.

TABLE 1.6.   Present and Estimated Future Fabrication
Capability for X-Ray Optical Components

Fabrication parameter	Present	Future
Microroughness	50 Å	10 Å
Zone-plate resolution	10^5 lines/cm	2×10^6 lines/cm
Thin-film thickness	1 Å	1 Å
Waveguide smoothness	Atomically smooth	Atomically smooth

Figure 1.35a shows the tunnel junction of the type originally proposed by Josephson. Here two superconducting films are coupled via coherent tunneling of bound electrons pairs through an insulating oxide film between the layers.

In a Dayem bridge (Fig. 1.35b), the two superconductors are coupled by a narrow constriction (weak link). The bridge must be extremely narrow ($\sim 1\ \mu$m) to be coupled weakly enough to exhibit Josephson behavior. Figure 1.35c shows the structure of a Dayem bridge fabricated by electron-beam lithography and ion implantation.[76] The implanted nitrogen ions lower the transition temperature (T_c) from ~ 9 to 2.7–4.1 K and increase the resistivity by a factor of 3 to 26, depending on the dose.

Since Josephson junctions can be switched between different voltage states in less than 10^{-9} s and can also be packed densely, they could conceivably form the

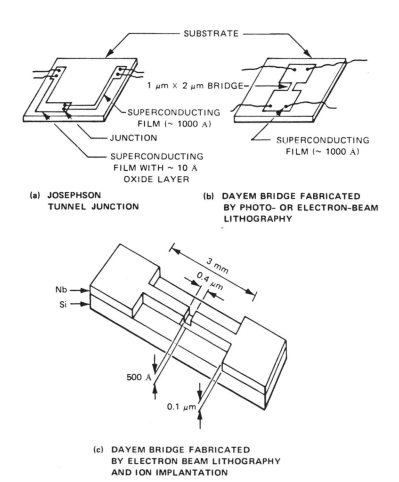

(a) JOSEPHSON TUNNEL JUNCTION

(b) DAYEM BRIDGE FABRICATED BY PHOTO- OR ELECTRON-BEAM LITHOGRAPHY

(c) DAYEM BRIDGE FABRICATED BY ELECTRON BEAM LITHOGRAPHY AND ION IMPLANTATION

FIGURE 1.35. Microfabricated Josephson junction structures. (a) Josephson tunnel junction; (b) Dayem bridge fabricated by photo- or electron-beam lithography; (c) Dayem bridge fabricated by electron-beam lithography and ion implantation.

basis for logic elements for a small, fast computer. As the speed of a computer increases, its dimensions must decrease, since the transmission velocity of signals is limited by the speed of light. For example, if a computer has a cycle time of 1 ns and the dielectric constant of its transmission line is 4, then it must have a linear dimension less than or equal to 0.5 μm. Because of this small size, the heat dissipation must be low so that the component (active or passive) density can be high. Josephson junctions satisfy these conditions and are thus suitable for the purpose. The advent of the new high-temperature superconducting materials has caused a resurgence of interest in these types of devices.

1.5.4. Vacuum Microelectronic Devices

Conventional point field-emission sources need a high operating voltage (2–10 kV) to produce the high fields (10^7–10^8 V/cm) at the tip necessary for field emission of electrons. The high voltages allow ions formed by electron impact with the residual gas to obtain sufficiently high energies so that they can sputter or erode the tip. By reducing the radius of the tip to 500 Å and bringing the anode to 0.5 μm from the tip, one can reduce the voltage to 50–100 V, that is, as close to or below the sputtering threshold for the tip material, and thus produce long-life devices.[77]

Figure 1.36 shows SEM photographs of microfabricated field-emission cathode sources.[78,79] Figure 1.36a shows part of a regular array of cathode tips arranged on 6.35-μm centers. Typical cathode arrays have over 20,000 tips in an area of about 1 mm². A magnified view of one of the field-emission tips is shown in Fig. 1.36b. The radius of the tip is about 300 Å, and the diameter of the hole in the anode film is slightly larger than 1 μm. The top anode film and cone are made of electron-beam-evaporated molybdenum, and the cone rests on a silicon substrate. Electrical insulation is provided by a thin film of thermally grown SiO_2.

The cutaway view in Fig. 1.36c shows the cone in the silicon (dark area) and the hole in the SiO_2. Emitter arrays have been operated at current densities as high as 1000 A/cm², averaged over the area occupied by the array, and have been life-tested at lower currents (3 A/cm²) for operating times exceeding 70,000 h.

Arrays of field-emission tips have been used as electron sources for traveling wave tubes and matrix addressed display tubes, and the application of individual tips to form microtriodes has been reported.[80]

1.5.5. Field Ion Sources

Structures in the form of miniature volcanoes have been fabricated[81] (Fig. 1.37). The edge of the volcano aperture is made extremely sharp (radius ~300 Å) and needlelike whiskers can be grown at the edge to augment the ionization efficiency. In this way, electric fields above 10^8 V/cm are generated with modest voltage. If a carrier gas containing organic molecules flows through the aperture into a vacuum, the organic molecules can be ionized by field ionization. The strong electric field first polarizes an organic molecule and the resulting dipole is drawn by the field gradient toward the tip. Close to the tip the electric field is sufficiently high to allow an electron to tunnel from the molecule into the metal of the tip, and the resultant ion is expelled into the vacuum. Since the molecular ion is formed without fragmentation, the

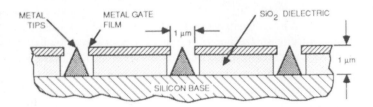

(a) SCHEMATIC OF A THIN-FILM FIELD-EMISSION CATHODE (TFFEC) ARRAY

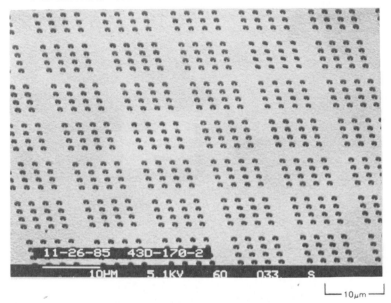

(b) SEM MICROGRAPH OF TFFEC ARRAY

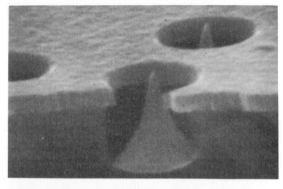

(c) SEM MICROGRAPH OF TFFEC CATHODE

FIGURE 1.36.   Spindt cathode arrays. (a) Schematic of a Spindt cathode array; (b) SEM micrograph of a spindt array; (c) SEM micrograph of a spindt cathode.

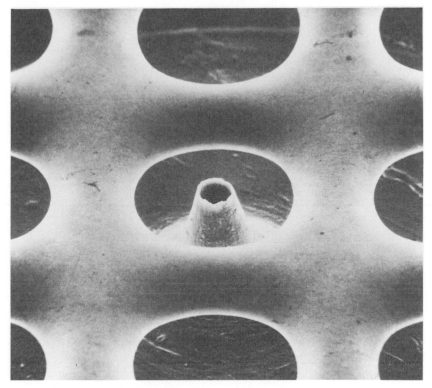

FIGURE 1.37. SEM micrograph of a single volcano in an array.

volcano makes an excellent source for the mass spectroscopy of organic and biological fluids.[82] It has been used for example to show changes in mouse urine content when the mice are infected with cancer.[83] Arrays of microvolcanoes have been built with apertures of less than 1 μm and require voltages of less than 100 V to produce ionization.

1.5.6. Micromechanical Devices

Gears and motor mechanisms have been fabricated on the micron scale.[84] Light modulation devices have been made by forming a thin metal membrane over a shallow well as shown in Fig. 1.38. The well is made by etching silicon out from

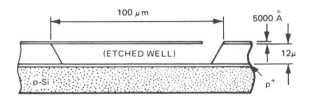

FIGURE 1.38. Metal-coated oxide membrane. From Ref. 63.

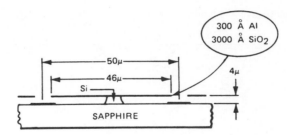

FIGURE 1.39. Light modulator (field controlled). From Ref. 63.

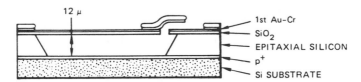

FIGURE 1.40. Micromechanical switch.

under the deposited film with a controlled etching procedure. Using an electric field, it is possible to move the film down toward the substrate or up away from it. An array of these membranes can function as light modulators (Fig. 1.39). The light is directed onto the thin film, and by changing its position the angle by which the light is deflected changes. Figure 1.40 shows a micromechanical switch that can be open or closed, depending on the electric field applied. Moving thin-film structures can also be used as accelerometers.[85]

In the fabrication processes for these devices, EDP silicon[86] etchant is used because of its unique capability as an anisotropic etchant of silicon. It has a very low etch rate for SiO_2 ($\sim$100 Å/h) and other insulators, compared to silicon ($\sim$50 μm/h). It also does not appreciably attack Cr or Au, so metallization can be patterned prior to the Si etching. The etching rate of Si highly doped with boron is reduced by at least a factor of 50.

1.5.7. Microsensors

With the availability of inexpensive computers, a demand has arisen for inexpensive sensing devices to directly input experimental data, so that computations based on these data may be made available in "real time." Such rapidly available computed results may then be fed back to control elements. Physical microsensors that measure temperature, pressure, strain, and other parameters are available, but microchemical sensors with good specificity have proved more elusive. A gas chromatograph built on a chip has been reported, and micro-electrochemical cells are under consideration.[60]

PROBLEMS

1. a. Sketch the cross section through a bipolar transistor and indicate the direction of current flow for both electrons and holes.
   b. Sketch a cross section through a MOSFET and describe how a channel is formed.

2. Discuss the differences and similarities of the following layering technologies:
   - Epitaxy
   - Cathodic sputtering
   - Vacuum evaporation

3. What are some of the potential advantages of X-ray lithography over conventional electron-beam and UV lithography? Discuss any limitations.

4. Microfabrication is normally applied to the fabrication of integrated circuits, but it has also been used successfully to make a range of microdevices.
   a. Give three examples of microdevices and describe how they are made.
   b. What are the advantages of microdevices over their macroscopic equivalents?

REFERENCES

1. R. P. Feynmann, in: *Miniaturization* (H. D. Gilbert, ed.), pp. 22–36, Reinhold, New York (1961).
2. A. R. Von Hippel (ed.), *The Molecular Designing of Materials and Devices*, MIT Press, Cambridge, Mass. (1965).
3. R. S. Becker, J. A. Golovchenko, and B. S. Schwartzentruber, Atomic scale surface modifications using a tunneling microscope, *Nature* **325**, 419 (1987).
4. I. Langmuir, Pilgrim trust lecture—Molecular layers, *Proc. R. Soc. London Ser. A* **170**, 1–39 (1939).
5. A. Y. Cho and J. R. Arthur, Molecular beam epitaxy, *Prog. Solid State Chem.* **10**(3), 157–191 (1975).
6. M. Von Ardenne, *Tabellen der Electronenphysik, Ionenphysik, und Ubermikroskopie*, Vols. I and II, Deutscher Verlag der Wissenchaften, Berlin (1956).
7. V. E. Coslett, *Practical Electron Microscopy*, Butterworths, London (1951).
8. C. W. Oatley, W. C. Nixon, and R. F. W. Pease, Scanning electron microscopy, *Adv. Electron. Electron Phys.* **21**, 181 (1965).
9. P. R. Thornton, *Scanning Electron Microscopy: Applications to Materials and Device Science*, Chapman & Hall, London (1968).
10. G. Binnig and H. Rohrer, Scanning tunneling microscopy, *Helv. Phys. Acta* **55**, 726–735 (1982).
11. J. C. Slater, in: *The Molecular Designing of Materials and Devices* (A. R. Von Hippel, ed.), pp. 7–8, MIT Press, Cambridge, Mass. (1965).
12. K. R. Shoulders, in: *Advances in Computers*, Vol. 2 (F. L. Alt, ed.), Academic Press, New York (1961).
13. B. E. Deal and J. M. Early, The evolution of silicon semiconductor technology: 1952–1977, *J. Electrochem. Soc.* **126**(1), 20C–30C (1979).
14. A. E. Anderson, Transistor technology evaluation, *West. Electr. Eng.* **3**(3), 3 (1959); **3**(4), 30 (1959); **4**(1), 14 (1960).
15. J. Bardeen and W. H. Brattain, Physical principles involved in transistor action, *Phys. Rev.* **74**(8), 1208 (1949).
16. W. Shockley, The path to the conception of the junction transistor, *IEEE Trans. Electron Devices* **ED-23**(7), 597–620 (July, 1976).
17. W. Shockley, The theory of p–n junctions in semiconductors and p–n junction transistors, *Bell Syst. Tech. J.* **28**, 435 (1949).
18. G. K. Teal and J. B. Little, Growth of germanium single crystals, *Phys. Rev.* **78**, 647 (1950).
19. W. G. Pfann, Principles of zone melting, *Trans. AIME* **4**(7), 747–753 (1952).
20. R. N. Hall and W. C. Dunlap, P–N junctions prepared by impurity diffusion, *Phys. Rev.* **80**(3), 467–468 (November, 1950).
21. J. A. Hoerni, Paper presented at the IEEE International Electron Devices Meeting, Washington, D.C. (October, 1960).
22. H. C. Theurer, J. J. Kleinmack, H. H. Loar, and H. Christensen, Epitaxial diffused transistors, *Proc. IRE* **48**, 1642–1643 (September, 1960).
23. S. M. Sze, *Physics of Semiconductor Devices*, 2nd ed., Wiley–Interscience, New York (1981).

24. C. Mead and L. Conway, *Introduction to VLSI Systems*, Addison–Wesley, Reading, Mass. (1980).
25. D. Kahng and M. M. Atalla, Paper presented at IRE International Solid-State Device Research Conference, Pittsburgh (June, 1960).
26. J. E. Lilienfeld, U.S. Patent 1,745,175 (1930).
27. R. N. Noyce, U.S. Patent 2,981,877 (1961).
28. F. M. Wanlass and C. T. Sah, Paper No. WPM 3.5, presented at the University of Pennsylvania IEEE International Solid-State Circuits Conference (February, 1963).
29. W. S. Boyle and G. E. Smith, Charge-coupled semiconductor devices, *Bell Syst. Tech. J.* **49**(4), 587–593 (April, 1970).
30. A. S. Grove, *Physics and Technology of Semiconductor Devices*, Wiley, New York (1967).
31. J. W. Mayer, L. Ericksson, and J. A. Davies, *Ion Implantation in Semiconductors: Silicon and Germanium*, Academic Press, New York (1970).
32. J. M. Meese (ed.), *Neutron Transformation Doping in Semiconductors*, Plenum Press, New York (1978).
33. J. F. Gibbons, Ion implantation in semiconductors—Part I: Range distribution theory and experiments, *Proc. IEEE* **56**(3), 295–319 (March, 1968).
34. L. I. Maissel and R. Glang (eds.), *Handbook of Thin Film Technology*, McGraw–Hill, New York (1970).
35. J. C. Vossen and W. Kern (eds.), *Thin Film Processes*, Academic Press, New York (1978).
36. W. S. DeForest, *Photoresist Materials and Processes*, McGraw–Hill, New York (1975).
37. E. Bassous, Fabrication of novel-three-dimensional microstructures by the anisotropic etching of $\langle 100 \rangle$ and $\langle 110 \rangle$ silicon, *IEEE Trans. Electron Devices* **ED-25**(10), 1178–1185 (1978).
38. R. M. Finne and D. L. Klein, A water–amine–complexing agent system for etching silicon, *J. Electrochem. Soc.* **114**(9), 965–970 (1967).
39. S. Somekh, Introduction to ion and plasma etching, *J. Vac. Sci. Technol.* **13**(5), 1003–1007 (September–October, 1976).
40. M. Cantagrel and M. Marchal, Argon ion etching in a reactive gas, *J. Mater. Sci.* **8**(12), 1711–1716 (December, 1973).
41. N. Hosokawa, R. Matsuzaki, and T. Asamaki, RF sputter-etching by fluorochloro-hydrocarbon gases, *Jpn. J. Appl. Phys. Suppl.* **2**, 435–438 (1974).
42. M. Cantagrel, Comparison of the properties of different materials used as masks for ion-beam etching, *J. Vac. Sci. Technol.* **12**(6), 1340–1343 (1975).
43. E. D. Wolf, The national submicron facility, *Phys. Today* **32**(3), No. (34), 34–36 (November, 1972).
44. D. A. Markle, A. New projection printer, *Solid State Technol.* **17**(6), 50–53 (1974).
45. M. C. King, Future developments for 1 : 1 projection photolithography, *IEEE Trans. Electron Devices* **ED-26**(4), 711 (April, 1979).
45. M. C. King, Future developments for 1 : 1 projection photolithogrpahy, *IEEE Trans. Electron Devices* **ED-26**(4), 711 (April, 1979).
46. J. Roussel, Step-and-repeat wafer imaging, *Solid State Technol.* **21**(5), 67–71 (May, 1978).
47. G. L. Resor and A. C. Tobey, The role of direct step-on-the-wafer in microlithography strategy for the '80s, *Solid State Technol.* **22**(8), 101 (August, 1979).
48. H. E. Mayer and E. W. Loebach, in: *Proc. SPIE* **221**, 9 (1980).
49. M. Lacombat, in: *Proc. Int. Conf. Microlithography*, Paris, p. 83, Comite du Colloque International de Microlithographie, Paris (1977).
50. R. E. Tibbetts and J. S. Wilczvnski, High-performance reduction lenses for microelectronic circuit fabrication, *IBM J. Res. Dev.* **13**(2), 192–196 (1969).
51. J. P. Scott, Recent progress on the electron image projector, *J. Vac. Sci. Technol.* **15**(3), 1016–1021 (1978).
52. P.R. Malmberg, T. W. O'Keefe, M. M. Sopira, and M. W. Levi, LSI pattern generation and replication by electron beams, *J. Vac. Sci. Technol.* **10**(6), 1025–1027 (November–December, 1973).
53. H. I. Smith, Fabrication techniques for surface-acoustic-wave and thin-film optical devices, *Proc. IEEE* **62**(10), 1361–1387 (1974).
54. P. A. Totta and R. P. Sopher, SSLT device metallurgy and its monlithic extension, *IBM J. Res. Dev.* **13**(3), 226–238 (May, 1969).
55. F. M. d'Heurle, Electromigration and failure in electronics: Introduction, *Proc. IEEE* **59**(10), 1409–1418 (1971).

56. F. Mohammadi, Silicides for interconnection technology, *Solid State Technol.* **24**(1), 65 (January, 1981).
57. L. V. Gregor, Thin-film processes for microelectronic application, *Proc. IEEE* **59**(10), 1390–1403 (1971).
58. J. E. Sitch, Microwave semiconductor devices, *Rep. Prog. Phys.* **48**, 277–326 (1985).
59. R. W. Whatmore, Pyroelectric devices and materials, *Rep. Prog. Phys.* **49**, 1335–1386 (1986).
60. M. J. Madou and S. R. Morrison, *Chemical Sensing with Solid State Devices*, Academic Press, New York (1989).
61. K. Board, New unorthodox semiconductor devices, *Rep. Prog. Phys.* **48**, 1595–1635 (1985).
62. T. H. O'Dell, Magnetic bubble domain devices, *Rep. Prog. Phys.* **49**, 589–620 (1986).
63. F. Sibille, Infrared detection & imaging, *Rep. Prog. Phys.* **49**, 1197–1242 (1986).
64. D. R. Scifres, R. D. Burnham, and W. Streifer, Heterojunctions in integrated optics, *J. Vac. Sci. Technol.* **14**(1), 186–194 (1977).
65. V. Evtuhov and A. Yariv, GaAs and GaAlAs devices for integrated optics, *IEEE Trans. Microwave Theory Tech.* **MIT-23**(1), 44–57 (January, 1975).
66. P. D. Townsend, Optical effects of ion implantation, *Rep. Prog. Phys.* **50**, 501–558 (1987).
67. D. R. Scifres, W. Steifer, and R. D. Burnham, Leaky wave room-temperature double heterostructure GaAs: GaAlAs diode laser, *Appl. Phys. Lett.* **29**(1), 23–25 (July, 1976).
68. H. L. Garvin, E. Garmire, S. Somekh, H. Stoll, and A. Yariv, Ion beam micromachining of integrated optics components, *Appl. Opt.* **12**(3), 445–459 (March, 1973).
69. C. V. Shank and R. V. Schmidt, Optical technique for producing 0·1 mm periodic surface structure, *Appl. Phys. Lett.* **23**(3), 154–155 (August, 1973).
70. W. W. Ng, C. S. Hong, and A. Yariv, Holographic interference lithography for integrated optics, *IEEE Trans. Electron Devices* **ED-25**(10), 1193–1200 (1978).
71. H. Koops, Verwendung der Kondensor-Objective-Einfelldinse zur Elektronenoptischen Microminiaturisation, *Optik* **29**(1), 119–121 (April, 1969).
72. T. Barbee, in: *Proceedings of the NSF Workshop on Opportunities for Microstructure Science, Engineering, and Technology*, Airlie, Var., p. 94 NSF, Washington, D.C. (November, 1978).
73. J. R. Schrieffer, *Theory of Superconductivity*, Benjamin, New York (1964).
74. J. Clarke, Electronics with superconducting junctions, *Phys. Today* **24**(8), 30–37 (August, 1971); D. G. MacDonald, Superconductive electronics, *Phys. Today* **34**(2), 36–37 (February, 1981).
75. P. W. Anderson and J. M. Rowell, Probable observation of the Josephson superconducting tunneling effect, *Phys. Rev. Lett.* **10**(6), 230–232 (March, 1963).
76. J. Gu, W. Cha, K. Gamo, and S. Namba, Properties of niobium superconducting bridges prepared by electron-beam lithography and ion implantation, *J. Appl. Phys.* **50**(10), 6437–6442 (1979).
77. I. Brodie, Bombardment of field emission cathodes by positive ions formed in the interelectrode regions, *Int. J. Electron.* **38**, 541–550 (1975).
78. C. A. Spindt, A thin-film field-emission cathode, *J. Appl. Phys.* **39**(7), 3504–3505 (1968).
79. C. A. Spindt, I. Brodie, L. Humphrey, and E. R. Westerberg, Physical properties of thin film field-emission cathodes with molybdenum cones, *J. Appl. Phys.* **47**(12), 5248–5263 (December, 1976).
80. H. F. Grey and C. A. Spindt (eds.), *IEEE Trans. Electron Devices* **36**(11) Part II, 2635–2747 (1989).
81. C. A. Spindt, Method of fabricating a funnel shaped miniature electrode for use as a field ionization source, U.S. Patent 4,141,405 (February 1979).
82. W. Aberth and C. A. Spindt, Characteristics of a volcano field ion quadrupole spectrometer, *Int. J. Mass Spectrom. Ion Phys.* **25**, (1977).
83. W. Aberth, R. Marcuson, R. Barth, R. Edmund, and W. B. Dunham, Profile analysis of volcano field ionization mass spectra of mice with sarcoma I transplanted by intraperitoneal inoculation, *Biomed. Mass Spectrosc.* **10**(2), 84–93 (1983).
84. K. E. Petersen, Dynamic micromechanics on silicon: Techniques and devices, *IEEE Trans. Electron Devices* **ED-25**(10), 1241–1250 (1978).
85. H. C. Nathanson and J. Guldberg, in: *Physics of Thin Films* (G. Haas, M. H. Francomb, and R. W. Hoffman, eds.), Academic Press, New York (1975).
86. R. M. Finne and D. L. Klein, A water–amine–complexing agent system for etching silicon, *J. Electrochem. Soc.* **14**, 965–970 (1975).

2

Particle Beams

Sources, Optics, and Interactions

2.1. INTRODUCTION

A key element in our ability to view, fabricate, and in some cases operate microdevices is the availability of tightly focused particle beams, particularly of photons, electrons, and ions. Consideration of diffraction effects leads to the general rule that if one wishes to focus a beam of particles into a spot of a given size, the wavelength associated with the particles should be smaller than the required spot diameter. In Table 2.1 are listed the wavelengths (in μm) of three particles (photons, electrons, and protons) at various energies.

From Table 2.1 we see that photons in the visible range (1.6–3.5 eV) are useful for resolutions down to perhaps a micron and for feature sizes of a few microns. This is improved by the use of photons in the UV to soft X-ray range (5–1000 eV). Photon energies greater than 1000 eV rapidly lose their usefulness as a result of the increasing range of scattered photons. We are obviously unconcerned about wavelength limitation to electron beams over the energy range in which they are normally used (10^2–10^5 eV), even if we are seeking to resolve atoms that are a few angstroms (10^{-10} m) in diameter. Again, a limitation lies in the range of the scattered electrons. With ions, even very-low-energy beams are not wavelength limited. Because they are of a size comparable to the atomic arrays on which they are impinging, their range is always quite small.

In this chapter we treat electron, ion, and photon beams from the viewpoints of particle sources, particle-beam formation, optics, focusing, deflection, and energy analysis. The interactions of particle beams with surfaces are discussed, since these form the basis of the microfabrication functions desired. Electrons are treated in greater detail because of their versatility and emerging importance in submicron lithography, surface analysis, structural analysis, and microscopy. The concepts of electron optics were derived by analogy with classical photon optics and carry over directly to ion optics simply by interchanging the mass and charge of the ionic particle for that of the electron in the relevant equations.

55

TABLE 2.1.   Particle Wavelengths (μm) at Various Particle Energies E_0 (eV)

Particle	Particle energy E_0 (eV)						
	1	10	10^2	10^3	10^4	10^5	10^6
Photons $\lambda = \dfrac{1.2399}{E_0}\ \mu\mathrm{m}$	1.24	1.24×10^{-1}	1.24×10^{-2}	1.24×10^{-3}	1.24×10^{-4}	1.24×10^{-5}	1.24×10^{-6}
Electrons $\dfrac{1.23 \times 10^{-3}}{(E_0 + 10^{-6}E_0^2)^{1/2}}$	1.23×10^{-3}	3.88×10^{-3}	1.23×10^{-4}	3.88×10^{-5}	1.22×10^{-5}	3.70×10^{-6}	8.7×10^{-7}
Protons $\lambda = \dfrac{28 \times 10^{-6}}{E_0^{1/2}}\ \mu\mathrm{m}$	2.87×10^{-5}	9.07×10^{-6}	2.87×10^{-6}	9.07×10^{-7}	2.87×10^{-7}	9.07×10^{-8}	2.87×10^{-8}

This chapter is fairly comprehensive, since the manipulation of particle beams has formed the basis of much of the microdevice technologies used today. Descriptions in the subsequent chapters rely heavily on the specifics discussed here.

2.2. ELECTRON SOURCES

2.2.1. Elementary Theory of Thermionic and Field Emission

The conductivity of a metal is due to electrons that move freely within the crystal lattice. At the surface of the crystal there are intense electric fields which, because of the asymmetric spatial distribution of the surface ions, form a potential barrier that prevents the conduction electrons from escaping. The energy distribution of the conduction electrons is given by Fermi–Dirac statistics. The energy required to remove an electron with the Fermi energy inside the conductor to a field-free region outside the conductor is termed the work function ϕ and is usually measured in electron volts. If an electric field E (volts per meter) is applied at the surface in a direction so as to help an electron escape, the work function is lowered by an amount

$$\Delta\phi = \left(\frac{eE}{4\pi\varepsilon_0}\right)^{1/2} eV = 3.79 \times 10^{-5}E^{1/2}\, eV \qquad (2.1)$$

where e is the electronic charge and ε_0 is the permittivity of free space. This is due to the reduction in the amount of work required to overcome the image force between the electron and the conductor surface and is termed the Schottky effect. The value of the work function is in part determined by the lattice structure and, as such, also changes with temperature. It is generally assumed, with good experimental evidence, that the work function is linearly dependent on the temperature with coefficient α; so in general we may write

$$\phi = \phi_0 + \alpha T - \left(\frac{eE}{4\pi\varepsilon_0}\right)^{1/2} \qquad (2.2)$$

where ϕ_0 is termed the zero-temperature, zero-field work function.

The work function may also be substantially modified by a single atomic layer of foreign atoms adsorbed on the metal surface. Chemisorbed atoms tend to give up or accept an electron from the substrate lattice to form a strong electrostatic bond and a dipole layer at the surface. Electropositive atoms such as thorium and barium are adsorbed as positive ions and make it easier for electrons to escape from the crystal; that is, they lower the work function. Electronegative atoms such as oxygen and fluorine are adsorbed as negative ions and make it more difficult for electrons to escape; that is, they increase the work function.

The number of conduction electrons packed per unit volume within a metal is extremely large, being on the order of 10^{22} per cm^3, making it an attractive possible source of electrons. To extract electrons from the potential well of the crystal lattice in which they are trapped, one must seek means to give sufficient energy to the conduction electrons so that they may overcome the potential barrier. Alternatively, by applying an intense external electric field to the surface, one can provide conditions for quantum-mechanical tunneling of electrons through the potential barrier.

In the first category we may simply heat the metal since the number of conduction electrons with energies sufficient to overcome the work-function barrier increases exponentially with temperature according to Fermi–Dirac statistics. This is the thermionic emission on which most practical vacuum electronic devices depend on as their source of electrons. Bombardment of the metal with energetic particles (photons, electrons, and ions (gives rise to an energy exchange with both conduction and core electrons in the lattice, whereby some may acquire sufficient energy to escape from the surface. Such effects are now rarely used as sources for electron guns, since the energy distribution of the emitted electrons is too broad for most practical applications. The second category, termed field emission, is very attractive in theory because very large current densities can be obtained with a small energy spread. However, it has proved to be most difficult to implement such sources because of practical difficulties that will be discussed in Section 2.2.3.2.

By heating a cathode to near thermionic temperatures, a substantial number of energy levels above the Fermi level are occupied by the conduction electrons. The tunneling probability of these electrons under the action of a strong surface electric field is substantially greater than that of electrons at the Fermi level or below, and the cathode is said to be operating in the thermal-field or TF mode.

The classical theory of thermionic emission has been well summarized by Nottingham,[1] and that of field emission by Good and Müller.[2] A brief outline is given below.

If the energy associated with the velocity component of an electron moving in a direction perpendicular to a surface is designated ε, then the current escaping from a unit area of the surface is given in general by

$$j = e \int_{-\infty}^{+\infty} N(T, \varepsilon) D(E, \varepsilon, \phi) \, d\varepsilon \qquad (2.3)$$

where $N(T, \varepsilon) \, d\varepsilon$ is the number of electrons incident upon the surface per unit area per second with velocity component perpendicular to the surface in the associated energy range ε to $\varepsilon + d\varepsilon$, when the crystal is at temperature T. $N(T, \varepsilon)$ is called the supply function. $D(E, \varepsilon, \phi)$ is the probability of an electron with energy ε penetrating a potential barrier of work function ϕ with external field E applied; $D(E, \varepsilon, \phi)$ is called the transmission function.

The supply function for conduction electrons is calculated from Fermi–Dirac statistics and is given by

$$N(T, \varepsilon) = \frac{4\pi m k T}{h^3} \ln\left[1 + \exp\left(\frac{-\varepsilon}{kT}\right)\right] \qquad (2.4)$$

where m is the mass of an electron, h is Planck's constant, and k is Boltzmann's constant. In the case of thermionic emission with small external electric fields, the transmission function $D = 0$ for

$$\varepsilon \leq \phi - \left(\frac{eE}{4\pi\,\varepsilon_0}\right)^{1/2} \tag{2.5}$$

and

$$D = 1 - r(\varepsilon) \tag{2.6}$$

for

$$\varepsilon > \phi - \left(\frac{eE}{4\pi\,\varepsilon_0}\right)^{1/2} \tag{2.7}$$

where $r(\varepsilon)$ is the reflection coefficient for electrons of energy ε arriving at the surface. Since $r(\varepsilon)$ is essentially unknown, it is usually assumed to be zero, or at least an average value r is assumed so that the integration may be carried out. This results in Richardson's equation for thermionic emission:

$$j = \left(\frac{4\pi m k^2 e}{h^3}\right) T^2 (1 - \bar{r}) \exp\left[-\left(\frac{\phi_0 + \alpha T - |eE/4\pi\,\varepsilon_0|^{1/2}}{kT}\right) \right] \tag{2.8}$$

Numerically,

$$j = 120.4(1 - \bar{r}) \exp\left(-\frac{\alpha}{k}\right) T^2 \times \exp\left[-\left(\frac{\phi_0 - 3.79 \times 10^{-5} E^{1/2}}{kT}\right) \right] \text{ A/cm}^2 \tag{2.9}$$

Figure 2.1 shows plots of j as a function of T for various ϕ, assuming that r, α, and $E = 0$.

For field emission at $T = 0$, it has been shown[2] that

$$D \approx \exp \frac{-4(2m\varepsilon^3)^{1/2} v(y)}{3\hbar e E} \tag{2.10}$$

where $y = (eE/4\pi\,\varepsilon_0)^{1/2}/\phi_0$ and $\hbar = h/2\pi$. Putting this value of D in Eq. (2.3) and integrating leads to the Fowler–Nordheim equation for field emission

$$j = \frac{e^3}{8\pi h} \frac{E^2}{\phi t^2(y)} \exp\left[-\left(\frac{4(2m)^{1/2}\phi^{3/2}}{3\hbar e E} v(y)\right) \right] \tag{2.11}$$

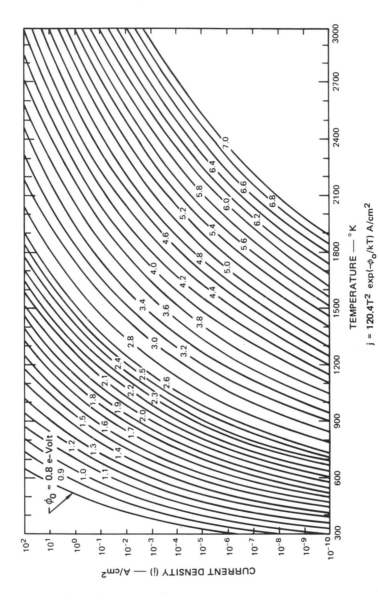

FIGURE 2.1. Plots of Richardson's equation for current density (j) as a function of temperature (T) with ϕ_0 as a parameter.

where $v(y)$ and $t(y)$ are the Nordheim elliptic functions, for which computed values are available. To a good approximation we may write $t^2(y) = 1.1$, and $v(y) = 0.95 - y^2$, leading to

$$j = 1.5 \times 10^{-6} \frac{E^2}{\phi} \exp\left(\frac{10.4}{\phi^{1/2}}\right) \times \exp(-6.44 \times 10^7 \phi^{3/2}/E) \; \text{A/cm}^2 \qquad (2.12)$$

Figure 2.2 shows plots of j as a function of E for various ϕ.

For field emission at moderate temperatures, Good and Müller[2] have shown that

$$j(T) = j(0) \frac{\pi kT/d}{\sin(\pi kT/d)} \qquad (2.13)$$

where $j(0)$ is given by Eq. (2.11), and

$$d = \frac{\hbar eE}{2(2m\phi)^{1/2}t(y)} \qquad (2.14)$$

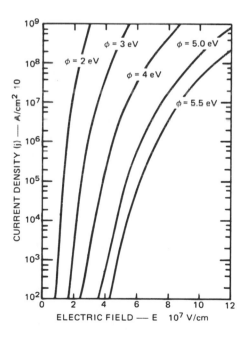

FIGURE 2.2. Plots of the Fowler–Nordheim equation for current density (F) as a function of energy (E) with ϕ as a parameter (Eq. 2.12).

For field emission at higher temperatures (TF emission), it has been shown [3] that

$$j = \frac{E}{2\pi}\left(\frac{kTt(y)}{2\pi}\right)^{1/2} \exp\left(-\frac{\phi}{kT} + \frac{E\theta(y)}{24(kT)^3}\right) \tag{2.15}$$

where

$$\theta(y) = \left(\frac{3}{t^2(y)} - \frac{2v(y)}{t^2(y)}\right) \cong 1.1 \tag{2.16}$$

Equation (2.15) gives thermionic and field emission as limiting cases.

2.2.2. Cathode Requirements

Electron sources for devices requiring a beam of electrons for their operation are called cathodes and are subject to a number of requirements depending on the device specifications. These relate to the following:

- The uniformity of emission over the cathode surface
- The current density required from the cathode
- The energy distribution of the emitted electrons
- The current fluctuations as a function of time
- The evaporation of material from the cathode surface
- The ability of the cathode to operate in a given environment
- The lifetime of the cathode under given operating conditions

For applications in microscience, we usually need to focus a beam of electrons of a given energy into the smallest possible spot and with the highest possible current. The Langmuir limit to the focused spot size in electron optics (see Section 2.4.3) requires that the current density drawn from the cathode be as large as possible.

Thermionic cathodes are usually operated space-charge limited for the following reasons:

- For practical polycrystalline cathodes, the work function over the emitting surface varies from point to point on a macroscopic scale, depending on the crystal face presented to the surface. The presence of space charge creates a potential minimum of value V_m a distance d_m in front of the cathode surface. If d_m is greater than the dimensions of the polycrystals forming the surface, the effects of different current densities emerging from the crystallites forming the surface and the small contact potential differences between them are blurred out. The emitting surface, as far as electron optics are concerned, appears as a uniformly emitting virtual cathode a distance d_m from the cathode surface and at a potential $-V_m$ below that applied to the cathode. The value of V_m is given by the increase in the average work function of the cathode that would be necessary to decrease the zero-field emission j_0 from the cathode to that required for space-charge-limited flow j_{scl}; that is,

$$j_{scl} = j_0 \, e^{-eV_m/kT} \tag{2.17}$$

The value of d_m can then be calculated from the full theory of space-charge-limited flow.[1] For a planar diode of interelectrode spacing d with applied voltage V, we may use the approximation

$$j_{scl} = \tfrac{4}{9}\varepsilon_0 \left(\frac{2e}{m}\right)^{1/2} \frac{(V - V_m)^{3/2}}{(d - d_m)^2} \tag{2.18}$$

By measuring j_0, j_{scl}, V, T, and d, an estimate of d_m can be made using Eqs. (2.17) and (2.18).

- The space-charge-limited current density is only slightly dependent on the cathode temperature, whereas the maximum available (saturated) thermionic emission with zero field at the cathode is exponentially dependent on temperature. The temperature stability requirements for constant emission are usually too stringent for the cathode to be used in the saturated mode. This effect is further aggravated by the fact that the work-function distribution of the emitting surface may change with time and with any changes in the vacuum environment in which the cathode is operating.
- The space charge serves to reduce—by several orders of magnitude—current fluctuations due to both the discrete nature of the electron (shot noise) and random changes in the emission capability of different points on the cathode surface (flicker noise).[4]

The energy distribution of thermionically emitted electrons is given by Maxwell–Boltzmann statistics. This gives rise to tails of electrons with energies both above and below the energy given to them by the electric fields through which they are accelerated. The mean spread, however, is very narrow, being on the order of kT. In field emission the maximum of the energy distribution is at the Fermi level. Since at normal temperatures there are very few electrons in states above the Fermi level, essentially a single tail exists, containing only electrons below the energy to which they are accelerated. By arranging to cut off the low-energy tail, a very narrow energy spread may be obtained. The sharp front of the low-temperature field-emission distribution is especially useful for measuring effects whose appearance occurs at sharply defined electron energies.

Flicker noise is prominent in thermionic cathodes in which the work function is lowered by an adsorbed monolayer of electropositive atoms. Such atoms move over the surface as a two-dimensional gas until they are lost by evaporation. Lost atoms may be replenished from the body of the cathode material by a chemical dispensing mechanism. An atomic site at the surface of the cathode spends some portion of its time in an emitting state when an electropositive atom is adsorbed on it, and the remaining time in a nonemitting state when there is no electropositive atom adsorbed on it. The constant on-and-off twinkling of the atomic sites gives rise to the flicker noise. The flicker noise spectrum depends on the mean times during which the sites are emitting and nonemitting.

If a high current density is required from a given thermionic cathode, its temperature of operation has to be increased. Natural limits occur when the cathode material melts or the vapor pressure of the evaporation products exceeds some tolerable limit for the device in which it is operated. Because of such limits, very few practical thermionic cathodes can be operated continuously at current densities exceeding 20 A/cm^2.

Another limit lies in the cooling of the cathode surface when high current densities are drawn from it. As the electrons leave, each carries away an average energy $\phi e + 2kT$. Thus, when drawing j A/cm^2, the number of watts per square centimeter that must flow to the surface to maintain its temperature of operation, assuming no conduction losses, is given by

$$w = j(\phi + 1.725 \times 10^{-5}T) + \xi\sigma T^4 \quad \text{W/cm}^2 \tag{2.19}$$

where ξ is the emissivity and σ is Stefan's constant. When the work-function cooling loss becomes substantially greater than the radiation loss ($\xi\sigma T^4$), it becomes increasingly difficult to supply sufficient heating power to maintain the surface temperature without melting or obtaining substantial evaporation from the cathode support structure. While continuous current densities from cathodes up to 10 A/cm^2 have been demonstrated, it is likely that an absolute limit to continuously operated thermionic cathodes lies between 20 and 100 A/cm^2. The problem can be somewhat alleviated by drawing current for very short times (ns to μs) and allowing relatively long times for the cathode to thermally recover, but this restricts such usage to a narrow range of applications.

Cathodes that depend on the dispensing of an active material to their surface for their operation can be rendered inactive (poisoned) by minute amounts of substances that change the chemical nature of the cathode surface or inhibit the flow of dispensed materials.

In field emission, it has proved possible to obtain current densities exceeding 10^{10} A/cm^2. However, such current densities can only be obtained over very small areas (less than a few μm^2). It has been found that with macroscopic surfaces, even those of a single-crystal face, cleaned and prepared by the best methods known, electrons are emitted from them when macroscopic fields above 10^5 V/cm are applied rather than the anticipated 10^7–10^8 V/cm required for field emission. These electrons have been shown to originate at relatively few tiny whiskers, or projections, where field-enhancement factors of over 100 are easily obtained. These whiskers range in height from 10 to 1000 atoms tall, with 1 to 100 atoms in their tips.[5]

Wires etched to extremely sharp points are erratic in their emission characteristics unless, as was first demonstrated by Muller in 1936, they are heat treated to form a single crystal at their tips. These crystals are usually 0.2–1 μm in radius and exhibit various crystal faces. The lowest-work-function faces emit more profusely, and in using such cathodes for an electron gun it is necessary to find means for selecting only electrons from a particular face. The currents are unstable and cathode lifetimes are short, mainly owing to sputtering of the surface by ions formed in the residual gas close to the cathode surface. It has been shown[6] that a residual pressure of 10^{-14} Torr would be necessary to avoid sputtering damage for long, although useful lives of a few hours may be obtained at 10^{-10} Torr to perform short-term experiments.

For these reasons, some workers have turned to using TF emission, since heating to these temperatures has the effect of allowing the tip to repair itself after damage by the sputtering action. The increased stability, however, is gained at the price of increasing the energy spread of the emitted electrons.

Field-emission electrons tunnelling from below the Fermi level give up energy to the lattice, thereby heating the surface (Nottingham effect). Those that tunnel from above the Fermi level cool it down. Hence, when field-emission current is drawn, the cathode surface will move toward the temperature at which the energy carried away by electrons from above the Fermi level is exactly compensated by the energy given up by electrons from below the Fermi level. This is known as the transition temperature and is given approximately[7] by

$$T_c = \frac{5.32 \times 10^{-5} E\phi^{1/2}}{t(y)} \qquad (2.20)$$

The actual surface temperature T_s will be given by a complex balance between the various forms of heat loss and generation,[8] but if T_s exceeds the melting point, a sudden disruption of the emitting tips may be expected.

2.2.3. Practical Cathodes

2.2.3.1. Thermionic Cathodes. Although there have been attempts to utilize an extraordinary variety of cathodes and cathode structures to overcome the difficulties associated with particular devices, relatively few cathodes have met with general acceptance.[9] Of the pure metals, tungsten is the favorite. Tantalum is less favored because although it operates at a slightly lower temperature than tungsten for a given emission, it also melts at a lower temperature (2996°C compared with 3410°C for tungsten).

Thoriated tungsten (containing about 2% by weight thorium oxide) may be considered the first dispenser-type cathode, since the work function of the tungsten is lowered by an adsorbed monolayer of thorium.[9] In order to dispense thorium to the surface at a rate sufficient to maintain the low work function at the operating temperature in a normal vacuum environment, the cathode is carburized; that is, a layer to tungsten carbide is formed at the surface by heating it to a high temperature in a hydrocarbon atmosphere or in contact with powdered carbon. In operation, the tungsten carbide reacts with the thorium oxide, forming free thorium and carbon monoxide. The presence of carbon also enables the cathode to operate at relatively high partial pressures of oxygen (10^{-5} Torr) without poisoning, since the oxides of carbon do not remain on the emitting surface. These cathodes can give up to 3 A/cm^2 when operated at 1720°C.[10] The main disadvantage of thoriated tungsten emitters lies in the fact that they constantly evolve carbon monoxide gas. This can cause problems in UHV devices and devices sensitive to the presence of ions. However, in cases in which a rugged cathode is necessary, such as a demountable system or electron-beam lithography systems used for mask or wafer mass production, this cathode performs extremely well.

Lanthanum hexaboride also relies on a dispensing mechanism for its operation.[11] The crystal structure is unusual in that the boron atoms are linked by strong valence bonds and form a three-dimensional cagelike structure around the lanthanum atom. The lanthanum, however, is not bonded to the boron. The strong boron bonds make the crystal refractory and the lanthanum valence electron makes it conductive. When heated, lanthanum diffuses to the surface, forming a low-work-function monolayer. Lanthanum lost by evaporation from the surface is replenished from the interior. The material, however, is difficult to support, since it reacts or alloys with most refractory metallic substrates at operating temperatures. Cathodes were originally made with sintered powders, but single crystals are currently favored. The crystal tips usually in the form of a pointed rod, can be heated by electron bombardment while the support structure and electrical contact are attached at the cooler base.[11] Such cathodes appear to be used in the saturated (Schottky) mode to obtain the highest brightness, since the emission is obtained in "lobes" that must be steered into the electron-optical path.

The oxide-coated cathode is attractive because of its low temperature of operation.[12] It consists of loosely packed crystals of the oxides of barium, strontium, and calcium on a nickel substrate. The oxide crystals, which are hygroscopic in air, have to be formed in the vacuum by heating the carbonates to drive off the excess carbon dioxide. Care must be taken so that the mixture of carbonates and oxides does not melt during this process, thereby losing cathode porosity. The nickel substrate contains an impurity such as aluminum or zirconium which at the operating temperature (750°C) reacts with the barium oxide, forming free barium. The free barium contributes both to increasing the conductivity of the crystals and in forming a low-work-function monolayer on the surfaces. In addition to electronic conduction through the crystals, charge is transported from the base nickel to the emitting surface by space-charge-limited flow in the interstices and electrolysis of the crystals. Continuous currents up to about $0.5 \, A/cm^2$ can be obtained with long life, and pulsed currents over $100 \, A/cm^2$ have been utilized.

Unfortunately, oxide cathodes are highly sensitive to their vacuum environment and their emission properties can easily be permanently destroyed (poisoned). Thus, they are unsuitable for the demountable, high-voltage systems usually associated with electron microscopy and lithography, although they are the first choice for cathode-ray display tubes and camera tubes. To overcome these effects, attempts have been made to enclose the cathodes in some form of nickel matrix. While such cathodes are more rugged, they operate at higher temperatures, and their performance is limited by the temperature at which the nickel evaporation rate is too high to tolerate (1000°C).

Tungsten-based barium-dispenser cathodes have been highly successful for high-current-density applications. They consist of a tungsten sponge with interconnected pores made by powder metallurgy. Barium is dispensed through the pores to the emitting surface, where it forms a low-work-function monolayer. In the preferred configuration the pores are filled with barium/calcium aluminate compounds by impregnation in the molten phase.[13] In operation the aluminates react with the tungsten, forming the free barium, which diffuses easily through the matrix. Current densities of $5 \, A/cm^2$, with the cathode operating at 1100°C, are routinely utilized in sealed-off power tubes. If the emitting surface is precoated with osmium[14] (M cathode) or iridium,[15] the work function is substantially lowered and the cathodes can deliver up to $20 \, A/cm^2$ before the barium evaporation rate becomes intolerable. While barium-dispenser cathodes are sensitive to their environment, they are not usually permanently poisoned by high pressures of ambient gases.

2.2.3.2. Field-Emission Cathodes. For the purposes of obtaining a small focused spot as is required in electron microscopy and lithography, the necessarily small emission area of a field emission cathode tip (radius in the range of 0.2–1 μm) is attractive, since the electron optics are required to produce an only slightly demagnified image of the tip. A crossover where the electron density is large enough to cause energy exchange between electrons that increases their effective temperature (see Section 2.4.4) is necessary for thermionic electron optics in which large demagnifications are required, but can be eliminated in field-emission electron optics, which utilizes a virtual crossover.

The major disadvantages of field emission are:

- The emission from the tip is emitted in "lobes" because of the various crystal faces present in the surface
- The instability of the emission

The crystal faces in the tip are usually symmetrically arranged around a central face, which usually has a high work function, i.e., is nonemitting, unless special steps are taken to ensure that this is not so. One approach is to use tungsten wire made from a single crystal oriented so that the low-work-function $\langle 110 \rangle$ face is axial. Another utilizes adsorbed material to lower the work function of the central face. However, such cathodes are only useful in devices for which the pressure of the vacuum environment can be maintained sufficiently low ($<10^{-10}$ Torr) to prevent rapid erosion of the tip from sputtering of the surface by ions formed by interaction of the electron beam with the residual gas. One approach to overcome this problem has been to operate the field emission tip at a temperature sufficiently high to enable it to continuously repair itself by recrystallization as it is damaged. This brings the cathode into the TF mode of operation. This heating usually causes the crystal to reorient itself continuously, presenting different faces to the electron-optical axis. Despite these difficulties, certain workers[16] have found that zirconium adsorbed on $\langle 100 \rangle$-oriented tungsten operated at 1350 ± 50 K can give 100-μm emission with 2500–3000 V applied for 1000 h, provided that the pressure is less than 2×10^{-8} Torr.

A radically different approach[17] has been to bring the anode so close to the tip that the applied voltage required to produce a field of 5×10^{7} V/cm at the cathode surface for field emission is sufficiently low that ions formed in the interelectrode region cannot gather sufficient energy to damage the tip. Using microfabrication technology, molybdenum tips of 50-nm radius set in the center of an anode hole of 0.6-μm radius have been fabricated. Currents in the range of 1–500 μA can be drawn with voltages in the range of 25–250 V, and cathodes can be operated in pressures of up to 10^{-5} Torr. Since the tips are formed by vacuum evaporation, they are not single crystals and their surfaces are not atomically clean. Emission appears to proceed from a few atomic sites on the surface that are constantly changing, giving rise to lobe and current fluctuations. Operating the cathodes at about 400°C substantially reduces these fluctuations.

Under the action of an intense electric field, liquid metals can be extracted from tiny nozzles. The tips of the liquid metal cones can be of extremely small radii, and when the field is in the correct direction electrons are field-emitted from the tips. It

TABLE 2.2. Cathodes and Their Properties

Type of emission	Type of cathode	Emission (A/cm^2)	Operating temperature (T_c) (K)	Upper pressure limit (Torr)	Brightness (β) $(A/cm^2 \cdot sr$ at 20 kV)
Thermionic	Tungsten	0.6	2470	10^{-4}	1.8×10^4
		7.3	2700		1.9×10^5
Thermionic	Tantalum	0.5	2300	10^{-5}	1.6×10^4
Thermionic	Thoriated tungsten	1–3	2000	5×10^{-6}	$3.75 \times 10^4 \rightarrow 1.1 \times 10^5$
Thermionic	Oxide coated	0.5	1050	10^{-6}	3.4×10^4
Thermionic	Barium dispenser	0.5–6	1150–1400	5×10^{-6}	$3.3 \times 10^4 \rightarrow 3.2 \times 10^5$
Thermionic	Lanthanum hexaboride	20.4	2100	10^{-6}	9.5×10^5
Field	Single-crystal tungsten	Up to 10^6	Room	10^{-10}	10^8
Temperature field	Zirconated tungsten		1400–1800 (1.5 eV)	10^{-9}	10^{10}
Photo	Pd	2×10^{-5}	Room (2 eV)	10^{-7}	2×10^{-1}
Photo	CsI	5×10^{-6}	Room (0.5 eV)	10^{-4}	2×10^{-1}

has been demonstrated[18] that with an indium–gallium alloy, pulsed currents as high as 250 A can be obtained (see Section 2.3.5).

It is clear that although substantial progress has been made in operating field-emission cathodes for particular devices, they cannot at the time of this writing be considered a fully satisfactory emission source for microdevice fabrication or usage.

Table 2.2 summarizes the properties of the more practical electron sources.

2.3. ION SOURCES

2.3.1. Background

In this section we are concerned with the creation and extraction of ions. We will first discuss the different ways in which ions can be created and then the properties of ion sources used in microelectronic applications. This will be followed by a discussion of space-charge effects that play an important role in ion-beam formation used for ion beam lithography, ion milling, ion implantation, and ion microscopy.[19,20]

In general, each application of ion beams requires a somewhat different source and set of beam parameters. The source performance parameters of greatest importance are the following:

- The ion species produced
- The ion current produced by the source
- The brightness of the source
- The energy spread

There are other parameters of lesser importance (e.g., ionization efficiency, source life, energy conversion efficiency), and discussions of these can be found in the specialist literature.[21] To illustrate the different characteristics of the ion sources, three classes are discussed: electron-impact, field-ion, and liquid-metal field-ion sources.

2.3.2. Electron-Impact Ionization Sources

Electron-impact ionization is the most used technique for generating ions for implantation and ion processing. In this ionization process the energy transferred to a molecule from an energetic electron exceeds the ionization energy eV_i, where V_i is the ionization potential for that molecule. Electrons for ionization can be created by thermionic or cold-cathode emission or can result from the discharge itself (see Chapter 3). These electrons are accelerated by the use of dc or rf fields and confined by the use of magnetic fields.

Energetic electrons can lose energy to gas particles by means of elastic collisions, the dissociation of gas particles, and the excitation of gas particles. The momentum transfer to the gas particles is usually negligible because of the large mass imbalance; thus, the increase in kinetic energy of the gas particles is negligible. When an incident (primary) electron has an energy in excess of the ionization energy, this excess energy can appear as the kinetic energy of a scattered incident electron, as the kinetic energy of an ejected (secondary) electron, as the multiple ionization of a gas particle, the excitation of a gas particle, or any combination of these effects.

The number of ionizing collisions suffered by an electron in passing through a gas per unit path length per unit pressure is called the differential ionization coefficient S_e; the value of this coefficient depends on the electron energy U_e and the gas species. The differential electron-impact ionization coefficients are shown for several gas species in Fig. 2.3. It is seen that electrons with energies much less than or much greater than the atomic and molecular ionization energies do not ionize these particles very effectively; electron energies a few times the ionization energies are optimum for electron-impact ionization.

The number of ions per square centimeter per second (n_-) produced by an electron current density $j-(A/cm^2)$ passing a distance l (cm) through a gas of pressure p (Torr) is given by

$$n_+ = \frac{j_-plS_e}{e} \tag{2.21}$$

As an energetic electron loses energy from ionizing and scattering collisions, its ionizing effectiveness changes. When its energy decreases below the lowset ionization energy of any atom in the volume, it no longer causes ionization and is called an ultimate electron. Energetic electrons can cause ionization efficiently only if they are maintained in the ionizing volume for a time long enough for them to lose their energy by ionizing collisions. A confining force resulting, for example, from a magnetic field

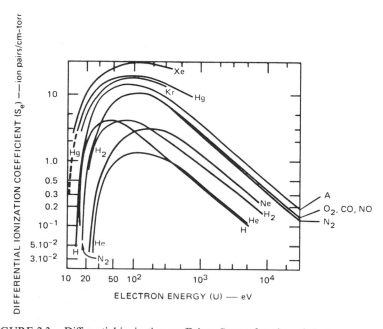

FIGURE 2.3.   Differential ionization coefficient S_e as a function of electron energy U.

or an oscillating rf field is often employed to increase path lengths and electron lifetimes. These concepts are illustrated in Fig. 2.4.

In the hot- and cold-cathode ion sources, the anode is maintained at a positive potential with respect to the two end plates so that electrons emitted by the cathode are injected through the plasma sheath in front of the cathode into a region where the potential varies with both the radial and axial distances. The application of an approximately axial magnetic field prevents the electrons from moving directly to the outer cylinder. In this crossed electric and magnetic field configuration, the electrons spiral around the magnetic field lines and oscillate longitudinally. As a result

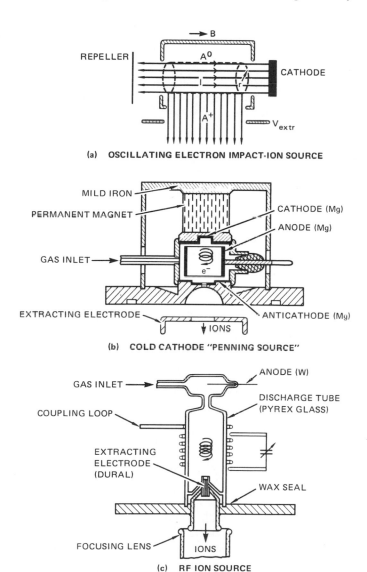

FIGURE 2.4. Electron-impact ionization sources. (a) Oscillating electron-impact ion source; (b) cold cathode "Penning source"; (c) rf ion source.

of ionizing collisions between the electrons and the atoms of the source gas, the plasma will be supplied continuously with ions and electrons. The primary electrons lose energy and diffuse across the magnetic field lines to the outer cylinder, where they are collected, causing an electric current to flow from the anode to the cathode. The plasma essentially fills the discharge chamber. In principle, this discharge mechanism can be used to ionize any gas.

In the rf ion source the ever-present electrons formed in the discharge tube fill the chamber and form a plasma sheath along the chamber wall and the extractor electrode. The wall potential is given by (see Section 3.4.4)

$$V_w = \frac{kT_r}{2e} \ln \frac{T_e m_i}{T_i m_e}$$ (2.22)

for an insulator surface, where T_e and T_i are the electron and ion temperatures and m_e and m_i are the electron and ion masses, respectively.[22] This wall potential causes the plasma boundary S_1 to move away from the wall surface S_2, as shown in Fig. 2.5a. the measure of the plasma withdrawal is given by λ_D, the Debye screening length (see Section 3.4.1)

$$\lambda_D = \left(\frac{kT_e}{4\pi n_e e^2}\right)^{1/2}$$ (2.23)

The following are useful forms of Eq. (2.23):

$$\lambda_D = 6.9(T/n)^{1/2} \text{ cm} \qquad (T \text{ in K}, n \text{ in number per cm}^3)$$

and

$$\lambda_D = 740(kT/n)^{1/2} \text{ cm} \qquad (kT \text{ in eV}, n \text{ in number per cm}^3)$$

when $n_e = n_i = n$ and $T_e = T_i = T$.

If an aperture larger than $2\lambda_D$ is made in the chamber wall, the plasma will balloon out through this opening, as shown in Fig. 2.5a. In the opening of the extractor electrode, the value of the negative potential applied to this electrode determines the shape of the plasma surface (Fig. 2.5b) because the plasma boundary must be an equipotential surface and the normal component of the electric field must be zero. This means that if the negative potential is increased, the plasma surface is forced back so that a curved surface will be the ion emitter forming a convergent beam of ions.

The shape and position of the ion-emitter surface of the plasma are thus dependent on the value of the negative potential applied to the extracting electrode. This means that the ion-optical system composed of the extracting electrode and the plasma surface can have different geometries determined by the value of the negative potential applied to the extracting electrode. This imposes limitations on the extracting voltage so that the beam intensity cannot be varied beyond a given value by varying the extracting voltage.

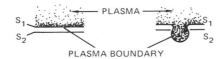

(a) SEPARATION OF PLASMA BOUNDARY S$_1$ FROM THE
WALL SURFACE S$_2$ CAUSED BY THE WALL POTENTIAL V$_W$

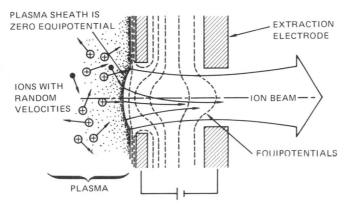

(b) PLASMA BOUNDARY IN AN ION SOURCE

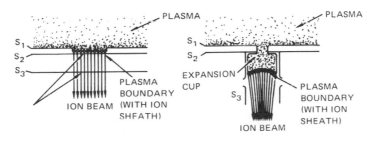

(c) ION EXTRACTION BY THE USE OF GRIDS   (d) ION EXTRACTION BY THE USE OF AN "EXPANSION CUP"

FIGURE 2.5. Beam extraction from plasma sources. (a) Separation of plasma boundary S$_1$ from the wall surface S$_2$ caused by the wall potential V$_W$; (b) plasma boundary in an ion source; (c) ion extraction by the use of grids; (d) ion extraction by the use of an "expansion cup."

Under normal operating conditions, the shape of the plasma boundary adjusts itself so that the ion extraction rate is equal to the supply rate from the plasma. If the extracted ion current density is space charge limited, it is given by the Child–Langmuir equation:

$$j_i = \frac{4\varepsilon_0}{9}\left(\frac{2e}{m_i}\right)^{1/2}\frac{V_{ext}^{3/2}}{d^2} \tag{2.24}$$

where V_{ext} is the extractor potential, d is the extractor electrode–plasma boundary distance, and m_i is the ion mass.

The current limitation can be overcome by keeping the shape and position of the ion emitter surface fixed. This can be achieved by placing a grid of high transparency in the extracting aperture, which ensures both the constant shape and fixed position of the plasma boundary independently of the extractor voltage (Fig. 2.5c).

For low-pressure (10^{-2}–10^{-3} Torr) and high-current operation, it is important to restrict the extraction aperture size but enlarge the ion-emissive surface. This can be achieved by the use of an "expansion cup" as shown in Fig. 2.5d. These expansion cups are very useful devices in ion sources such as plasmatrons, where high-current-directed ion beams are required.

2.3.3. Plasmatrons

The plasmatron is an arc-discharge type of ion source that differs from that of the electron-impact ionization ion source in that higher gas pressures (3×10^{-3} to 3×10^{-2} Torr) and higher electron currents are used. An arc discharge of a few hundred volts is maintained between a heated cathode and an anode. Electrons from the filament are accelerated toward the anode through the gas to be ionized. Near the anode the plasma is constricted by a conical intermediate electrode with an aperture (e.g., 5 mm) at a potential approximately halfway between those of the cathode and the anode (negative ~ 100 V) that creates a plasma bubble bounded by a charge double layer that focuses the electrons from the plasma on the cathode side in the region in front of the ion-extraction aperture. In the duoplasmatron shown in Fig. 2.6a the plasma is further compressed by a nonuniform magnetic field created by a magnetic-pole lens. The poles are insulated from one another; one can also be used as the intermediate electrode and one can be the anode.

The magnetic field confines the electrons so as to limit intense ionization to a small region around the anode aperture.[23] The primary advantage of the duoplasmatron is the increased efficiency obtained (50–95%). One disadvantage in some applications is that in extracting large ion currents the ions are accelerated to 70 kV. Thus, for low-energy applications a decelerating step is required which can reduce the current deliverable to the target. Proton current densities of up to 65 A/cm^2 have been obtained from a pulsed duoplasmatron, corresponding to a current of about 1 A.

The shape of the extracted ion beam can be modified by the use of an expansion cup. If the axial magnetic field is very high in the plasma, a "halo beam" will be formed, as shown in Fig. 2.6b, but upon insertion of a conical element into the cup the beam becomes homogeneous again,[24] as seen in Fig. 2.6c.

2.3.4. Field-Ionization Sources

As the electric field surrounding a neutral atom or molecule is increased, the particle first becomes polarized and ultimately the valence electron can tunnel from within the particle to a region just outside of it. Once the particle is ionized, the intense electric field rapidly separates the ion and electron. The tunneling mechanism is similar in principle to the field emission of electrons from a metal crystal, but the detailed theory involves only the escape of a single electron from the ground state of the atom or molecule.[2]

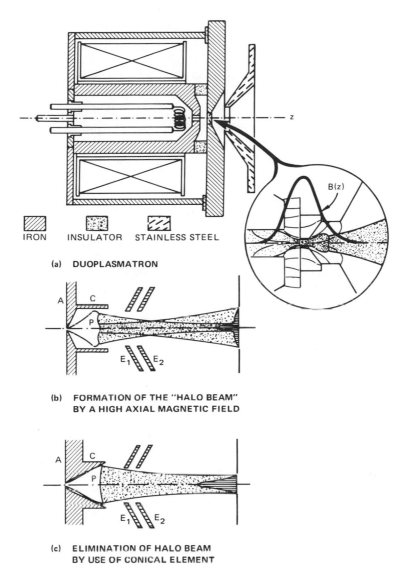

IRON   INSULATOR   STAINLESS STEEL

(a) **DUOPLASMATRON**

(b) **FORMATION OF THE "HALO BEAM"
BY A HIGH AXIAL MAGNETIC FIELD**

(c) **ELIMINATION OF HALO BEAM
BY USE OF CONICAL ELEMENT**

FIGURE 2.6.   Beam extractions from plasmatrons. (a) Duoplasmatron; (b) formation of the "halo beam" by a high axial magnetic field; (c) elimination of halo beam by use of conical element.

The theory predicts the mean lifetime τ for free atoms or small molecules in an electric field E to be given by

$$\tau \approx 10^{-16} \exp\left(\frac{0.68\phi_i^{3/2}}{E}\right) \tag{2.25}$$

where ϕ_i is the ionization energy. For larger molecules the calculations are more complex but are assumed to take on the same exponential form.

Since ionization energies of atoms and molecules are usually higher than the work functions of solid surfaces, the field required to extract the electron is usually an order of magnitude greater than for field emission from solid surfaces. Typically, fields of about 10^8 V/cm or 1 V/Å are required for field ionization. Such fields can be simply generated around sharp points of radius about 1000 Å. Since ionization occurs only an angstrom or so from the surface, the energy with which the electron strikes the tip is always small. The ions, on the other hand, are accelerated radially from a region close to the point, which forms a tiny virtual source. At very high field strengths, the ion current produced in a small volume element close to the tip is given by

$$i = en$$

where n is the number of particles arriving per second in the volume element. If A_0 is the area of the tip where the field is high, from kinetic theory we obtain

$$n_0 = A_0 \frac{P}{(2\pi m k T)^{1/2}} \tag{2.26}$$

However, because of the high field gradient around the point, molecules are drawn toward it from a relatively large distance, where the fields are sufficient to polarize the molecule. This gives rise to a much larger effective area of capture A given by

$$A/A_0 = 1 - \frac{2}{3} \frac{V_0}{kT} \tag{2.27}$$

where

$$-V_0 = \mu E_0 + \tfrac{1}{2}\alpha E_0^2 \tag{2.28}$$

is the potential energy of the particles just outside the tip, of average permanent dipole moment μ in the field direction, and of polarizability α, and E_0 is the field at the tip. The effective area of capture can exceed that of the tip by factors of 10–100, giving rise to correspondingly higher field-ionization currents.

For ion-beam lithography and microscopy, where a high brightness is required, the use of field-ionization (FI) sources is very attractive. With simple ion optical systems, a resolution <1000 Å can be achieved with monatomic gases (H, He, Ar, Xe).[25]

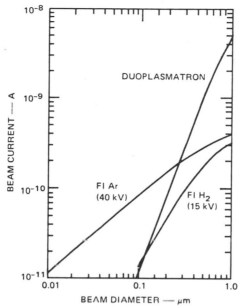

FIGURE 2.7. Limits on ion-beam current as a function of beam diameter for field ionization (FI) and duoplasmatron sources.

An FI source can generate a total current of 10^{-8}–10^{-7} A, with a virtual source size of 10 Å and brightness

$$\beta \approx 10^8 \, \frac{A}{cm^2 \cdot sr} \tag{2.29}$$

Figure 2.7 shows a comparison of currents as a function of beam diameter for two FI sources and a duoplasmatron source.[25] As seen from these curves, the FI source can provide more current in spots smaller than 2000 Å.

The source-extractor resolution for FI systems is limited by chromatic aberration due to the wide energy spectrum (4–15 V) of the extracted ions.

The ultimate limit of the emitted current density is set by the kinetic supply of ionizable gaseous particles to the high-field (>2.2 V/Å) region of the conically shaped emitter.[26] This limit can be removed if a liquid-film reservoir of the ionizable material is formed on the high-field region of the emitter.

2.3.5. Liquid Field-Ionization Sources

A liquid-metal ion source[27–30] typically consists of a fine tungsten capillary (0.02–0.002 cm in diameter) or of a capillary with a tungsten needle projection through it and an extraction electrode. As shown in Fig. 2.8,[18] the operation of the source involves feeding liquid metal to the tip of the capillary tube and applying a voltage between the tip and the extractor electrode. The interaction of the electro-static and surface-tension forces causes the liquid-metal meniscus to form a sharply peaked cone of small radius (Taylor cone).[31] The application of the critical Taylor voltage on a liquid-metal cone extracts ions or electrons, depending on the polarity. This critical cone-forming voltage is given by

$$V_c = 1.43 \times 10^3 \gamma^{1/2} R_0^{1/2} \tag{2.30}$$

where R_0 is the electrode spacing (cm) and γ is the surface tension (dyn/cm). To get some feeling for V_c let us assume that $R_0 \sim 0.1$ cm, and $\gamma = 700$ for liquid Ga; then we find that $V_c = 11$ kV.

It was shown by Taylor[31] that at the critical voltage, where the onset of cone formation occurs, the cone half-angle is 49.3°. As the cone forms, the apex radius decreases sufficiently to enable field emission. If negative field is applied to the emitter field, electron emission occurs when the local apex field reaches typical field-emission values (0.1–0.5 V/Å). At higher voltages, considerable heating takes place, resulting in the vaporization of the liquid metal followed by field ionization of the thermally evaporated atoms.[29]

Without a negative feedback mechanism, the liquid-cone radius r_a (ion source size) would decrease without obvious limit with increasing apex field. The apex field

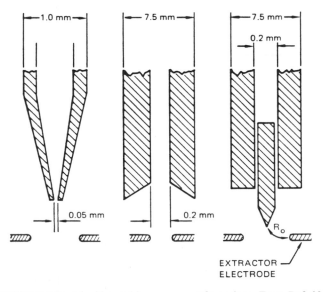

FIGURE 2.8.   Liquid-metal ion source configurations. From Ref. 30.

(E_a) is given by[27]

$$E_a \approx 3322\left(\frac{\gamma}{r_a}\right)^{1/2} \quad \text{V/cm} \tag{2.31}$$

The formation of Ga ions via field desorption requires a field of 1.6×10^8 V/cm. With this E_a and $\gamma = 700$ dyn/cm, Eq. (2.31) gives an r_a value of 30 Å.

The ion emission is extremely stable owing to space-charge effects in the emission process, even at low currents ($<1\ \mu$A).[29] In the case of electron emission from liquid Ga cathode, current pulses as high as 250 A with a 2- to 3-ns rise time have been observed at a repetition rate of 40–8000 pulses/s.[18] The mechanism involves the formation of a field-stabilized zone of the liquid cathode, which forms a sufficiently small apex radius (r_a) that a regenerative field electron current initiates an explosive emission process. Runaway is not observed for ion emission because there is always sufficient negative feedback present in all regimes of ion emission. This is provided by space charge and by the fact that a decrease in apex field leads to an increase in apex radius; hence, a further decrease in the field and decrease in emission, which in turn reduce the space charge, thus leading to an increase in the field, and so on. Space charge is effective in ion emission but not in electron emission because the primary charge carriers in ion generation are massive and the countercharge carriers (which reduce space charge) are light, whereas the opposite situation occurs in electron emission. Typical onset currents in liquid-ion sources are 10^{-5} A, which is a

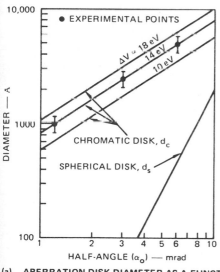

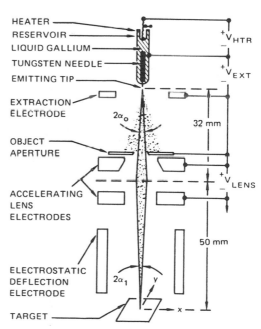

(a) ABERRATION DISK DIAMETER AS A FUNCTION OF BEAM HALF ANGLE

(b) ION-BEAM LITHOGRAPHY MACHINE

FIGURE 2.9. Limits on ion-beam spot as a function of α_0 for an ion-beam machine. (a) Aberration disk diameter as a function of beam half angle; (b) ion-beam lithography machine. From Ref. 32.

TABLE 2.3. Comparison of Ion-Source Performance Parameters

Source	Total current (A)	Current density (A/cm²)	Energy spread (eV)	Brightness (A/cm² · sr)	Angular brightness (A/sr)	Pressure (Torr)
Duoplasmatron	10^{-3}–10^{-1}	10^{-2}–1	10	$<10^3$	—	10^{-2}–10^{-1}
Rf	10^{-4}–10^{-2}	10^{-3}–10^{-1}	30–500	—	—	10^{-4}–10^{-3}
Penning (cold cathode)	10^{-5}–10^{-3}	10^{-4}–10^{-1}	50	—	—	10^{-1}–1
Hot cathode, electron impact	10^{-3}	$\sim 10^{-2}$	10–50	—	—	10^{-4}–10^{-3}
Field ion	10^{-9}–10^{-8}	10^{-3}	2–5	10^5	1.5×10^{-6}	$\sim 10^{-8}$
Liquid-metal, field ion	10^{-8}–10^{-4}	$\sim 10^5$	5–4	10^7	10^{-4}	10^{-7}

factor of 10^3 times as large as the equivalent current of the mass flow or the supply of liquid in the capillary.

Thus, the conclusion is that, in contrast to the case of gas FI sources, for liquid-ion sources the atom supply is adequate for the formation of a space-charge-stabilized ion current. Probe diameters obtained from liquid-ion sources are limited by the chromatic aberration at the extractor lens. Spherical aberration is negligible in the region of half-angles between 1 and 6 mrad. Figure 2.9a shows[32] the beam and aberration disk diameter as functions of the lens half-angle, where $V_{ext} = 57$ kV, $C_c = 33$ nm, $C_s = 396$ mm, and $\Delta U = 10$, 14, and 18 eV.

The liquid-metal source was a major breakthrough in high-brightness ion-source technology. Its superior performance can be used to form focused ion beams with probe sizes and currents comparable to those used for scanning electron microscopy, shown in Fig. 2.9b. The outlook for application to microcircuit fabrication will depend on the ability to generate other desired ion species (e.g., dopant ions) with liquid-metal sources. For diagnostics the impact will depend largely on the fundamental focusing limit. Extrapolating the probe size down to 100 Å or below by decreasing the lens half-angle is tempting, but perhaps a limiting virtual source size that is independent of α_0 will be reached. Ion-source performance parameters of greatest interest in microfabrication applications are compared in Table 2.3.

2.3.6. Microwave ECR Ion Sources

These sources[33] employ very-high-frequency power generators in the microwave region, typically 2.45 GHz, to generate the plasma, together with static magnetic fields in the discharge region to confine the electron motion and increase the

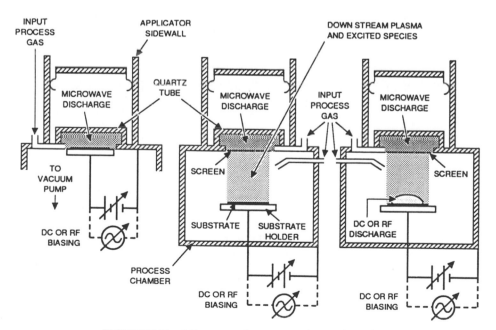

FIGURE 2.10.   Microwave plasma processing configurations.

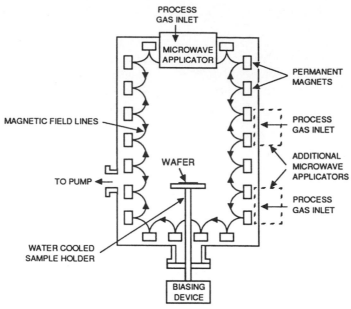

FIGURE 2.11.   Multipolar microwave plasma processing chamber.

plasma density. In the electron cyclotron resonance (ECR) mode, the value of a magnetic field is set so that the cyclotron frequency of an electron in this field, f_{ECR} (Section 3.5.5), is equal to the microwave frequency f. This means that an electron in this region is continuously accelerated (as in a cyclotron) until it either drifts out of the region of resonant magnetic field, or suffers a collision with a gas molecule. The microwave power can be increased until the electron/ion density in the plasma, n, is such that the plasma frequency f_p (Section 3.4.3) is also equal to the microwave frequency, in which case the maximum power is transferred to the plasma. Thus, the condition for the maximum possible electron/ion concentration n_M in the plasma is

$$f = f_{ECR} = \frac{eB}{m} = f_p = \left(\frac{n_M e^2}{4\pi^2}\right)^{1/2} \tag{2.32}$$

Clearly the higher the microwave frequency is, the higher n_M, provided the gas pressure in the discharge region is sufficiently high to maintain this value. In practice, 2.45 GHz is a preferred frequency of operation, requiring a magnetic field of about 900 gauss to meet the ECR condition. Gas pressures in the 10^{-5} to 10^{-1} Torr range are used and plasma densities up to 10^{12} electron/ions per cm^3 obtained.

For some applications the substrate may be immersed in the plasma, or the positive ions may be extracted through holes in a gridded wall of the chamber containing the plasma and accelerated. In this way, fluxes of up to 10^{16} ions/cm^2 per second (1 mA/cm^2) have been obtained. Figures 2.10 and 2.11 show various plasma processing configurations that have been used.

2.3.7. Cluster Ion Sources

Clusters are formed by condensation of supersaturated vapors. To realize this condition the vapor is adiabatically expanded through a nozzle at supersonic speeds into a high vacuum.[34] The dimensions and factors governing the design of a cluster source are shown in Fig. 2.12 and Table 2.4, where D is the nozzle diameter, λ is the mean free path of the atoms in the crucible, P_0 is the inner pressure of the crucible, P is the vapor pressure outside the crucible, and L is the thickness of the nozzle.[34,35] Figure 2.13 shows the process of formation of clusters during the expansion. When the vapor is ejected through a nozzle into a high-vacuum region, the vaporized atoms start to aggregate to form nuclei. The growth of the clusters occurs at the maximum rate near the nozzle and then slows down and finally ceases in the region where the pressure decreases. In the high-vacuum region, the sizes of the clusters remain constant because there are few collisions between clusters and atoms.

The flow characteristics can be treated quantitatively by condensation and thermodynamic models by assuming one-dimensional steady state and nonviscous flow. If the ejected vapor is continuous (i.e., if a sufficient number of collisions occur during the expansion), the temperature T and the pressure P of the vapor are given as functions of the Mach number, M, as follows[36]:

$$T = T_0\{1 + \tfrac{1}{2}(\gamma - 1)M^2\}^{-1} \tag{2.33}$$

$$P = P_0\{1 + \tfrac{1}{2}(\gamma - 1)M^2\}\gamma/\lambda' \tag{2.34}$$

where T_0 and P_0 are temperature and pressure inside the crucible. In accordance with the above equations, T and P of an expanding vapor change along the dry isentropic curve as shown in the P–T diagram in Fig. 2.14.[37] Since the expansion starts from near saturation which is shown by P_0 on the P_∞ curve, the vapor becomes rapidly supersaturated. Such a rapid supersaturation causes phase transition to the vapor by homogeneous nucleation, and clusters start to form. As clusters grow, the latent heat of condensation is released to increase their temperature.

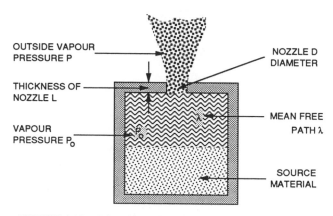

FIGURE 2.12.   Dimensions for the design of the cluster source.

TABLE 2.4.   Dimensions and Parameters for the Design of the Cluster Source Compared
with Those for a Molecular-Beam Source

Molecular beam	Cluster beam
Nonsupersaturated vapor	Supersaturated vapor in an adiabatic expansion
$D < \lambda$	$D \gg \lambda$ (e.g., $D = 0.1$–2 mm)
$P_0/P < 10^4$–10^5	$P_0/P \gtrsim 10^4$–10^5 (e.g., $P = 10^{-7}$–10^{-5} Torr, $P_0 = 10^{-2}$ Torr–several torr)
L/D not strongly restricted, tunnel-type nozzle also available ($L \gg D$)	$L/D \leq 1$ (for a cylindrical nozzle, experimentally $L/D = 1$)

According to classical nucleation theory, the change in the Gibbs free energy
(ΔG) in forming a cluster in the condensed phase is expressed by[38]

$$\Delta G = 4\pi r^2 \sigma - \tfrac{4}{3}\pi r^3 \left(\frac{kT}{V_c}\right) \ln S \qquad (2.35)$$

where r is the radius of the cluster, σ is the surface tension, V_c is the volume per
molecule in the condensed phase, and S is the saturation ratio. S is given by the ratio
of P/P_∞ by using the points indicated in Fig. 2.14. The first term on the right-hand
side is the surface free energy for the formation of a cluster. The second term is the

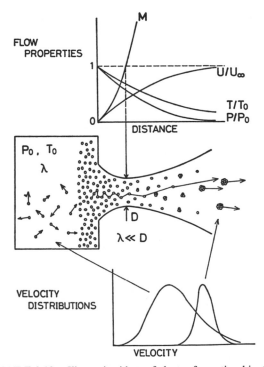

FIGURE 2.13.   Illustrative ideas of cluster formation kinetics.

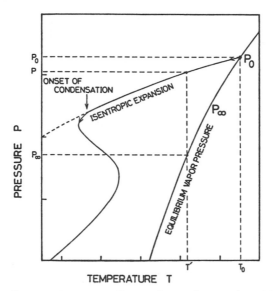

FIGURE 2.14.   Pressure–temperature diagram for nozzle expansion with condensation.

bulk free energy contribution to the excess free energy. The equation shows that r, T, S, and σ are important factors for cluster formation.

The change of the Gibbs free energy ΔG increases first with r, reaches a maximum at $r = r^*$, and then decreases continuously. The maximum value of ΔG ($= \Delta G^*$) can be calculated by differentiating Eq. (2.35) with respect to r and setting the derivative equal to zero:

$$\left(\frac{\partial \Delta G}{\partial r}\right)_{T,P} = \left(8\pi\sigma r - 4\pi r^2 \frac{kT}{V_c}\ln S\right)_{r=r^*} = 0 \qquad (2.36)$$

$$r^* = 2\sigma V_c/kT \ln S \qquad (2.37)$$

Substitution of Eq. (2.37) into Eq. (2.35) gives the maximum value of $\Delta G \equiv \Delta G^*$

$$\left(\frac{\Delta G^*}{kT}\right) = \frac{16\pi}{3}\left(\frac{\sigma}{kT}\right)^3\left(\frac{V_c}{\ln S}\right)^2 \qquad (2.38)$$

Several examples of the Gibbs free energy as a function of radius r are shown in Fig. 2.15, where values for surface tension for an infinite flat film were used. Pressure and temperature were determined by the condition for the isentropic expansion by using Eqs. (2.33) and (2.34). These results show that the nucleation barrier for metals and semiconductors is comparable to that for Ar, H_2O, or alkali metals even when infinite flat-film surface tension is used in the calculation. This is due to the fact that $\Delta G^*/kT$ is a function of σ/T, so that even when a metal has a high value of surface tension the ratio does not increase significantly because a metal has

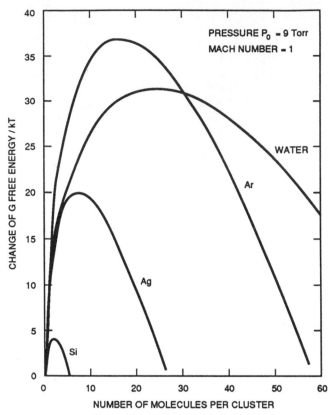

FIGURE 2.15.   Change of Gibbs free energy required to assemble a cluster as a function of number of atoms.

to be heated to a high temperature to get sufficiently high vapor pressure for expansion through the nozzle.

Since ΔG^* is influenced by the surface tension σ and saturation ratio S, lower barrier height for cluster formation is possible when the vapor material has a small value of σ at the condition of a large value of S. The surface tension of the cluster might differ significantly from the flat-film value which is usually referred to simply as the surface tension. This is easily understood by taking surface effects into account.[39] In clusters with less than 20–50 atoms, more than 70% of the constituent atoms are situated on the cluster surface. In clusters of 50–1000 atoms, more than half of the atoms are situated on the surface. Thus, the structure and properties of the cluster are strongly affected by surface forces. In clusters with more than a few thousand atoms, however, the surface contribution becomes less important, and the properties of the clusters come closer to those of the bulk. Several trials to estimate the surface tension have been performed. The Kirkwood–Duff equation is often used to relate the surface tension to the radius of curvature by[40]

$$\frac{\sigma}{\sigma_\infty} = \frac{1}{1 + \delta/r} \qquad (2.39)$$

where σ_∞ is the surface tension of an infinite flat film and δ is on the order of one molecular diameter. The steady-state homogeneous nucleation rate is expressed by[41]

$$J = k \exp(-\Delta G^*/kT) \tag{2.40}$$

$$K = (P/kT)^2 V_c \sqrt{2\sigma/\pi m} \tag{2.41}$$

A nucleation rate of $10^{20}–10^{24}/\text{m}^3$ sec is generally taken as a criterion for cluster formation. Comparison of the nucleation rate for Ar and Al as a function of Mach number, i.e., ratio of the flow velocity to the velocity of sound, is shown in Fig. 2.16.

The growth rate of the cluster is determined by the difference between the condensation and evaporation of atoms from the critical nucleus and is expressed by

$$\frac{dr}{dt} = \xi V_c \frac{1}{\sqrt{2\pi m kT}} \{p - p_c\} \tag{2.42}$$

where ξ is the condensation coefficient and p_c is the saturated vapor pressure for a cluster of radius r and given by the Helmholtz equation as

$$p_c = p_\infty \exp\left\{\frac{2V_c\sigma}{kTr}\right\} \tag{2.43}$$

The rate of condensation of clusters may be calculated if the flow through the nozzle is assumed to be one-dimensional. The continuity, momentum, and energy

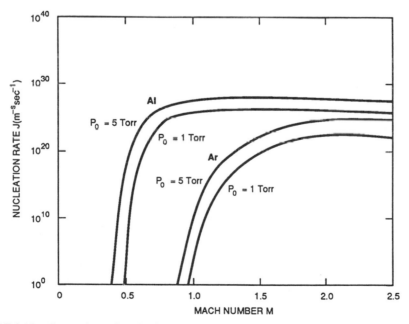

FIGURE 2.16. Comparison of nucleation rate of Al and Ar as a function of Mach number at $P_0 = 1$ Torr and 5 Torr.

equations can be used to compute the effect of the condensation process on gas dynamics[42]:

$$\dot{m} = \rho A V/(1 - \mu)$$

$$-A\, dp = \dot{m}\, dv \qquad (2.44)$$

$$C_p\, dT + V\, dV = h_{fg}\, d\mu$$

where m is the total flow rate, ρ the vapor density, A the cross-sectional area of the flow, μ the ratio by mass of the condensed and vapor phase, and h_{fg} the latent heat of condensation. The vapor is assumed to be a perfect gas, and the state equation,

$$P = (1 - \mu)\rho k T/m \qquad (2.45)$$

was also used.

The effect of the nozzle size on the cluster beam intensity is shown in Fig. 2.17. The results show that the larger size nozzle produces maximum intensity at lower source pressure.

The difference between a cluster beam and a molecular beam is characterized by their velocities. A molecular beam is formed by a so-called "Knudsen cell," i.e., an effusion source that looks similar in shape to what is used in the ICB deposition.

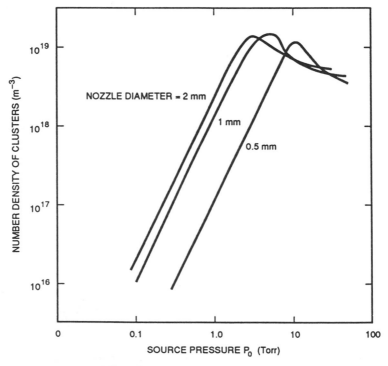

FIGURE 2.17.   Effect of nozzle size on the cluster beam density for silver.

The effusion source is operated under the condition that the orifice diameter d is much smaller than the mean free path of the atoms (λ) ($\lambda/D \ll 1$). The velocity distribution is expressed by[43]

$$n_{u,c(effusive)} = \frac{1}{2}\left(\frac{m}{kT_0}\right)^2 u^3 \exp\left(-\frac{mu^2}{2kT_0}\right)\Delta u \tag{2.46}$$

where u is the velocity of atoms and T_0 is the source temperature. Total intensity after collimation is given by

$$n_{c(effusive)} = \frac{S}{\pi l^2} \tag{2.47}$$

where l is the distance between the orifice and the collimator and S is the area of the small hole on the collimator. For the cluster beam the velocity distribution and the total intensity are given by

$$n_{u,c(nozzle)} = \left(\frac{m}{2\pi kT_1 l^2}\right)^{1/2} \exp\left(\frac{m(u-U)^2}{2kT_1}\right) \tag{2.48}$$

where U is the mass flow velocity and T_1 is the temperature of the particles at the skimmer and

$$n_{c(nozzle)} = \frac{mU^2 S}{2\pi kT_1 l^2} \tag{2.49}$$

The velocity is also given as a function of the Mach number as follows:

$$v = M\left\{\gamma\left(1 + \frac{\gamma-1}{2}M^2\right)^{-1}\frac{kT_0}{m}\right\}^{1/2} \tag{2.50}$$

The maximum attainable velocity occurs when all of the thermal energy is transformed to translational kinetic energy, which corresponds to the Mach number of infinity, and is given as

$$v_\infty = \left(\frac{2\gamma}{\gamma}\frac{kT_0}{m}\right)^{1/2} \tag{2.51}$$

Thus, the cluster beam so generated has considerably higher velocity than that of the molecular beam. In summary, the cluster beam has the following characteristics relative to a molecular beam:

- Operates at a much lower temperature
- Has a much more monochromatic velocity distribution
- Much higher beam intensities are possible

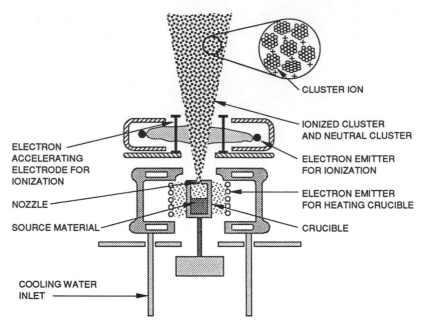

FIGURE 2.18.   Ionized cluster beam source.

Once the clusters have been formed, they are then ionized by electron impact so that the beam may be directed toward a target, with a specified energy, usually in the range of 1–10 kV. Figure 2.18 illustrates a simple cluster ion source. The collision cross section for ionization of a cluster by an electron will, of course, be much larger than for a single atom so that a large fraction of the clusters may be ionized. However, if more than one electron is removed from a cluster, there will be a strong repulsion between the associated ionized atoms that opposes the forces holding the cluster together. For this reason, most clusters are believed to be singly ionized. The energy per atom of the cluster is given by

$$E_u = \frac{meV}{n} \tag{2.52}$$

where V is the beam voltage, n is the number of atoms per cluster, and m is the number of electronic charges lost by the cluster. The application of this source to layer deposition is discussed in Section 4.13.5.

2.3.8. Space-Charge Effects

Whenever one attempts to obtain large currents (electron or ion) in a limited space, the mutual repulsion of the charged particles becomes evident and places definite limits on the particle densities obtainable. In diodes the space charge limits the current density that can be extracted with a given anode potential from both electron and ion sources and formed into beams. Space charge also limits the minimum spot size attainable in optical systems and the maximum current that can be passed through a drift tube. The best-known example of the influence of space charge is the limitation of the current density that can cross a simple infinite parallel-plane diode. In this case the maximum current density that can cross the diode with an electrode spacing d is given by Child's law:

$$j = \frac{4\varepsilon_0}{9}\left(\frac{2e}{m}\right)^{1/2}\frac{V_a^{3/2}}{d^2} \tag{2.53}$$

where V_a is the potential difference between the anode and cathode. This solution, which comes from a direct integration of the one-dimensional Poisson equation, also gives the potential and the space-charge distributions as

$$V = V_a(x/d)^{4/3} \tag{2.54}$$

and

$$-\rho(x) = \frac{4\varepsilon_0}{9}\left(\frac{V_a}{d^{4/3}}\right)x^{-2/3} \tag{2.55}$$

Equation (2.54) shows that the potential is lowered by the charge below what it would be (linear in x/d) were there no space charge. Near the cathode surface the potential depression is just sufficient to cancel any normal field there, clearly a stable self-limiting situation.

The 3/2 power dependence of j on V_a [Eq. (2.53)] may be shown to be generally true for any geometry in which the off-cathode gradient vanishes because of space charge. The ratio $j/V_a^{3/2}$ (perveance) is a geometric factor roughly analogous to conductance. Similar results are available for coaxial and cylindrical diodes.[44] These solutions, along with the parallel-plane case, are especially important in electron- and ion-beam work as the starting point in Pierce's method of high-current electron and ion gun design.

In the infinite-plane parallel diode the potential lowering resulting from space charge is in the direction of electron flow. However, in general, a potential profile transverse to the direction of electron flow can exist. The most important case of this sort is the dense charged-particle beam inside the fixed potential drift tube. Here the space charge results in a radial potential profile, with the beam axis at a lower potential than at the edges. In both the longitudinal and transversal cases we find that the space can support only a certain amount of current under given voltage conditions.

Consider a long charged-particle beam (electron or ion) carrying a current, with the particles moving in the $+z$ direction with a constant velocity given by

$$\dot{z} = \left(\frac{2q}{m}\right)^{1/2}V \tag{2.56}$$

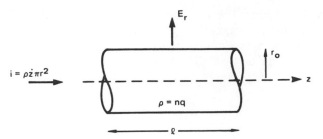

FIGURE 2.19.   Charge enclosed in a cylindrical volume.

The radial force giving rise to the spreading in a nonrelativistic case ($z \gg c$) comes from the charge enclosed in a cylindrical volume (see Fig. 2.19). Using Gauss's theorem,

$$2\pi r l E_r + \pi r^2 \int_0^l \frac{\partial E_z}{\partial z}\, dz = \pi r^2 l \frac{\rho}{\varepsilon_0} \tag{2.57}$$

where $\rho = nq$ is the space-charge density. Since in the drift space $E_z = 0$, the second term on the left-hand side of Eq. (2.57) is equal to zero. Then the space-charge density can be expressed as

$$\rho = \frac{j}{\dot{z}} = \frac{i}{\pi r_0^2 [(2q/m)V]^{1/2}} \tag{2.58}$$

and the radial component of the electric field intensity is

$$E_r = \frac{ir}{2\pi \varepsilon_0 [(2q/m)V]^{1/2} r_0^2} \tag{2.59}$$

At the boundary of the beam ($r = r_0$) the radial component of the field is

$$E_{r_0} = \frac{i}{2\pi \varepsilon_0 [(2q/m)V]^{1/2} r_0} \tag{2.60}$$

From this formula it is evident that at a constant current the radial field intensity increases with decreasing beam radius. In the case of laminar particle flow, this means that the beam does not intersect the axis (no real crossover).

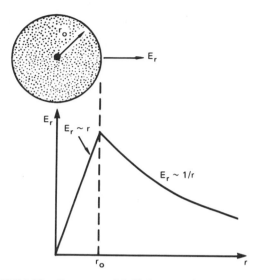

FIGURE 2.20. Currents and fields in space-charge-limited beams.

Equation (2.59) can also be used for the calculation of the field near the outer boundary of the beam ($r > r_0$) (see Fig. 2.20). We get

$$E_r(r > r_0) = \frac{i}{2\pi \varepsilon_0 [(2q/m)V]^{1/2} r} \qquad (2.61)$$

Now the envelope of the beam can be found from the equation of radial motion, $m\ddot{r} = qE_r$. By using the radial space-charge field at the beam edge and $\eta = q/m$, the equation of motion becomes

$$\frac{d^2 r}{dt^2} = \eta \frac{i}{2\pi \varepsilon_0 r (2\eta V)^{1/2}} \qquad (2.62)$$

With the assumption that the axial velocity is constant, the time derivative may be changed to one with respect to z:

$$\frac{d^2 r}{dt^2} = \dot{z}^2 \frac{d^2 r}{dz^2} \qquad (2.63)$$

which on substitution into Eq. (2.62) gives the final ray equation for this problem:

$$\frac{d^2 r}{dz^2} = \frac{1}{4\pi \varepsilon_0 (2\eta)^{1/2}} \frac{i}{V^{3/2}} \frac{1}{r} \qquad (2.64)$$

The first term on the right-hand side is a constant coefficient, which for electrons is given by

$$k_e = 1.52 \times 10^4$$

and for ions

$$k_i = 6.5 \times 10^5 M$$

where M represents the atomic weight of the ions in the beam. The second term is the perveance P. Thus, Eq. (2.63) can be written as

$$
\begin{aligned}
r'' - 1.52 \times 10^4 Pr^{-1} &= 0 \quad \text{for electrons} \\
r'' - 6.5 \times 10^5 \sqrt{M} Pr^{-1} &= 0 \quad \text{for ions}
\end{aligned}
\tag{2.65}
$$

where the primes denote the differentiation with respect to z. The solution of Eq. (2.65) is simplified by the introduction of dimensionless variables R and Z,

$$
\begin{aligned}
R &= \frac{r}{r_0} \\
Z &= \frac{z}{r_0}
\end{aligned}
\tag{2.66}
$$

where r_0 is the initial radius of the beam. With Eq. (2.66), Eq. (2.65) reduces to

$$R'' - \frac{k}{R} = 0 \tag{2.67}$$

where $k = k_e P$ for electrons and $k = k_i PM$ for ion beams.

Equation (2.67) can be solved with the following initial conditions:

$$
\begin{aligned}
Z &= 0 \\
R &= 1 \\
R' &= R_0'
\end{aligned}
\tag{2.68}
$$

If we multiply Eq. (2.67) by $2R'$ we obtain

$$\frac{d}{dZ}(R')^2 = \frac{2kR'}{R} \tag{2.69}$$

This equation can be integrated, taking into account the initial conditions in Eq. (2.68) as

$$R' = [2k \ln R + (R'_0)^2]^{1/2} \tag{2.70}$$

To find the entire beam envelope, $R = f(z)$, this equation must be integrated again.[44]

The beam, if initiated with a convergent slope, will decrease in radius until a certain minimum radius r_{min} is reached, beyond which it will expand again. The beam envelope will be symmetric around this minimum. At the position of minimum beam diameter $R' = 0$ in Eq. (2.70) and the minimum radius of the beam is

$$r_{min} = r_0 \, e^{-(R'_0)^2/2k} \tag{2.71}$$

Equation (2.71) shows that in order to achieve a small spot (i.e., a small minimum radius if the target is placed at the position $r = r_{min}$) a low-perveance beam with a high initial convergent slope should be used. However, a strongly convergent beam exhibits greater spherical aberration, which will limit the minimum spot size. Also, for any target location the smallest attainable radius, defined by the beam envelope, lies on the profile of a beam that has previously reached its waist and is expanding. For a minimum spot size, therefore, it is necessary to produce the beam waist a little before the target.[45,46]

In the case of electron beams the beam spread is also influenced by the residual gas pressure. The ionization of gas molecules by the beam produces positive ions that tend to accumulate along the beam axis.[47] This positive ion neutralization of the beam space charge reduces the spreading of the beam, but gives rise to other undesirable effects, such as beam attenuation from scattering, positive ion bombardment of the cathode, and beam oscillations.

In some applications (e.g., ion milling), neutralization of the ion beam is desired. In these machines, electrons are injected into the ion beam for space-charge neutralization.

2.4. ELECTRON GUNS

2.4.1. Gun Requirements and Limitations

An electron gun is a device that generates, accelerates, and focuses a beam of electrons into a small spot or parallel beam which is used as the object or "source" for a subsequent electron-optical column. In this section we discuss the requirements and limitations for the type of electron gun structure used in microfabrication equipment (electron-beam lithography) and instruments (electron microscopes).

In most electron-optical instruments the electron gun is required to give as high as possible a current density over a prescribed area. The angular divergence of electrons from the focal spot has to be small to minimize lens aberrations or to obtain electron-optical coherence. Thus, the electrons should travel through the focal spot with the highest possible current density per unit solid angle. We regard a gun as "good" if there is a high current density in the focal area, if the focal area is small, and if the aberrations from its optical elements are small.

The "brightness" of a gun is defined as the current density (j) per unit solid angle. The solid angle of a cone of semiangle θ is $2\pi(1-\cos\theta) \approx \pi\theta^2$ for small θ. Thus, the brightness denoted by β or R (from the German term *Richtstrahlwert*) is given by:

$$\beta = \frac{j}{\pi\theta^2} \tag{2.72}$$

As shown in Section 2.4.3, β has an upper limit (Langmuir limit) given by

$$\beta_{max} = \frac{jeV}{\pi kT} \tag{2.73}$$

where j is the current density at the cathode, T is the cathode temperature, $-e$ is the electronic charge, and k is Boltzmann's constant.

This theoretical limitation derives from the fact that electrons emitted from the cathode have a half Maxwellian energy distribution, with a most probable energy of kT and an average energy of $2kT$. Electrons emitted into a free-field space follow Lambert's law:

$$j(\theta) = \frac{j}{\pi}\cos\theta \tag{2.74}$$

where $j(\theta)$ is the emission current density in any direction making an angle θ with the normal and j is the total current density into the hemisphere (see Fig. 2.21). If the electrons are accelerated by a uniform electrostatic field parallel to the optical axis, their velocity component (v_z) remains unchanged. The semiangle of divergence is thus reduced from π to the ratio of perpendicular and parallel velocities $\theta(\cong v_r/v_z)$. If the electrons have an initial energy $eV_0 = kT$ and they are accelerated to an energy, eV, then the new semiangle θ' (Fig. 2.21) can be written in terms of energies; i.e.,

$$\theta' = \left(\frac{kT}{eV}\right)^{1/2} \tag{2.75}$$

and the brightness, using Eq. (2.73), is

$$\beta = \frac{j}{\pi\theta'^2} = \frac{jeV}{\pi kT} \tag{2.76}$$

This expression can be applied irrespective of any focusing system that may follow the gun and, in fact, no focusing system can increase the brightness beyond that of the electron-emitting surface.

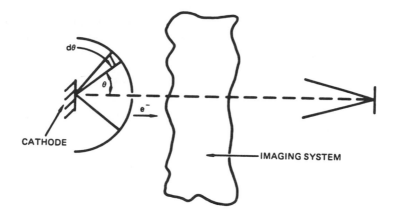

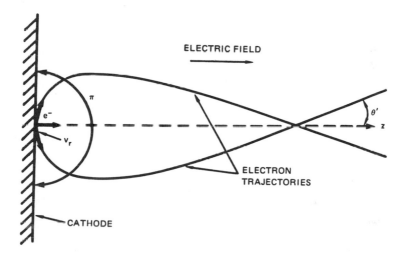

FIGURE 2.21.   Electron trajectories from a flat-surface emitter.

2.4.2. Source Formation

Fermat's law of geometrical optics states that among a number of possible paths between two points A and B, a light ray will take that path that has the shortest transit time. This statement can be expressed mathematically by requiring the variation of the transit time to be zero; i.e.,

$$\delta \int_{A}^{B} \frac{ds}{v} = \delta \int_{A}^{B} \frac{n}{c} ds = 0 \qquad (2.77)$$

where v is the velocity of propagation of the light wave, $v = c/n$ (where n is the index of refraction of the medium through which the light passes), and ds is an element of that path.

A corresponding principle governs the motion of material particles in conservative force fields. This is the principle of least action or the principle of Maupertuis:

$$\delta \int_A^B mv \, ds = 0 \tag{2.78}$$

Since $mv = (2 \, meV)^{1/2}$ from $1/2 \, mv^2 = eV$, this equation is identical with Eq. (2.77) if the refractive index is taken as

$$n = \sqrt{V} \tag{2.79}$$

assuming that V is zero at a point of zero electron velocity.

The use of Eq. (2.54) may be illustrated by the example of the refraction of an electron beam (see Fig. 2.22). Suppose that an electron traveling with uniform velocity v through a space of constant potential V passes a potential step into a space with another homogeneous potential V', so that the path of the electron suddenly changes its direction. Assuming that the potential $V' > V$, the normal velocity component v_y of the electron is increased, but the tangential component and v_x remains unchanged ($v_x = v_x'/x$). Moreover,

$$\sin \alpha = v_x/v \\ \sin \alpha' = v_x'/v \tag{2.80}$$

Therefore,

$$\frac{\sin \alpha'}{\sin \alpha} = \frac{\sqrt{V}}{\sqrt{V'}} \tag{2.81}$$

which is Snell's law for electron optics.

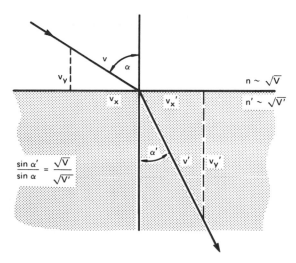

FIGURE 2.22.   Refraction of an electron beam.

In light optics the Abbe sine law states that

$$ny \sin \theta = n'y' \sin \theta' \qquad (2.82)$$

where y and y' are the distances from the axis of an object point and its corresponding image point, θ and θ' are the semiaperture of rays at the object and image (see Fig. 2.23), and n and n' are the refractive indices of the object and image space. In electron optics, when the object is in a region of potential V and the image is in a region of constant potential V', the sine law can be written as

$$\sqrt{V}y \sin \theta = \sqrt{V'}y' \sin \theta' \qquad (2.83)$$

Since $\sqrt{V} \sin \theta$ is proportional to the transverse velocity (v_y) of the electrons, this law is consistent with Liouville's theorem that states that the (yv_y) product is an invariant of the motion. In electron optics, just as in light optics, Abbe's sine law implies that the image formation is paraxial if θ is very small. Now since $\sin \theta = \theta - \theta^3/3! + \theta^5/5!$, Eq. (2.83) can be written for small θ as

$$\sqrt{V}y\theta = \sqrt{V'}y'\theta' \qquad (2.84)$$

When $\sin \theta$ can be approximated by the first term, it is called "Gaussian optics" or first-order theory, as distinguished from third-order theory, which includes the second term in the series.

Relation (2.83) is referred to as the Helmholtz–Lagrange theorem. If there is no loss of electron current in an axially symmetric optical system, and i represents the

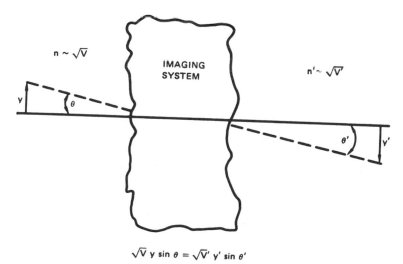

$$\sqrt{V}\, y \sin \theta = \sqrt{V'}\, y' \sin \theta'$$

FIGURE 2.23.   Image formation in electron optics.

total emission current of the source and j the current density, then the electron current density in the image is given by

$$\frac{j'}{j} = \frac{i}{\pi r'^2} \cdot \frac{\pi r^2}{i} = \frac{V' \sin^2 \theta'}{V \sin^2 \theta} \tag{2.85}$$

2.4.3. Beam Characteristics

Consider the characteristics of electron beams generated by guns restricted by the following conditions:

- They are of circular symmetry.
- The emitter has a Maxwellian velocity spectrum, both radially and axially.
- The image and object spaces are related by the Helmholtz–Lagrange relation.
- Space-charge forces are neglected.

Figure 2.24 shows a simplified gun configuration used in microfabrication equipment and instruments.[48] Electrons are extracted from the cathode when suitable potentials are applied to the succeeding electrodes (grid and anode), and these electrons pass through a "crossover," which serves as the "object" in the optical system. After passing through the crossover, the rays diverge into a roughly conical bundle. An aligned circular aperture serves to pass only a small central portion of the total ray cone. This accepted portion of the whole beam is then focused by the lens onto the target or fluorescent screen.

The cathode, grid, and anode collectively form a triode gun which is often referred to simply as "the triode." If the lens is assumed to be "thin," then the lens images the crossover with a magnification $M = b/a$. By simply varying a and b it is apparent that the final spot size can be controlled.

The distinctive feature of such guns is that the cross section of the beam at the screen is determined to a first order not by the area and shape of the emissive surface of the cathode, but by the crossover radius (r_c), which can be much smaller than the

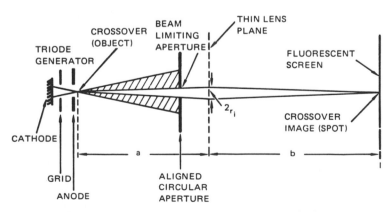

FIGURE 2.24.   Gun configuration for beam fabrication.

radius of the cathode. The first lens of the gun is an immersion objective that forms the crossover and controls the beam current. The crossover is the object for the main lens.

Figure 2.25 shows the configuration and electron trajectories in the near-cathode region. If the electrons leave the cathode with zero initial velocity, they would intersect the optical axis at the same point 0, i.e., the crossover radius r_c would be zero; electrons with initial velocities result in paths that intersect the axis farther than the point, depending on the direction of the initial velocity. [The electrons appear to originate from the surface of the virtual cathode (C′).]

To determine r_c for electrons emerging from the cathode with a velocity v_0 corresponding to an energy eV_0, we use the Helmholtz–Lagrange theorem in the following form:

$$r_0\sqrt{V_0}\sin\gamma_1 = r_i\sqrt{V}\sin\gamma_2 \qquad (2.86)$$

where r_0 and r_i are the radii of the cathode and image, respectively, V_0 and V are the potentials at the cathode and image sides, and γ_1 and γ_2 are the aperture angles.

From Fig. 2.25 we find that $r_c = b\tan\gamma_2 \cong b\sin\gamma_2$ for small γ_2, $r_i \cong b\tan\theta$, and using Eq. (2.86), we obtain

$$r_c = \frac{r_0}{\tan\theta}\left(\frac{V_0}{V}\right)^{1/2}\sin\gamma_1 \qquad (2.87)$$

If the direction of the initial velocity is assumed to be equiprobable in the whole range of γ_1 from 0 to 90°, the maximum value of $\sin\gamma_1$ is equal to unity and the maximum crossover radius (r_c) is given by

$$r_c = a\left(\frac{V_0}{V}\right)^{1/2} \qquad (2.88)$$

where $a = r_0/\tan\theta$ is the distance from the cathode surface to the crossover plane.

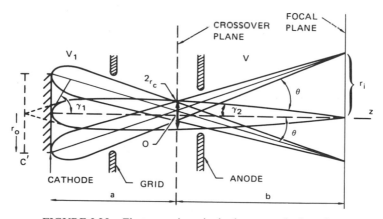

FIGURE 2.25. Electron trajectories in the near-cathode region.

Equation (2.88) shows that to a first approximation the crossover radius does not depend on the area of the emissive surface and is determined only by the ratio of the initial energy of the electrons (eV_0) to the energy of the electrons in the crossover region (eV).

Equation (2.88) was derived assuming that all of the electrons emerging from the cathode have the same initial energy (eV_0), resulting in a crossover having a sharp boundary radius r_c. However, this assumption is rather crude, since the electrons emitted by the cathode actually have a Maxwellian velocity distribution. Electrons having energies lower than eV_0 will intersect the crossover plane inside the crossover circle r_c, but electrons with higher initial energies may cross the plane outside of r_c. Each group of electrons with the same initial energy lies within a specific circle in the crossover plane, and for higher initial electron energies the radius of the crossover circle is greater. In the case of a Maxwellian distribution the electron density in the crossover circle will not have a sharp boundary but will decrease rapidly with the distance from the axis. For this reason the crossover radius usually defined as the radius of the circle that contains 90% of the electrons crossing the plane.

To estimate this crossover radius, one has to know the current distribution in the crossover as a function of the radial distance from the optical axis. According to Maxwell's energy distribution law, the number of electrons (N) emitted per unit area of cathode surface per unit time per unit solid angle and having an initial energy between eV_0 and $e(V_0 + dV_0)$ is given by

$$N(V_0)\, dV_0 = N_0 \frac{eV_0}{kT} e^{-eV_0/kT}\, d\left(\frac{eV_0}{kT}\right) \tag{2.89}$$

where N_0 is the number of electrons having all possible energies emitted by the cathode per unit area per unit time per unit solid angle. N_0 can be obtained from the emission current density of the cathode (j_c), since

$$j_c = e \int_{v_0 = 0}^{\infty} \int_{\gamma = 0}^{\pi} N(V_0)\, d\gamma = e\pi N_0 \tag{2.90}$$

The current emitted from the area A with electron energies in the range eV_0 to $e(V_0 + dV_0)$ into the solid angle from γ to $\gamma + d\gamma$ is given by

$$di_\gamma = 2\pi A N(V_0)\, dV_0\, e \sin \gamma \cos \gamma\, d\gamma \tag{2.91}$$

This current flows into an annular ring of radius r (see Fig. 2.26), where the current density is

$$j_r = \frac{di_\gamma}{d(\pi r^2)} = \frac{A N(V_0)\, dV_0\, e \sin \gamma \cos \gamma\, d\gamma}{r\, dr} \tag{2.92}$$

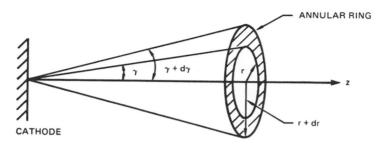

FIGURE 2.26.   Correlation between crossover radius and energy spread $(r \sim V_0)$.

Using Eq. (2.87) to calculate $r\, dr$, one obtains

$$j_0 = \frac{AN(V_0)\, dV_0\, e}{a^2} \frac{V}{V_0} \tag{2.93}$$

where V is the potential at the crossover plane. Equation (2.93) shows that in the annular ring $(r - r_c)$ the current density produced by the electrons having energies between eV_0 and $e(V_0 + dV_0)$ is independent of γ, the initial angle of the emitted electrons.

To calculate the total current density $j(r)$ at a radius r in the crossover plane, it is necessary to integrate the current densities for all annular rings formed by all electrons with energies larger than eV_0, for which $r_c > r$. Here we assume that all of the crossovers are formed in a single plane independently of the initial energy of the emitted electrons; i.e., there is no chromatic aberration in the gun's electron-optical system. Integrating Eq. (2.93), we see that the total current density as a function of the radial distance is given by

$$j(r) = \frac{Ae}{d^2} \int_{eV_0}^{\infty} \frac{V}{V_0}\, dV_0 \tag{2.94}$$

where eV_0 is the energy corresponding to a crossover radius of r. Using the Maxwellian energy distribution [Eqs. (2.88) and (2.89)], we obtain

$$j(r) = \frac{AeN_0}{a^2} \left(\frac{eV}{kT}\right) e^{-(eV/kT)(r^2/a^2)} \tag{2.95}$$

or

$$j(r) = j_0\, e^{-r^2/\rho_0^2} \tag{2.96}$$

where

$$j_0 = \frac{A N_0 e}{a^2} \frac{eV}{kT}$$ (2.97)

equals the current density at the crossover center ($r = 0$), and

$$\rho_0^2 = \frac{a^2 kT}{eV}$$ (2.98)

is a constant for a given operating condition of the gun.

The current density at the crossover center (j_0) is associated with the density of the cathode emission current j_c and leads to the Langmuir limit equation

$$j_0 = j_c \left(\frac{eV}{kT}\right) \sin^2 \theta$$ (2.99)

This is obtained from Eq. (2.96) by replacing a^2 by $r_0^2/\tan^2 \theta \cong r_0^2/\sin^2 \theta$ and taking into account that $A = \pi r_0^2$ and $j_c = e\pi N_0$ from Eq. (2.90).

From Eq. (2.97) it follows that an increased current density in the crossover center can only be obtained by increasing the cathode emission current density (j_c) or decreasing its operating temperature. It is clear that in the case of thermionic cathodes, these requirements are conflicting, since a decrease in the cathode temperature causes a decrease in the emission current density.

For small θ, Eq. (2.99) can be written as

$$j_0 = \pi \beta \theta^2$$ (2.100)

which is identical to Eq. (2.72). This relation is quite general; no assumptions concerning the configuration of fields or beam concentration system were made in its derivation.

The Langmuir limit [Eq. (2.99)] gives the maximum current density in the spot that is theoretically possible; it does not indicate that such a value is actually attainable. In practice, spot current density values, roughly 50% of those predicted by the Langmuir limit equation, can be obtained with well-designed optical systems.[48]

2.4.4. Space-Charge Effects

The simplified theory of forming a crossover is based on the assumption that the electrostatic field in the near-cathode region is uniquely determined by the potentials of the gun electrode configuration. The field created by the space charge of the electrons is not taken into account. The effect of space charge can be ignored only for low-perveance (P) guns where P is defined by

$$P \equiv \frac{i}{V^{3/2}} \, \text{perv} \tag{2.101}$$

for i in amperes and V in volts.

The influence of the internal electric field of the beam only has a perceptible effect on the focus of electrons when $P \geq 10^{-8}$ perv. For the electron beams used in microlithography, P is less than this value so that space-charge effects are not important in the image formation. However, for ion beams used for implantation or annealing, space-charge effects have to be taken into account.

In the near-cathode region where the electron velocity is low, and in the crossover region, where the current density is very high, the coulombic interaction between the electrons is much greater than in the final image spot. Such interactions can have a substantial effect on the object formation. Qualitatively, the presence of space charge in the cathode region results in a radial drift of electrons across the laminar trajectories and a change in the current density distribution.

Figure 2.27a shows the formation of a hollow beam due to the transverse drift of electrons away from the laminar trajectory. Figure 2.27b depicts the Gaussian beam formation in the same triode electron gun without any space-charge effect. We see, therefore, that the action of space charge results in an increase in the crossover radius and a reduction of current in the crossover center. The quantitative theory of space-charge effects will be discussed briefly in connection with ion extraction and ion-beam formation.

2.4.5. Aberrations

The size and shape of the electron beam are also influenced by imperfections in the electron-optical elements that make up the electron gun. We briefly summarize these effects here; more complete information is available in the electron-optics literature.[49–53]

Spherical aberration of the immersion objective (first lens next to the cathode) influences the crossover formation. Its effect can be observed even at low beam currents when electrons are taken only from the central part of the cathode. As the current increases (by increasing the temperature) and electrons are taken from larger areas of the cathode surface, the conditions for paraxiality are violated, and other kinds of aberrations become noticeable. Hence, an increase in beam current is generally accompanied by an increase in the size of the crossover.

The four types of electron-optical defects associated with the electron lenses used in electron-beam devices are listed below.

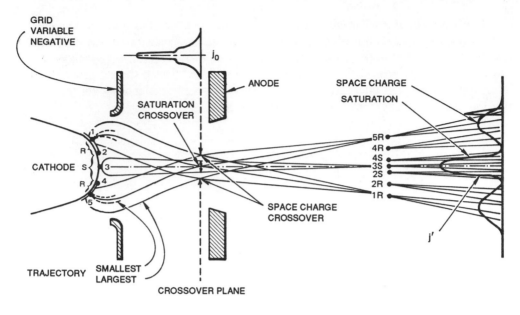

(a) HOLLOW BEAM

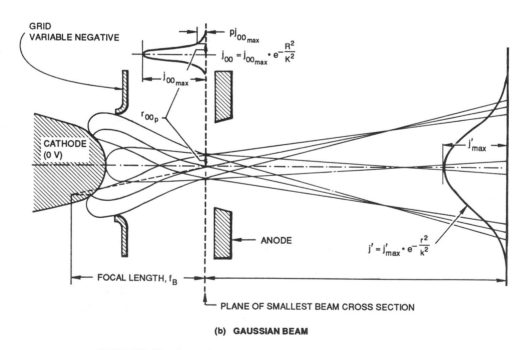

(b) GAUSSIAN BEAM

FIGURE 2.27.   Beam formation. (a) Hollow beam; (b) Gaussian beam.

2.4.5.1. Spherical Aberration.   Aberrations result in a minimum beam diameter, called the disk of least confusion, as shown in Fig. 2.28a. This minimum diameter d_s results from the crossing of several electron trajectories passing through the lens that do not come to a focus at the same axial position. The defect is caused by focusing

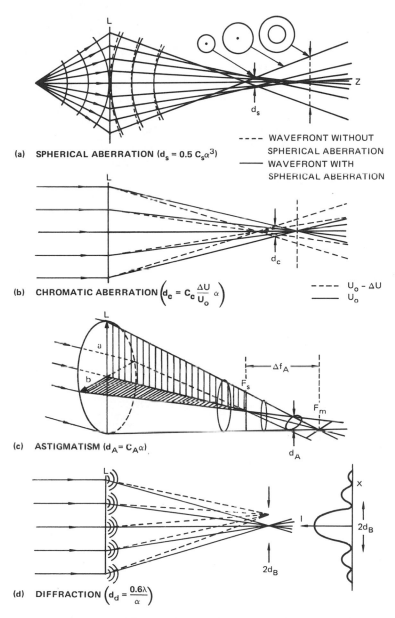

FIGURE 2.28.   Aberrations and diffraction. (a) Spherical aberration ($d_s = 0.5\ C_s a^3$) (– – –, wavefront without spherical aberration; ——, wavefront with spherical aberration); (b) chromatic aberration [$d_c = C_c(\Delta U/U_0)a$] (– – –, $U_0 - \Delta U$; ——, U_0); (c) astigmatism ($d_A = C_A \propto$); (d) diffraction ($d_d = 0.6\lambda/a$).

fields that are stronger near the electrodes that produce them than at the paraxial focal position. The diameter of the disk of least confusion, d_s, is

$$d_s = 0.5C_s\alpha^3 \tag{2.102}$$

where C_s is the spherical aberration constant, related to the paraxial focal length f by $C_s = K_s f$, K_s is a constant dependent on the lens geometry, and α is the convergence half-angle referenced to the image space. It can be related to the equivalent angle on the object side α_0 by the Lagrange–Helmholtz relation [Eq. (2.83)].

It is essentially impossible to compensate for the spherical aberration introduced into a beam by a lens by any subsequent electron-optical action. It is therefore of great importance to try to design each electron-optical element with a minimum C_s.

The effect of this aberration on the final image (spot) can be reduced, at the expense of beam current, by placing a limiting aperture in or before the lens. This aperture reduces the convergence angle α of the focused beam, and thereby the effect of spherical aberration. Such trade-offs between the spot diameter and the current density are part of the optimization of a system design.

The total aberration, resulting from two lenses in series spaced a distance L apart, with spherical aberration coefficient C_{s1} and C_{s2} and focal lengths f_1 and f_2, can be expressed in terms of $K_s = C_s/f$ as

$$K_s = \frac{K_{s1}f_2^2 + K_{s2}(L - f_1)^2}{(f_1 + f_2 - L)^2} \tag{2.103}$$

If the final lens is stronger, $f_1 \gg f_2$, and this expression reduces to

$$K_s = \left(\frac{f_2}{f_1}\right)^2 K_{s1} + K_{s2} \tag{2.104}$$

Thus, the aberration of the stronger lens exerts a greater effect on the aberration of the system.

2.4.5.2. Chromatic Aberration. Chromatic aberration in a lens refers to the sensitivity of the focal properties to the energy with which the particles enter the lens. A particle with a higher energy will come to a focus farther from the lens than a particle with a lower energy. This effect is illustrated in Fig. 2.28b. It is seen that a disk of least confusion exists, and its diameter is

$$d_c = C_c \frac{\Delta U}{U_0} \alpha \tag{2.105}$$

where C_c is the chromatic aberration constant, frequently expressed as $C_c = K_c f$, which indicates that chromatic aberration is also greater in lenses with longer focal lengths, and ΔU is the total energy spread about an average energy U_0.

2.4.5.3. Astigmatism. Astigmatism or the formation of an asymmetric focal spot results when the apertures of the optical elements are not circular, are displaced or tilted relative to the optical axis. The disk of confusion is given by (see Fig. 2.28c)

$$d_A = C_A \alpha \tag{2.106}$$

For a noncircular aperture, C_A is given by δ, where δ is the difference between the axes of the ellipse that forms a limiting diaphragm, $\delta = a - b$. Astigmatism can be corrected by the use of a stigmator, which in its simplest form consists of a multipole element of opposite electric fields arranged around the beam.

2.4.5.4. Diffraction. If a particle beam is passing through a limiting aperture, it will be diffracted to form a spot of diameter (see Fig. 2.28d)

$$d_d = \frac{0.6\lambda}{\alpha} \tag{2.107}$$

where λ is the (nonrelativistic) de Broglie wavelength of the particle and is given for electrons by

$$\lambda \, (\text{Å}) = \frac{12.26}{U_0^{1/2}} \quad (U_0 \text{ in eV}) \tag{2.108}$$

The crossover size, including all of the various aberrations, is usually assumed to be given by the rms sum of the diameters of the disks of confusion, i.e.,

$$2r_{\text{eff}} = [(2r_c)^2 + d_s^2 + d_c^2 + d_A^2 + d_d^2] \tag{2.109}$$

In practice, only d_s is significant for the triode; the others (d_c, d_A, d_d) are usually small enough to be neglected.

2.4.5.5. Defocusing Due to Increased Energy Spread. The thermal energy spread in the electrons emerging from the cathode can be increased by coulombic interaction between electrons.[52] The thermal spread with an initial Maxwellian energy distribution, broadens significantly, and changes to wider Gaussian distribution as the beam current density increases. This broadening, first reported by Boersch,[54] is now generally believed to be mainly a result of electron–electron interaction in the beam crossover(s), where the electrons are most tightly packed.

The total energy spread in the triode is given as

$$(\Delta U_{\text{triode}})^2 = (\Delta U_{\text{th}})^2 + (\Delta U_b)^2 \tag{2.110}$$

and in the complete instrument by

$$|\Delta U_{tot}|^2 = |\Delta U_{triode}|^2 + |\Delta U_{syst}|^2 \tag{2.111}$$

We can gain some physical insight into this effect by considering Fig. 2.29, which describes the interaction in terms of geometrical parameters.[55] A reference electron is considered to travel along the optical axis; a second electron travels along a general trajectory. The closeness of approach is measured by an impact parameter b.

Three interactions are considered. Interactions parallel to b_z, i.e., along the optical axis, lead to energy variations. Scattering along the direction of b_r leads to changes in trajectory displacement. Using Monte Carlo methods[56] the expected mean values for the three interactions have been found to be as follows. The interaction energy change is

$$\Delta U_i \approx |e^2/4\pi\varepsilon_0|F_1(\lambda r_c)/\alpha_o r_c \tag{2.112}$$

the angular displacement

$$\bar{\Delta\alpha} \approx |e/4\pi\varepsilon_0|F_2(\lambda r_c)/2V_b\alpha_o r_c \tag{2.113}$$

and the radial displacement

$$\bar{\Delta r} \approx |e/4\pi\varepsilon_0|F_3(\lambda r_c)/2V_b\alpha_0^2 \tag{2.114}$$

where r_c is the radius of the crossover, α_0 is the beam half-angle subtended at the crossover, V_b is the beam voltage, and λ is given by

$$\lambda = |I_b/V_b^{1/2}|(e^2/m)^{1/2} \tag{2.115}$$

where I_b is the beam current.

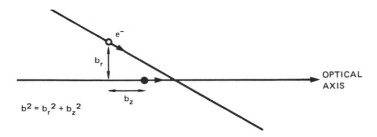

FIGURE 2.29. Geometry of the coulomb interaction between two electrons.

The functions F_1, F_2, and F_3 have the forms shown in Fig. 2.30 when plotted against λr_c. At high current levels, all three functions tend to a $(l_b r_c / V_b)^{1/2}$ dependence. At lower currents the dependence is steeper and more variable. In the high-current regime we can use the relationship between the beam current and the brightness β [Eq. (2.100)] to eliminate α_0 from Eq. (2.112) to obtain

$$\bar{\Delta}U \approx (\beta r_c)^{1/2} V_b^{-1/4} \tag{2.116}$$

This result has been confirmed experimentally.[57]

Since most beam paths in electron-optical columns can be considered as a sequence of crossovers, and since each crossover increases the energy spread, the number of these crossovers has to be minimized to achieve the smallest spot size. Coulombic interactions are "diffractionlike" in the sense that, unlike all other aberrations except for diffraction, they produce a loss of focusing ability that increases as the beam angle is decreased.

In microfabrication systems the ultimate in resolution is not usually required, so that resolution (spot size) can be traded for increases in current. The limitations and trade-offs due to aberrations are shown in Fig. 2.31 as a function of the beam half-angle. In general, system performance is limited by spherical aberration. To increase the spot current the angular aperture has to be increased to its maximum value, consistent with the required final image spot size (d_i), where

$$d_i^2 = d_0^2 + d_s^2 + d_c^2 + d_d^2 \tag{2.117}$$

Here d_0 is the demagnified image of the gun source and d_s, d_c, and d_d are the aberration disk sizes. In this case the optimization is obtained by increasing the current in the final image spot (d_i) by increasing the angular aperture α.

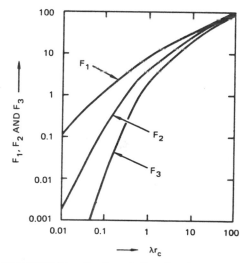

FIGURE 2.30. F_1, F_2, and F_3 as functions of λr_c.

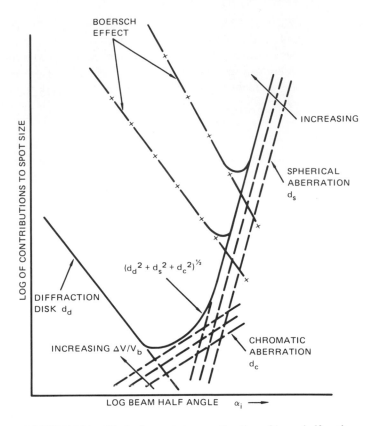

FIGURE 2.31.   Limits for spot size as a function of beam half-angle.

2.4.6. Gun Structures

To extract electrons from the cathode and accelerate and shape them into the desired beam profile, an arrangement of appropriately designed electrodes must be used. This electrode system must create the proper configuration of electric fields at the surface of the cathode and along the acceleration path and is often called the "gun." In this section the more common structures utilized in microelectronics applications are reviewed.

A general model[58] for a triode gun, in which the cathode is a wire filament, is shown in Fig. 2.32. The electrons are emitted from a small region around the tip of the filament. In the immediate vicinity of the tip their trajectories are sharply curved, but the paths soon straighten out and if we extrapolate back we find that the electrons appear to come from a virtual source which, in general, does not coincide with the real filament tip. It is the size of this virtual source that we attempt to keep small in designing the triode.

A triode gun is designed to produce a crossover or focus between the grid (or Wehnelt electrode) and the anode; the size of this crossover is limited by the spherical aberration and the size of the virtual source, and is typically 10–50 μm in diameter.

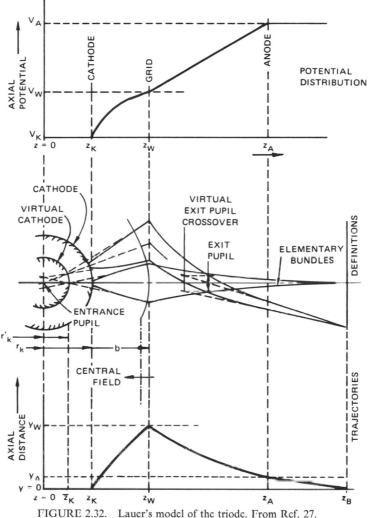

FIGURE 2.32. Lauer's model of the triode. From Ref. 27.

In quantitative terms the radius of the virtual cathode is given by

$$r'_k \approx r_k \left(\frac{V_k}{V_w - V_k} \right)^{1/2} \tag{2.118}$$

where eV_k is the kinetic energy of the electrons at the cathode, r_k is the cathode radius, and V_w is the potential on the axis at the position of the Wehnelt cylinder. For a tungsten filament $eV_k = 0.25$ eV and the radius of the virtual cathode is

$$r'_k \approx r_k \left(\frac{0.25}{V_w} \right)^{1/2} \tag{2.119}$$

hence, the crossover radius can be written as

$$r_c \approx r_k' \frac{1}{1 - (r_k + b)(V_A - V_w)/4(z_A - z_w)V_w} \qquad (2.120)$$

where V_A is the anode potential and B, z_A, and z_w are the position coordinates depicted in Fig. 2.32.

The telefocus gun (Fig. 2.33a) is designed to produce a beam focused at a relatively large distance away from the anode. The long-focus effect results from the hollow shape and negative bias of the Wehnelt electrode. Near the cathode the electric field is diverging so that emerging electrons are given an outward radial velocity. Between the Wehnelt electrode and the anode the equipotentials first become flat and then converge toward the anode. The beam acquires a net radial velocity inward that is of smaller magnitude than the initial outward velocity because of the higher energy of the electrons. Thus, the beam converges quite slowly and has a long focal length. An increase in the bias increases the focal length by increasing the curvature in the cathode region and starting the beam with more divergence, but it also has the effect of reducing the beam current. The spot size is further limited by the beam current owing to the space-charge forces that act over an increased path length.

The gradient gun shown in Fig. 2.33b has postacceleration. Its main advantage is that the beam current is independent of the final beam voltage V_5, since the current is controlled by the voltage V_1 on the first grid. Thus, the total beam power may be varied over a wide range with only a small variation in spot size. On the other hand, relatively large voltages must be used. To take full advantage of the guns' capabilities, the final voltage V_5 must be a great deal larger than V_1, which in turn must be high enough to draw adequate emission from the cathode.

Perhaps the most significant advance in the design of space-charge-dominated electron guns was made by Pierce,[44] who evolved a philosophy of design to which all high-current-density guns have been built in the past decades. He proposed to achieve a laminar flow in the emerging beam from the gun by forcing the electric field configuration external to the beam to be similar to that anticipated from an ideal diode. Pierce showed that the effect of space charge within a finite electron beam could be compensated for by a set of beam-forming electrodes that causes the field in the region external to the beam to satisfy the proper boundary conditions at the beam edge. By forcing this boundary condition, the gun can be designed as if the electron flow were space-charge limited between the cathode and anode of an ideal diode of planar, spherical, or cylindrical configuration (see Fig. 2.33c).

The current density that can be drawn with infinite plane parallel electrodes under space-charge-limited emission conditions in a diode of spacing d with a potential difference V is given by Child's law as

$$J = \frac{4\sqrt{2}\varepsilon_0}{9}\left(\frac{e}{m}\right)^{1/2}\frac{V^{3/2}}{d^2} = 2.33 \times 10^{-6}\frac{V^{3/2}}{d^2} \quad \text{A/cm}^2 \qquad (2.121)$$

where V is expressed in volts and d in centimeters.

The Pierce gun is designed to produce a uniform current density over the beam cross section. For converging guns the minimum beam diameter will lie well beyond the anode, and thereafter the beam will diverge uniformly. Aside from the design simplicity, the principal advantage of the Pierce gun is its high efficiency, which may range to 99.9% or more (i.e., less than 0.1% of the cathode current will be lost to the gun electrodes) is reasonable care is exercised.

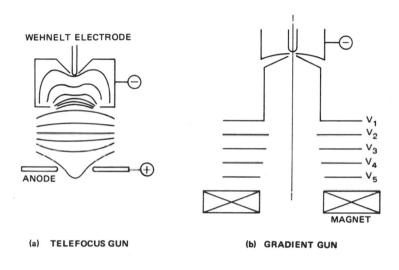

(a) **TELEFOCUS GUN**   (b) **GRADIENT GUN**

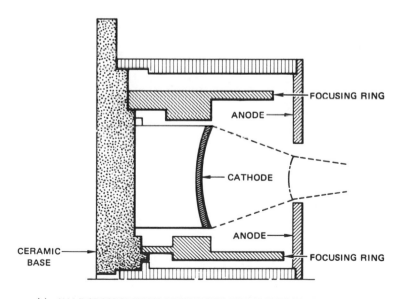

(c) **HALF CROSS SECTION OF MODIFIED PIERCE-TYPE TOROIDAL GUN**

FIGURE 2.33. Sample electron gun structures. (a) Telefocus gun; (b) gradient gun; (c) half cross section of modified Pierce-type toroidal gun.

Guns that produce a line source rather than a round source are also of interest. This is because most patterns are made up of orthogonal rectangular-shaped elements which can be imaged from a line source by independently varying the magnification of the length and width of the source. Much higher total currents can be put into a rectangular element with the same current density and resolution as a round source. This could result in higher throughputs in lithography and faster readouts for testing and microscopy.[59] The use of scanned line beams for rapid thermal annealing has proved to be of particular value.

2.4.7. Gun Optics

The factors that limit the crossover size and the maximum achievable current in the crossover are source brightness, lens aberrations, diffraction, path changes due to space charge, and increased energy spread due to coulomb interaction. In general, these factors cannot be eliminated. However, depending on the configuration and application, they may vary in significance from negligible to performance limiting.

As can be seen from Table 2.2, field-emission sources have a very high brightness as compared to thermionic sources. However, this high brightness can be achieved only over very small source sizes, resulting in a low total beam current. Hence, the advantage of the high brightness of field-emission sources over conventional thermionic sources can be utilized only for small beam sizes (10–1000 Å) such as are usually employed in electron microscopy. The optical properties of point-cathode field-emission guns have been studied in some detail.[60] We note that in field-emission guns the object is usually not the crossover but the point behind the tip from which the electrons appear to originate, i.e., the "virtual" cathode. Thus, the final spot size (d) is set by the aberrations of the optical column and the allowed half-angle of the lens system. In other words, the source size is so small that it cannot be resolved by the lens system of the optical column.

In the case of a thermionic cathode with a resolvable source size (d_0), the final spot (d) is set by the effective crossover size ($d_i = 2r_c$) and the aberrations of the lenses in the optical column. By the use of Eq. (2.109) the crossover size is given as

$$d_i^2 = (2r_c)^2 = d_g^2 + d_s^2 + d_c^2 + d_d^2 \tag{2.122}$$

where d_g^2 is the Gaussian spot size at the crossover (demagnified image of the source $d_g = m d_0$). But from Eq. (2.72) $j = \pi \beta \alpha^2$, and the current contained within the Gaussian spot

$$\text{area} = \pi \frac{d_g^2}{4} \tag{2.123}$$

is

$$i = \beta \frac{\pi^2}{4} d_g^2 \alpha^2 \tag{2.124}$$

and

$$d_g^2 = \left(\left|\frac{4i}{\pi^2\beta}\right|\right)\frac{1}{\alpha^2} = \frac{C_0^2}{\alpha^2} \qquad (2.125)$$

Then, using Eqs. (2.125) and (2.102)–(2.107), in Eq. (2.121) we get for the effective crossover size

$$d_{\text{eff}}^2 = (C_0^2 + (0.6\lambda)^2)\frac{1}{\alpha^2} + \tfrac{1}{4}C_s^2\alpha^6 + \left(C_c\frac{\Delta U}{U_0}\right)^2\alpha^2 \qquad (2.126)$$

Now we can calculate the maximum beam current that can be achieved for a given source size d_0, a lens system (C_s, C_c), and desired operating characteristics [λ (V_0), α].

First we consider a resolvable thermionic source, and we will calculate the maximum current in the crossover with spot size d_{eff}. Figure 2.34a shows the simplified optical system of this gun, which images on an axis source of size d_0 to a final spot d_{eff} at the crossover plane. The imaging system is a single lens characterized by a linear magnification m and aberration coefficients C_c and C_s.

Since this type of optical system is usually limited by spherical aberration, the effective spot size is given [from Eq. (2.126)] by

$$d_{\text{eff}}^2 \approx \frac{C_0^2}{\alpha^2} + \tfrac{1}{4}C_s^2\alpha^6 \qquad (2.127)$$

The spot size is minimized for a given current with respect to the beam half-angle if

$$\frac{\partial d_{\text{eff}}^2}{\partial\alpha} = -\frac{2C_0^2}{\alpha^3} + \tfrac{6}{4}C_s'^2\alpha^5 = 0 \qquad (2.128)$$

which gives the optimum half-angle for minimum spot size for a given current,

$$\alpha_{\text{opt}} = \left(\frac{4}{3}\right)^{1/8}\left(\frac{C_0}{C_s}\right)^{1/4} \qquad (2.129)$$

And with this

$$d_{\text{eff,min}} = (\tfrac{4}{3})^{3/4}(C_0^3 C_s)^{1/2} \qquad (2.130)$$

Using Eq. (2.125) in (2.130), the maximum current in the beam is given by

$$i = \frac{3\pi^2}{16}\beta C_s^{-2/3}d_{\text{eff}}^{8/3} \qquad (2.131)$$

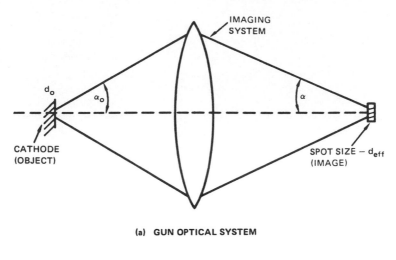

(a) GUN OPTICAL SYSTEM

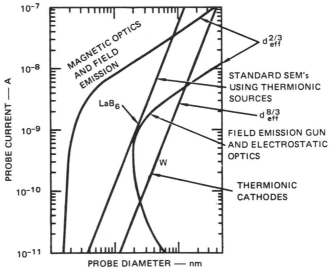

(b) PROBE CURRENT AS A FUNCTION OF PROBE DIAMETER

FIGURE 2.34. Imaging system and current relationships for electron guns. (a) Gun optical system; (b) probe current as a function of probe diameter.

as a function of the spot size d_{eff}. Figure 2.34b shows the maximum probe current as a function of the spot size for thermionic (W and LaB$_6$) and field-emission cathodes.[60]

For the extreme case of a nonresolvable source, consider again the optical system shown in Fig. 2.34a with our ideal point source ($d = 0$). For this condition Eq. (2.126) reduces to

$$d_{\text{eff}}^2 \approx d_g^2 + \tfrac{1}{4} C_s^2 \alpha^6 = (m d_0)^2 + \tfrac{1}{4} C_s^2 \alpha^6 \approx \tfrac{1}{4} C_s^2 \alpha^6 \qquad (2.132)$$

for a spherical-aberration-limited system. Because the source is an ideal point, the usual definition of source brightness (β) (in $A/cm^2 \, sr^{-1}$) is meaningless, since the source current density (j_c) cannot be defined. In this case the source parameter of interest is the angular brightness, Ω (A/sr). This source parameter is applicable to field-emission or thermal/field-emission sources that have very small size (0.1–1 μm). For these sources the beam current delivered into a solid angle defined by the source half-angle α_0 can be expressed in terms of the angular brightness Ω as

$$i = \Omega \pi \alpha_0^2 \tag{2.133}$$

Using Snell's law we can relate the half-angle in the object space α_0 to the half-angle in the image space by the equation

$$\alpha_0 = m\alpha \left(\frac{V}{V_0}\right)^{1/2} \tag{2.134}$$

where V and V_0 are the beam potentials in the image and object spaces, respectively. Substituting Eq. (2.134) into Eq. (2.133), we obtain for the beam current

$$i = \Omega \pi m^2 \left(\frac{V}{V_0}\right) \alpha^2 \tag{2.135}$$

The beam half-angle α in the image space is defined in terms of the final effective spot size for a spherical-aberration-limited system from Eq. (2.132) as

$$\alpha = \left(\frac{2d_{\text{eff}}}{C_s}\right)^{1/3} \tag{2.136}$$

and substituting Eq. (2.136) into (2.135) we get the maximum beam current in the imaged spot (d_{eff}):

$$i = \Omega \pi m^2 \left(\frac{V}{V_0}\right) \frac{2d_{\text{eff}}^{2/3}}{C_s^{2/3}} \tag{2.137}$$

A comparison of Eqs. (2.137) and (2.131) illustrates an important difference in the beam current capabilities of optical imaging systems with nonresolvable and resolvable sources. For the case of a resolvable source, Eq. (2.131) shows that the beam current is proportional to the $\frac{8}{3}$ power of the spot diameter (d_{eff}), but for a nonresolvable source the beam current is proportional to the $\frac{2}{3}$ power of the final spot diameter (d_{eff}), as shown by Eq. (2.137).

In practice, this means that more current at small spot sizes can be achieved with field-emission sources than with conventional thermionic sources, whereas the converse is true for larger spot sizes, as shown in Fig. 2.34.

In the case of small sources the current dependency on the spot size is generally given as

$$i = cd_{\text{eff}}^{2/3} \tag{2.138}$$

where the constant (c) depends strongly on the optical system constants (C_s) and on the electrode arrangement (diode, triode, tetrode) in the gun. With an electrostatic triode gun (Crewe gun[61]) a spot size of 100 Å has been achieved, which can be

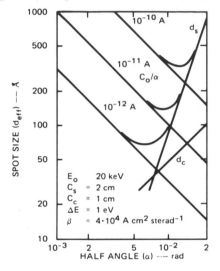

(a) SPOT SIZE AS A FUNCTION OF BEAM HALF ANGLE

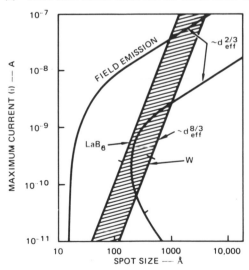

(b) MAXIMUM CURRENT AS A FUNCTION OF SPOT SIZE

FIGURE 2.35. Limits for spot size and maximum current as functions of beam half angle for LaB$_6$, W, and field-emission cathodes. (a) Spot size as a function of beam half angle; (b) maximum current as a function of spot size.

improved by the use of magnetic optical elements as shown in Fig. 2.34b. Generally, the spot (probe) current (i) is a function of the spot size (d_{eff}), the type of electron source, and the type of optical elements (magnetic or electrostatic). However, the spot size is limited by the beam half-angle α for a given optical system (see Fig. 2.31). Thus, for system specifications two sets of data $d_{\text{eff}}(\alpha)$ and $i(d_{\text{eff}})$ are required, as shown in Fig. 2.35a, b.

For example, it can be seen from these figures that for a scanning electron microscope operating at a probe current in the range $i = 10^{-12}$–10^{-11} A a probe size $d_{\text{eff}} = 100$ Å is practical with a thermionic cathode. The angular aperture of the optical system should be in the range 5×10^{-3}–10^{-2} rad. With the use of α the limiting aperture sizes can be calculated for a given electrode geometry.

2.5. COMPONENTS FOR ELECTRON AND ION OPTICS

2.5.1. Electrostatic Lenses and Mirrors

The earlier discussion (Section 2.4.2) of Snell's law introduced the analogy between electron dynamics and geometrical optics. From the discussion we saw that the quantity analogous to the optical index of refraction is $\sqrt{V}$. Thus, considering the equipotential surfaces as refractive surfaces of an optical medium, we can find the trajectories of the electrons in an electric field by using the laws of geometrical optics. In the following paragraphs we will use electrons as test particles; but, by changing e/m, all of the conclusions are also valid for ions.

In a magnetic field the force acting on a charge a particle depends on the magnitude and direction of the particle velocity. Using the analogy between geometrical optics and charged-particle motion, we observe that the presence of a magnetic field creates an anisotropic medium instead of an isotropic medium, as in the case of an electric field.

The refractive index in a magnetic field is expressed as[49,52,53]

$$n = \frac{e}{m}(A \cdot s) \tag{2.139}$$

where A is the vector potential and s is a unit vector tangent to the trajectory.

Although the analogy between light and charged-particle optics is close, it is important to bear in mind the distinctions between the propagation of light and the motion of charged particles. First, the energy of charged particles moving in an electric field varies continuously, but the energy of a photon does not change. Second, the refractive index in light optics can change discontinuously at the interface between two media with different refractive indices, but in electron optics the potential and therefore the refractive index varies continuously from point to point. Consequently, the path of a light ray usually consists of straight-line sections, while the trajectory of a charged particle is a smoothly varying curve. The third important difference between light and electron optics is the fact that in light optics the shape of refracting surfaces and the refractive index are not interconnected, but in electron optics the refractive index ($\sqrt{V}$) and the shape of the refracting (equipotential) surfaces are interdependent through the Laplace equation

$$\nabla^2 V = 0 \tag{2.140}$$

Finally, charged particles can interact through coulomb forces, whereas there is no interaction between light particles.

On the other hand, despite these differences, the analogy between light and electron optics is very useful for a large set of practical beam design problems that we are concerned with.

A bundle of electron paths passing through a common point near the axis of an axially symmetric field system can be made to pass through another common point (or if the paths diverge, to appear to have passed through such a point) through the action of a relatively limited region of field variation. It seems appropriate to call the first common point the object, the second the image, and the region of fields bringing this about the electron lens. Imperfections in this process are naturally called aberrations. In electron optics the potentials often differ on opposite sides of the lens, corresponding to different media on opposite sides of a light lens. Indeed, there may be a varying potential on one side corresponding to a medium of continuously varying index of refraction. Here we will first discuss purely electrostatic lenses and then consider the purely magnetic ones.[52,62,63]

In the following paragraphs we consider only the motion of bundles of electrons that are near the axis of an axially symmetrical system and whose paths make only small angles with the axis. The potential near the axis of a symmetrical system cannot depend on the angle ϕ and, moreover, must be an even function of r. If we assume a power-series expansion in r for the potential, Laplace's equation enables us to determine the coefficients of the series in terms of the potential and its derivatives on the axis only. Thus, we write

$$V(r, z) = V_0(z) + a_2(z)r^2 + a_4(z)r^4 + \cdots \tag{2.141}$$

where $V_0(z)$ is the potential on the axis.

Laplace's equation in cylindrical coordinates, with no ϕ dependence, is

$$\frac{1}{r}\frac{\partial}{\partial r}\left(r\frac{\partial V}{\partial r}\right) + \frac{\partial^2 V}{\partial z^2} = 0 \tag{2.142}$$

Performing the indicated operations on the first term of the power series, we obtain

$$\frac{1}{r}\frac{\partial}{\partial r}\left(r\frac{\partial V}{\partial r}\right) = 2^2 a_2 + 4^2 a_4 r^2 + 6^2 a_6 r^4 + \cdots = 0 \tag{2.143}$$

Similarly, we can find the expression for $\partial^2 V/\partial z^2$, which must equal the expression in Eq. (2.143),

$$-\frac{\partial^2 V}{\partial z^2} = -(V_0'' + a_2'' r^2 + a_4'' r^4 + \cdots) \qquad (2.144)$$

where the primes indicate the derivative with respect to z. Comparing these series term by term and solving successively for the coefficients, we finally find that

$$V(r, z) = V_0(z) - V_0''(z)\frac{r^2}{2^2} + V_0''''(z)\frac{r^4}{2^2 4^2} - \cdots \qquad (2.145)$$

This series converges quickly for small r and enables us to solve electron-optical problems after we have determined only the potential on the axis.

To take advantage of the convenience of Eq. (2.145), it is necessary to write the equations of motion in a particular form, which we will now derive. It is assumed, of course, that there is no motion in the ϕ direction; all electrons follow paths that lie in r–z planes. In this case the equations of motion are particularly simple:

$$\ddot{r} = \frac{e}{m}\frac{\partial V}{\partial r} \qquad (2.146)$$

and

$$\ddot{z} = \frac{e}{m}\frac{\partial V}{\partial r} \qquad (2.147)$$

where e is the charge and m is the mass of the particle. Time can be eliminated and the two equations combined into a single radial trajectory equation as follows:

$$\dot{r} = \frac{dr}{dz}\dot{z}, \qquad \ddot{r} = \frac{dr}{dz}\ddot{z} + \frac{d^2r}{dz^2}\dot{z}^2 \qquad (2.148)$$

and substituting from Eqs. (2.147) and (2.148) into (2.146) gives

$$-\frac{e}{m}\frac{\partial V}{\partial r} = -\frac{e}{m}\frac{dr}{dz}\frac{\partial V}{\partial z} + \frac{d^2r}{dz^2}(\dot{z})^2 \qquad (2.149)$$

But by using the energy conservation, $1/2m\,(\dot{r}^2 + \dot{z}^2) = eV$ and with Eq. (2.148) we can write

$$\dot{z}^2 = \frac{-2(e/m)V}{1 + (dr/dz)^2} \qquad (2.150)$$

Using Eq. (2.150) in (2.149) and rearranging gives us the general ray equation

$$2V\frac{d^2r}{dz^2} = \left(\frac{\partial V}{\partial r} - \frac{dr}{dz}\frac{\partial V}{\partial z}\right)\left[1 + \left(\frac{dr}{dz}\right)^2\right] \tag{2.151}$$

This is a nonlinear differential equation for r as a function of z, but it can be made linear by assuming that the paths never make large angles with the axis (paraxial rays). In this case $(dr/dz)^2 \ll 1$, and the equation, now called the paraxial ray equation, becomes

$$2V\frac{d^2r}{dz^2} = \frac{\partial V}{\partial r} - \frac{dr}{dz}\frac{\partial V}{\partial z} \tag{2.152}$$

It is now possible to substitute the series expansion [Eq. (2.145)] into this equation. Since it has already been assumed that r and dr/dz are small, it is reasonable to approximate V, $\partial V/\partial r$, and $\partial V/\partial z$ by the first terms of only their respective series. Thus,

$$V \approx V_0(z) \tag{2.153}$$

$$\frac{\partial V}{\partial r} \approx V_0''(z)\frac{r}{2} \tag{2.154}$$

$$\frac{\partial V}{\partial z} \approx V_0'(z) \tag{2.155}$$

These substitutions lead to the final form of the paraxial ray equation:

$$\frac{d^2r}{dz^2} + \frac{dr}{dz}\left(\frac{V_0'}{2V_0}\right) + \frac{r}{4}\frac{V_0''}{V_0} \tag{2.156}$$

This equation has several significant features. First, it is independent of the ratio e/m. Second, the derivatives V_0' and V_0'' are normalized with respect to V_0; hence, it is the field distribution or shape but not its intensity that governs the trajectories. Finally, the equation is unchanged in form if a scale factor is applied to r. It is this fact that allow us to talk of electron lenses, since it implies the same focus ($r = 0$) for all trajectories parallel to the axis and independent of the initial radius. Of course, this fortunate result is the outcome of neglecting all but the first terms of the potential and its axial derivatives. If an extra term is included in the analysis, the various aberrations of such lenses are revealed. These are discussed in Section 2.5.4. The paraxial ray equation can be rewritten in the form

$$\frac{d}{dz}\left(\sqrt{V_0}\frac{dr}{dz}\right) = -\frac{r}{4}\frac{V_0''}{\sqrt{V_0}} \tag{2.157}$$

And from this, after integration,

$$\sqrt{V_0}\,\frac{dr}{dz}\bigg|_1^2 = -\frac{1}{4}\int_1^2 \frac{rV_0''}{\sqrt{V_0'}}\,dz \qquad (2.158)$$

where 1 and 2 are as yet only arbitrary limit points. The integral will receive contributions to its value only in regions where V_0'' has some finite value. In many cases of interest, only a limited range of z contains such values of V_0''. Elsewhere the field $-V_0''$, or the potential V_0 are constant, making $V_0'' = 0$ there. The limits 1 and 2 are chosen at values of z that completely enclose this active lens region, as shown schematically in Fig. 2.36.

Equation (2.158) demands only that the active region of the lens be limited between points 1 and 2, but if this requirement is made even more stringent, the equation is easier to solve. The lens must be thin (or weak); i.e., its active region is short compared to its focal length, so r cannot change appreciably between points 1 and 2 (although dr/dz does). Now r can be taken out of the integral and we obtain

$$\sqrt{V_2}\left(\frac{dr}{dz}\right)_2 - \sqrt{V_2}\left(\frac{dr}{dz}\right)_1 = -\frac{r}{4}\int_1^2 \frac{V_0''}{\sqrt{V_0}}\,dz \qquad (2.159)$$

If $V_0(z)$ is known explicitly, the integral can be evaluated either analytically or numerically. The focal lengths f_1 for electrons moving parallel to the axis and to the right in region 1, and f_2 for those moving parallel to the axis and to the left in region

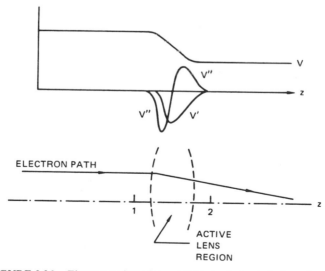

FIGURE 2.36. Electron trajectories and fields in electrostatic lens elements.

2, are found by setting $(dr/dz)_1 = 0$ and $(dr/dz)_2 = 0$ separately (see Fig. 2.37). This procedure results in

$$\frac{1}{f_2} = \frac{-(dr/dz)_2}{r_1} = \frac{1}{4\sqrt{V_2}} \int_1^2 \frac{V_0''}{\sqrt{V_0}} \, dz \tag{2.160}$$

and

$$\frac{1}{f_1} = \frac{-(dr/dz)_1}{r_2} = \frac{1}{4\sqrt{V_1}} \int_1^2 \frac{V_0''}{\sqrt{V_0}} \, dz \tag{2.161}$$

Thus,

$$\frac{f_2}{f_1} = \frac{\sqrt{V_2}}{\sqrt{V_1}} \tag{2.162}$$

which is again analogous to the corresponding law in light optics for a lens between two different media.

If f_2 is positive, then f_1 is negative, but both conditions imply convergence. The sign of the lens depends on the sign of V_0'', which determines the integral, since all other factors are positive. Thus, wherever the axial potential curves concavely upward, it is a converging region, and wherever it curves concavely downward, it is diverging.

To obtain more physical insight it is instructive to write $1/f_2$ in a different form. If the integral of Eq. (2.160) is integrated by parts, we obtain two terms contributing to $1/f_2$:

$$\frac{1}{f_2} = \frac{1}{4\sqrt{V_2}} \left(\frac{V_0'(z_2)}{\sqrt{V_2}} - \frac{V_0'(z_1)}{\sqrt{V_1}} \right) + \frac{1}{8\sqrt{V_2}} \int_1^2 \frac{(V_1')}{V_0^{3/2}} \, dz \tag{2.163}$$

If the axial fields on both sides of the lens are zero, then only the integral remains. It is clear from the squared factor in the integral that this must indicate a convergent lens. We are thus led to an important and useful result: *All electric lenses then lie between field-free regions are convergent.* Such lenses are formed by the field

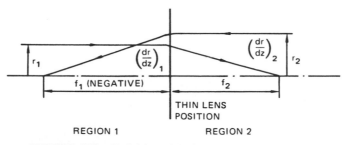

FIGURE 2.37.   Definition of the focal points for a thin lens.

between the ends of two long coaxial cylinders, as shown in Fig. 2.38a. These lenses, although always converging, are asymmetrical because of the differing potentials on each side. To explain the converging nature of these lenses, consider the influence of the longitudinal field component on the particle motion in the case of a symmetrical two-electrode (bipotential) lens, shown in Fig. 2.38a.

In passing from the first half of the lens to the second, the sense of the radial field is reversed; if the field had a radial component only, it would have no overall effect. There is, however, a longitudinal component that accelerates the electrons as they pass through the lens, which as a result are moving faster in the second half of the lens than in the first half. In the first half, therefore, the converging power is increased, and in the second the diverging power is diminished; i.e., the whole lens is convergent. In the three-electrode lens the action can be analyzed similarly (see Fig. 2.38b) into an assembly consisting of a convergent unit between two divergent units.

Two forms of these symmetrical "einzel" or unipotential lenses are shown in Fig. 2.38c. These lenses have the virtue of leaving the beam energy unchanged and

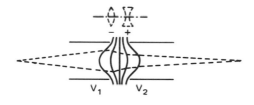

(a) BIPOTENTIAL LENS

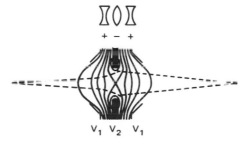

(b) THREE-ELECTRODE LENS

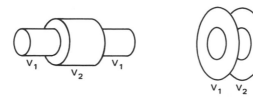

(c) TWO TYPES OF UNIPOTENTIAL (EINZEL) LENSES

FIGURE 2.38. Examples of electrostatic lens configurations. (a) Bipotential lens; (b) three-electrode lens; (c) two types of unipotential (Einzel) lenses.

may be "inserted" at any convenient position along the beam. Regardless of whether $V_2 < V_1$ or $V_2 > V_1$, the lens is convergent and its power is conveniently adjusted by varying V_2. In practice, V_2 is usually less than V_1 because a greater lens power can be achieved with smaller changes in voltage.

When either of the gradients $V_0'(z_2)$ or $V_0'(z_1)$ are not zero, the first term of Eq. (2.159) dominates; then, if the lens is thick, $V_1 \sim V_2$:

$$\frac{1}{f_2} \approx \frac{1}{4V_2}[V_0'(z_2) - V_0'(z_1)] \tag{2.164}$$

This well-known expression applies to the lens action of a round aperture in a plate with different longitudinal fields on either side. Such apertured plates, or their equivalent, are virtually unavoidable in electrons guns in which electron are accelerated to an anode through which they must pass. The lens action of these anodes must be taken into account by a suitable aplication of Eq. (2.164).

In the following paragraphs we summarize the optical properties of three types of electrostatic lenses: the aperture lens, the bipotential (immersion) lens, and the unipotential lens.

The aperture lens Fig. 2.39a is formed by a disk-shaped electrode with a circular aperture at a potential V_d. Fields with constant but different values, E_1 and E_2, act on both sides of the electrode. In some particular cases, one of these fields may be absent (e.g., E_1 or E_2 is along the axis and $V_0'' \neq 0$).

The optical power of an aperture lens can be easily calculated with some approximation. Since the potential near the aperture varies insignificantly, we may in a first approximation transfer the term V_0 in Eq. (2.160) from under the integral sign. In this case the aperture lens power is calculated by the following approximate formula:

$$\frac{1}{f} \approx \frac{1}{4V_0}[']_1^2 = \frac{1}{4V_d}(V_2' - V_1') \tag{2.165}$$

where V_1' and V_2' are potential gradients on both sides of the aperture lens. Taking into account that $V' = E_2$, Eq. (2.165) may be conveniently rewritten in the form

$$\frac{1}{f} = \frac{|E_2| - |E_1|}{4V_d} \tag{2.166}$$

where E_1 and E_2 are the field intensities on the two sides of the aperture.

Depending on the absolute values of the field intensities, the aperture lens can be converging (positive) or diverging (negative). In the case where an electron passes from the domain with a lower intensity to the domain of a higher intensity of the field, the lens is converging. The direction of the force acting on the electron in the lens domain is determined by the sign of the second derivative V'''. The aperture lens has a limited application as a self-contained focusing device, but it is often used as a component in electron-optical systems.

The bipotential lens is usually formed by two coaxial apertures, an aperture and a cylinder, or two coaxial cylinders at different potentials (Fig. 2.39b). The diagrams

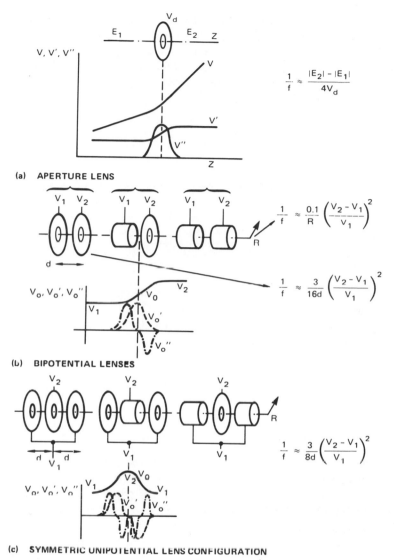

$$\frac{1}{f} \approx \frac{|E_2| - |E_1|}{4V_d}$$

(a) APERTURE LENS

$$\frac{1}{f} \approx \frac{0.1}{R} \left(\frac{V_2 - V_1}{V_1}\right)^2$$

$$\frac{1}{f} \approx \frac{3}{16d} \left(\frac{V_2 - V_1}{V_1}\right)^2$$

(b) BIPOTENTIAL LENSES

$$\frac{1}{f} \approx \frac{3}{8d} \left(\frac{V_2 - V_1}{V_1}\right)^2$$

(c) SYMMETRIC UNIPOTENTIAL LENS CONFIGURATION

FIGURE 2.39. Lens and field configurations. (a) Aperture lens; (b) bipotential lenses; (c) symmetric unipotential lens configuration.

of the variations of the potentials as well as their first and second derivatives are shown in Fig. 2.39b. When the electrons pass through a bipotential lens, their velocities vary; at $V_2 > V_1$ the lens is accelerating and at $V_1 > V_2$ it is decelerating.

From Fig. 2.39b it is clear that the bipotential lens has zones with a positive and a negative second derivative of the potential $V_0''(z)$; i.e., it has converging and diverging portions of the field. However, the particle velocity determined by the potential value is higher in that zone where $V_0''(z) < 0$; i.e., the diverging region is passed by the electron in a shorter time, so that an impulse imposed upon this electron radially toward the axis turns out to be greater than that directed from the axis. Thus, the converging action is predominant.

The focal lengths of the bipotential lenses cannot be calculated from the paraxial equation for every geometry; however, the focal length of the weak immersion lens formed by two apertures is given by the expression

$$\frac{1}{f} \approx \frac{3}{16d}\left(\frac{V_2 - V_1}{V_1}\right)^2 \tag{2.167}$$

where d is the distance between the apertures. This expression is valid with

$$\frac{V_2 - V_1}{V_1} \le 0.2 \tag{2.168}$$

From Eq. (2.167) it is evident that the bipotential aperture lens is always converging, regardless of the sign of the potential difference $V_2 - V_1$.

The approximate expression for the focal length of a bipotential lens formed by two cylinders of the same radius R is given as

$$\frac{1}{f} \approx \frac{k}{R}\left(\frac{V_2 - V_1}{V_1}\right)^2 \tag{2.169}$$

where $k \sim 0.1$.

The given Eqs. (2.167) and (2.168) show that the focal length of the bipotential lenses depends on the potential ratio V_2/V_1 of the electrodes. The optical properties of the lenses are also affected by the geometrical configurations, particularly the radius of the cylinder. A bipotential lens in which the object (usually the electron emitter) is positioned in the field of the lens (objective lens) is widely used in electron-beam lithography and electron microscopy.

The unipotential lens (Fig. 2.39c) is formed by three coaxial electrodes (apertures or cylinders), and outer electrodes being at a common potential V_1 and the center electrode at a different (lower or higher) potential V_2. If the outer electrodes are of the same shape and are located at the same distance from the center electrode, then the field of the lens is symmetrical relative to the midplane of the lens. Such a unipotential lens is called symmetrical. The electrode systems shown in Fig. 2.39c form symmetrical unipotential lenses. When the outer electrodes are not identical or are unequally spaced from the center electrode, the unipotential lens is asymmetrical. The unipotential lens is characterized by the equality of the potentials on the outer electrodes. As a result, the energy of electrons passing through this lens does not change, it is the velocity vector direction that changes. The equality of the potentials on the outer electrodes results in the electric fields on either side of the midplane (plane of symmetry) of the lens acting in opposite directions. For a weak symmetrical unipotential lens

$$\frac{V_2 - V_1}{V_1} \ll 1 \tag{2.170}$$

The approximate expression for the focal length can be written as

$$\frac{1}{f} \approx \frac{3}{8d}\left(\frac{V_2 - V_1}{V_1}\right)^2 \tag{2.171}$$

In the case of asymmetrical unipotential lenses the focal lengths have to be calculated numerically, since analytical expressions are not known.

The unipotential lenses can be used with both $V_1 > V_2$ and $V_2 > V_1$. However, the selection of $V_1 > V_2$ is more practical because the lens has lower aberrations and simpler voltage sources.

Unipotential electrostatic lenses with electron-transparent foils as outer electrodes have interesting optical properties, including negative values for both their focal lengths and a third-order spherical aberration coefficient.[64]

More complicated electrostatic electron-optical systems formed by several electrodes can be regarded as a combination of aperture lenses and bipotential or unipotential lenses. The total optical power of a complex electron-optical system can be approximated by the light-optical formula

$$\frac{1}{f_\Sigma} = \frac{1}{f_1} + \frac{1}{f_2} + \cdots + \frac{1}{f_n} \tag{2.172}$$

where $f_1, f_2, \ldots, f_n$ are the focal lengths of the lenses forming the electron-optical system.

The bipotential and unipotential electrostatic lenses with their electrodes at certain potential differences can be used for reflecting the electron flow; i.e., they can be converted into electron mirrors (Fig. 2.40). Depending on the sign of curvature

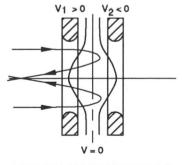
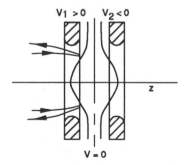

(a) CONVERGING ELECTRON MIRROR ($V_2 < 0$)   **(b)** DIVERGING ELECTRON MIRROR ($V_2 > 0$)

FIGURE 2.40.   Bipotential lenses used as electron mirrors. (a) Converging electron mirror ($V_2 < 0$); (b) diverging electron mirror ($V_2 > 0$).

of the equipotential surfaces in the reflecting area, the mirror can be either converging (concave) or diverging (convex). Shown in Fig. 2.40a is a bipotential lens with the right-hand electrode at a potential $V_2 < 0$. Of course, the electron beam near the reflecting surface cannot be regarded as paraxial, because the angles formed by the electron trajectories relative to the axis in the reflection area cannot be considered small, even as a rough approximation. Therefore, when determining the optical parameters of electron mirrors, the equations of paraxial optics may be used only for a preliminary estimation. Furthermore, electron mirrors have considerable chromatic aberration.

A unipotential lens with the central electrode potential at $V_2 < 0$ can also be used as an electron mirror. Varying this potential makes it possible to change the sign of the optical power of the mirror (to obtain converging and diverging mirrors) and control the optical power within a wide range. When the central electrode potential of a unipotential lens is reduced, the area having a potential below zero and reflecting the electrons is first formed near the electrode itself, while the potential along the lens axis remains positive. In this case the lens serves as a lens for near-axis electrons and changes into a mirror for peripheral electrons. This property of a unipotential lens can be used to control the intensity of the electron flow passing through the lens. By varying the potential of the central electrode from zero to a negative value such that the equipotential $V = 0$ lies at the saddle point of the field, it is possible to smoothly vary the electron-beam current from its maximum to zero. Under these conditions the unipotential lens acts as an "iris" diaphragm widely used in optical devices.

2.5.2. Magnetic Lenses

In addition to electrostatic lenses, electron-beam devices widely employ magnetic lenses. These lenses consist of a limited region of axially symmetric magnetic fields, usually produced by ring-shaped magnets, as shown in Fig. 2.41. As the following

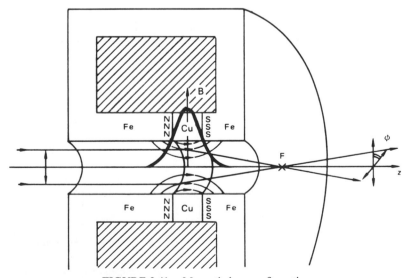

FIGURE 2.41. Magnetic lens configuration.

brief analysis shows, this arrangement has useful and attractive lens characteristics. The equation of motion for a charged particle of mass m and charge e in cylindrical coordinates in an axially symmetric magnetic field are

$$\ddot{r} = r\dot{\phi}^2 - \frac{e}{m}(B_\phi \dot{z} - B_z r\dot{\phi}) \qquad (2.173)$$

$$\ddot{z} = \frac{e}{m}(B_r r\dot{\phi} - B_\phi \dot{r}) \qquad (2.174)$$

$$r\ddot{\phi} + 2\dot{r}\dot{\phi} = \frac{e}{m}(\dot{r}B_z - \dot{z}B_r) \qquad (2.175)$$

where the dot notation refers to total time derivatives. The last equation may be rewritten as

$$\frac{1}{r}\frac{d}{dt}(\dot{\phi}r^2) = \frac{e}{m}(\dot{r}B_z - \dot{z}B_r) \qquad (2.176)$$

The difference between the flux linking a circular ring element of radius r and of width $dr(\phi_2)$ and the radial flux leaving through the curved surface is given by $\psi_2 - \psi_1 = 2\pi r(B_z\,dr - B_r\,dz)$. If this change of flux occurs in a time dt as a result of the electron motion in a magnetic field of axial symmetry from position 1 to position 2, then

$$\frac{d\psi}{dt} = 2\pi r(\dot{r}B_z - \dot{z}B_r) \qquad (2.177)$$

Substituting into Eq. (2.176) gives

$$\frac{1}{r}\frac{d}{dt}(\dot{\phi}r^2) = \frac{e}{2\pi rm}\dot{\psi} \qquad (2.178)$$

or

$$\frac{d}{dt}\left(\dot{\phi}r^2 - \frac{e}{2\pi m}\dot{\psi}\right) = 0 \qquad (2.179)$$

Integration with respect to time yields

$$m\dot{\phi}r^2 - \frac{e}{2\pi}\psi = \text{const} \qquad (2.180)$$

or

$$\dot{\phi} = -\frac{e}{m}\left(\frac{\psi_c - \psi}{2\pi r^2}\right) \tag{2.181}$$

where ψ_c is the flux when $\phi = 0$. If the magnetic field is confined to the lens region only, then $\psi_c = 0$, and

$$\dot{\phi} = -\frac{e}{m}\frac{\psi}{2\pi r^2} \tag{2.182}$$

This equation (Busch's theorem) tells us that the position of the electron in the initial and final radial planes are different because in the lens ϕ had a finite value; i.e., the image will be rotated away from the object in the ϕ direction.

As in the electrostatic case we can simplify the equations of motion by using an approximate expression for the magnetic field. Since the axial component of magnetic field B_z must satisfy Laplace's equation, it may be expanded in the same series as was the electrostatic potential, i.e.,

$$B(z) = B_0(z) - B_0''\frac{r^2}{2^2} + B_0''''\frac{r^4}{2^24^2} - \cdots \tag{2.183}$$

But the magnetic flux density B is a vector and it is necessary to include its other components. We assume that $B_\phi = 0$, and B_r may be determined in terms of B_z by means of the divergence equation, i.e.,

$$\frac{1}{r}\frac{\partial}{\partial r}(rB_r) + \frac{\partial B_z}{\partial z} = 0 \tag{2.184}$$

Substitution of Eq. (2.183) into (2.184) and solving for B_r gives

$$B_r = -B_0'\frac{r}{2} + B_0'''\frac{r^3}{2^24} - B_0''''\frac{r^5}{2^24^26} + \cdots \tag{2.185}$$

Thus, the complete magnetic field may be represented in terms of the axial component and its axial derivatives.

Neglecting all terms higher than the first power in r in Eqs. (2.183) and (2.185) therefore gives

$$B_z \approx B_0(z) \tag{2.186}$$

$$B_r \approx -B_0'(z)\frac{r}{2} \tag{2.187}$$

Putting these values into the equations of motion [Eq. (2.174)] gives

$$\ddot{r} = r\dot{\phi}^2 + \frac{e}{m}B_0 r\dot{\phi} \tag{2.188}$$

$$\ddot{z} = \frac{e}{m}B_0'\frac{r}{2}r\dot{\phi} \tag{2.189}$$

$$\dot{\phi} = -\frac{e}{m}\frac{B_0}{2} \tag{2.190}$$

Substituting the latter expression for $\dot{\phi}$ in the first two equations, we obtain

$$\ddot{r} = -\left(\frac{e}{m}\right)^2\frac{B_0^2}{4} \tag{2.191}$$

For paraxial rays r is small so that $z \cong 0$. Note that the radial acceleration is linear in r, a necessary condition for image formation. Now

$$\ddot{r} = \frac{d^2r}{dz^2}\dot{z}^2 = -\left(\frac{e}{m}\right)^2\frac{B_0^2}{4} \tag{2.192}$$

[from Eq. (2.191)] and

$$\dot{z}^2 = 2\left(\frac{e}{m}\right)V \tag{2.193}$$

where V is the electric potential energy referred to the cathode. Thus,

$$\frac{d^2r}{dz^2} = \frac{(e/m)rB_0^2}{8V} \tag{2.194}$$

This is the paraxial ray equation for the axially symmetric magnetic fields. In contrast to its counterpart for electric fields, this paraxial equation does contain the ratio e/m, so the magnetic-lens effect does depend on the particles involved. If B_0 is reversed in sign, the equation is unchanged; thus, these lenses are symmetrical. This is consistent with the constant potential throughout the system, which implies the same index of refraction on the two sides of the lens. The important feature of Eq. (2.194), however, is its invariance to the change of scale in r, which is an essential property for image formation. Again, this invariance appears only because we could approximate the fields near the axis to make the right-hand side linear in r.

The focal length is found as before by integrating once with respect to z between points 1 and 2, completely enclosing the active lens region:

$$\left(\frac{dr}{dz}\right)_2 - \left(\frac{dr}{dz}\right)_1 = \int_1^2 \frac{e}{m}\frac{B_0^2 r}{8V}\,dz \tag{2.195}$$

If the lens is sufficiently thin, r may be taken outside the integral. For focal length f_2 is then found by setting $(dr/dz)_1 = 0$. Thus,

$$\frac{1}{f_2} = \left.\frac{-(dr/dz)_2}{r_1}\right|_{dr/dz|_1 = 0} = \frac{-e/m}{8V}\int_1^2 B_0^2\,dz \tag{2.196}$$

Clearly f_2 is always positive, so the lens is always converging; and because of the symmetry already noted, f_1 equals $-f_2$. From Eq. (2.190)

$$\dot{\phi} = -\frac{e}{m}\frac{B_0}{2} \tag{2.190}$$

hence

$$\frac{d\phi}{dz} = -\frac{e}{m}\frac{B_0}{2\dot{z}} = -\frac{e}{2m}\frac{B_0}{[2(q/m)V]^{1/2}} \tag{2.197}$$

Thus, the image rotation is given by

$$\phi_2 - \phi_1 = \left(\frac{e}{8mV}\right)^{1/2}\int_1^2 B_0\,dz \tag{2.198}$$

The axial magnetic field of a short circular coil carrying a current i and having n turns and a mean radius R_m is given by

$$B(z) = \frac{\mu_0 R_m ni}{2(R_m^2 + z^2)^{2/3}} \tag{2.199}$$

Thus, using Eq. (2.196), the focal length of a thin magnetic lens is given by

$$f \approx 98\frac{VR_m}{(ni)^2}\quad \text{cm} \tag{2.200}$$

and the angle of image turn is

$$\Delta\phi = \phi_2 - \phi_1 = 10.7 \frac{ni}{\sqrt{V}} \quad \text{degrees} \tag{2.201}$$

The values of V, i, and R in Eqs. (2.200) and (2.201) should be in volts, amperes, and centimeters, respectively, to obtain f in centimeters and $\Delta\phi$ in angular degrees.

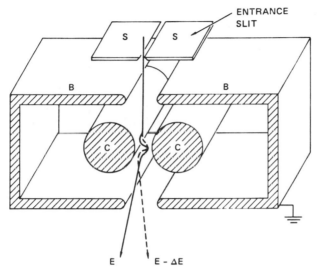

(a) CYLINDRICAL ELECTROSTATIC ENERGY ANALYZER

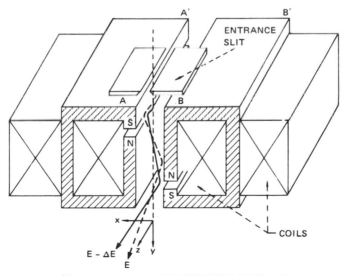

(b) CYLINDRICAL MAGNETIC ENERGY ANALYZER

FIGURE 2.42. Electrostatic and magnetic energy analyzers. (a) Cylindrical electrostatic energy analyzer; (b) cylindrical magnetic energy analyzer. From Ref. 65.

2.5.3. Energy-Analyzing and -Selecting Devices

Energy analysis of charged-particle beams is useful in two classes of instruments used in manufacturing, namely, electron microscopes and surface-analysis equipment [Auger spectroscopy and low-energy electron diffraction (LEED)].

In energy-analyzing microscopes the image is formed only by electrons with a selected energy. The spectral width (0.1–1 eV) is given by the product of the dispersion of the analyzer and the width of the selecting aperture. The energy-selecting microscope produces the electron-optical analogue of a monochromatic photograph.

The following analyzers are used in energy-selecting electron microscopes[65]:

- The cylindrical electrostatic analyzer
- The cylindrical magnetic analyzer
- The electrostatic mirror magnetic prism

The cross-sectional view of a cylindrical electrostatic analyzer is shown in Fig. 2.42a. A fine slit S a few microns wide and about 1 cm long apertures the incoming charged-particle beam. The apertured particles pass through regions of high-energy dispersion near the two electrodes C, which are biased at a potential near the cathode (or ion source). The energy spectrum is recorded on the exit side of the lens. At high accelerating voltages the problem of electrical breakdown limits the use of the electrostatic analyzer. A purely magnetic system (shown in Fig. 2.42b) avoids insulation problems. The magnetic analyzer is the magnetic analogue of the cylindrical electrostatic system discussed above.

The electron-optical properties of a double magnetic prism and an electrostatic mirror are shown in Fig. 2.43. Here the prism employs a uniform magnetic field and a concave electrostatic mirror that is biased to the source (cathode) potential.

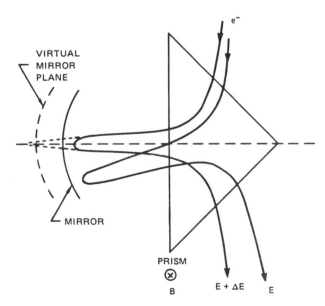

FIGURE 2.43.   Behavior of charged particles of energies E and $+E\Delta E$ in traversing a double magnetic prism and electrostatic mirror.

In Auger electron spectroscopy it is useful to measure the derivative of the secondary-electron energy distribution to display the effects more vividly. Energy analysis is obtained by using a retarding-field or cylindrical-mirror analyzer, shown in Fig. 2.44.

In a retarding-field analyzer the secondary electrons are collected by a fluorescent screen after passing the grid system (Fig. 2.44a). A retarding potential is applied to the two central grids, with a small modulating voltage added. The outer and inner grids are at ground potential. The modulated collector current that contains the information is amplified and demodulated using lock-in amplifier techniques.

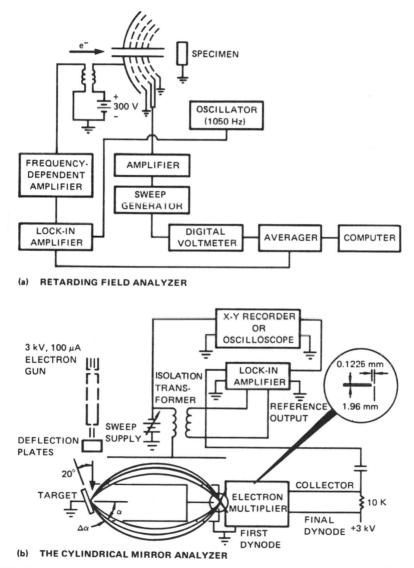

FIGURE 2.44.   Analyzers for recording the derivative Auger spectra. (a) Retarding field analyzer; (b) the cylindrical mirror analyzer. From Ref. 6.5.

The collector current is given by

$$I_{coll}(E) = I_p \int_E^{E_p} N(E) \, dE \qquad (2.202)$$

where $N(E)$ is the secondary emission distribution, E_p and E are the primary energy and the retarding potential, respectively, and I_p is the primary current. It can easily be shown that for a small modulating voltage the first harmonic coefficient in the Taylor expansion of the current is proportional to $N(E)$, the second harmonic coefficient to $dN(E)/dE$, and so on. In the second harmonic mode, the dc component is eliminated and electronic amplification can be considerably increased. The cylindrical-mirror analyzer is a deflection bandpass instrument with high transmission (10%) for moderate resolution (0.3%) (Fig. 2.44b).[66] Here the first harmonic coefficient is proportional to $E[dN(E)/dE]$ for small modulation voltages. If true energy distribution curves are desired, an integration has to be performed.

2.5.4. *Image Aberrations* (Ref. 67)

2.5.4.1. Geometric Aberrations. In the previous derivation of the basic equations of electron (ion) optics, we considered only the rays close to the optical axis, i.e., paraxial beams. The condition of paraxiality was expressed in the approximation that we used by taking into account only the first terms of the field expansion series [Eqs. (2.145) and (2.185)]. However, neglecting the subsequent terms in the expansion can result in considerable error. As seen from the radial motion equation in electric fields [Eq. (2.156)] and magnetic fields [Eq. (2.194)], one of the conditions for obtaining ideal images is linearity in r.

The use of subsequent terms in the field expansion results in terms proportional to the third, fifth, and so on, degrees of r or dr/dz in the equation of motion. If only the third-order terms (r^3, $r^2 \, dr/dz$, $r \, d^2r/dz^2$, and d^3r/dz^3) are taken into account in the equations of motion, the errors in image formation are called third-order aberrations.

In image formation, both r, the distance between the trajectory and the axis, and dr/dx, the inclination angle relative to the optical axis, can be expressed by r_a, the distance between the object point and the axis, and r_d, the radius of the limiting aperture for the lens (see Fig. 2.45). To classify the aberrations for wide beams we assume that a narrow paraxial beam nn is emitted from the point a of an object and creates an undistorted image of the point a in the plane z_b. For the wide beam mm emitted from the point a, a certain aberration figure is formed in the plane z_b rather than a point.

Since the aberration figures in the image plane are not circles, it is expedient to use Cartesian coordinates in the object and image planes to estimate the distortion of the image. The origin of these coordinates coincide with the point where the planes intersect the axis, and the aberrations are characterized by the values Δx and Δy of the deviation of the points of the aberration figure from the point of the image formed by the paraxial beam.

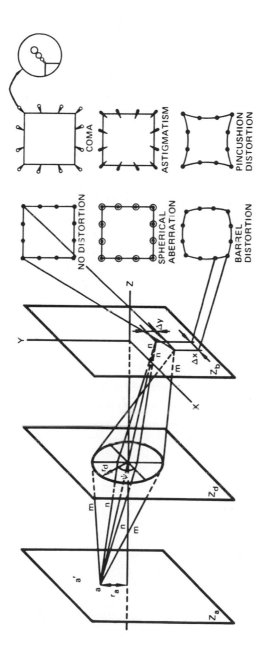

FIGURE 2.45. Image formation and examples of common aberrations.

Since the aperture limiting the beam must be circular, the coordinates of the aperture edges are preferably expressed in polar coordinates:

$$x_d = r_d \cos \psi, \qquad y_d = r_d \sin \psi \qquad (2.203)$$

With these coordinates the deviations in the image plane from the point of the image formed by the paraxial beam can be expressed as [52]

$$\Delta x = B r_d^3 \cos \psi + F(2 + 2\psi) r_d^2 x_a + (2C + D) \cos \psi r_d x_a^2 + E x_a^2 \qquad (2.204)$$
$$\Delta x = B r_d^3 \sin \psi + F \sin 2\psi r_d^2 x_a + D \sin \psi r_d x_a^2 + E x_a^2 \qquad (2.205)$$

where the coefficients B, C, D, E, and F are functions of the axial potential and its derivatives of the first, second, third, and fourth orders. These coefficients characterize the five geometrical aberrations of the third order.

From Eq. (2.205) it is clear that for points of the object lying on the system axis ($x_a = 0$), all of the terms except the first vanish and there remains only one aberration characterized by the coefficient B. This error is known as the spherical aberration and is very important for electron-beam devices. In the presence of spherical aberration the image of a point on the axis is a circle with radius Δr centered on the axis and given by

$$\Delta r = [(\Delta x)^2 + (\Delta y)^2]^{1/2} = B r_d^3 \qquad (2.206)$$

Since the radius of the circle for spherical aberration is proportional to r_d^3 this error can be reduced by decreasing the aperture radius, i.e., by using narrower beams. However, a decrease in the aperture radius also decreases the beam current.

The radius of the limiting aperture is proportional to the tangent of the aperture angle α. Since for small angles $\tan \alpha \cong \alpha$, Eq. (2.206) is usually written in the form

$$\Delta r = C_s \alpha^3 \qquad (2.207)$$

where C_s is the spherical aberration coefficient of the lens. C_s can be approximated for unipotential lenses by the empirical formula

$$C_s \approx k \frac{f^3}{R^2} \qquad (2.208)$$

where f is the focal length of the lens, R is the radius of the electrode, and $k = 2.5$–5 is an empirical constant.

The coefficient F in Eq. (2.205) characterizes the image error called the coma. The aberration figure produced by this error also has the form of a circle but with a center that does not coincide with the point of the image formed by the paraxial beam. Putting all of the coefficients in Eq. (2.205) except for F equal to zero and squaring and summing these equations gives

$$(\Delta x + 2Fx_a r_d^2)^2 (\Delta y)^2 = F^2 x_a^2 r_d^4 \tag{2.209}$$

Equation (2.209) is an equation of a circle with radius $\Delta r = Fx_a r_d^2$ and with the center at a distance $2Fx_a r_d^2$ from the image point.

Since the aberration figures are formed by the electrons passing through the entire aperture and not only by those close to the electrodes, we have in the image plane a combination of superimposed circles with gradually increasing radii and with centers located at a greater distance from the point of the image created by the paraxial beam.

The coefficients C and D characterize the image error known as astigmatism. Combining the terms with the coefficients C and D in Eq. (2.204), we obtain the ellipse equation

$$\frac{(\Delta x)^2}{[(2C + D)r_d x_a^2]^2} + \frac{(\Delta y)^2}{(Dr_d x_a^2)^2} = 1 \tag{2.210}$$

whose center coincides with the point of undistorted image and with the axis values proportional to $r_d x_a^2$. With $C = 0$ the ellipse degenerates into a straight line of length $4Cr_d x_a^2$. With $C = 0$ the aberration figure transforms into a circle.

The final geometric error determined by the coefficient E in Eqs. (2.204) and (2.205) is called distortion. This aberration is a result of nonlinear scaling, which results in a system magnification that depends on the distance of the object from the axis, $\Delta x = Ex_a^3$. This distortion for larger than Gaussian magnification is called barrel distortion, and for smaller than first-order magnifications, pincushion distortion. The aberration figures corresponding to the geometric image errors for electrostatic lenses are shown in Fig. 2.45.

Aberrations are also inherent in magnetic lenses. However, because of the "twisting" action of the magnetic field the magnetic lenses have additional "anisotropic" aberrations. This group of aberrations includes anisotropic coma, anisotropic astigmatism, and anisotropic distortion.

2.5.4.2. *Electronic Aberrations.* Geometrical aberrations depend on the fields forming the electron lenses and are independent of the electron parameters. So far we have assumed that all of the electrons at a given point in the system have the same energy that is uniquely determined by the potential of this point. This assumption is valid only for electrons entering the systems with zero initial velocity. Real electron beams enter the system with a finite energy distribution. Thus, an electron at potential $V(R_1, z)$ will have a kinetic energy given by

$$\frac{mv^2}{2} = e[V(R_1, z) + \Delta V] \tag{2.211}$$

where $e \Delta V$ is the initial kinetic energy of the electron.

When attempting to focus electrons having different initial kinetic energies, there arise additional errors called chromatic aberrations, by analogy to light optics. From the physical point of view, chromatic aberrations result from the dependence of the optical parameters of the electron lenses on the electron energy in the beam being focused. This dependency follows directly from Eqs. (2.211) and (2.196). Consider the trajectories of two electrons, one of which has an initial energy $e\,\Delta V$ and the other zero. If these electrons are emitted from point z_a (see Fig. 2.46) and pass through a lens, they will intersect the axis at $z_b + \Delta z$ and z_b. For electrons with larger initial energies, the value of Δz is greater. Generally, electrons with nonzero initial energies will create a circle of confusion in the image plane at z_b. The radius of this chromatic aberration circle is equal to

$$r_c = \Delta z \tan \gamma_2 \approx r_d \frac{\Delta z}{z_b} \tag{2.212}$$

where r_d is the radius of the aperture. Then for the two electrons the least and most energetic in the beam we can use the basic lens laws

$$\frac{1}{z_a} + \frac{1}{z_b} = \frac{1}{f}$$

$$\frac{1}{z_a} + \frac{1}{z_b + \Delta z} = \frac{1}{f + \Delta f} \tag{2.213}$$

where Δf is the increase in focal length corresponding to the most energetic electron in the beam.

Subtracting the first equation of system (2.213) from the second and neglecting Δf as compared to f and Δf as compared z_b we get

$$\frac{\Delta z}{z_b} = \frac{z_b}{f} \frac{\Delta f}{f} \tag{2.214}$$

From Eqs. (2.214) and (2.212) it is clear that the radius of the confusion circle (disk) for chromatic aberration is proportional to the ratio $\Delta f/f$. Using the formula

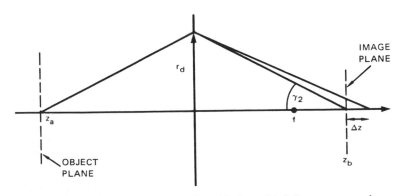

FIGURE 2.46.   Electron imaging in a thin lens with finite energy spread.

for the electrostatic lenses we can calculate the focal distances as

$$\frac{1}{f} = \frac{1}{4\sqrt{V}} \int_{-\infty}^{+\infty} \frac{V_0''}{V_0} \, dz$$

$$\frac{1}{f + \Delta f} = -\frac{1}{4(V + \Delta V)^{1/2}} \int_{-\infty}^{+\infty} \frac{V_0''}{\sqrt{V_0}} \, dz$$

(2.215)

Dividing the first equation by the second, we obtain

$$\frac{\Delta f}{f} = \frac{1}{2} \frac{\Delta V}{V}$$

(2.216)

Thus, the ratio $\Delta f/f$ is proportional to the ratio of the initial energy to the final energy. Finally, from Eqs. (2.212), (2.214), and (2.216) the radius of the chromatic aberration disk can be expressed as follows:

$$r_c = r_d \frac{z_b}{f} \frac{1}{2} \frac{\Delta V}{V} = C_c \frac{\Delta V}{V} \gamma_2$$

(2.217)

This expression shows that the chromatic aberration is proportional to the aperture radius r_d. To reduce this aberration, the diameter of the aperture should be reduced and the beam energy increased.

Other types of electronic aberrations include space-charge errors in intense electron beams and diffraction errors, which are the consequence of the wavelike properties of electron beams (see Section 2.3.5).

2.5.5. Multiple-Focusing Elements

In axially symmetric lenses the electron trajectories are approximately parallel to the field lines; hence, these lens elements are referred to as longitudinal systems.[52] In longitudinal systems, electron beams are focused by a small (compared to the longitudinal) transversal component of the field. Transversal systems, in which the lines of force of the field are directed across the beam, are more effective. Transversal electron-optical systems have gained wide use, particularly for focusing high-energy particles (strong focusing). Transversal systems are interesting in that they can be used to design aberration-free focusing systems.

Transverse focusing fields are usually created by four electrodes or four magnetic coils located about the axis of the system, the diametrically opposite electrodes or

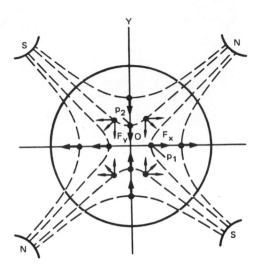

FIGURE 2.47.   Pole configuration for a magnetic quadrupole.

magnets being of similar polarity and the adjacent ones of different polarity (Figs. 2.47 and 2.48). Such four-pole systems having two symmetrical planes are known as quadrupole lenses. A specific feature of the quadrupole lenses is that the axial component of the field is equal to zero.

As an example, consider a quadrupole electrostatic lens[68] formed by four identical electrodes in the form of hyperbolic cylinders located symmetrically and equally spaced from the axis of potentials $\pm V_1$ (Fig. 2.48). The potential distribution in such a system is described by the equation

$$V_1(x, y) = \tfrac{1}{2}k(x^2 - y^2) \tag{2.218}$$

where $k = 2V_1/a^2$ and a is the distance from the optical axis to the electrode surface.

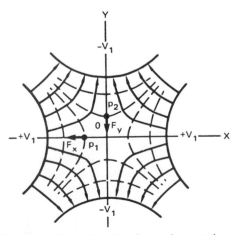

FIGURE 2.48.   Electrode configuration for an electrostatic quadrupole lens.

With the potential distribution described by Eq. (2.218), the field intensity components E_x and E_y are linear functions of the coordinates:

$$E_x = -\frac{\partial V_1}{\partial x} = -kx$$

$$E_y = -\frac{\partial V_1}{\partial y} = ky \tag{2.219}$$

$$(E_z = 0)$$

i.e., the gradients of the field intensity are constant. For this reason the quadrupole lens with hyperbolic electrodes is called a constant-gradient lens.

Consider the motion of an electron entering such a lens parallel to the axis 0Z. One can easily see that at the point p_1 (Fig. 2.48) the electron is acted on by the force $F_x = -eE_x = ekx$ directed from the lens axis, while at the point p_2 it is under the action of the force $F_y = -eE_y = eky$ directed toward the axis. Thus, in the plane X0Z the lens is diverging, whereas in the plane Y0Z it is converging. In any point of the field not lying in the symmetry planes the electron is acted on by a force having a component forcing the electron toward the plane X0Z and away from the plane Y0Z. The equations of motion in this quadrupole field can be written as

$$m\frac{d^2x}{dt^2} = -eE_x = \frac{2eV_1}{a^2}x$$

$$m\frac{d^2y}{dt^2} = -eE_y = -\frac{2eV_1}{d^2}y \tag{2.220}$$

Since there is no field on the lens axis, it is permissible to consider the longitudinal component of the electron velocity as constant for the region near the axis (paraxial approximation); hence,

$$m\left(\frac{2e}{m}V_0\right)^{1/2}\frac{d}{dz}\left[\left(\frac{2e}{m}V_0\right)^{1/2}\frac{dx}{dz}\right] = \frac{2eV_1}{a^2}x$$

$$m\left(\frac{2e}{m}V_0\right)^{1/2}\frac{d}{dz}\left[\left(\frac{2e}{m}V_0\right)^{1/2}\frac{dy}{dz}\right] = -\frac{2eV_1}{a^2}y \tag{2.221}$$

that is

$$x'' - k_E^2 x = 0$$

$$y'' - k_E^2 y = 0 \tag{2.222}$$

where $k_E = V_1/a^2 V_0$ and the primes indicate differentiation with respect to z.

For the symmetrical magnetic quadrupole lens (Fig. 2.47) we assume that the electrons enter the lens as a beam parallel to the z axis with velocity

$$v_z = \left(\frac{2e}{m} V_0\right)^{1/2} \tag{2.223}$$

At the point P_1 the electron is affected by a force $F_x = -ev_zB_y$ directed from the axis, while at the point P_2 the electron is affected by the force $F_y = -eV_zB_x$ directed toward the lens axis. At any point not lying in the symmetry planes the electron is acted on by a force having a component forcing the electron toward the plane X0Z and from the plane Y0Z. Thus, the magnetic quadrupole lens will be converging in the plane Y0Z and diverging in the plane X0Z.

The equations of motion in a quadrupole magnetic lens are

$$m\frac{d^2x}{dt^2} = ev_zB_y$$
$$m\frac{d^2y}{dt^2} = -ev_zB_x \tag{2.224}$$

The magnetic field components are similar to the axially symmetrical case; i.e.,

$$B_x(x, y, z) = F(z)y - \tfrac{1}{12}F''(z)(3x^2y + y^3) + \cdots$$
$$B_y(x, y, z) = F(z)x - \tfrac{1}{12}F''(z)(x^3 + 3xy^2) + \cdots \tag{2.225}$$

where

$$F(z) = \left(\frac{\partial B_x}{\partial y}\right)_0 = \left(\frac{\partial B_y}{\partial x}\right)_0 \tag{2.226}$$

Within the limits of paraxial approximation we use only first terms in Eq. (2.225) and the equations of motion become

$$\frac{d^2x}{dt^2} = \frac{e}{m}v_z\left(\frac{\partial B_y}{\partial x}\right)_0 x$$
$$\frac{d^2y}{dt^2} = -\frac{e}{m}v_z\left(\frac{\partial B_y}{\partial x}\right)_0 y \tag{2.227}$$

Assuming that in the paraxial region

$$v_z = \left(\frac{2e}{m} V_0\right)^{1/2} = \text{const} \tag{2.228}$$

we can change the differentiation with respect to t in Eq. (2.227) to differentiation with respect to z, giving

$$\frac{d^2x}{dz^2} - \left(\frac{e}{2mV_0}\right)^{1/2}\left(\frac{\partial B_y}{\partial x}\right)_0 x = 0$$

$$\frac{d^2y}{dz^2} - \left(\frac{e}{2mV_0}\right)^{1/2}\left(\frac{\partial B_y}{\partial x}\right)_0 y = 0 \tag{2.229}$$

or

$$x'' - k_m^2 x = 0$$
$$y'' - k_m^2 y = 0 \tag{2.230}$$

where

$$k_m^2 - \left(\frac{e}{2mV_0}\right)^{1/2}\left(\frac{\partial B_y}{\partial x}\right)_0 \tag{2.231}$$

Equations (2.230) are similar in structure to those describing the motion of electrons in an electrostatic quadrupole lens. Therefore, in the general case the equations of motion for electrons near the axial region of a symmetrical electrostatic or magnetic lens can be presented in the form

$$x'' - k^2 x = 0$$
$$y'' + k^2 y = 0 \tag{2.232}$$

where $k = k_E$ for the electrostatic lens and $k = k_M$ for the magnetic lens.

If the lens has a length L, the lens field is concentrated between the planes $z = +L/2$ and $z = -L/2$ and abruptly drops down to zero outside this region. In this case we can consider the field independent of z and the motion equations (2.232) have solutions of the form

$$x = A \cosh kz + B \sinh kz$$
$$y = C \cos kz + D \sin kz \tag{2.233}$$

Here the constants A, B, C, and D are determined by the initial conditions. For example, if in the plane $z = -L/1$ (in the "entrance" plane of the lens) $x = x_0$, $y = y_0$, $x' = x'_0$, and $y' = y'_0$, then

$$x = x_0 \cosh k\left(z + \frac{L}{2}\right) + \frac{x'_0}{k} \sinh k\left(z + \frac{L}{2}\right)$$

$$y = x_0 \cos k\left(z + \frac{L}{2}\right) + \frac{y'_0}{k} \sin k\left(z + \frac{L}{2}\right)$$

(2.234)

To explore the optical characteristics of the quadrupole lens we consider the path of an electron entering the lens parallel to the axis 0Z. In this case

$$x'_0 = y'_0 = 0$$

$$x = x_0 \cosh k\left(z + \frac{L}{2}\right)$$

$$y = x_0 \cos k\left(z + \frac{L}{2}\right)$$

(2.235)

The focal lengths are equal to

$$f_x = -\frac{x_0}{x'} \quad \text{and} \quad f_y = -\frac{y_0}{y'}$$

(2.236)

If we differentiate Eq. (2.235) and determine x' and y' when $z = L/2$ (in the "exit" plane of the lens), we obtain

$$x'\left(\frac{L}{2}\right) = x_0 k \sinh kL$$

$$y'\left(\frac{L}{2}\right) = -y_0 k \sin kL$$

(2.237)

Substituting Eq. (2.237) into (2.236) then gives

$$f_x = -\frac{1}{k \sinh kL}$$

$$f_y = \frac{1}{k \sin kL}$$

(2.238)

The expressions for the focal lengths show that, indeed, in the plane Y0Z the lens is converging ($f_y > 0$, real focus) and in the plane X0Z the lens is diverging

($f_x < 0$, imaginary focus). Thus, upon passing through the quadrupole lens, the electron beam parallel to the axis is deformed in such a way that a line focus is produced in the focal plane of the image space.

In the case of a thin, weak quadrupole lens, $kL \ll 1$, so that $\sin kL \cong \sinh kL \cong kL$, and the expressions for the focal lengths are simplified to

$$f_x \approx -\frac{1}{k^2 kL}$$

$$f_y \approx \frac{1}{k^2 kL}$$

(2.239)

i.e., for the thin, weak lenses the focal lengths are equal in magnitude and the principal planes coincide with the center plane of the lens. These equations are valid only for lenses with a constant field gradient. This condition is satisfied when using lens electrodes or pole pieces in the form of hyperbolic cylinders positioned symmetrically around the lens axis. In practice, cylindrical electrodes and pole pieces are usually used, since they are easier to make. To achieve an effectively constant gradient field a configuration in which the radius of the circular cylinder R is in the range $1.1a$ to $1.15a$ for the electrostatic lens is required, where a is the distance from the electrode axis. With magnetic lenses the optimum value for R is $1.15a$.

Similarly, the assumption of an abrupt drop of the field at the boundary of the electrodes or pole pieces is not a realistic approximation. Satisfactory results, however, may be obtained if the value L is regarded as an "effective length" including the stray fields. It has been found experimentally that good agreement of the design data with the experimental data is obtained when the effective length $L = l - 1.1a$, is used, where l is the length of the electrodes or pole pieces and a is the distance from the surface of the electrode (pole) to the axis.

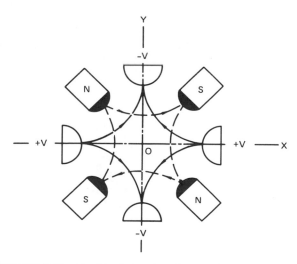

FIGURE 2.49. Combined electric and magnetic quadrupole lens.

Two quadrupole lenses located along an axis, called a doublet, produce a point image of a point object. Each lens of a doublet produces a linear image of a point object, but a doublet is not identical to an axially symmetric lens because the quadrupole lenses do not have the same magnification in the 0X and 0Y directions.

Quadrupole lenses have the same axially symmetric aberrations as the axially symmetric lenses, but in contrast to the other lenses compensation for spherical aberration can be made. It is also possible to build an achromatic combination quadrupole lens with electric and magnetic fields superimposed in one lens (Fig. 2.49).

2.5.6. Deflection Systems

In this section we are concerned with the aberrations caused when charged-particle beams are deflected. In many applications (electron microprobes, microscopes), deflection aberrations are of somewhat secondary importance because they limit the field size but not the basic probe size and beam current. Deflection aberrations distort the spot from its ideal round shape and distort the raster scan lines into curves. These subjects are treated in detail in the literature, and a brief discussion is given here.[69,70]

The deflection of a round beam of charged particles by a pair of electrostatic plates is illustrated in Fig. 2.50 where the distortions of the scan pattern and of the spot shape are shown. The displacement of the spot on the screen can be considered

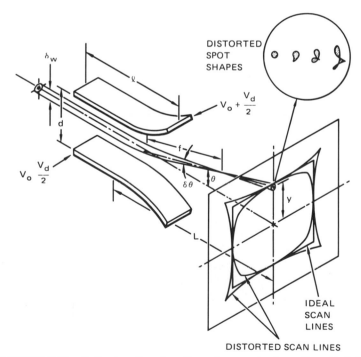

FIGURE 2.50.   Deflection distortions from electrostatic plates. From Ref. 69.

to be made up of linear terms x and y and error terms Δx and Δy, the latter being functions of the deflection angle, the entrance angle, and the deflection field configuration. The linear deflection term and some insight into the causes of beam distortion can be derived from the first-order deflection equation

$$m\frac{d^2y}{dt^2} = (-e)E_y$$

$$m\frac{d^2z}{dt^2} = 0$$

(2.240)

where $E_y = \partial V/\partial y$ and the initial conditions are given by

$$\left(\frac{d_y}{dt}\right)_0 = 0, \qquad \left(\frac{dz}{dt}\right)_0 = v = \left(\frac{2eV_0}{m}\right)^{1/2}$$

$$y_0 = 0 \qquad z_0 = 0$$

(2.241)

By integrating twice Eq. (2.240) we obtain

$$y = \frac{-e}{2m}E_y t^2$$

$$z = vt$$

(2.242)

Elimination of time t from Eq. (2.242) yields the equation of the particle trajectory:

$$y = \frac{-e}{2m}E_y\left(\frac{z}{v}\right)^2$$

(2.243)

This is the equation of a parabola. Now substituting $(2eV_0/m)^{1/2}$ for v we get

$$y = \frac{z^2}{V_0}\frac{V_d}{4d}$$

(2.244)

where d is the plate separation distance. The angle θ of deflection is obtained by differentiating Eq. (2.244), which yields

$$\tan\theta = \frac{dy}{dz} = \frac{z}{V_0}\frac{V_d}{2d} = \frac{1}{V_0}\frac{V_d}{2d}$$

(2.245)

This is the standard electrostatic deflection equation for charged particles (electrons here) passing between ideal plates. The deflection on a screen a distance L away will be

$$y \cong L\theta = \frac{V_d}{V_0} \frac{1}{2} \frac{L}{d} \tag{2.246}$$

This simple expression does not show the several optical defects present in a deflected beam. For example, the particles in a beam of finite thickness δw will attain different axial velocities in their transit through the plates. Depending on their transverse position, this effect results in different values of the deflection angle θ. This effect can be evaluated by differentiating Eq. (2.245):

$$\delta\theta = -\theta \frac{\delta V_0}{V_0} \tag{2.247}$$

where

$$\delta V_0 = \frac{V_d}{d} \tag{2.248}$$

thus

$$\delta\theta = -2\theta^2 \delta w \tag{2.249}$$

The electrons closest to the positive electrodes will move faster and will be deflected less; the two rays will therefore cross at a distance f away; i.e., the electrostatic deflection process acts as a convergent lens, f, and the focal length is given by

$$f = \frac{\delta w}{\delta\theta} = -\frac{1}{2\theta^2} \tag{2.250}$$

It is seen that the spot diameter at the target will vary with the deflection angle (unless the target is curved). To minimize this effect, the "focal length" f should be made equal to the distance from the deflection plate to the target and the beam diameter should be as small as possible. The deflection plates should always be operated in a "push–pull" mode so that the beam axis remains an equipotential independent of the deflecting voltage. Other types of deflection errors include distortion of the spot shape and of the scan line.

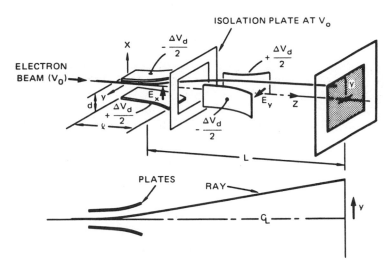

FIGURE 2.51. $x, -y$ electrostatic deflection system. From Ref. 69.

The simplest deflection or scanning system is a set of electrostatic deflection plates as illustrated in Fig. 2.51. This type of deflection does not depend on e/m and therefore can be used for electrons or ions. In the case of magnetic deflection systems, because the deflection is proportional to the particle velocity (and therefore to e/m), a very large magnetic field is required for ion deflection.

If a constant impingement angle is required on the target plane, a double-deflection method can be used (see Fig. 2.52). An interesting cylindrical octupole deflector configuration[71,72] is shown in Fig. 2.53a. In this system y deflection is obtained by applying the voltages $\pm V_y$ to the upper and lower pairs of plates, which can be called the "main" plates for this operation and by applying $\pm aV_y$ to the other or "auxiliary" plates in the pattern shown in Fig. 2.53b. The value of the parameter a is a constant less than unity that determines the degree of deflection-field shaping

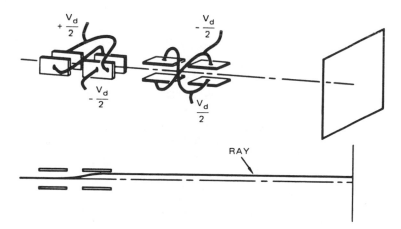

FIGURE 2.52.   Double $x, -y$ electrostatic deflection. From Ref. 69.

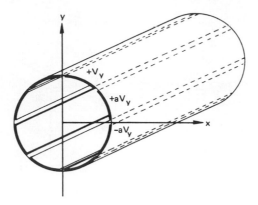

(a) OCTUPOLE DEFLECTION STRUCTURE

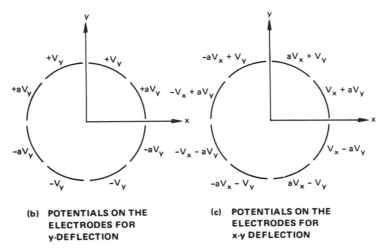

(b) POTENTIALS ON THE ELECTRODES FOR y-DEFLECTION

(c) POTENTIALS ON THE ELECTRODES FOR x-y DEFLECTION

FIGURE 2.53. Cylindrical octupole deflection system. (a) Octupole deflection structure; (b) potentials on the electrodes for y deflection; (c) potentials on the electrodes for x–y deflection.

provided by the auxiliary plates. For x deflection, the roles of the main and auxiliary plates are interchanged. In principle, combined x and y deflection can be obtained by summation (Fig. 2.53c).

The aberrations resulting from the deflection of two sets of electrostatic or magnetic deflections have been studied in the literature.[70–74] The equations of motion for electrons (or any other charged particles) in electrostatic and magnetostatic fields, characterized by $V(x, y, z)$ and the vector potential (x, y, z), can be derived from the variation of the action integral

$$\delta \int n \, ds = 0 \qquad (2.251)$$

where, using rectangular coordinates, $n(x, y, x', y')$

$$= \sqrt{V(x, y, z)} - \left(\frac{e}{2m}\right)^{1/2}\left(A_x\frac{dx}{ds} + A_y\frac{dy}{ds} + A_z\frac{dz}{ds}\right) \tag{2.252}$$

and

$$ds = dz(1 + x'^2 + y'^2)^{1/2}$$

$$x' = \frac{dx}{dz}, \qquad y' = \frac{dy}{dz} \tag{2.253}$$

The Lagrange equation associated with Eq. (2.251) can be written in the form

$$\frac{\partial n}{\partial x} - \frac{d}{dz}\left(\frac{\partial n}{\partial x'}\right) = 0$$

$$\frac{\partial n}{\partial y} - \frac{d}{dz}\left(\frac{\partial n}{\partial y'}\right) = 0 \tag{2.254}$$

Then, with Eq. (2.253) we get the general trajectory equations

$$\frac{1}{2\sqrt{V}}(1 + x'^2 + y'^2)^{1/2}\frac{\partial V}{\partial x} = \frac{d}{dz}\left[\frac{\sqrt{V}\,x'}{(1 + x'^2 + y'^2)^{1/2}} - \left(\frac{e}{2m}\right)^{1/2}A_x\right]$$

$$\times \left(\frac{e}{2m}\right)^{1/2}\left[x'\frac{\partial A_x}{\partial x} + y'\frac{\partial A_y}{\partial x} + z'\frac{\partial A_z}{\partial x}\right] \tag{2.255}$$

$$\frac{1}{2\sqrt{V}}(1 + x'^2 + y'^2)^{1/2}\frac{\partial V}{\partial y} - \frac{d}{dz}\left[\frac{\sqrt{V}\,y'}{(1 + x'^2 + y'^2)^{1/2}} - \left(\frac{e}{2m}\right)^{1/2}A_y\right]$$

$$\times \left(\frac{e}{2m}\right)^{1/2}\left[x'\frac{\partial A_x}{\partial y} + y'\frac{\partial A_y}{\partial y} + z'\frac{\partial A_z}{\partial y}\right] \tag{2.256}$$

The aberrations of a two-dimensional electrostatic or magnetic system can be derived by integrating these motion equations [see Eq. (2.256)]. The general technique used is to develop expansions for the potential $[V(x, y, z)]$ and the fields (E_x, E_y, H_x, H_y), using the symmetry of the system. The elimination of various cross terms is achieved by making the expansions consistent with Laplace's equation. Direct integration of the trajectory equations is then possible using a number of approximations.[72,73] The aberrations or deviations from first-order (Gaussian) deflection theory then appear as terms having two parts: an integral over some component or derivative of the deflection field, and a multiplier showing the dependence on Gaussian deflection and beam entry conditions (diameter and angle). The aberration integrals are dependent only on the deflection-field distribution.

2.5.7. Recent Developments in Electron and Ion Optics

In the recent years we witnessed an accelerated application of computer-aided design (CAD) in electron and ion optics. New computational techniques are used not just to design optical components but also to optimize electro-optical systems for microscopes, lithography machines, and analytical instruments. In addition to the conventional numerical methods (Runge–Kutta, multistep methods, etc.), new CAD optimization and synthesis of components have been invented (variation calculus, dynamic programming[66]).

These new computational techniques are applied to magnetic lens design,[75] for the optics of round and multipole electrostatic lenses,[76] and for trajectory computation in electron lenses.[77]

Many new system designs appeared also in the literature on thermal-field-emission electron optics,[78] on the design of electron deflection systems,[79] on ion optics,[80] on instruments,[81] and on electron beam testing.[82]

2.6. ELECTRON INTERACTIONS

2.6.1. Thin Films

The surface of a solid constitutes a region where the geometrical arrangement of atoms, the electronic structure, and the chemical composition may be quite different from the bulk. Under atmospheric conditions, all surfaces are contaminated by adsorbed layers. If an atomically clean surface is generated (by sputtering or fracture of a crystal) in a vacuum of 10^{-6} Torr, it will be covered by a complete monolayer (10^{15} atoms/cm^2) from the gas phase in about 1 s; therefore, an ultrahigh vacuum (10^{-10} Torr) is required to study atomically clean surfaces. In such an environment one can characterize a surface region by specifying the following[83]:

- The atomic and molecular species on the surface region (chemical characterization)
- The arrangements of these atoms and molecules on the surface
- The valence electron distribution on the surface

For chemical characterization, electron-, ion-, and photon-beam excitation processes are used. In this section the basic excitation processes will be discussed; applications to electron- and ion-beam microscopies are treated in Chapter 6.

Most of our knowledge of surface geometry has been obtained by electron diffraction and field-ion microscopy.[84,85] From these experimental studies we learned that the surface of a crystal can differ from the bulk structure in a number of ways (see Fig. 2.54). Foreign atoms may be chemisorbed into an overlayer (Fig. 2.54a) or the adsorption of reactive species may form a three-dimensional surface-layer compound (Fig. 2.54b). The surface atoms may be displaced, forming a new lateral periodicity (Fig. 2.54c), or at the surface of an alloy segregation of one of the components may occur, as shown in Fig. 2.54d.

The chemical composition of a surface is generally not the same as the bulk. For example, all metals have a native oxide on the surface that can vary in thickness

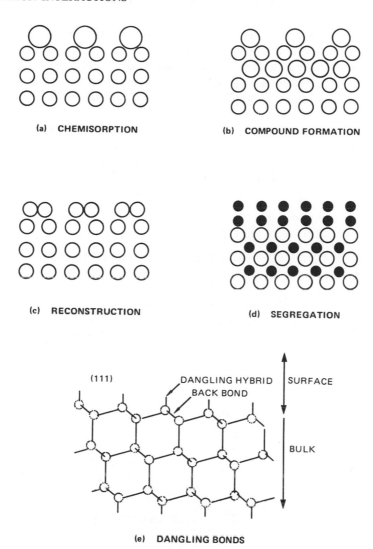

FIGURE 2.54. Examples of crystal surface structures. (a) Chemisorption; (b) compound formation; (c) reconstruction; (d) segregation; (e) dangling bonds.

from 1 to 10 nm. These oxide layers affect the surface mechanical and chemical properties if they are not removed by etching or chemical polishing.

Surface contamination normally consists of absorbed or adsorbed gases, organic materials, water, and inorganic ions.

The electronic structure at the surface of a crystal is usually different from that of the bulk. Some of the orbitals that form all of the bonds in the crystal remain unbonded at the surface. These orbitals, directed out of the surface, are called "dangling bonds," as shown in Fig. 2.54e. This is the case for many semiconductors (e.g., Si, Ge, GaAs) in which the bonds in the bulk are each made up of two electrons shared by adjacent atoms. At the surface, dangling bonds are left, each with only one electron. Thus, both filled and empty electron surface-state energy levels, which are different from the bulk energy levels, occur for these "dangling bonds." Such surface states play an important role in band bending and work-function-related effects. The differences between the bulk and surface properties of solids are important owing to their effects on surface bombardment by electron, ion, and photon beams.

2.6.1.1. Electron Transmission through Thin Films. If a beam of electrons with kinetic energy E_0 impinges upon a thin film sample (i.e., one whose thickness is much smaller than the range of the electrons in the target material), it is convenient to classify the transmitted electrons into three categories[88] (see Fig. 2.55):

- The unscattered beam
- The elastically scattered beam
- The inelastically scattered beam

Unscattered electrons pass through the sample without being deflected or suffering any energy loss.

2.6.1.2. Elastic Scattering. Elastic collisions are primarily due to the screened nuclear field and can result in large angle (90°) deflections, although most are smaller angle deflections. Some of the electrons are scattered on phonons, suffer small (1–10 meV) energy losses, and are difficult to distinguish from elastically scattered electrons.

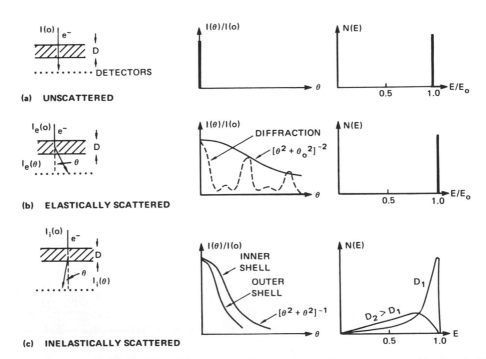

FIGURE 2.55. Angular and energy distributions of the electrons transmitted through thin films (arbitrary units). (a) Unscattered; (b) elastically scattered; (c) inelastically scattered.

The angular distribution of elastically scattered electrons can be calculated as Rutherford scattering screened by the electron cloud of the atom.[86] The incoming $I_e(0)$ and scattered $I_e(\theta)$ current ratios can be related to the angular distribution[87] by

$$I_e(\theta)/I_e(0) = (\theta^2 + \theta_0^2)^{-2} \tag{2.257}$$

where $\theta_0 = \lambda/2\pi a$ is the characteristic screening angle and a is the atomic radius given by

$$a = 0.9a_0 Z^{-1/4} \tag{2.258}$$

a_0 is the Bohr radius, Z is the atomic number, and λ is the incident electron wavelength. This angular distribution formula is approximately correct for amorphous samples, but in the case of crystalline films the angular distribution is modulated by diffraction effects, and peaks appear at the angles where the Bragg condition $\theta \approx \lambda/d$ is satisfied (see Fig. 2.55).

2.6.1.3. Beam Spreading.   The spatial distribution of electrons as a function of vertical penetration z and radial distance from the center of a fine (199 Å) incoming beam due to scattering has been calculated.[88] For film thicknesses of several thousand angstroms, the incident electrons of moderate energies (0.1–100 keV) suffer many collisions and the cumulative deflections can be determined statistically by using the Boltzmann equation.

Using Fermi's solution, the spatial probability distribution density of the scattered electrons is given as

$$H(r, z) = \frac{3\lambda}{4\pi z^3} \exp\left[-\frac{3}{4}\left(\frac{\lambda r^2}{z^3}\right)\right]$$

$$= \frac{1}{\pi r_0^2} \exp\left|-\frac{r^2}{r_0^2}\right| \tag{2.259}$$

where r is the radial distance measured from the axis of incoming electrons, r_0 is the beam diameter, z is the vertical distance measured along the axis of electron penetration, and λ is the scattering mean free path.

The normalization condition

$$\left(\frac{3\lambda}{4\pi z^3}\right) \int_{-\infty}^{\infty} \int_{-\infty}^{\infty} \exp\left(-\frac{3}{4}\frac{\lambda(x^2 + y^2)}{z^3}\right) dx\, dy = 1 \tag{2.260}$$

assures that at every thickness the spatial distribution is Gaussian. For nonrelativistic electrons

$$\lambda\,(\text{Å}) = \frac{5.12 \times 10^{-3} U^2 A}{\rho Z^2 \ln(0.725 U^{1/2}/Z^{1/3})} \tag{2.261}$$

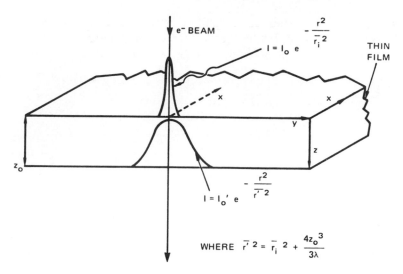

FIGURE 2.56.   Spatial distribution of electrons as a function of vertical penetration and radial distance from the beam center.

where A is the mass number, U is the electron energy (eV_0), ρ is the film density (g/cm^3), and Z is the atomic number. From these formulas [Eqs. (2.259) and (2.261)] the response density at any radius r_0 from the center of the Gaussian at penetration depth z_0 can be calculated. The effective beam radius r' after passing through the film of thickness z is related to the Gaussian radius of the incoming beam (r_i) by

$$\bar{r}'^2 = \bar{r}_i^2 + \frac{4z_0^3}{3\lambda} \qquad (2.262)$$

where the Gaussian radius r_i is for the incident beam at $z = 0$. The second term gives the spread in radius, as illustrated in Fig. 2.56. This calculation assumes no loss of energy of electrons in passing through the material, only elastic scattering.

2.6.1.4. Inelastic Scattering.   In the case of inelastically scattered electrons the angle of scattering θ depends on the energy lost by the incident electron. The angular distribution[89] in this case is given by

$$I_i(\theta)/I_i(0) = (\theta^2 + \theta_E^2)^{-1} \qquad (2.263)$$

where $\theta_E = E p v$ and E is the energy lost by an incident electron of velocity v and momentum p.

The energy loss in thin and thick targets resulting from inelastic scattering occurs as discrete events, with the creation of secondary electrons of low energy (up to 50 eV), excitation of density oscillations in metal electron plasmas, inner-shell ionization that leads to X-ray emission and Auger electron emission, creation of

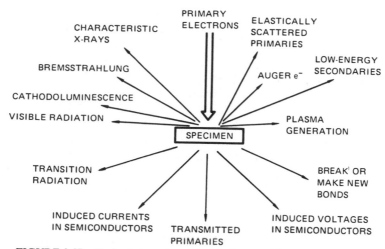

FIGURE 2.57.   Basic electron–target interactions used in microfabrication.

electron–hole pairs followed by photon emission (cathodoluminescence), transition radiation,[90] and lattice vibrations (phonon excitation) (see Fig. 2.57).

The energy losses from all inelastic processes are conveniently described by the Bethe continuous energy loss relation[91]:

$$\frac{dE}{ds} = \left(\frac{N_A e^4}{2\pi \varepsilon_0^2}\right)\left(\frac{Z\rho}{A}\right)\left(\frac{1}{E} \ln \frac{1.66E}{J}\right)$$

$$= -7.85 \times 10^4 \frac{\rho Z}{AE} \ln \frac{1.66E}{J} \quad \left[\frac{keV}{cm}\right] \tag{2.264}$$

where dE is the change in energy of the electron of energy E (keV) while traversing a path of length ds (cm), Z is the atomic number, A is the atomic weight, ρ is the density, and J is the mean ionization potential. J is an increasing function of atomic number and can be approximated as

$$J = 1.15 \times 10^{-2} Z \quad (keV) \tag{2.265}$$

The rate of energy loss with distance traveled is thus dependent on the atomic number because of the dependence on Z/A and on density. Thus, dE/ds generally increases with increasing Z. The rate of energy loss also increases with decreasing energy since the E term in the denominator of Eq. (2.264) dominates the ln E term in the numerator.

2.6.2. Energy-Loss Mechanisms

2.6.2.1. Electron Collisions in Solids.   Collisions are the most important processes used to analyze beam–target interactions on an atomic scale in microscience. To define a collision cross section, a monoenergetic particle beam is assumed to impinge on a target (Fig. 2.58). If the beam is uniform and contains n_i particles per unit volume, moving with velocity v with respect to the stationary target, the flux, F, is given by

$$F = n_i v \text{ particles crossing unit area per second} \qquad (2.266)$$

Particles scattered by the target are observed with a counter that detects all particles scattered by an angle θ into the solid angle $d\Omega$. The number dN_D recorded per unit time is proportional to the incident flux F, the solid angle, $d\Omega$, and the number N of independent scattering centers in the target that are intercepted by the beam, i.e.,

$$dN_D = FN\sigma(\theta) \, d\Omega \qquad (2.267)$$

The constant of proportionality is designated by $\sigma(\theta)$; it is called the *differential scattering cross section*. If $d\Omega$ is the solid angle about θ into which the particles are

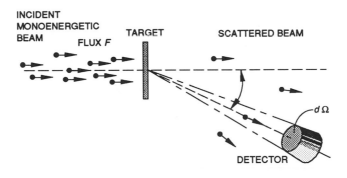

(a) FORWARD SCATTERING

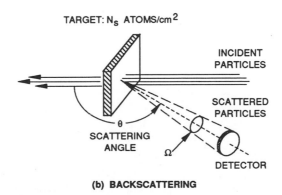

(b) BACKSCATTERING

FIGURE 2.58.   Geometries used in scattering experiments. (a) Forward scattering; (b) backscattering.

scattered, then

$$\sigma(\theta) \, d\Omega = d\sigma(\theta) \quad \text{or} \quad \sigma(\theta) = \frac{d\sigma(\theta)}{d\Omega} \tag{2.268}$$

The total number of particles scattered per unit time is obtained by integrating over all solid angles, i.e.,

$$N_D = FN\sigma_{tot} \tag{2.269}$$

where

$$\sigma_{tot} = \int \sigma(\theta) \, d\Omega \tag{2.270}$$

is called the total scattering cross section.

The total cross section has the dimension of an area, and it is customary to quote subatomic cross sections in barns (b) or decimal fractions of barns, where

$$1 \text{ b} = 10^{-24} \text{ cm}^2 = 100 \text{ fm}^2 \tag{2.271}$$

Finally, we note that if n is the number of scattering centers per unit volume, d the target thickness, and a the area intercepted by the beam, N is given by

$$N = and \tag{2.272}$$

If the target consists of nuclei with atomic weight A and has a density p, n is given by

$$n = \frac{N_0 p}{A} \tag{2.273}$$

where $N_0 = 6.0222 \times 10^{23}$ molecules/mole is Avogadro's number.

In the case of electron–electron or ion–electron scattering the number for the atomic (nuclear) scattering centers has to be multiplied by the number of electrons per atom.

The scattering cross section can be calculated from the force that acts during the collision between the projectile and target atom. The scattering cross section for central force scattering can be calculated for small deflections from the impulse imparted to the particle as it passes the target atom. As the particle with charge $Z_1 e$ approaches the target atom, charge $Z_2 e$, it will experience a repulsive force that will cause its trajectory to deviate from the incident straight-line path. If this momentum transfer is small relative to the momentum of the incoming particle, the scattering can be treated as a one-body process where the scattering center is fixed in space (no recoil).

The deflection of the particles in this one-body formulation is treated as the scattering of particles by a center of force in which the kinetic energy of the particle is conserved. The impact parameter b is defined as the perpendicular distance between the incident particle path and the parallel line through the target nucleus (Fig. 2.59a). Particles incident with impact parameters between b and $b + db$ will be scattered through angles between θ and $\theta + d\theta$. With central forces, there must be complete symmetry around the axis of the beam so that

$$2\pi b \, db = -\sigma(\theta) \cdot 2\pi \sin \theta \, d\theta \qquad (2.274)$$

In this case the scattering cross section $\sigma(\theta)$ relates the initial uniform distribution. The minus sign indicates that an increase in the impact parameter results in less force on the particle so that there is a decrease in the scattering angle.

If the scattering center is not fixed but recoils from its initial position as the result of scattering, then the process should be treated as a two-body central force problem (see Fig. 2.59b).

In this case the measured scattering angle θ differs from the angle θ_c calculated in the center of mass system, unless the mass of the scattered particle M_1 is much smaller than the mass of the scattering center M_2.

Therefore, electron and light ion scattering on nuclei can be treated as one-body scattering but medium and heavy ion scattering requires two-body kinematical treatment. As an important example, let us calculate the scattering cross section for a central force (coulomb force) in the one-body approximation.

When a particle with charge $Z_1 e$ approaches the target atom, charge $Z_2 e$, the repulsive force will cause its trajectory to deviate from the incident straight-line path (Fig. 2.60a). The value of the coulomb force F at a distance r is given by

$$F = \frac{Z_1 Z_2 e^2}{r^2} \qquad (2.275)$$

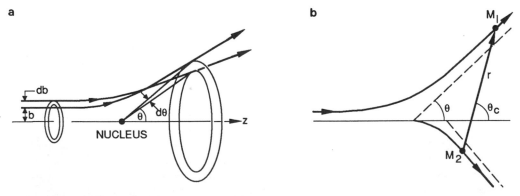

FIGURE 2.59. (a) Illustration of the one-body scattering process and definition of the scattering geometry. (b) Illustration of the two-body scattering process. θ_c is the scattering angle in the center of mass system.

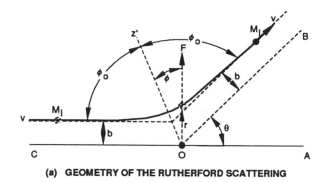

(a) GEOMETRY OF THE RUTHERFORD SCATTERING

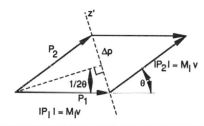

(b) MOMENTUM TRANSFER DIAGRAM FOR RUTHERFORD SCATTERING

FIGURE 2.60. Kinematics for Rutherford scattering. (a) Geometry of the Rutherford scattering; (b) momentum transfer diagram for Rutherford scattering.

Let p_1 and p_2 be the initial and final momentum vectors of the particle. From Fig. 2.60b it is evident that the total change in momentum $\Delta p = p_2 - p_1$ is along the z' axis. In this calculation the magnitude of the momentum does not change. From the isosceles triangle formed by p_1, p_2, and Δp one gets

$$\frac{\frac{1}{2}\Delta p}{M_1 v} = \sin\frac{\theta}{2} \qquad (2.276)$$

or

$$\Delta p = 2M_1 v \sin\frac{\theta}{2} \qquad (2.277)$$

We now write Newton's law for the particle, $F = dp/dt$, or

$$\Delta p = F\,\Delta t \qquad (2.278)$$

To obtain Δp one needs to integrate along the z' direction as the force F is given by Coulomb's law, and is in the radial direction. Taking components along the z'

direction, and integrating to obtain Δp, as:

$$\Delta p = \int (dp)_{z'} = \int F \cos \phi \, dt = \int F \cos \phi \, \frac{dt}{d\phi} \, d\phi \tag{2.279}$$

since the conservation of angular momentum (torque is zero for central forces) gives

$$M_1 r^2 \frac{d\phi}{dt} = M_1 v b \tag{2.280}$$

or

$$\frac{d\phi}{dt} = \frac{r^2}{vb} \tag{2.281}$$

and with this the momentum change

$$\Delta p = \frac{Z_1 Z_2 e^2}{r^2} \int \cos \phi \, \frac{r^2}{vb} \, d\phi = \frac{Z_1 Z_2 e^2}{vb} \int \cos \phi \, d\phi \tag{2.282}$$

or

$$\Delta p = \frac{Z_1 Z_2 e^2}{vb} (\sin \phi_2 - \sin \phi_1) \tag{2.283}$$

But $\phi_1 = -\phi_0$ and $\phi_2 = +\phi_0$, where $2\phi_0 + \theta = 180°$. Then $\sin \phi_2 - \sin \phi_1 = 2 \sin(90° - 1/2\theta)$, and Δp is

$$\Delta p = 2 M_1 v \sin \frac{\theta}{2} = \frac{Z_1 Z_2 e^2}{bv} 2 \cos \frac{\theta}{2} \tag{2.284}$$

This gives the relationship between the impact parameter b and scattering angle:

$$b = \frac{Z_1 Z_2 e^2}{M_1 v^2} \cot \frac{\theta}{2} = \frac{Z_1 Z_2 e^2}{2E} \cot \frac{\theta}{2} \tag{2.285}$$

and using the definition of cross section

$$\sigma(\theta) = \frac{-b}{\sin \theta} \frac{db}{d\theta} \tag{2.286}$$

and from the geometrical relations $\sin\theta = 2\sin(\theta/2)\cos(\theta/2)$ and $d\cot(\theta/2) = -1/2\, d\theta/\sin^2(\theta/2)$,

$$\sigma(\theta) = \left(\frac{Z_1 Z_2 e^2}{4E}\right)^2 \frac{1}{\sin^4 \theta/2} \tag{2.287}$$

This is the Rutherford formula for elastic scattering.

Electron scattering experiments fully confirm the theoretical results for electrons in the energy range used in microscience.

For heavier particles (ions) one has to take into account the nuclear recoil and with the two-body kinematics the Rutherford cross section can be written as

$$\sigma(\theta) = \left(\frac{Z_1 Z_2 e^2}{4E}\right)^2 \frac{4}{\sin^4 \theta} \frac{(\{1 - [(M_1/M_2)\sin\theta]^2\}^{1/2} + \cos\theta)^2}{\{1 - [(M_1/M_2)\sin\theta]^2\}^{1/2}} \tag{2.288}$$

which can be expanded for $M_1 \ll M_2$ in a power series to give

$$\sigma(\theta) = \left(\frac{Z_1 Z_2 e^2}{4E}\right)^2 \left[\sin^{-4}\frac{\theta}{2} - 2\left(\frac{M_1}{M_2}\right)^2 + \cdots\right] \tag{2.289}$$

where the first term omitted is on the order of $(M_1/M_2)^4$.

2.6.2.2. *Inelastic Electron–Electron Collisions.* In the case of small-angle electron–electron scattering, the momentum transfer Δp from the moving electron to the target electron is given by the following equation (see Section 2.6.2.1):

$$\Delta p = \frac{2e^2}{bv} \tag{2.290}$$

where b is the impact parameter and v is the electron velocity. Here the small-angle approximation is used where

$$\phi_1 = \phi_2 \approx 90 \tag{2.291}$$

and

$$M_1 = M_2 = m \tag{2.292}$$

The transferred energy can be expressed as

$$T = \frac{(\Delta p)^2}{2m} = \frac{e^4}{Eb^2} \tag{2.293}$$

where $E = 1/2mv^2$ is the energy of the scattered electron. Using the definition of differential cross section as function of the energy transferred

$$d\sigma(T) = -2\pi b \, db \tag{2.294}$$

one can get the differential cross section as

$$d\sigma(T) = -\frac{\pi e^4}{E} \frac{dT}{T^2} \tag{2.295}$$

The cross section for an electron to transfer an energy between T_{min} and T_{max} is

$$\sigma_e = \int_{T_{min}}^{T_{max}} d\sigma(T) \tag{2.296}$$

$$\sigma_e = \pi \frac{e^4}{E}\left(\frac{1}{T_{min}} - \frac{1}{T_{max}}\right) \tag{2.297}$$

For energetic electrons with energy E of several hundred electron volt or greater, the maximum energy transfer ($T_{max} = E$) is very much greater than T_{min}; therefore,

$$\sigma_e \cong \frac{\pi e^4}{E} \frac{1}{T_{min}} = \frac{6.5 \times 10^{-14}}{E T_{min}} \quad \text{cm}^2 \tag{2.298}$$

with E and T_{min} in electron volts and using $\sigma_e = 14.4 \, \text{Å}^2$.

Using the expression for the inelastic cross section, the electron impact ionization cross section can be estimated with the assumption that the minimum energy transfer T_{min} is equal to the electron binding, E_B, energy in the atom. Thus,

$$\sigma = \frac{\pi e^4}{E E_B}$$

2.6.2.3. Plasmon Excitation.   In addition to the individual inelastic scattering event, energetic electrons suffer a continuous energy loss to surface and bulk plasma excitations (plasmons). The plasma frequency is determined by oscillations of the valence electrons in a metal with respect to the positively charged cores. Consider the fluctuation δr in radial distance r from a positive core of a free electron gas containing a concentration n of electrons. If the gas expands from its equilibrium radius δr, the number of electrons in the shell, $\delta n = 4\pi n r^2 \delta r$, establish an electric field E

$$E = 4\pi n e \delta r \tag{2.299}$$

The retarding force F created by the expansion is

$$F = -eE = -4\pi e^2 \delta rn \tag{2.300}$$

The solution for the frequency of a harmonic oscillator is given as

$$\omega_p = \left(\frac{4\pi e^2 n}{m}\right)^{1/2} \tag{2.301}$$

m is the mass of the electron. For metals, the value of $n \cong 10^{23}/\text{cm}^2$ gives an oscillator frequency $\omega_p = 1.8 \times 10^{16}$ rad/s and an energy $\hbar\omega_p = 12$ eV.

In addition to the bulk plasma oscillations, surface plasma oscillations are also excited by electrons impinging on the target. The surface plasmon frequency is given as

$$\omega_p(s) = \frac{1}{\sqrt{2}} \, \omega_p(\text{bulk}) \tag{2.302}$$

This equation is found to hold for many metals and semiconductors. The calculated plasmon energies of Si and Ge are 16.0 eV based on four valence electrons per atom with the entire valence electron sea oscillating with respect to the ion cores. The measured values for Si are 16.4–16.9 eV and for Ge, 16.0–16.4 eV.

2.6.2.4. Bremsstrahlung. Bremsstrahlung is an always present continuous energy loss mechanism for accelerated charged particles.

Bremsstrahlung (braking radiation) yields a continuum of photons up to the energy of the incident electron. To derive the photon energy spectrum and the energy loss due to the Bremsstrahlung, recall the elastic (Rutherford) cross section

$$\frac{d\sigma}{d\Omega} = \left(\frac{Z_1 Z_2 e^2}{4E}\right)^2 \frac{1}{\sin^4 \theta/2} \tag{2.303}$$

where θ is the scattering angle. It is convenient to express this cross section in terms of the associated momentum transfer Δp

$$\Delta p = 2p \sin \theta/2 \tag{2.304}$$

and

$$d\Omega = 2\pi \sin \theta \, d\theta = \frac{2\pi \, \Delta p \, d\Delta p}{p^2} \tag{2.305}$$

Then the cross section for a momentum transfer Δp is given by

$$\frac{d\sigma}{d\Delta p} = 8\pi \left(\frac{Z_1 Z_2 e^2}{v}\right)^2 \frac{1}{(\Delta p)^3} \tag{2.306}$$

Using the formula for the total energy radiated per unit frequency interval per collision

$$\frac{dI}{d\omega} = \frac{2}{3\pi} \frac{(Z_2 e)^2 \Delta p^2}{m^2 c^3} \tag{2.307}$$

where m is the mass of the deflected particle. Then the differential radiation cross section is

$$\frac{d^2\chi}{d\omega \, d\Delta p} = \frac{dI}{d\omega} \frac{d\sigma}{d\Delta p} \tag{2.308}$$

or explicitly

$$\frac{d^2\chi}{d\omega \, d\Delta p} = \frac{16}{3} \frac{Z_2^2 e^2 (Z_1^2 e^2)^2}{m^2 v^2 c^3} \frac{1}{\Delta p} \tag{2.309}$$

after integration over all possible energy transfers Δp_{min} to Δp_{max} to find the frequency spectrum,

$$\frac{d\chi}{d\omega} = \frac{16}{3} \frac{Z_2^2 e^2 (Z_1^2 e^2)^2}{m^2 v^2 c^3} \ln \frac{\Delta p_{max}}{\Delta p_{min}} \tag{2.310}$$

To determine the ratio, $\Delta p_{max}/\Delta p_{min}$, one can use the energy and momentum conservation laws in the form of

$$E = E' + h\omega \tag{2.311}$$

and

$$(\Delta p)^2 = (p - p' - k)^2 \cong (p - p')^2 \tag{2.312}$$

where E, p and E', p' refer to the energy and momentum of the particle before and after the collision and hk is the momentum of the emitted Bremsstrahlung photon. Neglecting the photon momentum, one gets

$$\frac{\Delta p_{max}}{\Delta p_{min}} = \frac{p + p'}{p - p'} = \frac{(\sqrt{E} + \sqrt{E - h\omega})^2}{h\omega} \tag{2.313}$$

so that

$$\frac{d\chi}{d\omega} = \frac{16}{3} \frac{Z_2^2 e^2 (Z_1^2 e^2)^2}{m^2 v^2 c^3} \ln \left(\frac{2 - \hbar\omega/E + 2\sqrt{1 - \hbar\omega/E}}{\hbar\omega/E} \right) \qquad (2.314)$$

which is the Bethe–Heitler formula for the cross section for Bremsstrahlung. The functional form shows that the cross section falls as $\hbar\omega/E$ with a cutoff at $E = \hbar\omega$. The cross section is also proportional to $Z_2^2 Z_1^2/m^2$ expressing the fact that strong radiation comes from light particles (electrons) scattered in high-Z targets.

The total energy lost by Bremsstrahlung per incident particle in the target having N nuclei per unit volume is

$$\frac{dE_{rad}}{dx} = N \int_0^{\omega_{max}} \frac{d\chi(\omega)}{d\omega} \, d\omega \qquad (2.315)$$

Letting $x = \hbar\omega/E$ and noting that

$$\left(\frac{1 + \sqrt{1 - x}}{\sqrt{x}} \right)^2 = \frac{2 - x + 2\sqrt{1 - x}}{x} \qquad (2.316)$$

one gets for the energy loss

$$\frac{dE_{rad}}{dx} = \frac{16}{3} \frac{NZ_2^2 e^2 (Z_1^2 e^2)^2}{c^3 h} \int_0^1 \ln \left(\frac{1 + \sqrt{1 - x}}{\sqrt{x}} \right) dx \qquad (2.317)$$

2.6.2.5. Continuous Electron Energy Loss. An electron moving through matter loses energy by electromagnetic interaction. In these interactions the target electrons are excited to a higher energy level or an atomic electron is raised into the continuum level, in which case the target atom is ionized. In both cases the incremental energy is taken from the kinetic energy of the electron. The energy loss suffered by these excitations was first calculated by Bohr using only classical (non-quantum-mechanical) considerations.

To calculate the energy loss per unit distance $(-dE/dx)$, consider an electron of mass m at distance b from the path of the incident electron having a charge $Z_1 e$, mass m, and velocity v. Here again one can use the small-angle ($\theta \sim 0$) approximation for the momentum transfer (see Section 2.6.2.1)

$$\Delta p = \frac{2Z_1 e^2}{bv} \qquad (2.318)$$

and the kinetic energy is

$$\frac{\Delta p^2}{2m} = \frac{2Z_1^2 e^4}{b^2 m v^2} = T \qquad (2.319)$$

where T is the energy transfer in the collision.

The differential cross section, $d\sigma(T)$, for an energy transfer between T and $T + dT$ is

$$d\sigma(T) = -2\pi b\ db \qquad (2.320)$$

and the energy loss per unit path length, dE/dx, is

$$-\frac{dE}{dx} = n \int_{T_{\min}}^{T_{\max}} T\ d\sigma \qquad (2.321)$$

where n is the number of electrons per unit volume. In terms of impact parameter b,

$$-\frac{dE}{dx} = n \int_{b_{\min}}^{b_{\max}} T\ 2\pi b\ db \qquad (2.322)$$

and after integration

$$-\frac{dE}{dx} = \frac{4\pi Z_1^2 e^4}{mv^2} \ln \frac{b_{\max}}{b_{\min}} \qquad (2.323)$$

Using one-body kinematics it is assumed that the target electron is fixed; therefore, the maximum momentum transferred is $2mv$ or

$$b_{\min} \approx \frac{Ze^2}{mv^2} \qquad (2.324)$$

To estimate $b_{\max}$ it is assumed that the minimum energy transfer to the atomic electron is sufficient to raise the atomic electron to an excited state, i.e., $\Delta pv + 2T = 2I$, where $I = T_{\min}$ represents the exictation energy of an electron.
With these

$$\frac{b_{\max}}{b_{\min}} = \frac{mv^2}{I} \qquad (2.325)$$

and with this the Bohr continuous energy-loss formula is given as

$$-\frac{dE}{dx} = \frac{4\pi Z_1^2 e^4 n}{mv^2} \ln \frac{mv^2}{I} \qquad (2.326)$$

A more precise derivation of the energy loss is carried out by Bethe which gives

$$-\left(\frac{dE}{dx}\right)_{\text{ion}} = \frac{2\pi e^4 n}{v^2 m} \left[\ln \frac{v^2 mT}{2\bar{I}^2(1 - \beta^2)} - \ln 2(2\sqrt{1 - \beta^2} - 1 + \beta^2) + 1 - \beta^2 \right] \qquad (2.327)$$

where $\bar{I}$ is the average ionization potential of the atoms of the absorber and T is the relativistic kinetic energy of the electron.

Now one can take the ratio between the radiative loss for electrons (Bethe–Heitler) to nonradiative loss (Bohr energy loss) as

$$\left(\frac{dE}{dx}\right)_{rad}\bigg/\left(\frac{dE}{dx}\right) = \frac{2}{\pi}\left(\frac{e^2}{c\hbar}\right)\left(\frac{\sigma}{c}\right)^2 = \frac{2}{\pi 237}\frac{V^2}{c} \tag{2.328}$$

In the electron energy range used in microscience, V is less than c and the radiative energy loss is less than 1% at 100 keV.

2.6.3. Solid Surfaces

A schematic spectrum of electrons emitted from a surface under electron bombardment is shown in Fig. 2.61. Electrons of different origin are observed in the four energy ranges (I–IV in Fig. 2.61). The relative intensity of the various contributions to the spectrum depends on the primary energy E_0 (5–50 keV), the angle of incidence, and the angle of emission.[92] An actual elastic energy spectrum obtained under particular experimental conditions may therefore differ significantly in shape from Fig. 2.61. The narrow peak at the incident energy E_0 is made up of elastically scattered electrons plus diffracted electrons if the thin film is crystalline. Energy loss in range I due to photon excitations (see insert a in Fig. 2.61) can be resolved only by the most sophisticated energy analyzers.

In metals and semiconductors, most of the energy lost in range II (due to electronic excitations and ionization losses) is transferred to the conduction or valence electrons through individual or collective excitations. In metals the free electrons can be excited in unison by the incident electron beam, resulting in a collective or "plasma" oscillation of the valence electrons. Typical energy losses for plasma excitation are around 10–20 eV (see insert b in Fig. 2.61).

Above 50-keV energy losses (range III), the energy spectrum is shaped by various atomic excitation mechanisms. The most striking is the appearance of sharp edges corresponding to the excitation of inner-shell atomic electrons into the vacuum continuum (see insert d in Fig. 2.61). These edges correspond to the energy necessary to ionize the atoms, and the energy loss at which they occur is related to the atomic number. In these processes a core hole is generated by a primary or secondary electron and filled by another core or valence electron. In the Auger process an electron drops down to fill the hole and a second electron (Auger electron) is emitted (see insert c in Fig. 2.61). In the case of X-ray emission (X-ray fluorescence), instead of the second electron a photon is emitted. These processes (Auger, X-ray emission) are shown in Fig. 2.62. Finally, range IV (0–50 eV) contains the "true" secondary electrons, which are the result of cascade collision processes of the primary and target electrons. The secondary-electron distribution (dashed line in Fig. 2.61) shows a most probable energy at 3–5 eV. One cannot find a clear upper energy limit for secondary electrons, but the use of 50 eV is the present convention.

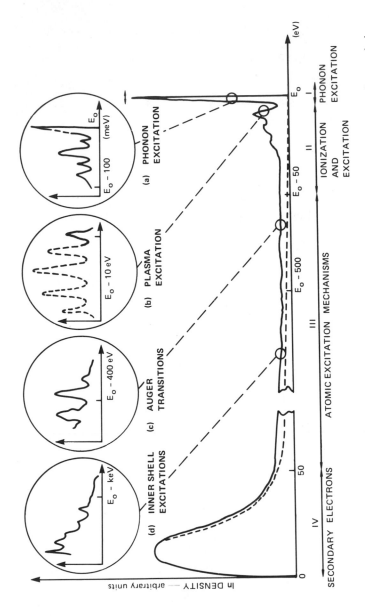

FIGURE 2.61.   Spectrum of inelastic electrons. ———, elastically scattered electrons; – – –, true (scattered) secondaries.

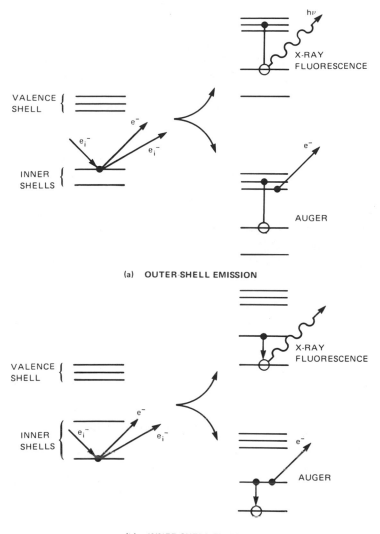

FIGURE 2.62. Two-step Auger and X-ray fluorescence processes. (a) Outer-shell emission; (b) inner-shell emission.

2.6.3.1. Electron Range. The penetration depth of electrons in a solid and the distribution of energy loss as a function of penetration depth are of great importance in microfabrication.

The rate of change of the electron kinetic energy E along the path s is described by the Bethe energy loss formula:

$$\frac{dE}{ds} = \left(\frac{N_A e^4}{2\pi\varepsilon_0^2}\right)\left(\frac{Z\rho}{A}\right)\left(\frac{1}{E}\ln\frac{1.66E}{J}\right) \tag{2.329}$$

which consists of three terms: The first is a constant, the second depends only on the material, and the third depends on both the energy and the material [$J = 1.15 \times 10^{-2}Z$ (keV)]. Since Z/A is nearly constant, the second term varies with the density, which changes with the position of the element in the periodic table. However, if the distance is measured in terms of ρs (g/cm^2), the change with density is accounted for.

To calculate the electron range, one has to integrate Eq. (2.329) starting with $E = E_0$ at $s = 0$ (electron entrance):

$$R_B = \int_{E_0}^0 \frac{dE}{dE/d(\rho s)} \tag{2.330}$$

The maximum value of ρs for complete slowing down ($E = 0$) is called the Bethe range R_B. R_B can be larger than the practical range R because of elastic scattering of the electrons in the target.

The practical electron range R, obtained by measuring the electron transmission of thin films, can be approximated by

$$R = 10E_0^{1.43} \left[\frac{\mu g}{cm^2} \right] \tag{2.331}$$

where E_0 is in deV and R is in μg/cm. This formula is useful in the range of $E_0 = 1$–100 keV.

For low-Z materials, R is of the same order as R_B. The probability for large-angle elastic scattering is small owing to the Z^2 in the Rutherford formula. More specifically, the inelastic/elastic total scattering cross sections can be expressed as

$$\frac{\sigma_i}{\sigma_e} \frac{1}{Z} \ln \left(\frac{4E_0}{\bar{E}} \right) \tag{2.332}$$

where $E \approx 5Z$, and

$$\sigma_e = \frac{1.5 \times 10^{-4}}{(v/c)^2} Z^{3/2} \quad \text{(in Å}^2) \tag{2.333}$$

where v is the velocity of the electron. From Eq. (2.332) it is clear that the ratio σ_i/σ_e increases slowly with increasing voltage and with decreasing Z.

In low-Z materials, most of the electrons pass through the target with multiple inelastic small-angle scattering events. For example, for carbon

$$\sigma_i \cong 3\sigma_e$$

and for platinum

$$\sigma_i \cong 0.4\sigma_e$$

In high-Z targets, the electron trajectories are bent more strongly and no electron can penetrate to a depth of R_B.

Many theoretical models[93-95] for electron penetration have been built on variations of Eq. (2.330). In these models the electrons penetrate to a depth z_d into the material, losing energy according to the Bethe law and scattering according to a screened Coulomb cross section. At depth z_d, owing to multiple scattering, the electron paths become isotropic, traveling in all directions until they reach their total range (R_B) in the target material. The "depth of complete diffusion"[94] is given by

$$z_d \approx R_B \left(\frac{40}{7Z} \right) \tag{2.334}$$

Based on this equation one can characterize the shape of the electron diffusion cloud with the shape of a fruit. For low-Z materials, the diffusion cloud resembles the shape of a pear, and for high-Z targets, that of an apple.

The energy dissipated $dE/d(\rho z)$ by an electron beam is used mainly for X-ray production, secondary emission, and electron–hole pair production. A typical plot of $dE/d(\rho z)$ as a function of ρz is shown in Fig. 2.63. The extrapolation of the linear portion of the curve defines the so-called Grün range $R_G(E)$, which can be expressed for low-Z targets[92] by

$$R_G = 4.0 E_0^{1.75} \tag{2.335}$$

where E_0 is the beam energy in keV and the range is expressed in $\mu g/cm^2$ (see Fig. 2.64). Knowledge of the energy dissipation $dE/d(\rho z)$ profile as a function of distance ρz ["depth–dose distribution" $\phi(z)$] from the point of entry of the primary-electron

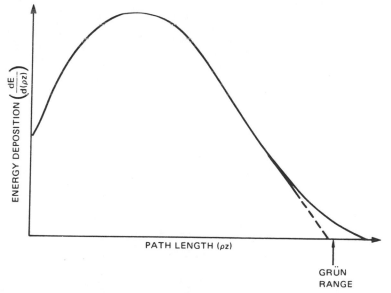

FIGURE 2.63. Energy deposition as a function of path length.

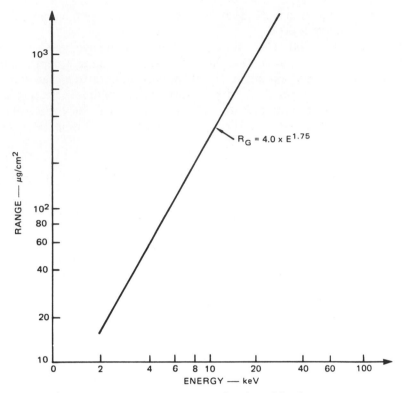

FIGURE 2.64.   The Grün range as a function of the electron energy.

beam is necessary for calculating secondary yields.[93,94] Figure 2.65 shows the depth–dose curves[93] for carbon and gold calculated by the Monte Carlo method.

2.6.3.2. Backscattered Electrons. When an electron beam impinges on a plane target, the electrons emerging from the target may be divided into two groups: backscattered electrons and "true" secondary electrons. Backscattered electrons leave

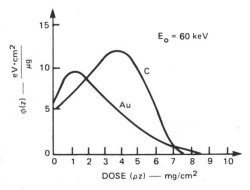

FIGURE 2.65.   Depth–dose distribution $\phi(z)$ generated by 60-keV electrons in carbon and gold.

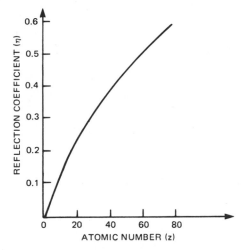

FIGURE 2.66. Reflection coefficient η as a function of atomic number Z,

the target with energies greater than 50 eV. These can be subdivided into the following:

- Elastically reflected primaries, which leave the target with only a small loss of energy
- Diffused electrons, which leave the target with a greater loss of energy
- Auger electrons, which have an energy related to the target material

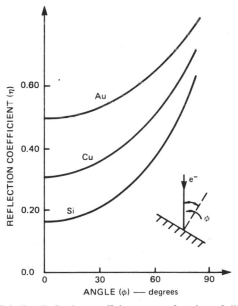

FIGURE 2.67. Reflection coefficient η as a function of tilt angle ϕ.

True secondary electrons typically have energies of only a few volts. However, the secondary-emission coefficient δ is customarily defined as the ratio of the total electron current leaving the target to the primary-electron current striking the target. Thus, δ has two components: η, the ratio of reflected-electron current leaving the target to the primary current striking the target, and δ_s, the ratio of the secondary-electron current leaving the target to the primary-electron current striking the target; i.e.,

$$\delta = \eta + \delta_s \tag{2.336}$$

From the integration of the Rutherford scattering for large scattering angles[96] one can define η as

$$\eta = \frac{a - 1 + 0.5^a}{a + 1} \tag{2.337}$$

where $a = 0.045Z$. Figure 2.66 shows η as a function of Z. This expression predicts that η is independent of the primary-electron energy and depends only on the atomic number of the target material and is in good agreement with the experimental data. Note that η increases with increasing tilt angle ϕ,[97] as shown in Fig. 2.67. The angular distribution of the backscattered electrons at normal incidence ($\phi = 0$) can be approximated by a $\sin \Omega$ law[98] where Ω is the takeoff angle for backscattered electrons (see Fig. 2.68). The exit depth of the backscattered electrons is on the order of half the electron range for normal incidence.[99]

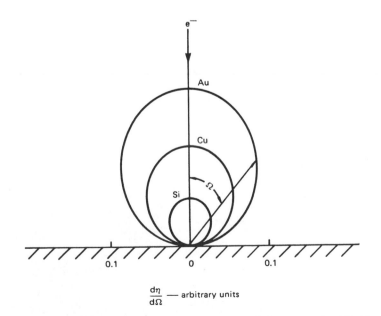

$$\frac{d\eta}{d\Omega} \text{ — arbitrary units}$$

FIGURE 2.68. Angular distribution of backscattered electrons at normal incidence.

2.6.3.3. Secondary Electrons. The energy distribution of the secondary electrons shows (Fig. 2.69) that the most probable energy is between 2 and 5 eV.

The secondary-electron yield δ_s, defined as the average number of emitted secondaries per incident primary, does not show a monotonic increase with increasing Z as does the backscattered electron yield, but it is influenced by the variation of the work function and surface conditions.[100] δ_s shows a maximum larger than unity at primary energies in the range $E_0 = 300$–800 eV.

The observed behavior of the secondary emission can be explained in terms of the production and escape mechanisms.[101] Let $n(z) \, \Delta z$ be the number of secondary

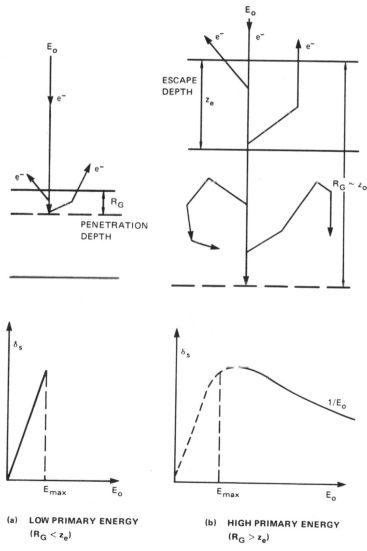

(a) LOW PRIMARY ENERGY
$(R_G < z_e)$

(b) HIGH PRIMARY ENERGY
$(R_G > z_e)$

FIGURE 2.69. Secondary electron field yield (δ_s) as a function of the energy of primary electron. (a) Low primary energy ($R_G < z_e$); (b) high primary energy ($R_G < z_e$).

electrons produced between z and $z + \Delta z$; let $g(z)$ be the average probability of escape of a secondary electron produced at a depth z. Then δ_s is given by

$$\delta_s = \int_0^{z_0} n(z)g(z) \, dz \tag{2.338}$$

where z_0 is the range of the primary electrons in the target material ($z_0 \sim R_G$). At very small primary energies the penetration depth is small so that $g(z) = g_0$ (see Fig. 2.69) and

$$\delta_s = g_0 \int_0^{z_0} n(z) \, dz = g_0 N \tag{2.339}$$

where N is the total number of secondary electrons produced at a given primary energy E_0. If on the average an energy E_s is needed to produce a secondary electron, then $N = E_0/E_s$ and

$$\delta_s = g_0 \frac{E_0}{E_s} \tag{2.340}$$

i.e., δ_s is roughly proportional to E_0 at low energies.

At high primary energies ($R_G > z_e$) the energy loss [Eq. (2.329)] can be approximated by

$$\frac{dE_0}{dz} = -\frac{a}{E_0^\alpha} \tag{2.341}$$

where α is a very slow function of E_0 starting at $\alpha < 1$ and approaching unity for large primary energies. Equation (2.341) can be integrated if we treat α as a constant, with the condition $E_0 = E_{00}$ at $z = 0$, obtaining the extended Whiddington law as

$$E_0^{1+\alpha} = E_{00}^{1+\alpha} - (1 - \alpha)az \tag{2.342}$$

From this the range $Z_0 \sim R_G$ is found by setting $E_0 = 0$. This yields, from Eq. (2.342),

$$z_0 = \frac{E_{00}^{1+\alpha}}{(1 + \alpha)a} \tag{2.343}$$

Now we assume that $n(z)$, the number of secondary electrons produced per unit length, is proportional to $-dE_0/dz$. That is, if b is a proportionality factor,

$$n(z) = \frac{ba}{E_0^\alpha} \tag{2.344}$$

and the escape function has the form

$$g(z) = g_0 \exp(-\beta z) \tag{2.345}$$

Substituting Eqs. (2.344) and (2.345) into Eq. (2.338) yields

$$\delta_s = bg_0 \int_0^{z_0} \frac{a}{E_p^\alpha} \exp(-\beta z) \, dz \tag{2.346}$$

Then using Eqs. (2.341) and (2.343) one obtains

$$\delta_s = bg_0 \int_0^{E_{00}} \exp\left[\frac{\beta}{(1+\alpha)a} (E_{00}^{1+\alpha} - E_0^{1+\alpha})\right] dE_0 \tag{2.347}$$

This integral goes through a maximum of $E_{00} = E_{max}$ and decreases for $E_{00} > E_{max}$. If E_{00} is larger than a few times E_{max}, then the integral yields

$$\delta_s = \frac{bg_0 a}{\beta E_{00}^\alpha} = \frac{const}{E_{00}^\alpha} \tag{2.348}$$

This result agrees with measurements above 10 kV, where α approaches unity.
     The exit depth (z_e) for secondary electrons is on the order of 1–10 nm.[102] The variation of the secondary-electron yield with tilt angle ϕ can be approximated by

$$\delta_s(\phi) = \frac{\delta_s(0)}{\cos \phi} \tag{2.349}$$

2.6.3.4. *X-ray Emission.*   When electrons excite atoms in a solid, the excitation energy can be transferred to other electrons in an Auger process (see Fig. 2.57) or it can be emitted in the form of characteristic photons (X-ray fluorescence). At low energies the transfer rate is much greater for the electron process than for photon emission. The photon process generally becomes more probable with increasing energy.[103] The interaction of an electron with a solid from the standpoint of X-ray emission can be divided into two distinct modes: emission resulting from the radiative capture of energetic electrons (normal Bremsstrahlung) and emission resulting from inelastic collisions with core electrons (Auger electrons or characteristic X rays). The fraction of incident electron intensity that is converted into Bremsstrahlung is found experimentally to vary linearly with both the Z of the material and the incident-electron energy:

$$f_B \approx 7 \times 10^{-10} Z E_0 \tag{2.350}$$

where Z is the atomic number of the target and E_0 is given in eV.[104]

The separation of channels producing X rays and electrons from an electron-bombarded target is shown in Fig. 2.70. The channel on the right side involves the dissipation of energy by the nuclear and valence band electron–electron scattering, and the left channel includes the energy dissipation involving atomic core-level excitation.[105] Here again, if one wants to calculate the number of emitted X-ray quanta from the solid, one has to calculate the number of ionization processes along the electron trajectory, using the Bethe formula.

The number of emitted X-ray quanta per path element $d(\rho s)$ per unit of solid angle for an element concentration c_a can be written [106] as

$$dN_K = \frac{1}{4\pi} \omega_K P_K \frac{c_a N_A}{A} \frac{\sigma_K}{dE/d(\rho s)} dE \tag{2.351}$$

where ω_K is the K-shell fluorescence yield, $\omega_K \sim 2 \times 10^{-2}$ for low Z and ~ 0.97 for gold, P_K is the ratio of intensity of the observed characteristic line to all lines of the K series, and σ_K is the total cross section of K-shell ionization.[107]

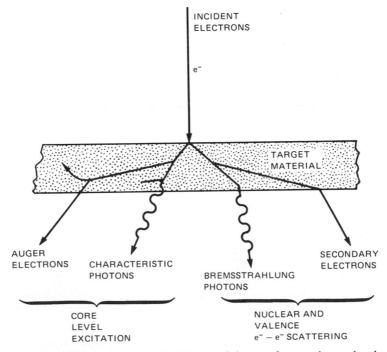

FIGURE 2.70.   Separation channels producing X rays and electrons from an electron-bombarded target.

The total number of X-ray quanta generated from the slowdown of an electron from energy E_0 to the average K-shell ionization energy, E_1, can be obtained from

$$N_K = \frac{1}{4\pi} \omega_K P_K \frac{c_a N_A}{A} \int_{E_K}^{E_0} \frac{\sigma_K}{dE/d(\rho s)} \, dE \qquad (2.352)$$

Analogous to the depth distribution for ionization, there is a depth distribution of X-ray emission $\phi_x(z)$ with a similar shape but reduced dimensions.

Now we can summarize (Fig. 2.71) the wide range of depths from which the secondary particles (e^-, X ray) originate. Figure 2.71 illustrates the range and spatial resolution of various secondaries that are generated when an electron beam strikes a solid target.[108]

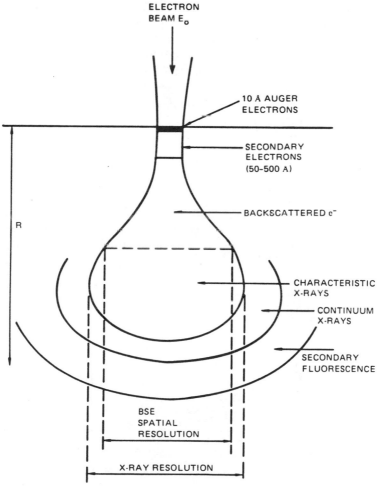

FIGURE 2.71. Range and spatial resolution of secondaries generated when an electron beam strikes a solid target. From Ref. 108.

2.6.4. Physical Effects in the Target

2.6.4.1. Polymerization: Bond Breaking. When energetic (1–100 keV) electrons strike a target made of organic molecules, the molecules may either break down to smaller fragments or link together to form larger molecules. Usually both types of change occur simultaneously, but for a given material one of these processes dominates and determines the net effect upon the material. For polymers it is known that if every carbon atom in the main chain is directly linked to at least one hydrogen atom, larger molecules are generally formed on irradiation by the process known as cross-linking.

If the polymer is cross-linked, it forms a network and becomes relatively insoluble and infusible. This type of polymer forms the basis of negative electron resist, which can be dissolved when it is not irradiated but cannot be removed when irradiated. Thus, a residual film pattern remains on the irradiated areas after development.

Another group of polymers is known in which irradiation causes a break in the main polymer chain, resulting in a lower average molecular weight in the irradiated material than in the unirradiated polymer. Since the lower-molecular-weight material can be dissolved selectively, the residual film pattern forms a positive working electron resist. Polymethyl methacrylate (PMMA) is the best known high-resolution long-chain polymer resist. Figure 2.72 is a schematic representation of the basic mechanisms of resists, showing the cross-linking of negative resists and the chain scission of positive resists.

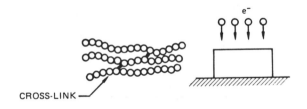

(a)   NEGATIVE RESIST (e.g., PS)

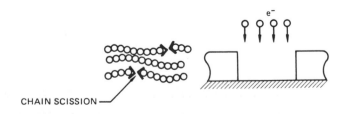

(b)   POSITIVE RESIST (e.g., PMMA)

FIGURE 2.72.   Schematic representation of basic resist mechanisms. (a) Negative resist (e.g., PS); (b) positive resist (e.g., PMMA).

2.6.4.2. Thermal Effects. Electron beams have many special properties that make them particularly well suited for thermal processing and machining.[109]

The first of these is a very high resolution. The second feature is the large energy densities that can be obtained ($\leq 10^{10}$ W/cm^2). Third, the dissipation of this power in surfaces can be used to alloy and anneal thin films (thermal lithography), catalyze chemical reactions, and selectively remove materials. Finally, electron beams are easy to control.

The distribution of the input electron-beam power, P_e, striking a target can be described by the following equation:

$$P_e = P_1 + P_2 + P_3 + P_4 \tag{2.353}$$

where P_1 is the power utilized in vaporizing the material, P_2 is the average power lost as a result of heat conduction, P_3 is the average power lost as a result of radiation, and P_4 is the power lost to ionization and excitation of the evaporated atoms in the path of the beam.

P_3 and P_4 are usually negligible. The efficiency of the evaporation is very low; much less than 1% of the beam power is used for evaporation.

Thus, the major effect of the beam is to cause the temperature to rise in the target area. The temperature rise caused by the beam is given as

$$\Delta T = \frac{3}{2} \frac{V_0 I_0}{2\pi \lambda R} \tag{2.354}$$

where V_0 is the beam voltage, I_0 the beam current, λ the heat conduction coefficient (J/m per s per degree) and R the range of the electrons in the material.

For metals ΔT is very small, but in plastics and insulators with low heat conductivities the temperature rise can cause the melting of the material. The heating effect of an electron beam will be discussed in more detail in connection with electron-beam annealing (Section 4.12.1).

2.6.4.3. Current Injection. If an electron beam impinges on a semiconductor target, collisions with the valence electrons produce electron–hole pairs. The energy necessary to create one electron–hole pair, E_p, is the same order of magnitude as the band-gap energy E_g. $E_g = 3.6$ eV in silicon and 2.8 eV in Ge. The number of electron–hole pairs created by one electron with energy E_0 is

$$N = \frac{E_0}{E_p} (1 - k\eta) \tag{2.355}$$

where η is the backscattering coefficient and $k = E_{BS}/E_0$, where E_{BS} is the mean energy of backscattered electrons.

To calculate the number of electron–hole pairs as a function of the penetration depth, one has to assume that the carrier generation is proportional to the main energy loss.[93] Measurement of the electron–hole generation profile (conductivity modulation) forms the physical basis for many electron-beam techniques used for device characterization.

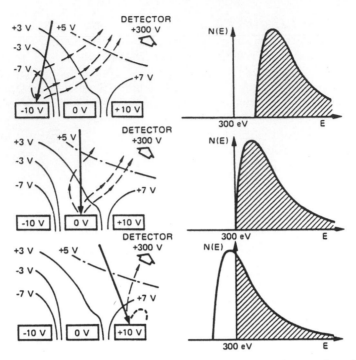

FIGURE 2.73. Change of the energy spectrum of the secondary electrons in voltage contrast measurements.

2.6.4.4. Field Effects. Local fields created by "energized" integrated-circuit electrodes change the energy spectrum of the secondary electrons when the electrodes are bombarded by a primary electron beam (e.g., in scanning electron microscopy). The gain or loss of the secondary electron kinetic energy with applied specimen voltage is the basis of the quantitative voltage measurement. As shown in Fig. 2.73, the whole energy spectrum shifts linearly with applied specimen voltage. The linear shift of the maximum of the spectrum is thus a direct measure of the applied specimen voltage. For quantitative measurement, an energy analysis of the secondary electrons is required in addition to the suppression of other field effects, except the field created by the tested electrode.[110]

2.7. ION INTERACTIONS

2.7.1. Introduction

In this section we consider the fate of energetic ions (1–100 keV) incident on a solid surface. The ten ways in which ions can interact with a surface are illustrated in Figure 2.74. An incoming ion can be backscattered by an atom or group of atoms in the bombarded sample (1). The backscattering process generally results in a deflection of the ion's incident path to a new trajectory after the encounter and an exchange of energy between the ion and the target atom. The energy exchange can

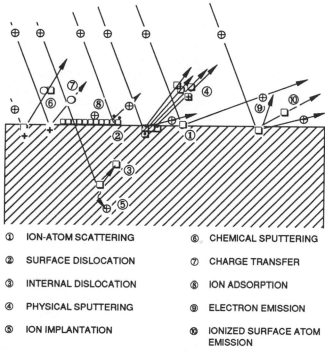

①	ION-ATOM SCATTERING	⑥	CHEMICAL SPUTTERING
②	SURFACE DISLOCATION	⑦	CHARGE TRANSFER
③	INTERNAL DISLOCATION	⑧	ION ADSORPTION
④	PHYSICAL SPUTTERING	⑨	ELECTRON EMISSION
⑤	ION IMPLANTATION	⑩	IONIZED SURFACE ATOM EMISSION

FIGURE 2.74. Ion–solid interactions. 1, ion–atom scattering; 2, surface dislocation; 3, internal disloca-tion; 4, physical sputtering; 5, ion implantation; 6, chemical sputtering; 7, charge transfer; 8, ion adsorp-tion; 9, electron emission; 10, ionized surface atom emission.

be either elastic or inelastic, depending on the constituent particles and the energy of the ions. The momentum of an ion can be sufficient to dislodge a surface atom from a weakly bound position on the sample lattice and cause its relocation on the surface in a more strongly bound position (2). This process is called atomic disloca-tion. Ions with greater energies can cause internal dislocations in the bulk of the sample (3). Physical sputtering (4) results when ions strike the sample surface and transfer enough momentum to entirely free one or more atoms. Ions can penetrate into the lattice and become trapped as their energy is expended (ion implantation) (5). Chemical reaction of the ions with the surface atoms can occur, and, as a result, new compounds can be formed on the sample surface or the outermost layer of atoms may leave as a gas (chemical sputtering) (6). The bombarding positive ion can gain an electron from the surface by Auger neutralization and be reflected as a neutral atom (7). Ions can become bound to the surfaces of the sample as adsorbed ions (8). Secondary electron emission (9) occurs under suitable conditions of ion bombardment of metal surfaces. Secondary ion emission (10) results when surface atoms are excited to ionized states and emitted from the sample.

In slowing down, the incident ion transfers energy to the solid. In analyzing the energy loss, it is convenient to distinguish between two major energy loss processes: electronic collisions and nuclear collisions.

The first process is the interaction of the fast ion with the lattice electrons, resulting in excitation and ionization. Since the density of the electrons in the target is high and the collisions are so numerous, the process similar to the electron energy loss can be regarded as continuous.[111]

The nuclear loss results from collisions between the screened nuclear charges of the ion and the target atoms. The trepidation of these collisions is lower, so they may be described by elastic two-body collisions. At high energies they are accurately described by Rutherford scattering, at medium energies by screened Coulomb scattering, while at low energies the interaction becomes even more complex.[112]

In addition, there is a contribution to the energy loss resulting from the charge exchange between the moving ion and target atom[113]; this energy loss reaches a maximum when the relative velocity is comparable to the Bohr electronic velocity $(2 \times 10^6 \text{ ms}^{-1})$.

The total energy loss $-dE/dz$ can thus be regarded as the sum of three components, nuclear, electronic, and charge exchange,

$$\frac{dE}{dz} = \left(\frac{dE}{dz}\right)_n + \left(\frac{dE}{dz}\right)_e + \left(\frac{dE}{dz}\right)_{ch} \tag{2.356}$$

At low ion energies, nuclear stopping dominates and is responsible for most of the angular dispersion of an ion beam. At high energies, electronic collisions are more important. A useful rule of thumb is that the energy lost to the lattice by nuclear collisions becomes predominant at energies less than A keV, where A is the atomic weight of the incident ion. At intermediate energy regions, the charge-exchange contribution may rise to roughly 10% of the total. The variation of the energy loss with ion energy is shown in Fig. 2.75.

Inelastic interactions with the target electrons lead to secondary electron emission, emission of characteristic X rays, and optical photon emission; elastic interactions lead to the displacement of lattice atoms, the formation of defects, and surface sputtering as illustrated schematically in Fig. 2.76.

The energy spectrum of ions[114] with initial energy E_0 scattered from a solid target is represented schematically in Fig. 2.77. There is a broad low-energy hump

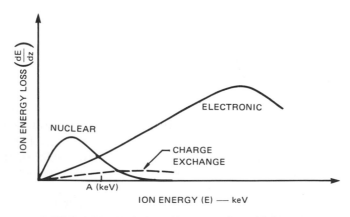

FIGURE 2.75. Variation of ion energy loss with ion energy.

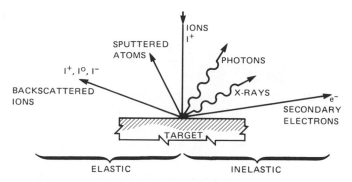

FIGURE 2.76.   Schematic representation of ion–target interactions.

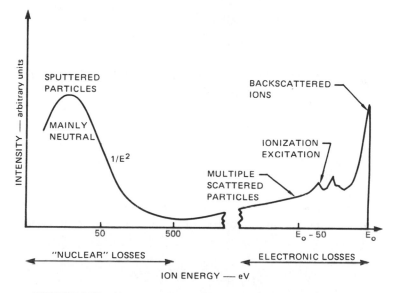

FIGURE 2.77.   Energy spectrum of ions scattered from a solid target.

(10–30 eV) corresponding to the emitted neutral atoms (sputtered atoms) and a high-energy hump distributed about the incident ion energy E_0 (elastically scattered ions).

  2.7.1.1. *Kinematics and Scattering of Ions.*   In this section the kinematics and scattering of ions by stationary atoms will be reviewed.[115]

  To describe ion–atom collisions the Rutherford–Bohr model of the atom will be used. This model of the atom is that of a cloud of electrons surrounding a positively charged central core—the nucleus—that contains Z protons and $A-Z$ neutrons, where Z is the atomic number and A the mass number. Outside the electron cloud the nuclear charge $(Ze+)$ is completely shielded by the orbiting electrons $(Ze-)$. However, if the ion penetrates the electron cloud it will experience a force (and potential) which depends on the interatomic separation and on the effective charges of the ion $(Z_{eff}e)$ and the atom $(Z_{2,eff}e)$. Then the interatomic force law is a repulsive coulomb force

$$F(r) = \frac{Z_{1,\text{eff}}Z_{2,\text{eff}}e^2}{r^2} \qquad (2.357)$$

and the potential interaction is the spatial integral of the force,

$$V(r) = \frac{Z_{1,\text{eff}}Z_{2,\text{eff}}e^2}{r} \qquad (2.358)$$

In ion implantation and ion scattering (Rutherford) experiments the closest approach between the atoms/ions never becomes less than about 10^{-1} Å which is smaller than the Bohr radius

$$a_0 = \frac{\hbar^2}{m_e e^2} = 5.3 \times 10^{-1} \text{ Å} \qquad (2.359)$$

In the case of Rutherford scattering the distance of closest approach, d, for 2-MeV He ions ($Z_1 = 2$) incident on Ag ($Z_2 = 47$), is

$$d = \frac{(2)(47)(14.4 \text{ eV Å})}{(2 \times 10^8 \text{ cV})} = 6.8 \times 10^{-4} \text{ Å} \qquad (2.360)$$

a value much smaller than the Bohr radius $a_0 = \hbar^2/m_e e^2 = 0.53$ Å, and the K-shell radius of Ag, $a_0/477 \cong 10^{-2}$ Å. Thus, the use of unscreened charges (Z_1, Z_2) is justified.

If $d > a_0$, the effect of the electron shielding of the positive nuclear charge has to be taken into account and the interaction potential has to be modified:

$$V(r) = \frac{Z_1 Z_2 \exp(-r/a)}{r} \qquad (2.361)$$

where a is a constant of the order of

$$a \approx \frac{a_0}{(Z_1 Z_2)^{1/6}} \qquad (2.362)$$

The general form of the repulsive component of the potential can thus be written as

$$V(r) = \frac{Z_1 Z_2 e^2}{r} f(Z_1 Z_2 r) \qquad (2.363)$$

where $f(Z_1 Z_2 r)$ is a screening function dependent on the calculation model used and the separation.

As a general rule, the potential for collisions between heavy atoms and ions in the low-energy region (0.1–100 keV) should include the effect but for light ions in the high-energy region (≥ 1 meV) the Coulomb potential is adequate.

Since the Coulomb force and potential decrease rapidly with increasing atomic separation, a first-order approximation to the potential can be found assuming that $V(r) = a$ constant from $r = 0$ to some distance $r = R_0$ and $V(r) = 0$ if $r > R_0$. The implication of this approximation is that each atom is treated as a perfectly elastic hard or rigid sphere of radius R_0 so that a collision takes place only at the separation $r = R_0$ and no interaction occurs for $r > R_0$. Since the spheres (atoms) are rigid and perfectly elastic, the kinetic energy and the momentum are conserved in the collision process.

As an example, let us consider the kinematics for Rutherford backscattering[116,117] (see Fig. 2.78). In Rutherford backscattering spectrometry, monoenergetic particles in the incident beam collide with target atoms and are scattered backwards into the detector–analysis system which measures the energies of the particles. In the collision, energy is transferred from the moving particle to the stationary target atom; the reduction in energy of the scattered particle depends on the masses of incident and target atoms and provides the signature of the target atoms.

The kinematics in elastic collisions between two isolated particles can be solved fully by applying the principles of conservation of energy and momentum. For an

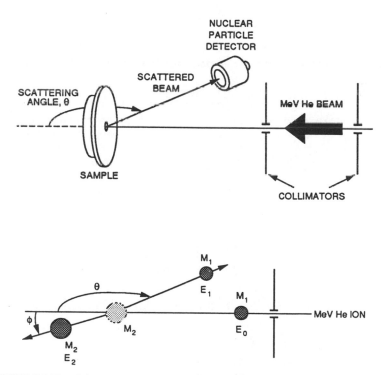

FIGURE 2.78.   Schematic and kinematics for Rutherford backscattering experiment.

incident energetic particle of mass M_1, the values of the velocity and energy are v and E_0 ($E_0 = 1/2 M_1 V^2$) while the target atom of mass M_2 is at rest. After the collision, the values of the velocities v_1 and v_2 and energies E_1 and E_2 of the projectile and target atoms are determined by the scattering angle θ and recoil angle ϕ.

Conservation of energy and momentum parallel and perpendicular to the direction of incidence gives

$$\tfrac{1}{2} M_1 v^2 = \tfrac{1}{2} M_1 v_1^2 + \tfrac{1}{2} M_2 v_2^2$$
$$M_1 v = M_1 v_1 \cos \theta + M_2 v_2 \cos \phi \qquad (2.364)$$
$$0 = M_1 v_1 \sin \theta - M_2 v_2 \sin \phi$$

Eliminating ϕ first and then v_2, one finds the ratio of particle velocities

$$\frac{v_1}{v} = [\pm (M_2^2 - M_1^2 \sin^2 \theta)^{1/2} + M_1 \cos \theta]/(M_2 + M_1) \qquad (2.365)$$

The ratio of the projectile energies for $M_1 < M_2$, where the plus sign holds, is

$$\frac{E_1}{E_0} = \left[\frac{(M_2^2 - M_1^2 \sin^2 \theta)^{1/2} + M_1 \cos \theta}{M_2 + M_1} \right]^2 \qquad (2.366)$$

The energy ratio, called the kinematic factor $K = E_1/E_0$, shows that the energy after scattering is determined only by the masses of the particle and target atom and the scattering angle.

For direct backscattering through 180°, the energy ratio has its lowest value given by

$$\frac{E_1}{E_0} = \left(\frac{M_2 - M_1}{M_2 + M_1} \right)^2 \qquad (2.367)$$

and at 90° is given by

$$\frac{E_1}{E_0} = \frac{M_2 - M_1}{M_2 + M_1} \qquad (2.368)$$

For $\theta = 180°$, the energy E_2 transferred to the target atom has its maximum value given by

$$\frac{E_2}{E_0} = \frac{4 M_1 M_2}{(M_2 + M_1)^2} \qquad (2.369)$$

with the general relation given by

$$\frac{E_2}{E_0} = \frac{4M_1 M_2}{(M_2 + M_1)^2} \cos^2 \phi \tag{2.370}$$

2.7.1.2. Energy Loss and Backscattering of Energetic Light Ions. In this section we will consider the interactions of light ions (H^+, d^+, and He^+) at high energies (0.5–5 MeV) during their passage through solids.

In Fig. 2.79 we follow the trajectory of a light ion passing into a solid target and suffering inelastic (excitation, ionization) energy losses and in depth t a Rutherford backscattering. The energy lost in penetration is directly proportional to the depth so that a depth scale can be assigned directly from the energy spectra of the backscattered particles. The energy loss suffered in the collision at depth t is given by the kinetic factor K. For large ($\sim180°$) scattering angle and light ions ($M_1 < M_2$) the kinetic factor

$$K \approx \left[\frac{M_2 + M_1 \cos \theta}{M_1 + M_2} \right] \tag{2.371}$$

The ion which penetrates a distance t loses ΔE_{in} energy by electronic processes and arrives at a depth t with an energy $E(t)$ where

$$E(t) = E_0 - \frac{dE}{dx} \cdot t \bigg|_{in} \tag{2.372}$$

At depth t an elastic (Rutherford) scattering occurs and the particle energy is reduced by the kinematic factor, i.e.,

$$E_1 = KE(t) \tag{2.373}$$

The ion loses further energy (ΔE_{out}) by electronic processes as it leaves the solid target. Thus,

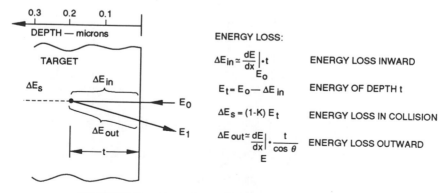

FIGURE 2.79. Energy losses for ions scattered from depth t.

$$E_1 = KE(t) - \frac{1}{|\cos \theta|} \left. \frac{dE}{dx} \right|_{\text{out}}$$

$$= K \left(E_0 - \left. \frac{dE}{dx} \right|_{\text{in}} t \right) - \frac{t}{|\cos \theta|} \left. \frac{dE}{dx} \right|_{\text{out}} \tag{2.374}$$

$$= KE_0 - t \left(K \left. \frac{dE}{dx} \right|_{\text{in}} + \frac{1}{|\cos \theta|} \left. \frac{dE}{dx} \right|_{\text{out}} \right)$$

The energy difference between particles scattered at the surface and at a depth t is

$$\Delta E = KE_0 - E_1 = t \left(K \left. \frac{dE}{dx} \right|_{\text{in}} + \frac{1}{|\cos \theta|} \left. \frac{dE}{dx} \right|_{\text{out}} \right) \tag{2.375}$$

so that the measured energy difference is converted into depth. The energy spectrum of Rutherford backscattering yield versus particle energy can consequently be converted into a spectrum showing yield as a function of depth into the target.

In backscattering measurements the energy-sensitive detector extends over a solid angle Ω so that the total number of detected particles, or yield Y from a thin layer of atoms Δt is

$$Y = \sigma(\theta) \cdot \Omega \cdot Q \cdot N \cdot \Delta t \tag{2.376}$$

where Q is the number of incident particles and $N \cdot \Delta t$ is the number of target atoms per square centimeter in the layer of thin film.

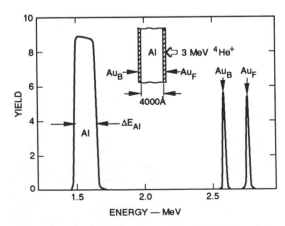

FIGURE 2.80.   A Rutherford backscattering spectrum of 3-Mev He ions for an Au–Al–Au thin film.

Using the Rutherford cross section for scattering, the yield from a slice of film width Δt at a depth t is given by the following equation for $\theta \cong 180°$:

$$Y(t) = \left(\frac{Z_1 Z_2 e^2}{4E(t)} \right)^2 N \cdot Q \cdot \Omega \cdot \Delta t \qquad (2.377)$$

where $E(t)$ is the ion energy at depth t and N is the atomic density.

A backscattered spectrum is shown in Fig. 2.80 for 3-MeV He ions for an Au–Al–Au thin-film structure.[116]

2.7.2. Energy-Loss Mechanisms

Microscopically, energy losses due to excitation and ionization (electronic losses) and due to nuclear collisions (nuclear losses) are discrete processes. However, macroscopically, the continuous energy-loss model is a more convenient description. The energy loss in a solid target depends on the distance Δt that the ions travel in the solid and the energy loss ΔE in this distance. The mass density ρ or the atomic density N are frequently combined with the distance, in the form $\rho \Delta t$ or $N \Delta t$, to express the amount of material per unit area or the number of atoms per unit area that the projectiles have traversed in losing energy ΔE to the target material. Energy loss can be expressed in several different ways. Some frequently used units are

$$dE/dx: \qquad eV/\text{Å}$$
$$(1/\rho)\, dE/dx: \qquad eV/(\mu g/cm^2)$$
$$S \equiv \varepsilon = (1/N)\, dE/dx: \qquad eV/(atoms/cm^2),\ eV\ cm^2$$

Recently, most authors have adopted $(1/N)\, dE/dx$ (eV cm^2) as the stopping cross section ε.

The ions in the solid-state target suffer inelastic "slow" and "fast" collisions compared to the orbital velocity of electrons in the target.

An estimate for the low-energy limit for fast collision is given by comparing the ion velocity with the Bohr orbital velocity V_0, where

$$V_0 = \frac{e^2}{h} = \frac{c}{137} = 2.2 \times 10^8\ cm/sec \qquad (2.378)$$

This velocity is that of 0.1-meV H or 25-keV H ions. Figure 2.81 shows values of the energy loss of the He or H ions in Al. The stopping cross section ($S \equiv \varepsilon$) decreases with increasing velocity because the interaction time is shorter. In this "fast" regime the Bohr theory gives a good description of the stopping power.

According to Bohr's theory of the stopping power, the collision between the ions and target electrons produces an energy transfer from the fast-moving ion. The energy loss from the fast particle to the stationary electrons or nucleus can be calculated from the scattering theory in the central-force field. This Bohr formula already

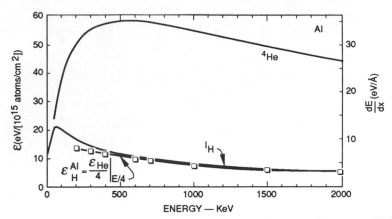

FIGURE 2.81. Stopping cross section, $\varepsilon = 1/N \cdot dE/dx$, and energy loss rate, DE/dx $N_{Al} = 6 \times 10^{22}$ atoms/cm^3.

derived (Section 2.6.2.5) and the energy loss is given including the distant resonance energy transfer, as

$$-\frac{dE}{dx} = \frac{2\pi Z_1^2 e^4}{E} \cdot NZ_2 \cdot \left(\frac{M_1}{m}\right) \ln \frac{2mv^2}{I} \tag{2.379}$$

where $E = M_1 v^2/2$ and $n = NZ_2$ and N is the atomic density. The average excitation energy I for most elements is about $10Z_2$ in eV, where Z_2 is the atomic number of the target atoms.

Fast ions also transfer energy to the nuclei of the solid through small-angle scattering. This nuclear energy loss is much smaller than the electronic loss and is given by Bohr's theory as

$$-\frac{dE}{dx}\bigg|_n = \frac{4\pi Z_2^2 e^4 N}{M_2 v^2} \ln \frac{b_{max}}{b_{min}} \tag{2.380}$$

where b_{max} represents the maximum energy transfer to a target atom in a head-on collision:

$$T_{max} = \frac{4M_1 M_2}{(M_1 + M_2)^2} E \tag{2.381}$$

so that b_{min} becomes

$$b_{min} = \frac{2Z_1 Z_2 e^2}{vp} \tag{2.382}$$

and

$$b_{max} = \frac{2Z_1 e^2}{(2mv^2 I)^{1/2}}$$

(2.383)

The ratio of the nuclear to the electronic energy loss per atom is

$$\frac{dE/dx|_n}{dE/dx|_e} \cong N \frac{Z_2^2}{M_2} \times \frac{m}{n} = \frac{Z_2}{M_2} m \cong \frac{1}{3600}$$

(2.384)

2.7.2.1. Nuclear Stopping. The calculation of elastic stopping requires a screened nuclear (Thomas–Fermi) potential[118]; the energy loss can be obtained from:

$$\left(\frac{dE}{dz}\right)_n = 27.8 \frac{Z_1 Z_2}{(Z_1^{2/3} + Z_2^{2/3})^{1/2}} \frac{M_1}{M_1 + M_2} N \quad (eV/\text{Å})$$

(2.385)

Here Z_1 and M_1 and Z_2 and M_2 are the atomic numbers and masses of the ion and lattice atoms, respectively, and N is the atomic density (particles/Å^3). This expression gives the magnitude of the energy loss rate as 10–100 eV/Å for the majority of ion–target atom combinations.

The energy E of an ion of incident energy E_0 after scattering through a given angle θ in any given system is determined by the laws of conservation of energy and momentum[119] and can readily be shown to be

$$E = \frac{E_0 [M_1 \cos \theta + (M_2^2 - M_1^2 \sin^2 \theta)^{1/2}]^2}{(M_1 + M_2)^2}$$

(2.386)

This expression is valid for $M_2 > M_1$ only when the positive root is chosen, but for $M_2 \le M_1$ both roots are valid, provided that $\sin \theta \le M_2/M_1$. The maximum energy that can be transferred to a target atom in a single collision is, from Eq. (2.386), for $\theta = \pi$:

$$E_{max} = \frac{4M_1 M_2}{(M_1 + M_2)^2} E_0$$

(2.387)

If $M_1 < M_2$, backscattering can occur in a single collision, whereas if $M_1 > M_2$, the scattering must be in the forward direction.

Since the stopping power in Eq. (2.385) is independent of energy, the ion range R_n is

$$R_n = \frac{E_0}{(dE/dz)_n} = 2kE_0$$

(2.388)

where

$$k = \frac{0.018}{N} \frac{(Z_1^{2/3} + Z_2^{2/3})^{1/2}}{Z_1 Z_2} \frac{M_1 + M_2}{M_1} \quad (\text{Å}/\text{eV}) \tag{2.389}$$

As expected, the range decreases when the atomic number Z or the density N increases.

2.7.2.2. Electronic Stopping. At ion velocities greater than the electron velocity in the K shell, an ion will have a high probability of being fully stripped of its electrons. The energy loss under these circumstances will be described by the Bethe formula, Eq. (2.329). However, for low ion velocities and high atomic numbers (Z_1, Z_2) the Bethe formula is invalid because it does not take into account the charge fluctuations, excitation of plasma resonance, and the charge-exchange effect.

An alternative approach has been formulated[120, 121] that accounts for the electronic stopping behavior at low intermediate energies for heavy ions. According to the Lindhard, Scharff, and Schiott (LSS) model,[118,120] the energy loss is proportional to the velocity of the ion,

$$\left(\frac{dE}{dz}\right)_e = k'E^{1/2} \tag{2.390}$$

The value of k' depends on both the ion being used and the target material, and is given[121] by

$$k' = 0.328(Z_1 + Z_2)M_1^{-1/2}N \quad (\text{Å}^{-1} \cdot \text{eV}^{1/2}) \tag{2.391}$$

2.7.2.3. Range. To calculate the range,[122] it is customary to assume that the nuclear and electronic losses are independent of each other, as follows:

$$R = \int_0^R dz = \int_0^{E_0} \frac{dE}{(dE/dz)_n + (dE/dz)_e}$$
$$= \int_0^{E_0} \frac{dE}{1/2k + k'E^{1/2}} = \frac{2E_0^{1/2}}{k'} - \frac{1}{kk'^2}\ln(1 + 2kk'E_0^{1/2}) \tag{2.392}$$

And using the approximation

$$\ln(1 + x) \approx x - (1/2x^2) + (1/3x^3)\cdots \tag{2.393}$$

we get

$$R \cong 2kE_0(1 - 4/3kk'E_0)^{1/2} \tag{2.394}$$

It is convenient to define a stopping cross section (stopping power), S, as

$$S = \left(\frac{1}{N}\right)\frac{dE}{dz} = \frac{1}{N}[S_n(E) + S_e(E)] \tag{2.395}$$

Generally, when $S_n(E)$ and $S_e(E)$ are known, Eq. (2.395) can be integrated to give the total distance R that an ion with energy E_0 will travel before coming to rest:

$$R = \int_0^R dz = \frac{1}{N}\int_0^{E_0} \frac{dE}{S_n(E) + S_n(e)} \tag{2.396}$$

In amorphous targets, R is called the average total range.

2.7.2.4. Channeling. In crystalline materials the target atoms are arranged symmetrically in space so that an incident energetic ion may have a correlated collision with these atoms. These correlated deflections (collisions) along a crystallographic direction can channel the energetic ions (see Fig. 2.82).

The channeling property of the crystalline substrate depends on the angle of entrance ψ of the individual ions and on the ion and substrate characteristics. If the entrance angle ψ is large, then the amplitude of oscillation will be large and the ion will not be channeled (trajectory a in Fig. 2.82). There will be a maximum permissible amplitude at which the ion will remain channeled (trajectory b in Fig. 2.82); ψ_c is the critical angle that produces this maximum amplitude trajectory.

This critical angle for channeling can be approximated[123] by

$$\psi_c = \left(\frac{2Z_1Z_2e^2}{4\pi\varepsilon_0 Ed}\right)^{1/2} \tag{2.397}$$

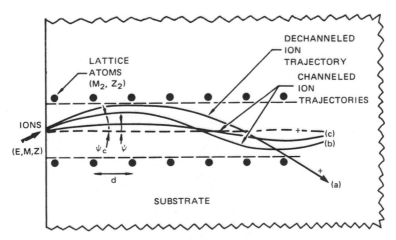

FIGURE 2.82.   Trajectories of channeled particles in a crystallographic direction.

It is clear from this formula that as the ion energy E is increased, ψ_c decreases and it becomes more difficult to remain in a stable trajectory. On the other hand, for closely packed atomic rows, d decreases and channeling is more likely.

2.7.2.5. Secondary Processes

Sputtering. In the sputtering process, an energetic ion incident on the target surface creates a cascade of atomic collisions in the bulk material. Some atoms in this cascade may be on the surface and escape. The sputtering yield S is defined as the number of atoms (mostly neutral) ejected per incident ion. The principal features of the sputtering of metals were explained by the first experimental investigators.[119] They demonstrated that the sputtering yield was a function of the atomic number and that the yield showed a periodicity that was closely correlated with that of the periodic table.

At low ion energies, there is a threshold for sputtering to occur. Above the threshold, the sputtering yield rises to a maximum and eventually at very high energies, decreases again as the ion energy is deposited so far into the solid that it cannot reach the surface.

A comprehensive and detailed theory of sputtering has been formulated in Ref. 124. This model is based on random collision processes, applying Boltzmann's equation and general transport theory. When the energy of the primary ion E is sufficient to transfer an energy greater than the displacement energy of a lattice atom, then a collision cascade may be initiated. If this cascade intersects the surface with atoms whose energy is greater than the surface-binding energy (U_0), then sputtering takes place.

For low energies up to 1 keV, the expression for the sputtering yield is given by[124]

$$S(E) = \frac{3}{4\pi^2} \alpha \frac{4M_1 M_2}{(M_1 + M_2)^2} \frac{E}{U_0} \tag{2.398}$$

where α depends on M_2/M_1, given in Table 2.5 and plotted in the original paper, and U_0 is the surface binding energy. For keV energies and heavy to medium-mass ions

$$S(E) = 0.0420\alpha \frac{4\pi Z_1 Z_2 e^2 a}{U_0} \left(\frac{M_1}{M_1 + M_2}\right) S_n(\varepsilon) \tag{2.399}$$

TABLE 2.5. α as a Function
of Mass Ratio

M_2/M_1	α
0.01	0.17
0.5	0.20
1.0	0.23
5.0	0.98
10.0	5.0

TABLE 2.6. $S_n(\varepsilon)$ as a
Function of Ion Parameters
(E, M_1, M_2, Z_1, Z_2)

ε	$S_n(\varepsilon)$
0.01	0.211
0.1	0.372
1.0	0.356
10.0	0.128
40.0	0.0493

where $\varepsilon = [M_2 E/(M_1 + M_2)]\,(Z_1 Z_2 e^2/a)$ and $a = 0.8853\,a_0(Z_1^{2/3} Z_2^{2/3})^{-1.2}$, a_0 is the Bohr radius, and $S_n\varepsilon)$ is a universal function tabulated in Table 2.6 and discussed in more detail in the original paper.

A statistical model of sputtering[125] predicts yields as a function of ion energy, mass, and incident angle. According to this model, the probability of displacing an atom in the target depends only on the distance of that atom from the point of impact.

The model assumes that the yield is proportional to the area defined by the intersection of the "cascade volume" with the target surface (see Fig. 2.83). At low energies the yield varies as

$$S(E, \theta) \approx E^{2/3}/\cos\theta \qquad (2.400)$$

The sputtering process widely used in microfabrication can be divided into the following application areas:

- Ion-beam machining or etching
- Production of atomically clean surfaces in vacuum
- Surface analysis
- Deposition of thin films

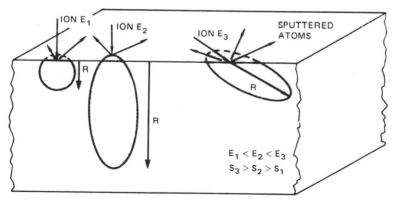

FIGURE 2.83. Sputtering model predicting yield as a function of ion energy, mass, and incident angle. From Ref. 123.

Inelastic Processes. The interaction of ions with atoms has been discussed in the previous sections on scattering and sputtering. The interactions with electrons do not result in any appreciable scattering of the incident ion, since momentum transfer is very small, but causes excitation and ionization of the electron shells of both the incident and target atoms. When this excitation occurs near the surface, it results in the emission of electrons, photons, and X rays.[112, 119]

Electron Emission. The ion striking the target surface has both potential energy because of its ionized state, and kinetic energy because of its relative velocity to the target. Electron emission can occur as a result of the rearrangement of either potential or kinetic energies in the ion–metal system. The secondary-electron yield (γ) is proportional to the energy loss of the incident ion and is observed experimentally to be the same for all metals.

X-ray Emission. X-ray emission results from the production of an inner electron shell vacancy as a result of a collision by the incident ion with a target atom. Once the inner shell vacancy is produced, Auger electron emission and X-ray emission compete to deexcite the atom. Both have energies characteristic of the excited atom.

2.7.3. Ion Interaction at Ultrahigh Energies

Heavy ion beams at ultrahigh energies ($>10^2$ MeV) serve as a unique tool for fabricating statistically distributed holes of predictable cross section (10–1000 nm) and usable density ($>10^{10}$ holes/cm^2), and is termed the nuclear track technique. This technology has decreased the smallest attainable diameter to the nanometer range for target thicknesses larger than 100 μm.

The formation of nuclear tracks is very complex but a simple three-step model explains the main features of the latent track formation. The energetic heavy ion in the thin-film target has a very short primary interaction time in the vicinity of the atoms of the solid (on the order of 10^{-17} s for a 5-MeV/nucleon ion which has a velocity of about 10% of the speed of light). The primary interaction is followed by an electronic collision cascade spreading out rapidly from the particle trajectory which lasts more than 1000 times longer than the primary interaction ($\geq 10^{-14}$ s) and leaves behind a highly positively charged cylindrical plasma zone. In the case of organic polymers, in addition, many chemically activated molecules are produced outside this zone. In dielectric solids, the remaining charged-plasma cloud "explodes" via the electrostatic repulsion of the formed positive ions. This process is often referred to as the "Coulomb explosion." This triggers an atomic collision cascade, which in turn comes to a halt on a time scale that again can be stretched by a factor of about 100 ($\cong 10^{-12}$ s), leaving behind a cloud of interstitial atoms and vacancies. The radial range of the electronic collision cascade ($\cong 100$–1000 nm) is one or two orders of magnitude greater than that of the atomic collision cascade ($\cong 10$ nm).[126–128]

Table 2.7 shows the temporal and spatial developmental steps in latent nuclear track formation. For hole fabrication the ion-beam exposure should be followed by etching to yield a variety of different hole sizes. The length of the nuclear track depends on the stopping power of the energetic ions. The energy loss increases with increasing ion mass.

TABLE 2.7. Temporal and Spatial Developments in Latent Nuclear Track Formation

Event	Time scale	Spatial extension	Results
Primary interaction	10^{-17} s	Few Å	Ionization excitation
Electronic collision cascade	10^{-14} s	10 nm	Cone formation
Track formation	10^{-12} s	100–1000 nm	Halo formation
			Broken bonds in the halo
			Radical formation in the latent nuclear tracks

To obtain an approximate value for the stopping power S in units of MeV/nucleon for a target of an "areal mass density" of 1 mg /cm^2, one can use

$$S \cong 0.14(Z_p^{*2})(Z_t/A_t) \ln(250E/Z_t) \qquad (2.401)$$

where the effective charge Z_p^* is

$$Z_p^* = Z_p[1 - \exp(-6E^{1/2}/A_p^{2/3})] \qquad (2.402)$$

and S is the stopping power of the projectile in MeV/nucleon per mg/cm^2, E is the projectile energy in MeV/nucleon, Z is the atomic charge number, A is the atomic mass number, p denotes the projectile, and t denotes the target. This equation gives values which are correct to 10% for light ions and target atoms and can deviate by a factor of three for arbitrary ion–target combinations.

2.8. PHOTON SOURCES AND INTERACTIONS

2.8.1. X-ray Sources for Lithography

There are several different types of photon sources used in microfabrication processes. X-ray sources are used for the mask replication of submicron features and for the fabrication of microdevices (optical elements, X-ray gratings, and microanalyzers). (See Sections 5.6 and 5.7).

The requirements for an ideal X-ray source for lithography can be described as follows: The source should be either very small (<1 mm) or highly parallel and very intense for short exposure times (<1 min) in order to cast sharp shadows. The X-ray energy should exceed 1 keV in order to be able to penetrate the mask support but should be absorbed in the resist (<10 keV). Using these selection rules, the following X-ray sources can be considered:

- Electron bombardment sources
- Synchrotron radiation
- Plasma sources—laser-generated plasma, plasma discharges
- X-ray lasers
- Transition radiation

2.8.1.1. Electron Bombardment Sources. In operation of this source, electrons in the kiloelectron volt energy range are focused onto a metallic target where they excite a spectrum of X-ray radiation consisting of a discrete line spectrum, character-istic of the target material, superimposed on a continuous spectrum (Bremsstrahlung) extending, in terms of energy, from zero to a cutoff energy of value equal to the bombarding energy. The characteristic radiation is produced when inner-shell elec-tron vacancies produced by the bombarding electrons are filled with electrons from outer shells.

To generate the characteristic radiation, the bombarding electrons with energy E_0 must excite inner-shell (K, L, or M) transitions as shown in Fig. 2.62(b). The result is a photon with energy E_x or an Auger electron. The wavelength of the photon related to its energy E_x can be expressed as

$$\lambda \, [\text{nm}] = hc/E_x = 1{,}238/E_x \quad [\text{keV}] \tag{2.403}$$

The emitted photon energy E_x is always smaller or equal to the energy of the bombarding electrons, $E_x \leq E_0$. The atomic cross section for ionization in the inner shell with quantum numbers n, l can be expressed as[122]

$$\sigma_{nl} = \frac{e^4 Z_{nl}}{16\pi \, \varepsilon_0^2 E_0 |E_{nl}|} \, b_{nl} \ln \frac{4E_0}{B_{nl}} \tag{2.404}$$

where Z_{nl} is the number of electrons in the inner shell, E_{nl} is the energy of the inner shell state and for the K shell $b_k = 0.35$, $B_k = 1.65 \, E_k$. Using the condition $E_0 = E_k$ and $B = 4E_k$ the cross section for the K-shell transitions can be expressed as

$$\sigma_K = \frac{e^4}{8\pi \, \varepsilon_0^2 E_0 |E_K|} \, b_K \ln \frac{4E_0}{B} = \frac{e^4}{8\pi \, \varepsilon_0^2 u |E_K^2|} \, b_K \ln \frac{4uE_K}{B} \tag{2.405}$$

where $B = [1.65 + 2.35 \exp(1 - u)]E_K$ and $u = E_0/E_K$. With these $\sigma_K E_K^2$ is a univer-sal function for each element as shown in Fig. 2.84 for Ni ($E_K = 8.3$ keV) and Ag ($E_k = 25.5$ keV).[129]

Using the atomic cross section and the number of target atoms in the electron interaction volume, the intensity of the characteristic radiation can be estimated.

The continuous part of the radiation in the case of electron bombardment comes from "Bremsstrahlung" (braking radiation). Because the electrons scattered in the Coulomb field of the target atoms' nuclei accelerate and therefore radiate, the maxi-mum energy of the radiated photons $E_{x\,\text{max}}$ is smaller or equal to the energy of the bombarding electrons.

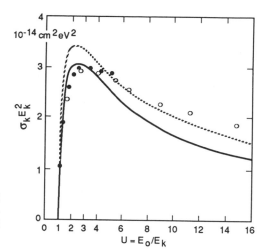

FIGURE 2.84. The dependence of the K-shell ionization cross section on the energy of the bombarding electrons E. Here $E_K = 8.33$ keV for Ni and $E_K = 25.5$ keV for Ag.

In the case of a thin film with mass thickness $dx = \rho\, ds$ and with the total number of atoms per unit film thickness $= N_L\, ds/A$, the radiation intensity per electron is given as[130]

$$I(E_x)\, dx\, dE_x = \frac{1}{12\pi\sqrt{3}\, h\varepsilon_0^3}\, \frac{Z^2 e^6}{mc^3 F_0}\, \frac{N_L}{A}\, dx\, dE_x \quad \text{for } E_x < E_0 \qquad (2.406)$$

and $I(E_x) = 0$ for $E_x > E_0$.

According to Eq. (2.406) the emitted radiation intensity is independent of the photon energy E_x and it is proportional to Z^2/E_0.

The energy distribution, the number of photons between the energy interval E_x and $E_x + dE_x$, can be obtained by dividing Eq. (2.406) by $E_x - hv$. Then

$$N(E_x)\, dE_x \approx -\frac{Z^2}{vE_0}\, dE_x \qquad (2.407)$$

The radiation emitted in this process has a dipole character having maxima perpendicular to the direction of the incidence of the electron. For relativistic electrons the radiation maximum is shifted to the direction of incidence. In the case of ultrarelativistic electrons, the radiation maximum is in the direction of incidence and this gives a searchlight character to the synchrotron radiation.

In the case of solid targets, the number of photons in the energy interval between E_x and $E_x + dE_x$ is given as[131]

$$N(E_x)\, dE_x = KZ\frac{E_0 - E_x}{E_x}\, dE_x \qquad (2.408)$$

where $K = 2.2 + 10^{-6}$ gives a good approximation for solid targets.

the rate of production of X rays increases with electron energy but so does the penetration depth of the electrons into the target. X rays generated inside the target

TABLE 2.8. Target Materials and Their Characteristic Wavelengths

Material	Atomic No.	Type of line	Wavelength (Å)	Energy (keV)
Pd	46	L	4.4	2.83
Rh	45	L	4.6	2.69
Mo	42	L	5.4	2.29
Si	14	K	7.1	1.74
Al	13	K	8.3	1.49
Cu	29	L	13.3	0.93
C	6	K	44.8	0.28

can be reabsorbed on their way out of the target. Because of this target reabsorption, there is an optimum voltage for each target material and source angular parameters (incidence angle of electrons and X-ray takeoff angle).

The coninuous X-ray spectrum intensity [formula (2.408)] is proportional to the atomic number (Z) of the target material. In the case of lithography applications, this radiation provides a background of unwanted high-energy X rays which tends to reduce the contrast of the X-ray exposure and should therefore be kept as low as possible by using low accelerating voltages and low-Z target materials. However, the choice of target materials is limited in the X-ray wavelength range between 4 and 50 Å to a few materials producing K and L lines, listed in Table 2.8.

X rays have conventionally been generated by accelerating electrons and allowing them to hit a solid target (W, Au, Pt, Cu, Al). The efficiency of this method is

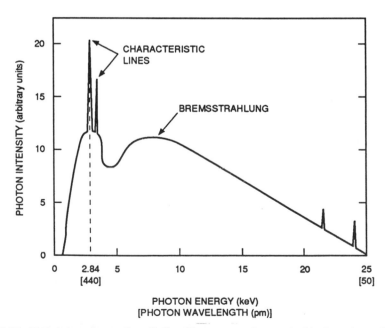

FIGURE 2.85 Emission spectrum of a palladium X-ray target, characterized by intensity peaks superimposed on background radiation (Bremsstrahlung).

quite low; typically $<10^{-3}$ of the electron energy is converted to X rays. It is even lower for soft X rays (≤ 1 keV) because of the absorption in the target. An advantage of these sources is that the strong characteristic emission lines from the target atoms can be isolated by simple filters. Al K (0.83 nm), Cu L (1.33 nm), and CK (4.48 nm) radiation sources have been specifically developed[130] for use in X-ray microlithography.

For the more powerful of such sources the source size is typically 1 mm or larger because of space-charge problems (especially at low-voltage operation) and heat load. The X rays are emitted in all directions and only a small fraction can be collected on the target. (See Section 5.6.2.)

Selection of the mode materials is one of the most critical factors in X-ray source design, since it influences a significant part of overall system design. The energy emission spectrum of a conventional source consists of one or more sharp peaks or characteristic lines, superimposed on a background of Bremsstrahlung radiation, as illustrated in Fig. 2.85. The lines are generated by electrons moving from one energy level to another in the atoms of the target material, while Bremsstrahlung radiation is produced by accelerating or decelerating incident electrons. The absorption characteristics of the various exposure system materials must be matched to the emission spectrum of the source so that maximum X-ray reaches the resist.[132]

2.8.1.2. Synchrotron Radiation. Synchrotron radiation is the electromagnetic radiation emitted by electrons due to the radial acceleration that retains them in orbit in an electron synchrotron or storage ring.[133,134]

Because of its intensity, tunability, small source size, and small divergence, synchrotron radiation comes close to being the ideal universal source.[133,134] Beam lines for X-ray lithography were recently designed for electron storage rings and synchrotrons. The spectral and temporal characteristics of the various X-ray sources are shown in Fig. 2.86 (see Section 5.7). The origin of *synchrotron radiation* can be

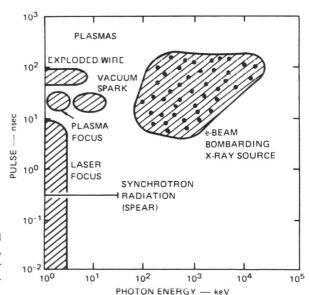

FIGURE 2.86. Pulse length and spectral energy of electron-impact, synchrotron-radiation, and high-temperature plasma flash X-ray sources. From Ref. 152.

explained on the basis of classical electrodynamics. Maxwell's equations predict that any accelerated charge particle radiates. The power radiated by a particle with charge e moving with velocity $v = \beta c$ on a circular path of radius R is given by

$$P = \frac{2e^2c}{3R^2} \frac{\beta^4}{(1-\beta^2)^2} \qquad (2.409)$$

If the velocity of the orbiting electron is close to the velocity of light $\beta \sim 1$, then

$$P \approx \frac{2e^2c}{eR^2} \gamma^4 = \frac{2e^2c}{3R^2} \left(\frac{E}{m_0c^2}\right)^4 \qquad (2.410)$$

by using the well-known relations

$$m = \frac{m_0}{\sqrt{1-\beta^2}} = \gamma m_0 \qquad (2.411)$$

and

$$E = mc^2 \qquad (2.412)$$

The time, T, for one revolution is given by

$$T = \frac{2\pi R}{v} \approx \frac{2\pi R}{c} \qquad (2.413)$$

and the energy lost in one revolution is

$$-\Delta E = PT \approx \frac{4\pi e^2}{eR} \left(\frac{E}{m_0c^2}\right)^4 \qquad (2.414)$$

In practical units (E in GeV, $R \equiv \rho$ in meters, I in amperes) for highly relativistic electrons

$$\Delta E \text{ (keV)} = 88.47 \frac{E^4}{\rho} \qquad (2.415)$$

Multiplying Eq. (2.415) by the current gives the total radiated power

$$P \text{ (kW)} = \frac{88.47E^4I}{\rho} \qquad (2.416)$$

or in terms of the magnetic field (B in kilogauss)

$$P \text{ (kW)} = 2.654BE^3 I \qquad (2.417)$$

These equations are valid also when the orbit is not circular but consists of arcs of bending radius ρ and field-free straight sections.

Synchrotron radiation is emitted in a cone with an opening angle θ, given by

$$\theta \approx \frac{1}{\gamma} \frac{mc^2}{E} = \frac{0.5}{E \text{ (GeV)}} \quad \text{(mrad)} \qquad (2.418)$$

The spectral distribution of the synchrotron radiation extends from the microwave region through the infrared, visible, ultraviolet, and into the X-ray region, decreasing in intensity below a critical energy E_c or, alternatively, the critical wavelength λ_c, where

$$E_c \text{ (keV)} = \frac{2.218E^3 \text{ (GeV)}}{\rho \text{ (m)}} \qquad (2.419)$$

or since

$$\lambda \text{ (Å)} = \frac{12.4}{E \text{ (keV)}}$$

then $\qquad\qquad\qquad\qquad\qquad\qquad\qquad\qquad\qquad\qquad\qquad$ (2.420)

$$\lambda_c \text{ (Å)} = \frac{5.59\rho \text{ (m)}}{E^3 \text{ (GeV)}}$$

For SPEAR operating at 3.5 GeV, $E_c = 7.4$ keV and $\lambda_c = 1.7$ Å. The spectrum of the synchrotron radiation from SPEAR[135] is shown in Fig. 2.87. The high polarization and the pulsed time structure available from synchrotron radiation, however, are relatively unimportant for X-ray lithography.

The spectral emission of the synchrotron radiation is a broad spectrum without characteristic peaks or line enhancements.

Insertion devices (ID) have been designed specifically as radiation sources for characteristic (broad line) radiation.[134] An insertion device is a device which causes the trajectories of electrons passing through it to describe periodic oscillations. As a result of these oscillations, the electrons emit electromagnetic radiation. Insertion devices can be classified into two types:

- Undulators
- Wigglers

Whether a device is an undulator or a wiggler depends on the so-called "wiggler parameter," usually denoted by K. When $K \leq 1$, one says that the device is an undulator, while for $K \geq 1$ one talks about wigglers. (K will be defined below.)

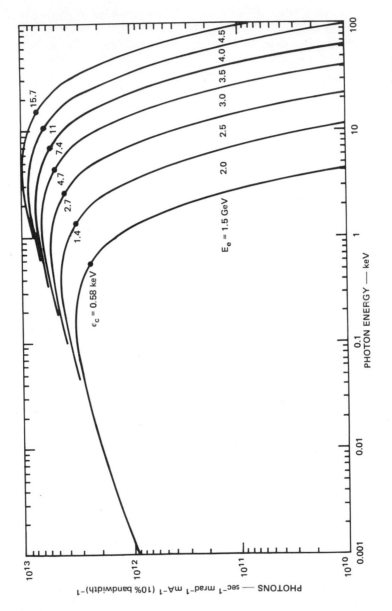

FIGURE 2.87.   Spectral distribution of synchrotron radiation from SPEAR ($\rho = 12.7$ m). From Ref. 133.

When $K < 1$, electrons will radiate predominantly into the first harmonic frequency of the device, i.e., emit close to monochromatic radiation. The smaller the value of K, the more monochromatic the radiation will be.

When $K > 1$, electrons will radiate a substantial fraction of their energy into harmonics higher than the first. The higher the value of K, the wider will be the spectral range into which the electrons will radiate while passing through the device.

The principle of an insertion device is to enhance the synchrontron radiation emission as much as possible without perturbing the stored beam. Usually, fields in these devices are periodic with a field integral along s (the main direction of propagation of the electron, see Fig. 2.88) of zero. The most conventional undulator has a planar sinusoidal field:

$$B(s) = B \cos(2\pi/\lambda_u)$$

This field is usually generated by an array of permanent magnets with period λ_u.

The magnetic field of a planar undulator is shown schematically in Fig. 2.88. The period of the undulator magnetic field is denoted by λ_u, the total length of the undulator by T_u, the spatially oscillating magnetic field strength vector by B, its y-component (vertical) peak amplitude by B_0, and the energy of the electrons by E_e. The electron charge and mass are e and m_e, respectively, c is the speed of light in vacuum, and $\gamma = E_e/m_ec^2$. It is customary to define a "wiggle" parameter K as

$$K = \frac{eB_0\lambda_u}{2\pi m_ec^2} = 0.934 B_0 \text{[tesla]} \cdot \lambda_u \text{ [cm]} \tag{2.421}$$

When $K \ll 1$, radiation emerging from the undulator in the direction that makes an angle θ with the undulator axis consists overwhelmingly of photons whose wavelength can be expressed by

$$\lambda = \frac{\lambda_u}{2\gamma^2} (1 + \tfrac{1}{2}K^2 + \gamma^2\theta^2) \tag{2.422}$$

These photons belong to the so-called "first harmonic" of the emitted radiation. When $K > 1$, the contribution of other harmonics is significant. As a result of their short wavelength, micropole undulators are ordinarily expected to have low K values (because $K > 1$ would require excessively high B_0 values).

FIGURE 2.88. Schematic of the conventional insertion device. The magnetic field stays parallel to the vertical direction (Oz). It is a sinusoidal function of the longitudinal coordinate s. The trajectory of an electron submitted to this field is a sinusoid in the horizontal plane (Os, Ox).

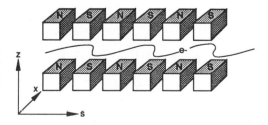

When photons are emitted in the forward direction, $\theta = 0$, then

$$\lambda = \frac{\lambda_u}{2\gamma^2}(1 + K^2/2) \tag{2.423}$$

When $K = 1$, higher harmonics also contribute. Their wavelength is simply λ/m, where m is any integer. In the forward direction, only odd values of m can contribute for ideal insertion devices. Or equivalently

$$E_\gamma = hc/\gamma = [2hc\gamma^2/\gamma_u](1 + K^2/2)^{-1} \tag{2.424}$$

where h is Planck's constant. In practical units, for low-K devices,

$$\lambda \, [\text{angstroms}] = \frac{13.056\lambda_u \, [\text{cm}]}{E^2 \, [\text{GeV}]} \tag{2.425}$$

and

$$E_\gamma \, (\text{eV}) = \frac{950E^2 \, [\text{GeV}]}{\lambda_u \, [\text{cm}]} \tag{2.426}$$

The spectral purity of this radiation is given by

$$\lambda/\Delta\lambda = l_u/\lambda_u = N \tag{2.427}$$

and its associated coherence length is

$$l_c = M\lambda = \lambda^2/\Delta\lambda \tag{2.428}$$

where N is the number of complete cycles described by the magnetic field in the undulator, commonly referred to as the "number of undulator periods."

The emitted radiation intensity has a maximum at $\theta = 0$. The half width of the first harmonic intensity peak in the vicinity of $\theta = 0$ can be written as

$$\Delta\theta = \left[\frac{1}{\gamma^2} \frac{(1 + 1/2K^2)}{N} \right]^{1/2} \tag{2.429}$$

The total power radiated by an electron traversing the undulator is

$$P_{\text{tot}} = \frac{(2\pi)^2}{3} e^2 \frac{\gamma^2}{\lambda_u^2} K^2 l_u I \tag{2.430}$$

or, in practical units,

$$P_{tot} \text{[watts]} = 633E^2 \text{[GeV]} B_0^2 \text{[tesla]} l_u \text{[m]} I \text{[A]} \qquad (2.431)$$

where I is the ring current.

The dependence on electron energy is quadratic. The radiated power, for example, for a simple insertion device where $B = 0.5$ tesla, $\lambda_u = 5$ cm, and $L = 1.5 \mu m$ with 0.8-GeV electrons and 0.5-A current, is 3×10^{-2} kW in the fundamental band and 7.6×10^{-2} kW in the whole spectral range. The number of photons per second, per 0.1% bandwidth contained in the central cone of the first harmonic is, for low-K devices.

$$N_\sigma = 1.25 \times 10^{12} l_u \text{[m]} \lambda_u \text{[cm]} B_0^2 \text{[tesla]} I \text{[A]} \qquad (2.432)$$

In a storage ring with emittance parameters ε_x [mm-mr] and ε_y [mm-mr], the on-axis output brightness of a low-K undulator is given by

$$B = \frac{N_\sigma}{4\pi^2 \varepsilon_x \varepsilon_y} \text{ photons/sec mr}^2 \cdot \text{mm}^2 \cdot 0.1\% \text{ bandwidth} \qquad (2.433)$$

2.8.1.3. Transition Radiation. In addition to Bremsstrahlung and synchrotron radiation, electrons can produce transition radiation when they move across an interface between two dielectrics.[136] The Poynting vector flux per unit solid angle due to transition radiation for a particle passing from vacuum to a medium characterized by a complex dielectric constant $\varepsilon = \varepsilon_1 + i\varepsilon_2$:

$$\frac{dW}{d\Omega} = \frac{e^2 \sin^2 \theta \cos^2 \theta}{\pi^2 c} \frac{\beta^2}{(1 - B^2 \cos^2 \theta)^2}$$

$$\times \int_0^\infty \left| \frac{(\varepsilon - 1)[1 - \beta^2 + \beta(\varepsilon - \sin^2 \theta)^{1/2}]}{[\varepsilon \cos \theta + (\varepsilon - \sin^2 \theta)^{1/2}][1 + \beta(\varepsilon - \sin^2 \theta)]} \right| d\omega \qquad (2.434)$$

In the formula θ is the angle between the direction of the incoming charged particle and the direction of the outgoing photon (see Fig. 2.89), β is the particle velocity, c is the velocity of light, and ω is the angular frequency of the radiation. This equation can be reexpressed in terms of the photon flux by using the fact that for photons of angular frequency ω, the energy dW associated with dN photons is

$$dW = (\hbar\omega) \, dN \qquad (2.435)$$

The result is

$$\frac{dN}{d\Omega} = \frac{\alpha\beta^2}{\pi^2} \frac{\sin^2 \theta \cos^2 \theta}{(1 - \beta^2 \cos^2 \theta)^2}$$

$$\times \int_0^\infty \left| \frac{(\varepsilon - 1)[1 - \beta^2 + \beta(\varepsilon - \sin^2 \theta)^{1/2}]}{[\varepsilon \cos \theta + (\varepsilon - \sin^2 \theta)^{1/2}][1 + \beta(\varepsilon - \sin^2 \theta)^{1/2}]} \right|^2 \frac{d\omega}{\omega} \qquad (2.436)$$

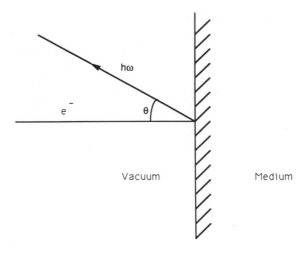

FIGURE 2.89.   Geometry illustrating production of transition radiation.

where α is the fine-structure constant. If the dielectric constant is known as a function of frequency over the range of interest, then the integral of Eq. (2.436) can be numerically evaluated.

The photon yield is small for a single interface between two dielectrics; however, a summation from each of a large number of interfaces is possible whenever the distance between interfaces exceeds a "formation" length. If the radiator is striated (composed of two alternating media), then the formation lengths of the two materials are given by

$$Z_{1.2} = \frac{2c}{\omega[1 - \beta(\varepsilon_{1.2} - \sin^2 \theta)^{1/2}]} \tag{2.437}$$

where β = ratio of electron velocity, v, to the velocity of light, c; ω = frequency of emission; θ = angle of emission relative to the direction of particle motion; $\varepsilon_{1.2} = 1 + \omega_{1.2}^2/\omega^2$ are the dielectric constants for the two media; and $\omega_{1.2}$ are their respective plasma frequencies. If medium 1 is replaced by a vacuum, then $\varepsilon_1 = 1$.

Including absorption, the spectral density (photons emitted per unit frequency) for M foils is given by

$$\frac{dN}{d\omega} = \frac{4\alpha[1 - \exp(-M\sigma)]}{\omega\sigma(1 + k)} \sum_n \theta_n \left(\frac{1}{\rho_1 + \theta_n} - \frac{1}{\rho_2 + \theta_n}\right)^2$$
$$\times [1 - \cos(\rho_2 + \theta_n)] \tag{2.438}$$

where l_1 and l_2 are the foil thicknesses with mass absorption coefficients μ_1 and μ_2, respectively, and:

$$\rho_1 = l_i/Z_i, \qquad i = 1, 2; \; \theta = 0 \tag{2.439}$$

$$\theta_n = \frac{2\pi n - (\rho_2 + k\rho_1)}{1 + k} > 0 \tag{2.440}$$

$$k = l_1/l_2 \tag{2.441}$$

$$\sigma = \mu_1 l_1 + \mu_2 l_2 \tag{2.442}$$

$$\alpha \cong 1/137 \tag{2.443}$$

There is a foil thickness that maximizes Eq. (2.438) for a particular photo energy. Setting $\mu_1 = 0$ for the case of a foil separated by a vacuum and taking the first derivative with respect to l_2, the optimum thickness is found to be

$$l_{opt} = \frac{\eta}{M\mu_2} \tag{2.444}$$

where η is given by

$$\frac{1}{\eta} + \frac{e^{-\eta}}{1 - e^{-\eta}} - \frac{2\eta}{3Z_{med}^2 M^2 \mu_2^2} = 0 \tag{2.445}$$

assuming $l_1 \gg l_2$. The formation length Z_{med} is evaluated at $\theta = \theta_{opt}$, the angle that maximizes the radiation from a single interface:

$$\theta_{opt}^2 = \frac{-(\delta_2 + \delta_2) + [(\delta_1 + \delta_2)^2 + 12\delta_1\delta_2]^{1/2}}{3} \tag{2.446}$$

where

$$\delta_{1.2} = 1/2\gamma^2 + \omega_{1.2}^2/2\omega^2 \tag{2.447}$$

and

$$\gamma^2 = 1/(1 - \beta^2) \tag{2.448}$$

For 25 foils of Be and a photon emission energy of 900 eV, the optimum thickness is 1 μm; with 25 foils of Mylar and a photon energy of 2.5 keV, the thickness is 2.5 μm; at 3.7 keV, the thickness is 3.5 μm.

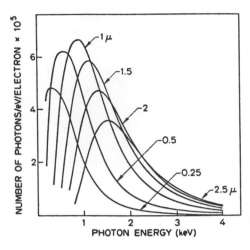

FIGURE 2.90. Theoretically predicted photon production from 25 foils of various thicknesses of Be. The electron beam energy is 100 MeV; the spacing between foils is 0.51 mm.

The effect of foil thickness is shown in Fig. 2.90 for 25 Be foil using 100-MeV electrons.[137] The observed spectra agree well with theory.

Using a linear accelerator as the source of relativistic electrons ($E \sim 100$ MeV), transition radiation in the soft X-ray range (0.6–1.2 nm) can be generated.[138] The generated photon flux is suitable for X-ray microlithography.

2.8.1.4. Plasma Sources. High-temperature plasmas are generated in galactic space, thermonuclear devices, electrical discharges, and by interaction of high-intensity laser beams with materials. They radiate infrared, optical and X-ray photons.

The radiation spectrum from a high-temperature plasma consists of a combination of discrete lines, recombination radiation, and Bremsstrahlung continuum. The spectral shape and the relative importance of these components depend on the gas species used, their relative concentrations, and the plasma temperature in the source volume. The plasma X-ray sources designed for lithography should produce a spectrum which has a maximum near 1-keV photon energy. The continuum part of the spectrum originates from electron–ion Bremsstrahlung. The radiated power from this interaction can be derived from the classical expression for the rate P_e at which energy is radiated by an accelerated electron, namely,

$$P_e = \frac{2e^2}{3c^3} a^2 \tag{2.449}$$

where e is the electronic charge, c is the velocity of light, and a is the acceleration. Suppose an electron moves past a relatively stationary ion of charge Z_e with an impact parameter b, as indicated in Fig. 2.91. The coulomb force between the charged particles is then $Ze^2/b^2 m_e$, and the rate of energy loss as radiation is

$$P_e \approx \frac{2e^6 Z^2}{3m_e^2 c^3 b^4} \tag{2.450}$$

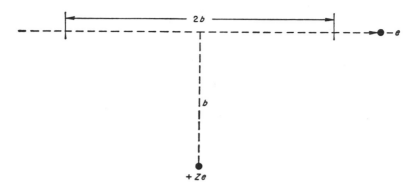

FIGURE 2.91. Coulomb interaction of electron with a nucleus.

The electron path length over which the coulomb force is effective is $2b$, and if the velocity is v, the time during which acceleration occurs is $2b/v$. If the acceleration is assumed to be constant during this time, the total energy E_e radiated as the electron moves past an ion with an impact parameter b is then

$$E_e \approx \frac{4e^6 Z^2}{3 m_e^2 c^3 b^3 v} \tag{2.451}$$

If this is multiplied by n_e and n_i, the numbers of electrons and ions, respectively, per unit volume, and also by v, the result is the rate of energy loss P_a per unit impact area for all ion–electron collisions occurring in unit volume at an impact parameter b; hence,

$$P_a \approx \frac{4e^6 n_e n_i Z^2}{3 m_e^2 c^3 b^3} \tag{2.452}$$

The total power P_{br} radiated as Bremsstrahlung per unit volume is obtained upon multiplying by $2\pi b \, db$ and integrating over all values of b from b_{min}, the distance of closest approach of an electron to an ion, to infinity; thus,

$$P_{br} \approx \frac{8\pi e^6 n_e n_i Z^2}{3 m_e^2 c^3} \int_{b_{min}}^{\infty} \frac{db}{b^2} = \frac{8\pi e^6 n_e n_i Z^2}{3 m_e^2 c^3 b_{min}} \tag{2.453}$$

An estimate of the minimum value of the impact parameter can be made by utilizing the uncertainty principle relationship,

$$\Delta x \Delta p \approx \frac{h}{2\pi} \tag{2.454}$$

when Δx and Δp are the uncertainties in position and momentum, respectively, and h is Planck's constant. The uncertainty in the momentum may be set equal to the momentum $m_e v$ of the electron and Δx may then be identified with b_{min}, so that

$$b_{min} \approx \frac{h}{2\pi m_e v} \tag{2.455}$$

Furthermore, assuming a Maxwellian distribution of velocity among the electrons it is possible to write

$$\tfrac{1}{2} m_e v^2 \times \tfrac{3}{2} k T_e \tag{2.456}$$

where T_e is the kinetic temperature of the electrons; hence,

$$b_{min} \approx \frac{h}{2\pi (3kTm_e)^{1/2}} \tag{2.457}$$

Upon inserting this result into Eq. (2.453), it is found that

$$P_{br} \approx \frac{16\pi^2}{3^{1/2}} \frac{(kT_e)^{1/2} e^6}{m_e^{3/2} c^3 h} n_e n_i Z^2 \tag{2.458}$$

The foregoing derivation has referred to a system containing a single ionic species of charge Z. In the case of a mixture of ions (or nuclei), it can be readily seen that the quantity $n_i Z^2$ should be replaced by $\Sigma(n_i Z^2)$, where the summation is taken over all of the ions present.

A more precise treatment, assuming a Maxwellian distribution of electron velocities, gives for the rate of Bremsstrahlung energy emission per unit volume[139]:

$$P_{br}\left(\frac{W}{cm^3}\right) = 1.69 \times 10^{-38} \Sigma(Z_i^2 n_i) n_e T_e^{1/2} \tag{2.459}$$

The line intensity for recombination radiation is given as

$$P_{rec}\left(\frac{W}{cm^3}\right) = 2.49 \times 10^{-39} \Sigma(E_Z n_i) n_e T_e^{-1/2} \tag{2.460}$$

where E_Z is the ground-state ionization energy of charge state Z, and n_i is the number density of ions with charge $Z + 1$.

In lithography a monoenergetic or a narrow band spectrum is preferred, hence the source emission should be line radiation. However, line radiation, being stimulated by electron bombardment, is necessarily accompanied by Bremsstrahlung. When ions are used to produce line radiation, recombination radiation may also be present. The task, then, is to create a plasma with an appropriate composition, temperature, and density to emit intense line radiation, and a relatively small amount of Bremsstrahlung and recombination radiation. One must be especially careful to suppress the emission, above a few kiloelectron volts, that would be less efficiently absorbed in the resist and would therefore penetrate to the substrate.

The above considerations lead to the consideration of low-atomic-number elements that have X-ray lines near 1 keV. These lines are produced by electronic transition from the $n = 2$ to $n = 1$ level in the atom or ion. The $n = 3$ to $n = 2$ transitions in higher-atomic-number elements can be ruled out because higher plasma temperatures would be required, resulting in increased Bremsstrahlung. Although several elements are suitable candidates for plasma sources, only one, neon, will be evaluated in more detail.[140]

The spectral characteristics of neon are given in Table 2.9. Ne^{9+} and Ne^{8+} are hydrogen-like and helium-like ions that have principal resonance lines at 12.1 and 13.5 Å on the $n = 2$ to $n = 1$ transition.

The emission power density in a hydrogen-like plasma is given by

$$P_{line}\left(\frac{W}{cm^3}\right) = 5.05 \times 10^{-31} f_{12} n_{ig} n_e T_e^{-1/2} \exp\left(-\frac{E_{21}}{T_e}\right) \tag{2.461}$$

where

$$E_{21} \text{ (eV)} = \text{photon energy}$$

$$f_{12} = \text{absorption oscillator strength of the } E_{21} \text{ transition}$$

$$n_{ig} = \text{ground-state density of ions with charge state } Z$$

$$n_e \text{ (m}^{-3}) = \text{electron density}$$

$$T_e \text{ (eV)} = \text{electron temperature}$$

To determine the parameters for this equation, one must first find the electron temperature that would produce a plasma containing predominantly Ne^{9+} and Ne^{8+} ions. This may be done by examining the ratio of ion densities as a function of temperature. The steady-state plasma model is appropriate for the temperatures and densities of interest. It gives

$$\frac{n_i^2}{n_i^{Z+1}} = 7.7 \times 10^{-9} E_Z^2 \left(\frac{E_Z}{T_e}\right)^{3/2} \exp\left(\frac{E_Z}{T_e}\right) \tag{2.462}$$

where E_Z (eV) = ionization potential of an ion species in charge state Z, and n_i^{Z+1} (m^{-3}) = density of an ion species in charge state $Z + 1$.

TABLE 2.9. Spectral Characteristics of Neon

	Lines		Lower bound of recombination spectrum	
Charge state	$E_v(2p\text{–}1s)$ (keV)		Recombination transition	E_v (min) (keV)
0	0.849		10+ → 9+	1.362
7+	0.906		9+ → 8+	1.196
8+	0.922		8+ → 7+	0.239
9+	1.022		7+ → 6+	0.207

Using $E_9 = 1362$ eV, $E_8 = 1196$ eV, and $E_7 = 239$ eV, one finds that the temperature range for which the density of Ne^{8+} or Ne^{9+} ions exceeds that of Ne^{7+} and Ne^{10+} is $38 < T_e < 410$ eV. Also, the temperature for which there would be equal densities of Ne^{8+} and Ne^{9+} is $T_e \sim 340$ eV, at which these ions would comprise about 82% of the Ne ions. A more accurate calculation shows that the maximum emission occurs at a temperature $T_e \approx 0.28$ keV as shown in Fig. 2.92.[141] At $T_e = 0.28$ keV, the

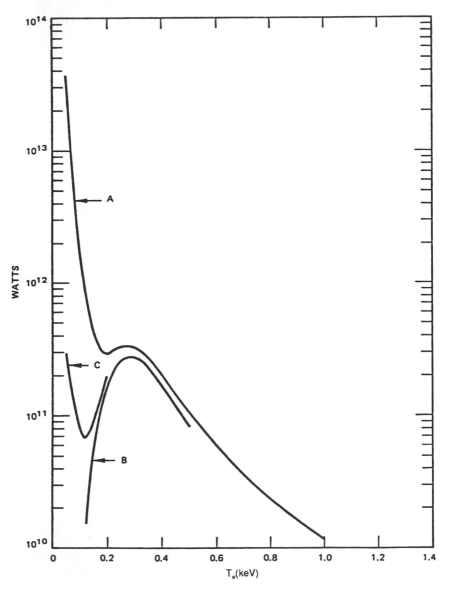

FIGURE 2.92.   Power radiated by neon plasma. Plasma focus: current = 10^6 A, radius = 1 mm, length = 1 cm. Curves: A = total power radiated by $Ne^{7+} + Ne^{8+} + Ne^{9+}$ ions; B = power radiated in lines near 1 keV by Ne^{8+} and Ne^{9+} ions; C = total power radiated by Ne^{8+} and Ne^{9+} ions.

recombination and Bremsstrahlung radiation amounts to $\approx 17\%$ of the total power emitted in lines near 1 keV.

The plasma density is needed to calculate the intensity of line emission. This is obtained by noting that, at maximum compression, the magnetic pressure caused by the driving current is balanced by the plasma pressure. This expression for pressure balance is called the Bennett relation[136]:

$$\mu_0 I^2/8\pi = (n_e + n_i)kT(\pi r^2) \qquad (2.463)$$

where I = discharge current, n_e and n_i = electron and ion densities, respectively, T = plasma temperature, and r = plasma radius.

The power radiated in lines can now be expressed in terms of the plasma current as follows:

$$P_{line} \text{ (watts)} = 0.28 \times 10^{-12} I^4$$

for a new plasma focus with $T_e = 0.28$ keV and $r = 1$ mm.

Using the typical pulse width of 50 ns for this type of system,[140] we find that the plasma energy is rapidly radiated away in the form of line radiation when the discharge current is large. A minimum capacitor bank storage capacity is about 1 kJ; above this energy the plasma density is high enough to radiate efficiently during the pulse.

2.8.1.5. *Plasma Focus and Related Sources.*   Several plasma sources have been developed for generating soft X rays. One type,[140,142] called a "plasma focus," uses the interaction of a tightly pinched electron beam with a dense plasma formed by a sliding spark capillary discharge (see Fig. 2.93a). This type of spark discharge through a capillary is a highly reproducible ($\pm 4\%$) radiant source of both vacuum ultraviolet and soft X-ray emissions. When a fast (100 ns), energetic (40 J) reproducible discharge occurs through a small-diameter (1 mm) hole in polyethylene under vacuum, reproducible dense carbon plasmas are created.

Another X-ray source (z-pinch of a puffed gas)[143] is shown in Fig. 2.93b. This device contains a cone-shaped cathode and a flat anode with a hole in the center. A fast valve filled with gas at a pressure of ~4 atm is activated, and the gas flows through a nozzle and a diaphragm (the anode) to form a collimated jet between the two electrodes. At this point, a capacitor bank charged to 30 kV is connected across the experiment via a high-voltage switch. As a result, the gas breaks down and current starts to flow. The gas burst is sensed by a miniature spark gap, located below the nozzle, which breaks down when the gas front reaches it. The hot plasma created is a useful source of X rays emitted by highly ionized elements such as neon, argon, krypton, or xenon. A somewhat similar puffed-gas system has been used recently in lithography studies,[144] for exposing submicron features.

In an imploding wire array system (see Fig. 2.93c),[145] a plasma is created by driving a 1-MA, 10,000-ns current pulse through a cylindrical array of 12 fine wires. The resulting implosion of each individually formed plasma toward the axis forms a single hot, dense plasma. The implosion occurs within 20 ns of the current peak, and the emission is characteristic of the wire materials used. The system as presently

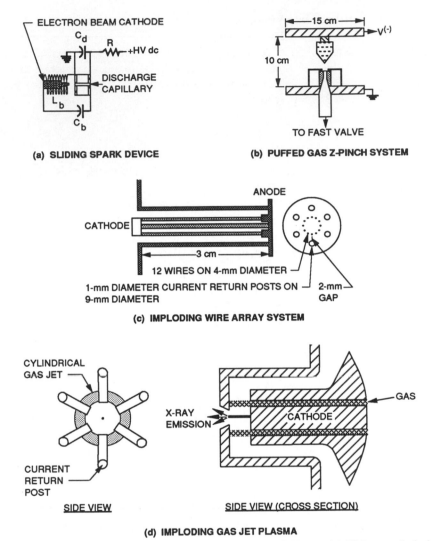

FIGURE 2.93. Configurations of various plasma sources for soft X rays. (a) Sliding spark device; (b) puffed gas Z-pinch system; (c) imploding wire array system; (d) imploding gas jet plasma.

constructed is quite massive, however, and cannot be fired with a reasonable repetition rate.

A more compact system is shown in Fig. 2.93d, where the source plasma is a high-density z-pinch produced by magnetically imploding a hollow, cylindrical gas jet.[146] The dynamics involved are quite similar to those observed with wire arrays. The gas jet is produced by opening a fast valve and allowing gas to pass through an annular nozzle. The discharge of a capacitor bank through the jet causes the gas to implode onto its axis. The resulting pinched plasma is about 1 to 3 mm in diameter, has a soft-X-ray spectrum characteristic of the gas used, and has a pulse length of 20 to 40 ns.

TABLE 2.10. Comparison of the Various X-ray Sources

Parameter	Storage ring	X-ray tube	Plasma source	Laser plasma
Spectrum	Wideband, typically 4–40 Å	Monochromatic 8.3 Å (Al) 6.7 Å (Si) 4.4 Å (Pd)	Wideband with monochromatic lines 7 Å, 12 Å	Mostly lines
Focus spot	0.5 mm	5 mm	≤ 1 mm	$\cong 0.5$ mm
Distance source/ mask	10 m	30 cm	>50 cm	>30 cm
Blurring ($\sim d$ 50 μm)	<0.01 μm	$\cong 0.5$ μm	$\cong 0.1$–0.2 μm	$\cong 0.2$ μm
Homogeneity	Horizontal: Homogeneous Vertical: Gaussian	Homogeneous	Homogeneous	Inhomogeneous
Pattern distortion	Fresnel	Blurring (Fresnel)	Fresnel blurring	Blurring
Intensity	10^{-1} W/cm^2	10^{-3} W/cm^2	10^{-2} W/cm^2	10 W/cm^2
Time dependence	Pulsed with high repetition rate (MHz). Vertical scan: 1 Hz to 100 kHz	Constant	Pulses with 10^5 to 10^6 W/cm^2	Pulse width ions 100 Hz

In Table 2.10 the relative characteristics of the various systems are compared, including a dense plasma focus system.

Plasma focus devices (PFD) have been studied extensively for applications such as fusion neutron production and X-ray generation for weapon simulation or structure studies.[147] Most of the experience has been with deuterium gas, although X rays have been produced with several other gases, such as argon, krypton, and neon. Through the process of collisional excitation, these ions will radiate most of the plasma energy in hydrogen- or helium-like transition lines.

The PFD produces a small volume of hot, dense plasma that is formed by pinched compression in a coaxial-electrode, pulse-discharge device. A number of different electrode geometries can be used to produce the dense plasma.[148] The coaxial plasma gun has been most successful in producing intense bursts of X rays and has received the most attention experimentally. The X-ray tube consists of two coaxial electrodes separated at the breech end by an insulator. The tube is connected by fast-acting switches to a low-inductance, high-voltage capacitor bank. The electrodes are located in a vacuum chamber filled to a pressure of 0.1 to 1 Torr. For X-ray production, the capacitor bank is charged to its operating voltage and the switch is triggered. The gas discharge initiates uniformly throughout the interelectrode region, but quickly becomes axially symmetric. Figure 2.94 shows the development of the density of the plasma calculated after breakdown.[149]

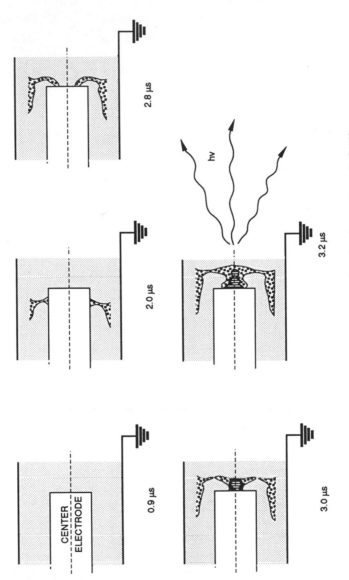

FIGURE 2.94.   Development of the density of a plasma. From Ref. 148.

The moving plasma sheet ionizes the gas it encounters and sweeps it up as it moves along. When the current sheet reaches the end of the anode (interior electrode), it collapses radially to the axis. This results in the formation of a dense, hot, pinched plasma, with a radius and lifetime of about 0.2 cm and 0.1 μs, respectively. The plasma density and temperature depend strongly upon the initial conditions of electrode geometry, capacitor band voltage, type of gas, and filling pressure. With the PFD source, the heat released in the focal plasma is either radiated away or transported to a large electrode area. Thus, much higher intensities are attainable than in accelerated electron-beam/target sources, where target cooling problems limit the maximum intensity achievable.

   2.8.1.6. Laser Plasma Sources.[150] The hot plasma that is generated in the focus of a high-power laser (≥ 100 mW/cm^2 power) represents a bright source of X rays.[151] Laser characteristics required for the X-ray source are

- Several joules total X-ray output
- Greater than 10^9 watts peak power during a few nanosecond pulse
- Less than 100 ns spot size on target
- At least 10% energy conversion efficiency yielding 10 mJ/cm^2 · sec for resist exposure

Solid Nd lasers with low beam divergence or amplified excimer lasers (KrF) with high power output are candidates for these X-ray sources. For greater than 10% energy conversion efficiency, the power density must be larger than 10^{13} W/cm^2 for nanosecond pulses. For a given spot size a minimum pulse length is required.[152] For example, for a 100-μm spot size the pulse length τ is related to the laser pulse energy E_L as

$$\tau \text{ (ns)} \approx 1.27 \times E_L \quad \text{(J)} \tag{2.464}$$

for a Cu target.

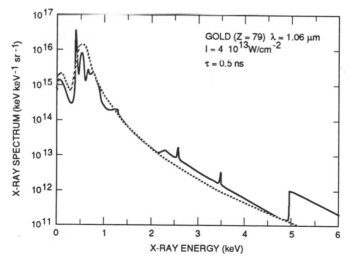

FIGURE 2.95. Experimental (– – – –) and theoretical (——) spectra of Au $\lambda = 1.06\,\mu$m, $I = 4 \times 10^{13}$ W/cm^2. From Ref. 150.

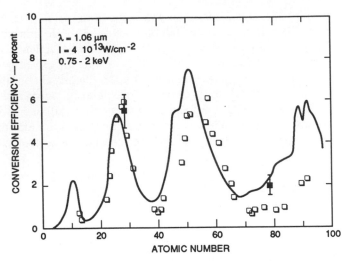

FIGURE 2.96.   Experimental and theoretical keV X-ray conversion efficiency as a function of atomic number. From Ref. 150.

The energy conversion efficiency does not change in the pulse length range of 0.3–30 ns if the intensity is greater than 5×10^{12} W/cm.

The X-ray emission from laser plasmas range from 0.1 keV to about 3 keV. Many lines from various transitions in several states of ionization appear on top of a continuum radiation, which is due to recombination and Bremsstrahlung. Figure 2.95 shows the experimental spectrum of a gold target irradiated by a glass phosphate laser ($\lambda = 1.06 \ \mu m$) at the intensity level of 4×10^{13} W/cm^2. The conversion efficiency depends on the transitions in the target material involving the K, L, M, and N shells. Therefore, it is expected that the efficiency will change with the atomic number of the target element as shown experimentally in Fig. 2.96.

The size of the X-ray emitting region in a laser-generated plasma is usually larger than the focal spot due to thermal conduction and plasma blowoff. Focal

TABLE 2.11.   Pulsed Plasma Sources

		PFD	Gas jet	Puffed gas	Sliding spark	Imploding wire
Energy	In	20 kJ	13 kJ	7.5 kJ	40–50 J	100 kJ
			20 J (6–8 Å)	50 J (12 Å)	0.44 J (184 Å)	30 J
	Out	25 J (12 Å)	500 J (30–300 Å)	100 J (10–13 Å)	1.6 J	
					(10–1000 Å)	
Emitting volume		2 × 20 mm	1–3 × 10 mm	1 × 10 mm	1 × 20 mm	1 × 30 mm
Pulse rate		1/two min	1/min	1/few min	seconds	? hours
Charging voltage		40 kV	50 kV	?	90–100 kV	50 kV
Current		450 kA	700 kA	?	6–9 kA	1 MA
Pulse length		20 ns	20–40 ns	20 ns	100 ns	100 ns
Resists exposed		FBM	COP	FBM	PMMA	COP
		EK-88	EK-88			
		PMMA				

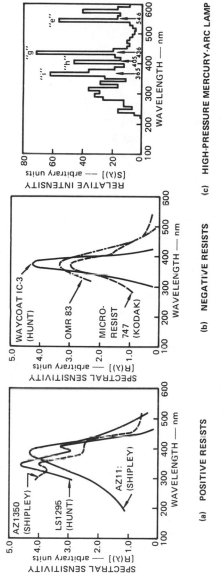

FIGURE 2.97. Comparison of spectral sensitivity curves for several positive and negative resists with spectral output of a high-pressure mercury-arc lamp. (a) Positive resists; (b) negative resists; (c) high-pressure mercury-arc lamp. From Ref. 151.

diameters fall into the 30–100 μm range, while the X-ray emission comes from a region up to ten times larger on the target surface.

The emission of X rays from the plasma is anisotropic but the maximum intensity is emitted normal to the target with relatively decreased line radiation due to self-absorption. Table 2.11 summarizes the typical characteristics of the X-ray sources discussed in this section. Figure 2.97 shows the spectral and temporal characteristics of the various X-ray sources used in microscience.

2.8.2. Light Sources for Microfabrication

2.8.2.1. Arc Lamps for Photolithography.   High-pressure mercury-arc lamps in the 200–600 nm range are used almost universally as light sources in optical lithography. Figure 2.97 shows the spectral output $S(\lambda)$ of a high-pressure mercury-arc lamp with the spectral sensitivity curves $R(\lambda)$ for several positive and negative resists.[153] For printers using reflective optics, the light spectrum of the wafer is essentially $S(\lambda)$. Printers employing refractive optics corrected for printing at one or two wavelengths ("g" and "h") must employ sharp cutoff filters to remove other emissions during exposure. If $F(\lambda)$ is the transmission characteristic of the printer, then the light spectrum reaching the photoresist-coated wafer is $S(\lambda)F(\lambda)$. As far as the photochemical reactions in the resist are concerned, the effective light intensity at the resist surface in the wavelength range $\Delta\lambda$ is proportional to $S(\lambda)F(\lambda)R(\lambda)\,\Delta\lambda$.

2.8.2.2. Laser Beam Characteristics.   The usefulness of lasers in microfabrication derives from the unique properties of laser light properties that offer distinct advantages over light obtained from conventional radiation sources. While not all of the unique properties of laser radiation may be important in a particular application, it is always found that the advantages of laser processing in a specific instance are due to one or more of the following properties:

- Monochromaticity
- Low beam divergence
- Coherence
- High brightness

These parameters influence the practical application of laser light by determining the extent to which laser radiation can be focused to a small spot (lithography, pantography) and directed to remote locations (alignment, testing). They also determine the spread of wavelengths emitted from a laser source, i.e., the laser linewidth. This is of fundamental importance in spectroscopic and photochemical applications.

Laser linewidth is a function of both the intrinsic width of the laser transition and the existence of a range of longitudinal cavity modes. These cavity modes arise because the wavelength λ_n of standing electromagnetic waves in the laser cavity is quantized with

$$\lambda_n = \frac{2L}{n} \qquad\qquad (2.465)$$

where L is the length of the cavity and n is an integer. The spectrum of these modes consists of a series of sharp resonances at frequencies $v v_n = nc/2L$ where c is the speed of light. The frequency interval between these modes is $\Delta v = c/2L$. For a 1-m cavity this corresponds to $\Delta v = 0.15$ GHz. The result is an oscillation which occurs alternatively on each of the cavity modes that overlap the fluorescent line. Frequency stabilization is needed for single frequency operation.

The spatial profile of a laser beam is determined by the geometry of the laser cavity. The shape of the laser cavity in a direction transverse to the optical axis provides boundary conditions on the wave equation that determines which config-urations of electromagnetic field will be allowed in the cavity. The notation for these transverse electromagnetic modes is taken from the theory of waveguides. Particular modes are labeled TEM_{mn} where m and n are the number of modes in two orthogonal directions. Both cylindrical and rectangular geometries have the TEM_{00} mode as the transverse mode of highest symmetry. Other higher-order modes are labeled TEM_{01}, TEM_{11}, TEM_{12}, and so on. Operation in the fundamental mode provides true diffraction-limited beam divergence. This can be achieved by reducing the cavity diameter to a critical size. Fundamental mode operation implies an intensity distribu-tion across the beam which can be expressed as

$$I(e) = I_0 \exp\left[\frac{2r^2}{W^2}\right] \tag{2.466}$$

where W is the radius of the beam at the radial distance for which $I = I_0 e^{-z}$, r is the radial coordinate, and I_0 is the intensity at $r = 0$.

A Gaussian intensity distribution is usually maintained at a distance z from the laser. The beam divergence θ can be defined through comparison at some distance z of conventional and laser light sources.

For the same total power P, one has

$$I\,(\text{conventional}) = \frac{P}{4\pi r^2}$$

$$I\,(\text{laser}) = \frac{P}{\pi z^2 \theta^2} \tag{2.467}$$

$$\frac{I\,(\text{laser})}{I\,(\text{conventional})} = \frac{4}{\theta^2}$$

Typical beam divergence is in the 0.1–10 mrad range for visible and infrared lasers. The intensity enhancement due to this small divergence is on the order of 10^6 for a laser light source. Since $\sin \theta \approx \lambda/l$, the enhancement is even larger for shorter wavelengths.

The brightness (radiance) is defined as power per unit area per unit solid angle:

$$B = \frac{P}{\pi r_0^2 \theta^2} \tag{2.468}$$

TABLE 2.12. Brightness, B, and Spectral Radiance, B_v, for Various Laser Sources and a 1-kW Arc Lamp

Source	B (W cm^{-2} sr^{-1})	B_v (W cm^{-2} sr^{-1} Hz^{-1})
1-kW arc lamp	10^3	10^{-10}
10-W Ar$^+$	4×10^9	1
100-W cw CO$_2$	10^8	10
1-J Nd-glass 10-ns pulse	10^{16}	10^3

where r_0 is the beam radius. In the diffraction-limited case

$$Q \approx \frac{\lambda}{r_0^2 \theta^2} \qquad (2.469)$$

thus,

$$B = \frac{P}{\lambda^2} \text{ W cm}^2 \text{ sec}^{-1} \qquad (2.470)$$

and the spectral radiance $B_v = B/\Delta v$ where Δv is the linewidth in hertz.

For comparison, Table 2.12 shows B and B_v for several sources.

2.8.2.3. Practical Lasers.   Lasers have been used in microelectronics for material processing (annealing, heat treatment) in addition to welding, cutting, drilling, and engraving. Table 2.13 shows the characteristics for each laser used in these applications.[154,155] For most of these applications, the lasers are used as high-power focused light sources to change surface or bulk material properties. In annealing ion–implant damage in semiconductors by means of pulsed lasers, the defect layer, which is generally amorphous, must be melted by the laser. The melt depth must extend slightly into the single-crystal layer to achieve optimum annealing and substitution for the implanted ions in the lattice structure. For Si implanted with As, B, or P, a Q-switched Nd:glass, Nd:YAG, or ruby laser can be used with a pulse range of 10- to 100-ns duration. The energy density of irradiation on the material is in the range of 0.5–10 J/cm^2. While pulsed laser annealing restores the electrical characteristics of the ion-implanted semiconductors, it causes the dopant to redistribute itself deeper into the host material. Another approach to laser annealing is the use of continuous-wave (cw) lasers, such as Ar lasers, the Kr-ion laser, or the CO$_2$ laser. The scan rate of the cw beam is usually about 0.5–10 cm/s, giving a dwell time of about 10–160 ms. The typical fluence is about 200 J/cm^2.

"Excimer" is a contraction of "excited dimers"—a diatomic molecule that has been raised to an excited energy state. In an excimer laser, such diatomic molecules comprise an atom each of a noble gas such as argon, krypton, or xenon and a halogen such as fluorine or chlorine.[151] Excimer lasers are pulsed, deep-ultraviolet lasers that produce high average power. The wavelength range within which excimers emit and their high power combine to induce the phenomenon of "ablative photo-decomposition" of materials, resulting in clean cuts with very little collateral heating.

TABLE 2.13. Characteristics and Applications of Material-Processing Lasers

Laser type	Wavelength (μm)	Operating mode	Typical power	Pulse repetition rate (pulses/s)	Pulse length	Typical application areas
Nd:YAG	1.06 (near infrared)	cw	2 W to 1 kW	N/A	N/A	Seam welding, sealing, reflow soldering, marking, surface hardening, and thin-metal cutting
		Q-switched cw	600 kW (peak) 300 W (avg)	500 to 50,000	100 to 600 ns	Scribing, marking, drilling, diamond cutting, trimming, metal cutting
		Pulsed	20 kW (peak) 400 W (avg)	1 to 300	0.6 to 10 ms	Drilling, cutting, welding, heat treating, and sealing
CO_2	10.6 (far infrared)	cw	50 W to 25 kW	N/A	N/A	Welding, cutting, heat treating, and sealing
		Pulsed	50 W to 2 kW (avg)	200 to 50,000	40 ms	Welding and cutting thick material
		TEA	Up to 300 W (avg)	Up to 100	10 μs	Marking
Ruby	0.6943 (visible red)	Pulsed	100 kW (peak)	Single pulse	0.2 to 5 ms	Spot welding and single-hole drilling
Nd:glass	1.06 (near infrared)	Pulsed	1 MW (peak)	Single pulse	0.5 to 10 ms	Spot welding and single-hole drilling

TABLE 2.14.   Typical Excimer Laser Specifications

Gas	ArF	KrF	XeCl	XeF	KCl
Wavelength (nm)	193	248	308	351	222
Average power (W)	20	60	30	20	30
Energy per pulse (mJ)	150	300	150	100	100
Repetition rate (Hz)	200	200	200	200	200

This characteristic makes excimers an excellent tool in many applications calling for high-precision removal of materials—particularly when the process requires heat. Promising application areas include micromachining, integrated optics, photolithography, gas immersion, and laser doping (gilding).

While the noble (inert) gases do not usually combine chemically, it is possible to induce dimer formation in halogens by supplying high-energy electrons, usually in the form of a high-voltage discharge, to a dilute noble–halogen mixture. The molecules thus formed, which have no stable ground state, quickly dissociate back into atomic rare gas and molecular halogen or halide, emitting ultraviolet radiation

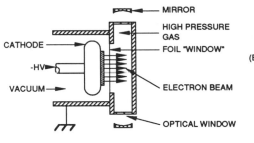

(a) SCHEMATIC DIAGRAM OF A LASER
PUMPED BY AN ELECTRICAL DISCHARGE

(b) SCHEMATIC DIAGRAM OF A LASER PUMPED
BY A HIGH INTENSITY ELECTRON BEAM

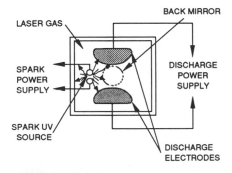

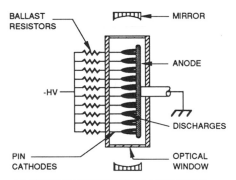

(c) SCHEMATIC DIAGRAM (VIEWED ALONG THE LASER
AXIS) OF A LASER PUMPED BY A TRANSVERSE
DISCHARGE USING UV RADIATION FROM A SPARK
SOURCE TO PRE-IONIZE THE GAS

(d) SCHEMATIC DIAGRAM OF A LASER PUMPED
BY CORONA POINTS

FIGURE 2.98.   Pumping systems for excimer lasers. (a) Schematic diagram of a laser pumped by an electrical discharge; (b) schematic diagram of a laser pumped by a high-intensity electron beam; (c) schematic diagram (viewed along the laser axis) of a laser pumped by a transverse discharge using UV radiation from a spark source to preionize the gas; (d) schematic diagram of a laser pumped by corona points.

TABLE 2.15.   Ablation Parameters for Hole Drilling in Polymer Foils

Type of foil	Wavelength (mn)	Useful energy density (J/cm^2)	Ablation depth per pulse (μm)
Polycarbonate	248	4.0	0.4
Polyester	248	4.0	0.8
Polyethylene	248	3.7	1.0
Silicone rubber	308	10.0	1.5
	248	30.0	0.17
Kapton	308	10.0	1.2
Plexiglas	193	1.0	0.3
Hostaform	248	2.8	0.57

in the process. Table 2.15 lists the various lasing media and some other characteristics of the excimer lasers.

The basic components of an excimer laser include a cavity containing a gas mixture at a pressure of 2 to 3 atm (100 to 300 kPa). Halogen atoms/molecules account for only about 1% of this total pressure, while the noble gas makes up about 5 to 15% and the remaining partial pressure is that of the buffer gas, usually neon or argon. Electrodes provide the high-voltage discharge (about 20 kV) that induces the emission of a pulse of ultraviolet light with a duration of 15 to 50 ns and energy ranging between 10 mJ and 2 J. Such pulses may be repeated at frequencies (repetition rates) up to 1 kHz. Other pumping systems such as electron beam, arc discharge, and corona are also used for excitation[156] (see Fig. 2.98). Also within the main cavity are resonator optics. The rear mirror fully reflects the pulse, while the output mirror provides partial transmission to extract the beam. The output beam is usually rectangular in shape, and about 1 cm^2 in area. Peak power is in the vicinity of tens of megawatts, with typical average power from about 5 W to 100 W.[157]

The high photon energies of ultraviolet excimer light can directly break bonds in the substrate by ablation rather than through such thermal reactions as melting, evaporation, or vaporization. Most of the UV energy is absorbed in a very thin (typically 0.5 to 2.0 μm) surface layer, and ablation starts within 4 to 6 ns after the start of the laser pulse. Consequently, very little energy remains in the substrate and thermal diffusion to the surrounding area is negligible.

The ablation dynamics during an excimer laser pulse have been described in the literature.[158] The variation in etch depth per pulse as a function of incident energy density (fluence) for three different excimer wavelengths are shown in Table 2.15. These data were obtained by recording the number of pulses required to completely etch through 25-μm-thick polyimide film. The depth of ablation depends on the particular polymer type, the fluence, and the area being exposed (Table 2.16).

The ability to selectively ablate thin sections of material to a controlled depth with minimal thermal diffusion into the substrate gives excimer lasers an operating advantage over such thermal sources as CO_2 and Nd:YAG, which produce a large heat-affected zone envelope. Recently, experimental X-ray lasers produced wavelengths in the 2–5 nm range. Soft X-ray radiation produced by a Nd:glass laser causes recombination in aluminum plasmas. The Al target is typically a 10-cm-long, 1-mm-wide slab. Table 2.16 shows the result of X-ray laser recombination experiments on lithium-like aluminum.

TABLE 2.16.   X-ray Laser Recombination Experiments on Lithium-like Aluminum

Transition	Wavelength (nm)	Target	λ (μm)	θ (ns)	I (W/cm^2)	Gain/length	Gain
Al 4f–3d	15.47	Fiber	0.53	0.12	1.7×10^{14}	3.0	2.1
5f–3d	10.57					3.0	2.1
Al 4f–3d	15.47	Mag.	1.06	50	2×10^{13}	3.5	3.5
Al 5f–3d	10.57	Foil	1.06	2.0	2×10^{12}	0.5	3.2
Al 5f–3d	10.57	Foil	0.53	0.15	1.3×10^{14}	0.8	1.0
5d–3p	10.38						
Al 5f–3d	10.57	Slab	1.06	5	2.5×10^{11}	3.4	4.1
4f–3d	15.47					4.5	5.4

The free electron laser (FEL) is a tunable, very powerful light source which converts the kinetic energy of a relativistic electron beam in a synchrotron or a storage ring, into coherent electromagnetic radiation.[159] The FEL is an insertion device (see Section 2.8.1.2) plus a pair of mirrors forming a resonant cavity. The mirrors provide a positive feedback resulting in stimulated emission at a wavelength

$$\lambda = \frac{\lambda_u}{2\gamma^2} (1 + \tfrac{1}{2}K^2 + \gamma^2\theta^2) \qquad (2.471)$$

The quantities are defined in Section 2.8.1.2. Figure 2.99 shows the FEL incorporated into an electron storage ring.

The potential advantages of the FELs are the following:

- It has the potential to operate at any wavelength including the XUV spectral range.
- It is tunable over a wide range of wavelengths. A given FEL should be tunable over typically at least a decade of wavelengths.
- It may be very powerful depending only on the intensity of the electron beam which feeds it.

The average power conversion from electron beam to photon beam is typically 0.1%; however, for peak power it is much higher, up to 1–5%.

In an undulator the electron transverse speed and the radiated electric field of the wave oscillate at different frequencies so that the net energy transfer is zero:

$$\int \overline{EV_\perp} \, dt \equiv 0 \qquad (2.472)$$

However, for the resonant wavelength and its harmonics, this integral does not vanish. The resonant wavelength, λ_R, is such that when the electron has traveled

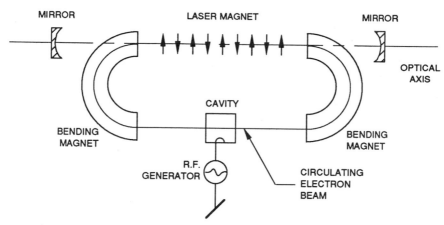

FIGURE 2.99. Schematic of free electron laser incorporated into an electron storage ring. From Ref. 159.

along one period $\lambda\mu$ of the undulator, the light has traveled the same distance plus an integer number of wavelengths λ_R (since the light travels slightly faster than the ultrarelativistic electrons considered). Therefore:

$$\frac{\lambda u}{v} = \frac{\lambda u + n\lambda_R}{c} \tag{2.473}$$

and the wavelength at $\theta \cong 0$ can be written as

$$\lambda_K(n) - \frac{\lambda u}{2n\gamma^2}[1 + \tfrac{1}{2}K^2] \tag{2.474}$$

Thus, the spectral lines can be in the optical region when γ is large (10^2–10^3). This radiation is called "undulator radiation," an analogue to the atomic laser's "spontaneous radiation." Integrating over θ (direction of photon emission) the spectrum is almost continuous. The electric field of the radiation interacts with the electrons in the undulator and "bunches" the electrons over certain positions. Once the electrons are bunched, the radiation becomes coherent. (The intensity is proportional to N_e^2 where N_e is the number of electrons in the bunch.) This increased radiation can be amplified further by the use of an optical cavity (two mirrors) or one can use an external source for radiation and amplify it. The optical cavity contains the undulator, the magnetic dipoles, the electron beam extractor, possibly some magnetic quadrupoles, and the light extraction device. Its length has to be matched with the electron bunch-to-bunch separation and to be adjusted within a few microns on pulsed accelerators.

The superbright FEL is expected to be an important future source for UV and X-ray lithography.

2.9. PHOTON INTERACTIONS

2.9.1. X-ray Photon Interactions

2.9.1.1. Interaction Processes. X-ray photons interact with matter chiefly via three processes: photoelectron ejection, the Compton effect, and pair production. A complete treatment of the three processes is rather complicated and requires the tools of quantum electrodynamics. The essential facts, however, are simple. In the photoelectric effect the photon is absorbed by an atom and an electron from one of the shells is ejected. In the Compton effect, the photon scatters from an atomic electron. In pair production, the photon is converted into an electron–positron pair. This process is impossible in free space because energy and momentum cannot be conserved simultaneously when a photon decays into two massive particles; however, it can occur in the coulomb field of a nucleus, where energy and momentum can be balanced.

The energy dependence of these processes are very different. At low energies below a few kiloelectron volts, the photoelectric effect dominates, the Compton effect is small, and pair production is energetically impossible. At an energy of 1 MeV, pair production becomes possible, and it soon dominates completely. Two of the three processes, the photoelectric effect and pair production, eliminate the photons undergoing interaction. In Compton scatterings, the scattered photon loses energy.

The characteristics of the transmitted beam are described by an exponential law:

$$N(z) = N(0) \, e^{-\mu z} \tag{2.475}$$

where the absorption coefficient μ is the sum of three terms,

$$\mu = \mu_{\text{photo}} + \mu_{\text{Compton}} + \mu_{\text{pair}} \tag{2.476}$$

The number of transmitted particles decreases exponentially. No range can be defined, but the average distance traveled by a particle before undergoing a collision is called the mean free path and it is equal to $1/\mu$.

Figure 2.100 shows the photon absorption coefficient μ (in cm^{-1}) for single-crystal silicon as a function of photon energy from 10^{-2} to 10^8 eV. Optical photons at the lowest energies can be absorbed by lattice phonons; since the silicon lattice is not infrared active, this absorption requires multiphonon processes. Absorption due to impurities can also occur in the energy range below 1 eV, but such absorption is not indicated in Fig. 2.100. Photoelectric absorption begins at the energy corresponding to excitation across the forbidden gap (E_g). For energies near E_g, the photoexcited electron remains in the solid, and this process is called the internal photoeffect. At higher energies (at approximately the ionization energy of the free atom), the photoelectron can leave the solid, and this process is called the external photoeffect. For energies near E_g (1.2 eV), the magnitude of the absorption cross section depends sensitively on the band structure of the solid. At higher photon energies, photoelectron absorption is the dominant process, until Compton scattering and pair production take over.

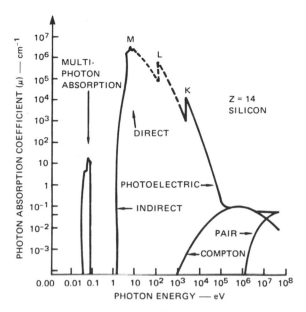

FIGURE 2.100. X-ray absorption coefficient as a function of X-ray energy.

2.9.1.2. Photoelectric Effect. In the photon energy range of 1–50 keV used in microscience, the most important photon interaction is the photoelectric effect. The transmitted X-ray intensity follows an exponential attenuation law expressed as

$$I = I_e \, e^{-\mu x} = I_0 \exp\left(-\frac{b}{\rho}\rho x\right) \tag{2.477}$$

where ρ is the density of the solid (g/cm²), μ is the linear attenuation coefficient, and μ/ρ is the mass attenuation coefficient. In this energy range, the attenuation is mostly due to the photoelectric effect $\mu \approx \mu_{photo}$ and shows a strong energy dependence due to the photon absorption in the K, L, and M shells.

The mass absorption coefficient μ/ρ for electrons in a given shell or subshell can be calculated from the photoelectric cross section σ:

$$\frac{\mu}{\rho} = \frac{\sigma \; (\text{cm}^2/\text{electron}) N \; (\text{atoms}/\text{cm}^3) n_s \; (\text{electrons}/\text{shell})}{\rho \; (\text{g}/\text{cm}^3)} \tag{2.478}$$

where ρ is the density, N the atomic concentration, and n_s the number of electrons in a shell. The photoelectric cross sections $\sigma_{photo} = \sigma(E)$ can be calculated approximately using first-order perturbation theory and solutions of the Schrödinger equation.

In this case, the Hamiltonian for the interacting system (photon + K-shell electron) can be written as

$$\hat{H} = \hat{H}_0 + \hat{H}'$$

where H_0 is a Hamiltonian for which the Schrödinger equation can be solved, and H' contains an additional applied electric field. Using H' the transition probability per unit time can be calculated from Fermi's Golden Rule as

$$W = \frac{2\pi}{\hbar} \rho(E) |\langle \psi_k | \hat{H}' | \psi_m \rangle|^2 \qquad (2.479)$$

In this equation $\rho(E)$ is the density of final states per unit energy,

$$\langle \psi_k | \hat{H}' | \psi_m \rangle = \int \psi_k^* | \hat{H}' | \psi_m \, d\tau = |\hat{H}'_{km}| \qquad (2.480)$$

where ψ^* is the complex conjugate of ψ, $d\tau$ is a three-dimensional volume element (i.e., $r^2 \, dr \sin \theta \, d\theta \, d\phi$), and the wave function ψ_m is

$$\psi_m = e^{it\omega_m} U_m \qquad (2.481)$$

ψ_k is the final wave function and U_m is a solution for H_0, the unperturbed Hamiltonian (i.e., $\hat{H}_0 U_m = E_m U_m$). The dimension of W is (time)$^{-1}$ since

$$W = \frac{1}{\text{energy} \cdot \text{time}} \frac{1}{\text{energy}} \cdot (\text{energy})^2 = \frac{1}{\text{time}} \qquad (2.482)$$

To calculate the transition probability and the cross section for the photoeffect, let us consider a one-dimensional potential well with length L where the electron is bound with binding energy E_B.

Using periodic boundary conditions, the normalized final states can be written as

$$\psi_E = \frac{1}{\sqrt{L}} e^{ikx} \qquad (2.483)$$

where $kL = (2mE)^{1/2} L / \hbar = 2\pi N$ (i.e., periodic boundary conditions).

   The density of states is the number of states with energy between E and $E + \Delta E$. Then

$$\rho(E)\,\Delta E = \Delta N = \frac{L}{2\pi}\,\Delta k$$

$$\rho(E) = \frac{L}{2\pi}\frac{\Delta k}{\Delta E}$$

(2.484)

For a free particle, $E = \hbar^2 k^2/2m$ and $\Delta E = (\hbar^2/m)k\,\Delta k$ so that

$$\rho(E) = \frac{L}{2\pi\hbar}\left(\frac{m}{2E}\right)^{1/2} 2$$

(2.485)

Here a factor of 2 is introduced to allow for positive and negative values of N. Then the matrix element for the transition probability is

$$H'_{fi} = \int \frac{1}{\sqrt{L}}\,e^{ikx}ex\psi_i(x)\,dx$$

(2.486)

where $\hbar k$ is the momentum of the final state.
   The initial state of the weakly bound electron in the one-dimensional well is described by the solution of the unperturbed Hamiltonian in the form

$$U_0 = C\,e^{\pm k_i x}$$

(2.487)

where k_i is the momentum associated with the bound state, $k_i = (2mE_B)^{1/2}/\hbar$ and $C = \sqrt{k_i}$. To simplify the calculation, the exterior wave function is extrapolated into the potential well as far as the origin, then the normalized initial wave function ψ_i is

$$\psi_i(x) \approx [\sqrt{k_k}]\,e^{-k_k|x|}$$

(2.488)

With this simplification one gets

$$\langle\psi_f|x|\psi_i\rangle = \frac{1}{\sqrt{L}}\int_{-\infty}^{\infty} e^{ikx}x\psi_i(x)\,dx$$

$$= \frac{1}{\sqrt{L}}\,k_i^{1/2}\left[\frac{4ikk_i}{(k^2 + k_i^2)^2}\right]$$

(2.489)

The transition rate is then

$$W = \frac{2\pi}{\hbar} \rho(E) |H'_{\text{fi}}|^2$$

$$= \frac{4e^2 E^2 \hbar}{m} \left(\frac{E_B}{E}\right)^{1/2} \frac{EE_b}{(E + E_B)^4} \tag{2.490}$$

where E is the external electric field. When $\hbar\omega \gg E_B$, so that $E \cong \hbar\omega$, then

$$W = \frac{4e^2 E^2 \hbar}{m} \frac{E_B^{3/2}}{E^{7/2}} \tag{2.491}$$

To calculate the cross section from the transition probability, one needs the formula for the power density for the field which is $cE^2/2$. The power density is the energy per area per second. Thus, the flux F, the number of photons per area per second, is given by $cE^2/2\hbar\omega$. Dimensionally the cross section is

$$\sigma = \frac{W}{F} = \frac{\text{Transition probability/time}}{\text{No. of photons/area/time}} \tag{2.492}$$

or

$$\sigma = \frac{8e^2\hbar}{mc} \frac{E_B^{3/2}}{E^{5/2}} \tag{2.493}$$

The cross section decreases with increasing photon energy ($E = \hbar\omega$) as $E^{5/2}$.

In the case of three-dimensional hydrogen-like atoms the photoelectric cross section σ_{ph} can be calculated as

$$\sigma_{\text{ph}} = \frac{288\pi}{3} \frac{e^2\hbar}{mc} \frac{E_B^{5/2}}{E^{7/2}} \tag{2.494}$$

For $\hbar\omega \gg E_B$ the cross section is given as

$$\sigma_{\text{ph}} = \frac{128\pi}{3} \frac{e^2\hbar}{mc} \frac{E_B^{5/2}}{E^{7/2}} \tag{2.495}$$

The value of $e^2\hbar/mc = 5.56 \times 10^{-2}$ eV Å^2, so that, for convenience, we write

$$\sigma_{\text{ph}} = \frac{7.45 \text{ Å}^2}{\hbar\omega} \left(\frac{E_B}{\hbar\omega}\right)^{5/2} \tag{2.496}$$

by setting the incoming photon energy $\hbar\omega$ (in eV) equal to the energy E of the outgoing electron since $E_B \ll \hbar\omega$ (i.e., $E = \hbar\omega - E_B \cong \hbar\omega$). Using the photoelectric cross section σ, the atomic density in the target N, and the number of electrons in a shell n, the attenuation coefficient for X rays can be written as

$$\frac{\mu}{\rho} = \frac{\sigma N n}{\rho} \approx \frac{1}{E^{7/2}} \tag{2.497}$$

2.9.2. Laser-Generated Photon Interactions

In this section we discuss the theory of laser interactions with a semiconductor substrate, the laser decomposition processes in a gas-phase thin film over the substrate,[160] and the laser/substrate interactions. These laser reactions are commonly referred to as "laser microchemistry." More precisely, "laser microchemistry" can be defined as a process in which energy supplied by a finely focused laser beam drives a chemical reaction in a localized area.[161] High-energy laser photons with energy ranging from 0.1 eV in the infrared to 6 eV in the ultraviolet initiating chemical reactions have been studied for a variety of gas–solid systems.[162] The interaction of laser photons can occur with: (1) the gaseous species, (2) the species adsorbed on the solid surfaces, and (3) the solid substrate. In processes (1) and (2), both the electronic and vibrational excitation of the gaseous molecules can be important. In process (3), both the electronic and lattice phonon (thermal) excitation of a solid may be involved.

UV lasers dissociate the reacting molecules while IR lasers are used primarily to raise the temperature of the reacting gases. Although UV lasers have been used to successfully deposit metal and insulator films, these films have generally not been of high quality because of the difficulty in controlling the reactions of the highly reactive photolyzed species. By using a pulsed IR CO_2 laser, however, high-quality gallium arsenide thin films have been grown in a process that raises the temperature of the reacting gas immediately above the substrate.[163]

It has been observed that SF_6 molecules, inert to Si at 25°C, can be reduced to react with the solid by multiple photon excitation with a pulsed CO_2 laser to form SiF_4 and SF_4 products.[164] Three or four photons are involved to vibrationally excite SF_6 molecules via process (1). At 90 K, SF_6 is adsorbed on the surface of Si with one or two monolayers of coverage. When irradiated by CO_2 laser pulses in resonance with SF_6 vibrations, a substantial number of SiF molecules are formed [process (1)] and desorb into the gas phase where they can be detected with a mass spectrometer.

Laser interaction with a solid surface generally results in localized heating that promotes surface reactions. A focused Ar laser in the UV range has been shown to excite and heat a spot on a Si wafer to its melting temperature and thus induce etching by Cl_2 or HCl gas at a rate approaching 10 μm/s.[165]

It has been shown that the reaction rate of a microchemical process induced by a focused laser beam can significantly exceed those of conventional diffusion-limited reactions for an extended heat source, in which reaction rates at a gas–solid interface are generally limited by the gas-phase diffusion of reactants and products, and by the reaction rates on the surface itself.[166] The reactant flux channeled into the reaction zone increases as the dimension of the reacting area decreases to a value small compared with the gas diffusion distance due to the enhancement of mass transport via three-dimensional gas-phase diffusion. In comparison, the reaction flux on a planar surface is limited by one-dimensional gas-phase diffusion.

Reactions (1) and (2) are commonly referred to as laser pyrolysis and laser photolysis, respectively, as illustrated in Fig. 2.101. With laser pyrolysis, the chemical reaction occurs at the surface while the volatile components remain in the vapor phase. Using this technique, features as small as 1 μm have been deposited. This method has a number of advantages:

- Depending on the optical delivery system, a compact 50-mW air-cooled argon-ion laser suffices as a source.
- Argon-ion laser output is tunable and stable over long periods of time.
- since the pyrolytic deposition utilizes a visible laser, no expensive UV or IR optics are required.

In laser photolysis, a laser is used to decompose an organometallic molecule such as $Cd(CH_3)_2$ in the vapor phase. Although studies have shown that submicron processing is possible, this method has several disadvantages:

- Second harmonics that are generated lead to inefficiencies in the process requiring a large, high-power argon laser.
- These second harmonics exhibit an output that fluctuates with time, requiring periodic adjustments.
- An argon laser operating at 257 nm requires costly, high-numerical-aperture, long-working-distance quartz optics.

The absorption of laser radiation by a solid (type 3 process) is equivalent to a source of heat in or on the solid. The response of a material to this source can be calculated by solving the three-dimensional heat transfer equation

$$\rho C \frac{\partial T}{\partial t} = \frac{K\nabla^2 T}{\frac{\partial}{\partial x}\left(K\frac{\partial T}{\partial x}\right) + \frac{\partial}{\partial y}\left(K\frac{\partial T}{\partial y}\right) + \frac{\partial}{\partial z}\left(K\frac{\partial T}{\partial z}\right)} + A(x, y, z, t) \qquad (2.498)$$

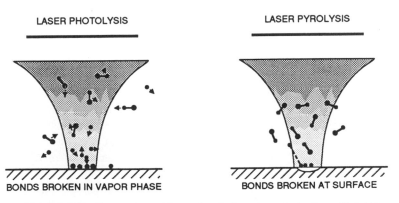

LASER PHOTOLYSIS   LASER PYROLYSIS

BONDS BROKEN IN VAPOR PHASE   BONDS BROKEN AT SURFACE

FIGURE 2.101.   Comparison of laser photolysis and pyrolysis. From Ref. 166.

While the form of the solution to this equation, $T(x, y, z, t)$, is determined by the source distribution function $A(x, y, z, t)$ and the boundary conditions pertaining to the problem, the amplitude of the solution is controlled by the magnitude of the thermal constants, C and K. These constants are

$$K = \text{thermal conductivity (W cm}^{-1} \, {}^{\circ}\text{C}^{-1})$$
$$C = \text{heat capacity (J g}^{-1} \, {}^{\circ}\text{C}^{-1}) \tag{2.499}$$

The magnitudes of C and K determine the rate at which thermal equilibrium is achieved in response to the sudden application of the heat source. This can be seen most easily from the thermal diffusivity, k, defined as

$$k = \frac{K}{\rho c} \tag{2.500}$$

k has the units cm^2/s and ρ is the sample density (g/cm^3). The distance, x, that the heat wave has penetrated the sample after time t can then be found approximately from

$$x = (Kt)^{1/2} \tag{2.501}$$

The solution to the heat transfer equation can be expressed in analytical form only when the system possesses certain symmetrical boundary conditions. When this is not the case, the heat transfer equation can be solved numerically. Detailed analysis of almost any practical heating problem requires that a numerical approach be adopted. However, considerable physical insight into laser heating mechanisms can often be obtained from approximate solutions expressed in analytical form.[167] Many applications in microfabrication involve several processes such as heating, phase changes (melting, evaporation), ablation, crystallization and molecular decomposition.

Table 2.17 shows the physical processes used in laser fabrication. In most of the applications, the lasers are used as spatially selective constants of pulsed heat sources.

TABLE 2.17. Physical Processes in Laser Fabrication

Laser processes	Physical phenomena
Drilling	Heating, evaporation
Welding	Heating, melting
Cutting	Heating, melting
Micromachining	Ablation, melting, recrystallization
Trimming	Evaporation
Crystal growth	Heating
Pyrolysis	Heating, bond breaking
Photolysis	Bond breaking (ionization)
Annealing	Melting, recrystallization
Diffusion	Heating, enhanced diffusion
Implantation	Heating, increased diffusion

In some other applications, the laser initiates other effects such as phase changes or electronic or vibrational transitions when the photon energies are higher.

These higher energy (3 to 4 eV) laser interactions form the basis of laser-driven metal deposition (laser pantography) processes.[160] Two different types of laser-induced metal deposition processes are candidates for this role: pyrolysis and photolysis. The pyrolytic processes are based on thermal decomposition reactions which occur due to laser-induced heating of the substrate, while the photolytic processes entail gas-phase photolytic decomposition reactions occurring in the laser-irradiated gas-filled region above the substrate. The ultimate limit on the rate of metal deposition by either pyrolytic or photolyic processes is determined by the maximum possible rate of mass transport of the reactant gas to the substrate.

In order for photolysis to occur in unit time, the photon fluence must equal the reciprocal of the cross section for the metal-bearing molecule to photolytically absorb the photon. Since the unit of time is that for the molecule to transit the reaction zone—about 10 ns for molecular weights on the order of 100 amu and reaction zone scales on the order of 1 μm—and since the largest photolytic cross sections of interest are less than 10^{-17} cm^2, the photon fluxes required for mass transport-limited photolysis are on the order of 10^{25} photons/cm^2 per s, or about 10 MW/cm^2, which implies a minimum input of 0.1 W of hard-won 4 to 6-eV laser photons. By way of comparison, the laser intensity required for pyrolysis of metal-bearing gases at a comparable mass-processing rate is just that needed to maintain a square micron of substrate at a temperature on the order of 1000 K, which is also on the order of megawatts/cm^2, but now involves an input of 0.1 W of readily available, wavelength-insensitive optical laser spectral power.

Using pyrolitic mechanisms for metal micropatterning, or interconnect formation, various metals (Al, Cd, Zn, Sn, Fe, W, Cr, and Mo) are deposited from the gas phase.[168] The laser-induced pyrolytic metal deposition processes are based on thermal decomposition reactions of the form:

$$[MX_n] \rightarrow [M] + n[X] \tag{2.502}$$

where MX_n is the reactant gas molecule (e.g., a metal alkyl or carbonyl), M is the metal atom being deposited, and X is a product gas of the reaction. The process can be thought of as involving five consecutive steps:

a. Diffusion of the reactant gas to the heated portion of the substrate
b. Adsorption of the reactant
c. Decomposition of the reactant
d. Desorption of the volatile product
e. Diffusion of the product gas away from the substrate

At lower temperatures, the rate-determining step of the process is usually (c); at higher temperatures, where the decomposition reaction rate becomes substantial, (a) and (e) become the rate-limiting steps.

TABLE 2.18. Characteristics of Particle Beams Used in Microfabrication

Beam type	Wave length λ (M)	Range	Deflection properties	Interactions with matter used in microfabrication
Photon ($h\nu$)	2000–4000 Å	$I = I_0 e^{-mz}$	Optical elements (lenses, mirrors)	Bond breakage and polymerization; photoelectron production, local heating (annealing); plasma generation; geometrical alignment; diffusion
X ray ($h\nu$)	2–20 Å	$R \sim$ range of photoelectrons	Crystals	Bond breakage and polymerization; material analysis; alignment
Electron (e^-, m_0)	~0.1 Å	$R[\frac{\mu g}{cm}] = 10 E_0^{1.43}$ E_0 in keV	E,H fields	Bond breakage and polymerization; secondary-e production; Bremsstrahlung (X-ray production); plasma generation; induced current, voltage production; alignment diagnostics (electron microscopies); annealing
Ion (e^+, M)	0.01 Å	$R \cong 2kE_0(1 - \frac{4}{3}kk'E_0)$	E,H fields	Sputtering; deposition; etching; ion implantation (surface treatment); diagnostics (microscopies)
Atom (A, Z)	0.01 Å	$R \cong 2kE_0(1 - \frac{4}{3}kk'E_3)$	Gradient fields when $\mu, p \neq 0$	Sputtering; crystal growing; molecular beam epitaxy

Typical laser-driven material deposition reactions are listed for

- Tungsten: $WF_6 + 3H_2 \rightarrow W + 6HF$
- Nickel: $Ni(CO)_4 \rightarrow Ni + 4CO$
- Silicon: $SiH_4 \rightarrow Si + 2H_2$
  $Si_2H_6 \rightarrow 2Si + 3H_2$
  $Si_3H_8 \rightarrow 3Si + 4H_2$

The reaction rates are generally given in the following form: $R \sim A \exp[-T_A/T]P^n$, where T is the substrate temperature, P is the reactant gas pressure, and T_A is the activation temperature.

The dimension of the deposit (ρ_0) relative to the laser spot size (r_0) can be estimated as

$$\frac{\rho_0}{r_0} \approx \sqrt{\frac{T}{T_A}} \approx 0.3\text{–}0.4 \tag{2.503}$$

for the reactions of interest. Experiments confirm that the diameter of deposition can be made smaller than the laser spot size by at least a factor of 2.

2.10. SUMMARY

In this chapter we discussed the source properties for generating particle beams (electron, ion, photon), the formation of microbeams, and beam–target interactions. Table 2.18 summarizes the most important characteristics of these beams and those interactions that are critically important in microscience.

PROBLEMS

1. a. Discuss the requirements for an electron source to function as a cathode in practical applications.

   b. What is the effect of coating a metal cathode with a monomolecular layer of (i) electronegative species, and (ii) an electropositive species?

   c. Discuss the characteristics of metal cathodes, and coated thermionic cathodes.

   d. What are the advantages of a field emission cathode?

2. The field emission currents obtained from a molybdenum microtip cathode at particular voltages are given below.

			V (volts)			
208	197	176	160	135	127	112
I (amps) 4.0×10^{-5}	3.2×10^{-5}	1.6×10^{-5}	8×10^{-6}	1.6×10^{-6}	8×10^{-7}	1.6×10^{-7}

Plot these data assuming the Fowler–Nordheim relationship.

Assuming that the work function of the cathode material is 4.5 eV, calculate the emission area and the beta factor, using $t^2(y) = 1.1$, $v(y) = 0.95 - y^2$, and $y = (eE/4\pi\varepsilon_0)^{1/2}/\phi$.

3. a. Why is space charge an important consideration in ion guns?

   b. Sketch the equipotentials in a plasma ion source.

   c. Discuss the formation and stability of a Taylor cone in a liquid metal ion gun.

4. a. What is the difference between a cluster ion source and a molecular ion source or "Knudsen cell"?

   b. Sketch the pressure–temperature diagram for the adiabatic expansion in a cluster ion source. Indicate (i) equilibrium vapor pressure, (ii) onset of condensation, and (iii) the isentropic expansion.

5. a. Describe how Snell's law of refraction for light optics may be applied to electron optics.

   b. Explain how space-charge effects limit the minimum spot size.

   c. Discuss the major types of aberrations which limit the spot size at a given half-angle.

   d. What are the advantages of the Pierce electron gun over other designs

6. a. Show that the focal length of a biopotential lens depends on the potential ratio V_2/V_1 of the electrodes using the weak immersion lens approximation.

   b. With suitable modifcations, how can this lens be used as (i) a mirror, (ii) an iris diaphragm?

   c. Discuss the advantages of magnetic lenses and as an example consider the quadrupole.

   d. What are the aberrations that occur in an electron deflection system?

REFERENCES

1. W. B. Nottingham, in: *Handbuch der Physik* (S. Flugge, ed.), Vol. 21, pp. 1–175, Springer-Verlag, Berlin (1956).
2. R. H. Good, Jr., and E. W. Müller, in: *Handbuch der Physik* (S. Flugge, ed.), Vol. 21, pp. 176–231, Springer-Verlag, Berlin (1956).
3. E. L. Murphy and R. H. Good, Jr., Thermionic emission, field emission, and the transition region, *Phys. Rev.* **102**, 1464–1473 (1956).
4. L. D. Smullin and H. A. Haus (eds.), *Noise in Electron Devices*, MIT Press/Wiley, New York (1959).
5. I. Brodie, Studies of field emission and electrical breakdown between extended nickel surfaces in vacuum, *J. Appl. Phys.* **35**, 2324–2322 (pages reverse numbered) (1964).
6. A. Van Oostrom, Field emission cathodes, *J. Appl. Phys.* **33**, 2917–2922 (1962).
7. F. M. Charbonnier, R. W. Straeger, L. W. Swanson, and E. E. Martin, Nottingham effect in field and T-F emission, *Phys. Rev. Lett.* **13**, 397–401 (1964).
8. I. Brodie, The temperature of a strongly field emitting surface, *Int. J. Electron.* **18**, 223–233 (1965).
9. G. A. Haas, *Electron Sources: Thermionic Methods of Experimental Physics* (C. Marton, ed.), Vol. 4, Part A, pp. 1–38, Academic Press, New York (1967).
10. R. G. Murray and R. J. Collier, Thoriated tungsten hairpin filament electron source for high brightness applications, *Rev. Sci. Instrum.* **48**(7), 870–873 (1977).
11. A. N. Broers, Thermal cathode illumination systems for round beam electron pulse systems, *Scanning Electron Microsc.* **1971**(I), 1–9 (1971).
12. G. Herrmann and S. Wagener, *The Oxide Coated Cathode*, Vols. I and II, Chapman & Hall, London (1951).

13. R. Levi, Improved barium dispenser cathode, *J. Appl. Phys.* **26**, 639 (1955); I. Brodie and R. O. Jenkins, Impregnated barium dispenser cathodes containing strontium or calcium oxide, *J. Appl. Phys.* **27**, 417 (1956).
14. P. Zalm and A. J. A. van Stratum, Osmium dispenser cathodes, *Philips Tech. Rev.* **27**, 69 (1966).
15. R. E. Thomas, T. Pankey, and G. A. Haas, Thermionic properties of BaO on iridium, *Appl. Surface Sci.* **2**, 187–212 (1979).
16. L. W. Swanson and N. A. Martin, Zirconium/tungsten thermal field cathode, *J. Appl. Phys.* **46**, 2029–2050 (1975).
17. C. A. Spindt, I. Brodie, L. Humphrey, and E. R. Westerberg, Physical properties of thin film field emission cathodes, *J. Appl. Phys.* **47**, 5248–5262 (1976).
18. L. W. Swanson and G. A. Schwind, Electron emission from a liquid metal, *J. Appl. Phys.* **49**(11), 5655–5662 (1978).
19. R. G. Wilson and G. R. Brewer, *Ion Beams: With Applications to Ion Implantation*, Wiley, New York (1973).
20. G. Carter and W. A. Grant, *Ion Implantation of Semiconductors*, Halstead Press, New York (1976).
21. L. Valyi, *Atom and Ion Sources*, Wiley, New York (1978).
22. L. B. Loeb, *Basic Processes of Gaseous Electronics*, University of California Press, Berkeley (1960).
23. M. von Ardenne, *Tabellen der Elektronenphysik, Ionenphysik, und Ubermikroskopie*, Vols. I and II, Deutscher Verlag der Wissenschaften, Berlin (1956).
24. N. B. Brooks, P. H. Rose, A. B. Wittkower, and R. P. Bastide, Production of low divergence positive ion beams of high intensity, *Rev. Sci. Instrum.* **35**(7), 894 (July, 1964).
25. J. Orloff and L. W. Swanson, A scanning ion microscope with a field ionization source, *Scanning Electron Micros.* **1977**(I), 57–62 (1977).
26. J. H. Orloff and L. W. Swanson, Study of a field-ionization source for microprobe applications, *J. Vac. Sci. Technol.* **12**(6), 1209–1213 (November–December, 1976).
27. L. W. Swanson, G. A. Schwind, and A. E. Bell, Emission characteristics of a liquid gallium ion source, *Scanning Electron Microsc.* **1979**(I), 45–51 (1979).
28. R. Clampitt, K. L. Aitken, and D. K. Jefferies, Intense field emission ion source of liquid metals, *J. Vac. Sci. Technol.* **12**(6), 1208 (November–December, 1976).
29. R. Gomer, On the mechanism of liquid metal electron and ion sources, *Appl. Phys.* **19**, 365–375 (1979).
30. L. W. Swanson, G. A. Schwind, A. E. Bell, and J. E. Brady, Emission characteristics of gallium and bismuth liquid metal field-ion sources, *J. Vac. Sci. Technol.* **16**(6), 1864–1867 (1979).
31. G. I. Taylor, Disintegration of water drops in an electric field, *Proc. R. Soc. London Ser. A* **280**, 383 (1964).
32. R. L. Seliger, J. W. Ward, V. Wang, and R. L. Kubena, A high-intensity scanning ion probe with submicrometer spot size, *Appl. Phys. Lett.* **34**(5), 310 (1979).
33. J. Asmussen, Electron cyclotron resonance microwave discharges for etching and thin film deposition, *J. Vac. Sci. Technol.* **A7**(3), 883–893 (1989).
34. T. Takagi, I. Yamada, M. Kunori, and S. Kobiyama, in: *Proc. 2nd Int. Conf. Ion Sources*, Vienna, 790 (1972).
35. I. Yamada, H. Takaoka, H. Inokawa, H. Usui, S. C. Cheng, and T. Takagi, Vaporized-metal cluster formation and effect of kinetics energy of ionized clusters on film formation, *Thin Solid Films* **92**, 137 (1982).
36. G. J. Van Wylen and R. E. Sonntag, *Fundamentals of Classical Thermodynamics*, p. 595, Wiley, New York (1976).
37. O. Hagena, in: *Molecular Beams and Low Density Gasdynamics* (P. P. Wegener, ed.), p. 93, Dekker, New York (1974).
38. G. S. Springer, Homogeneous nucleation, *Adv. Heat Transfer* **14**, 281 (1978).
39. M. R. Hoare, Structure and dynamics of simple microclusters, *Adv. Chem. Phys.* **40**, 49 (1979).
40. J. D. Hirschfelder, C. F. Curtiss, and R. B. Bird, *Molecular Theory of Gases and Liquids*, p. 351, Wiley, New York (1954).
41. P. G. Hill, H. Witting, and E. P. Demetri, Condensation of metal vapors during rapid expansion, *Trans. Am. Soc. Mech. Eng.* **85**, 303 (1963).
42. H. Usui, M.S. thesis, Kyoto University (1981).
43. H. Inokawa, M.S. thesis, Kyoto University (1981).

44. J. R. Pierce, *Theory and Design of Electron-Beams*, Van Nostrand, Princeton, N.J. (1954).
45. B. J. Thompson and L. B. Headrich, Space-charge limitation on the focus of electron beams, *Proc. IRE* **28**(7), 318 (July, 1940).
46. J. W. Schwartz, Space-charge limitation on the focus of electron beams, *RCA Rev.* **18**(1), 3 (1957).
47. E. L. Ginzton and B. H. Wadia, Positive ion trapping in electron beams, *Proc. IRE* **42**(10), 1548 (October, 1954).
48. H. Moss, *Narrow Angle Electron Guns and Cathode Ray Tubes*, Academic Press, New York (1968).
49. O. Klemperer and M. E. Barnett, *Electron Optics*, Cambridge University Press, London (1971).
50. P. Grivet, *Electron Optics*, Pergamon Press, New York (1972).
51. A. Septier, *Focusing of Charged Particles*, Vols. I and II, Academic Press, New York (1967).
52. W. Glaser, *Grundlagen der Electronenoptik*, Springer-Verlag, Berlin (1952).
53. V. K. Zworykin, G. A. Morton, E. G. Ramberg, J. Hillier, and A. W. Vance, *Electron Optics and the Electron Microscope*, Wiley, New York (1945).
54. H. Boersch, Experimentelle Bestimmung der Energieverteilung in thermisch ausgelosten Electronenstrahlen, *Z. Phys.* **139**, 115–146 (154).
55. K. H. Loeffler, Energy-spread generation in electron-optical instruments, *Z. Angew. Phys.* **27**(3), 145 (1969).
56. K. H. Loeffler and R. M. Hudgin, in: *7th Proceedings of the International Congress on Electron Microscopy*, p. 67, Grenoble, France (1970).
57. H. C. Pfeiffer, in: *IEEE 11the Symposium on Electron, Ion, and Laser Beam Technology*, p. 239, San Francisco Press, San Francisco (1971).
58. R. Lauer, Ein einfaches Modell fur Elektronenkanonen mit gekrummter Kathodenoberflache, *Z. Naturforsch.* **23a**(2), 100–109 (January, 1968).
59. A. D. Brodie and W. C. Nixon, An electron optical line source for microelectronic engineering, *Microelectron. Eng.* **6**, 111–116 (1987).
60. T. E. Everhart, Simplified analysis of point-cathode electron sources, *J. Appl. Phys.* **38**, 4944 (1967).
61. A. V. Crewe, J. Wall, and L. M. Welter, A high resolution scanning transmission electron microscope, *J. Appl. Phys.* **39**(13), 5861 (1968).
62. E. M. Yakushev and L. M. Sekunova, Theory of electron mirrors and lenses, *Adv. Electron. Electron Phys.* **68**, 337–416 (1986).
63. L. A. Harris, in: *Electron Beam Technology* (R. Bakish, ed.), Wiley, New York (1962).
64. N. D. Wittels, Unipotential lens with electron-transparent electrodes, *J. Vac. Sci. Technol.* **12**(6), 1165–1168 (November–December, 1976).
65. A. J. F. Metherell, in: *Advances in Optical and Electron Microscopy* (R. Barer and V. E. Cosslett, eds.), Vol. 4, Academic Press, New York (1971).
66. J. C. Tracy, in: *Electron Emission Spectroscopy* (W. Dekeyser, L. Fiermans, G. Vanderkelen, and J. Vennick, eds.), p. 331, Reidel, Dordrecht (1973).
67. X. Jiye, Aberration theory in electron & ion optics, *Adv. Electron. Electron Phys.* Suppl. 17 (1986).
68. M. Szilagyi, *Electron and Ion Optics*, Plenum Press, New York (1988).
69. R. G. E. Hutter, in: *Advances in Image Pickup and Display*, Vol. I, pp. 163–224, Academic Press, New York (1974).
70. C. C. T. Wang, Analysis of electrostatic small-angle deflection, *IEEE Trans. Electron Devices* **ED-18**(4), 258 (April, 1971).
71. L. N. Heynick, High-information-density storage surfaces, Research and Development Technical Report No. ECOM-01261-F, Stanford Research Institute, Menlo Park, Calif. (January, 1970).
72. J. Kelly, Recent advances in electron beam memories, *Adv. Electron. Electron Phys.* **43**, 43–135 (1977).
73. C. C. T. Wang, Computer calculations of deflection aberrations in electron beams, *IEEE Trans. Electron Devices* **ED-14**(7), 357 (July, 1967).
74. C. C. T. Wang, Two-dimensional small-angle deflection theory, *IEEE Trans. Electron Devices* **ED-15**(8), 603 (August, 1968).
75. B. Lencová and M. Lenc, in: *Scanning Electron Microscopy*, Vol. III, pp. 897–915, AMF O'Hare, Chicago (1986).
76. L. A. Baranova and S. Y. Yavor, The optics of round and multipole electrostatic lenses, *Adv. Electron. Electron Phys.* **76**, 3–207 (1989).
77. B. Lencová and M. Lenc, Paraxial trajectory computation in magnetic electron lenses, *Optik* **82**(2), 68–72 (1989).

78. M. Gesley, Thermal-field-emission electron optics for nanolithography, *J. Appl. Phys.* **65**(3), 914 (February, 1989).
79. B. Lencová, On the design of electron beam deflection systems, *Optik* **79**(1), 1–12 (1988).
80. D. Ioanoviciu, Ion optics, *Adv. Electron. Electron Phys.* **73**, 1–92 (1989).
81. O. Bostanjoglo, Electron microscopy of fast processes, *Adv. Electron. Electron Phys.* **76**, 209–279 (1989).
82. K. Ura and H. Fujioka, Electron beam testing, *Adv. Electron. Electron Phys.* **73**, 234–317 (1989).
83. C. B. Duke and R. L. Park, Surface structure—An emerging spectroscopy, *Phys. Today* **25**(8), 23–28 (August, 1972).
84. P. F. Kane and G. B. Larrabee (eds.), *Characterizations of Solid Surfaces*, Plenum Press, New York (1974).
85. E. W. Müller and T. T. Tsong, *Field Ion Microscopy*, American Elsevier, New York (1969).
86. V. E. Cosslett and R. N. Thomas, Multiple scattering of 5–30 keV electrons in evaporated metal films, I: Total transmission and angular distribution, *Br. J. Appl. Phys.* **15**, 883 (1964).
87. J. P. Langmore, J. Wall, and M. S. Isaacson, Collection of scattered electrons in dark field electron microscopy, I: Elastic scattering, *Optik* **38**(4), 335–350 (September, 1973).
88. R. W. Nosker, Scattering of highly focused kilovolt electron beams by solids, *J. Appl. Phys.* **40**, 1872–1882 (March, 1969).
89. M. Isaacson, All you might want to know about ELS (but were afraid to ask): A tutorial, *Scanning Electron Microsc.* **1978**(I), 763–776 (1978).
90. F. W. Inman and J. J. Muray, Transition radiation from relativistic electrons crossing dielectric boundaries, *Phys. Rev.* **142**(1), 272 (February, 1966).
91. H. A. Bethe, M. E. Rose, and L. P. Smith, Multiple scattering of electrons, *Am. Philos. Soc. Proc.* **78**(4), 573–585 (1938).
92. H. Raether, Electron excitations in solids, *Springer Tracts Mod. Phys.* **38**, 85 (1965).
93. T. E. Everhart and P. H. Hoff, Determination of kilovolt electron energy dissipation vs. penetration distance in solid materials, *J. Appl. Phys.* **42**(13), 5837–5846 (December, 1971).
94. L. Reimer, Electron specimen interactions, *Scanning Electron Microsc.* **1979**(II), 111–124 (1979).
95. D. B. Brown and R. E. Ogilvie, An evaluation of the Archard electron diffusion model, *J. Appl. Phys.* **35**(10), 2793–2795 (1964).
96. T. E. Everhart, Simple theory concerning the reflection of electrons from solids, *J. Appl. Phys.* **31**(8), 1483 (August, 1960).
97. H. Kanter, Zur Ruckstreuung von Elektronen im Energiebereich van 10 bis 100 keV, *Ann. Phys. (Leipzig)* **20**, 144–166 (1957).
98. L. Reimer, W. Poepper, and W. Broeker, Experiments with a small solid angle detector for BSE, *Scanning Electron Microsc.* **1978**(I), 705–710 (1978).
99. H. Seiler, Determination of the information depth in the SEM, *Scanning Electron Microsc.* **1976**(I), 9–16 (1976).
100. S. A. Blankenburg, J. K. Cobb, and J. J. Muray, Efficiency of secondary electron emission monitors for 70 MeV electrons, *Nucl. Instrum. Methods* **39**, 303–308 (1966).
101. O. Hachenberg and W. Brauer, Secondary electron emission from solids, *Adv. Electron. Electron Phys.* **11**, 413 (1959).
102. H. Seiler, Einige aktuelle Probleme der Sekundarelektronenemission, *Z. Angew. Phys.* **22**, 249 (1967).
103. W. Heitler, *The Quantum Theory of Radiation*, Oxford University Press, London (1954).
104. R. D. Evans, *The Atomic Nucleus*, McGraw–Hill, New York (1955).
105. J. T. Grant, T. W. Haas, and J. E. Houston, Quantitative comparison of Ti and TiO surfaces using Auger electron and soft x-ray appearance potential spectroscopies, *J. Vac. Sci. Technol.* **11**(1), 227–230 (January–February, 1974).
106. S. J. B. Reed, *Electron Microprobe Analysis*, Cambridge University Press, London (1975).
107. C. R. Worthington and S. G. Tomlin, The intensity of emission of characteristic X-ray radiation, *Proc. Phys. Soc. London* **69A**(5), 401 (1956).
108. P. F. Kane and G. B. Larrabee (eds.), *The Characteristics of Solid Surfaces*, Plenum Press, New York (1974).
109. C. K. Crawford, in: *Introduction to Electron Beam Technology* (R. Bakish, ed.), Wiley, New York (1962).
110. L. J. Balk, H. P. Feuerbaum, E. Kubalek, and E. Menzel, Quantitative voltage contrast at high frequencies in the SEM, *Scanning Electron Microsc.* **1976**, 615–624 (1976).

111. J. Lindhard and M. Scharff, Energy dissipation by ions in the keV region, *Phys. Rev.* **124**, 128 (October, 1961).

112. G. M. McCracken, The behavior of surfaces under ion bombardment, *Rep. Prog. Phys.* **38**(2), 241–327 (February, 1975).

113. G. Dearnaley, Ion bombardment and implantation, *Rep. Prog. Phys.* **32**(4), 405–492 (August, 1969).

114. R. J. MacDonald, The ejection of atomic particles from ion bombarded solids, *Adv. Phys.* **19**(80), 457–524 (July, 1970).

115. R. Boako and D. M. Newns, Theory of electric processes in atomic scattering from surfaces, *Rep. Prog. Phys.* **52**, 655–697 (1989).

116. W. K. Chu, J. W. Mayer, and M. A. Nicolet, *Backscattering Spectrometry*, Academic Press, New York (1978).

117. L. C. Feldman and J. W. Mayer, *Fundamentals of Surface and Thin Film Analysis*, North-Holland, Amsterdam (1986).

118. J. Lindhard, M. Scharff, and H. Schiott, Range concepts and heavy ion ranges, *K. Dan. Vidensk. Selsk. Mat. Fys. Medd.* **33**(14), 39 (1963).

119. G. Carter and J. S. Colligon, *Ion Bombardment of Solids*, Elsevier, Amsterdam (1968).

120. J. Lindhard and A. Winther, Stopping power of electron gas and equipartition rule, *K. Dan. Vidensk. Selsk. Mat. Fys. Medd.* **34**(4), 1 (1964).

121. O. B. Firsov, A qualitative interpretation of the mean electron excitation energy in atomic collisions, *J. Exp. Theor. Phys.* (USSR) **36**(5), 1517–1523 (May, 1959).

122. N. F. Mutt and H. S. W. Massey, *The Theory of Atomic Collisions*, Oxford University Press (Clarendon), London (1949).

123. J. Lindhard, Influence of crystal lattice on motion of energetic charged particles, *K. Dan. Vidensk. Selsk. Mat. Fys. Medd.* **34**(14), 64 (1965).

124. P. Sigmund, Theory of sputtering, I: Sputtering yield of amorphous and polycrystalline targets, *Phys. Rev.* **184**(2), 184 (August, 1969).

125. S. A. Schwarz and C. R. Helms, A statistical model of sputtering, *J. Appl. Phys.* **50**, 5492 (1979).

126. B. E. Fischer and R. Spohr, Production and use of nuclear tracks, imprinting structure in solids, *Rev. Mod. Phys.* **55**(4), 907 (October, 1983).

127. R. L. Fleischer, P. B. Price, and R. M. Walker, in: *Nuclear Tracks in Solids: Principles and Applications*, University of California Press, Berkeley (1975).

128. J. F. Ziegler, in: *Handbook of Stopping Cross-sections for Energetic Ions in All Elements*, Pergamon Press, New York (1980).

129. L. Reiner and G. Pfefferkorn, *Raster Electronen Microskopie*, Springer-Verlag, Berlin (1977).

130. E. Spiller and R. Feder, X-ray lithography, *Top. Appl. Phys.* **22**, 35 (1977).

131. M. Green, *X-ray Optics and X-ray Microanalysis*, Academic Press, New York (1963).

132. S. Harrel, X-ray lithography for VLSI production, *Microelectronic Manufacturing and Testing* (January, 1984).

133. H. Winick and A. Bienenstock, Synchrotron radiation research, *Annu. Rev. Nucl. Part. Sci.* **28**, 33 113 (1978).

134. P. Ellcaume, Special synchrotron radiation sources, Part I: Conventional insertion devices, *Synchrotron Radiation News* **1**(4) (1988); P. L. Csonka, *Part. Accel.* **7**, 21 (1977); Enhancement of Synchrotron Radiation by Beam Modulation, University of Oregon Preprint No. N.T. 059/75 (1975); Some coherence effects with wigglers and their applications, in: *Wiggler Magnets* (H. Winick and J. Knight, eds.), SSRL Report 77/05 (1977); R. O. Tatchyn and P. L. Csonka, Submillimeter period undulators: New horizons in insertion device magnet technology, Lecture delivered at the Adriatico Research Conference on Undulator Magnets for Synchrotron Radiation and Free Electron Lasers, Trieste, Italy (June, 1987); R. O. Tatchyn, P. L. Csonka, and T. Cremer, Note on the Possible Generation of 1–4 MeV Gamma Rays in the 10+ Kilowatt Range at PEP with a Novel Laminar/Superconducting Micropole Undulator Design, *Proceedings of the Workshop on PEP as a Synchrotron Radiation Source*, Stanford (October, 1987); R. Tatchyn and P. L. Csonka, Attainment of submillimeter periods and a 0.3-T peak field in a novel micropole undulator device, *Appl. Phys. Lett.* **50**(7), 377 (February, 1987).

135. R. Z. Bachrach, I. Lindau, V. Rehn, and J. Stohr, Report of the beam line III, Stanford Synchrotron Radiation Laboratory Report, SSRL 77/14 (1977).

136. M. L. Ter-Mikaelian, *High-Energy Electromagnetic Processes in Condensed Media*, Wiley–Interscience, New York (1972).

137. A. N. Chu, M. A. Piestrup, T. W. Barbee, Jr., R. H. Pantell, and F. R. Buskirk, Observation of soft x-ray transition radiation from medium energy electrons, *Rev. Sci. Instrum.* **51**(5), 597–601 (May, 1980).

138. M. J. Moran and P. J. Ebert, Transition radiation as a practical x-ray source, *Nucl. Instrum. Methods Phys. Res.* **A242**, 355 (1986).

139. R. W. P. McWhirter, in: *Plasma Diagnostic Techniques* (R. H. Huddlestone and S. L. Leonard, eds.), pp. 201–264, Academic Press, New York (1965).

140. D. C. Gates, in: *IEEE 1977 Int. Conf. Plasma Sci.*, Catalog No. 77CH1205-4 NPS, p. 221 (1977); D. C. Gates, Studies of a 60 kV Plasma Focus, *Second International Conference on Energy Storage, Compression and Switching*, Venice, Italy (December, 1978); R. A. Gutcheck and J. J. Muray, An intense plasma source for x-ray microscopy, *Proceeding of the Society of Photo-Optical Instrumentation Engineers* (1981).

141. R. L. Kelly and L. J. Palumbo, Atomic and Ionic Emission Lines Below 2000 Angstroms—Hydrogen through Krypton, NRL Report 7599, U.S. Govt. Printing Office, Washington, D.C. (1973); W. L. Wiese, M. W. Smith, and B. M. Glennon, Atomic Transition Probabilities: Vol. I, Hydrogen Through Neon, NSRDS-NBS4, U.S. Govt. Printing Office, Washington, D.C. (1966).

142. R. A. McCorkle and H. J. Vollmer, Physical properties of an electron beam-sliding spark device, *Rev. Sci. Instrum.* **77**, 1055–1063 (1977).

143. J. Shiloh, A. Fischer, and N. Rostoker, Z pinch of a gas jet, *Phys. Rev. Lett.* **40**, 515–518 (1978).

144. S. M. Mathews, R. Stringfield, I. Roth, R. Cooper, N. P. Economou, and D. C. Flanders, Pulsed plasma source for x-ray lithography, *Proc. SPIE* **275**, 52–54 (1981).

145. J. C. Riordan, J. S. Pearlman, M. Gersten, and J. E. Rauch, Sub-Kilovolt X-ray Emission from Imploding Wire Plasma, presented at the Topical Conference on Low Energy X-ray Diagnostics, Monterey, Calif. (June, 1981).

146. J. S. Pearlman and J. D. Riordan, X-ray Lithography Using a Pulsed Plasma Source, presented at the 16th Symposium on Electron, Ion and Photon Beam Technology, Dallas (May, 1981).

147. D. J. Johnson, Study of the x-ray production mechanism of a dense plasma focus, *J. Appl. Phys.* **45**, 1147–1153 (1974); H. L. L. van Paassen, R. H. Vandre, and R. S. White, X-ray spectra from dense plasma focus devices, *Phys. Fluids* **13**, 2606–2612 (1970); G. Herzinger, H. Krompholz, L. Micheal, and K. Schoenbach, Collimated soft x-rays from the plasma focus, *Phys. Lett.* **64A**, 390–392 (1978).

148. G. Decker and R. Wienecke, Plasma focus devices, *Physica* **82C**, 155–164 (1976); J. W. Mather, Investigation of the high-energy acceleration mode in the coaxial gun, *Phys. Fluids* (Suppl.) **7**, 528–540 (1964).

149. B. Küyel, Prospects of a plasma focus device as an intense x-ray source for fine line lithography, *Proc. SPIE* **275**, 44–51 (1981).

150. I. Toubhans, R. Fabbro, B. Faral, M. Chaker, and H. Pepin, X-ray lithography with laser-plasma sources, *Microelectron. Eng.* **6**, 281–286 (1987); E. Turcu, G. Davis, M. Gower, F. O'Neill, and M. Lawless, X-ray lithography using a KrF laser-plasma source at hv ≈ 1 keV, *Microelectron. Eng.* **6**, 287 (1987).

151. D. J. Nagel, Laser plasma sources for x-ray lithography, *Microcircuit Eng.* **85**(3), 557 (1985).

152. D. J. Nagel, R. R. Whitlock, J. R. Greig, R. E. Pechacek, and M. C. Peckerar, Laser-plasma source for pulsed X-ray lithography, *Proc. SPIE* III, **135**, 46 (1978).

153. J. D. Cuthbert, Optical projection printing, *Solid State Technol.* **20**(8), 59 (August, 1977).

154. W. W. Duley, *Laser Processing and Analysis of Materials*, Plenum Press, New York (1983).

155. W. E. Bushor, Industrial lasers—Versatile tools for microelectronics processing applications, *Mircoelectronic Manufacturing and Testing* (September, 1983).

156. C. K. Rhodes (ed.), *Excimer Lasers*, Vol. 30, Springer-Verlag, Berlin (1984).

157. D. C. Ferranti, The excimer laser: Characteristics and applications, *Microelectronic Manufacturing and Testing* (October, 1988); W. Shumay, Jr., New laser applications in microelectronics, *Advanced Materials and Processes* (October, 1986); L. Austin *et al.*, Industrial excimer lasers: Continuous operation and proven applications, *O-E/Laser SPIE* (January, 1988); R. M. Cleary *et al.*, Performance of an KrF excimer laser stepper, *Proc. SPIE* (March, 1988); B. Rückel, P. Lokai, H. Rosenkranz, B. Nikolaus, H. J. Kahlert, B. Burchardt, D. Basting, and W. Mückenheim, Wavelength stabilized and computer controlled 248.4 nm excimer laser for microlithography, *Proc. SPIE* (March, 1988).

158. L. Austin, High resolution lithography with excimer lasers, *Microelectronic Manufacturing and Testing* (April, 1989); T. A. Znotins, Excimer lasers: An emerging technology in semiconductor processing, *Solid State Technol.* **29**(9) (September, 1986); B. J. Garrison and R. Srinivasan, Laser ablation of organic polymers: Microscopic models for photochemical and thermal processes, *J. Appl. Phys.* **57**(8), 2909 (April, 1985); D. J. Erlich, Early applications of laser direct patterning: Direct writing and excimer projection, *Solid State Technol.* **28**(12) (December, 1985); D. J. Elliott, B. P. Piwczyk, and K. J. Polasko, VLSI Pattern Registration Improvement by Photoablation of Resist-Covered Alignment Targets, 1987 IEEE Triple Beams Conference, Woodland Hills, Calif. (1987).

159. D. A. G. Deacon, L. R. Elias, J. M. J. Madey, G. J. Ramian, H. A. Schwettman, and T. I. Smith, First operation of a free-electron laser, *Phys. Rev. Lett.* **38**(16), 892 (April, 1977); L. R. Elias, W. M. Fairbank, J. M. Madey, H. A. Schwettman, and T. I. Smith, Observation of stimulated emission of radiation by relativistic electrons in a spatially periodic transverse magnetic field, *Phys. Rev. Lett.* **36**(13), 717 (March, 1976).

160. I. P. Herman, R. A. Hyde, B. M. McWilliams, A. H. Weisberg, and L. L. Wood, Wafer-scale laser pantography: I. Pyrolytic deposition of metal microstructures, UCRL 8831, Lawrence Livermore National Laboratory (November, 1982); B. M. McWilliams, I. P. Herman, R. A. Hyde, F. Mitlitsky, and L. L. Wood, Wafer-scale laser pantography: II. Laser-induced pyrolytic creation of MOS structures, UCRL 8853, Lawrence Livermore National Laboratory (May, 1983); B. M. McWilliams, I. P. Herman, F. Mitlitsky, R. A. Hyde, and L. L. Wood, Wafer-scale laser pantography: III. Fabrication of n-MOS transistors and small-scale integrated circuits by direct-write laser induced pyrolytic reactions, UCRL 8839, Lawrence Livermore National Laboratory (June, 1983); I. P. Herman, B. M. McWilliams, F. Mitlitsky, H. W. Chin, R. A. Hyde, and L. L. Wood, Wafer-scale laser pantography: IV. Physics of direct laser-writing micron-dimension transistors, UCRL 89350, Lawrence Livermore National Laboratory (November, 1983); B. M. McWilliams, H. W. Chin, I. P. Herman, R. A. Hyde, F. Mitlitsky, J. C. Whitehead, and L. L. Wood, Wafer-scale laser pantography: VI. Direct-write interconnection of VLSI gate arrays, UCRL 90295, Lawrence Livermore National Laboratory (January, 1984); I. I. Burns and A. R. Elsea, Laser processing of semiconductors—A production machine, *SPIE* **945**, 97 (1988).

161. M. M. Oprysko, M. W. Beranek, D. E. Ewbank, and A. C. Titus, Repair of clear photomask defects by laser-pyrolytic deposition, *Semicond. Int.* **9**(1), 90–100 (1986).

162. T. J. Chuang, Laser-enhanced gas-surface chemistry: Basic processes and applications, *J. Vac. Sci. Technol.* **21**(3), 798–806 (1982).

163. K. A. Jones, Laser assisted MOCVD growth, *Solid State Technol.* **28**(10), 151–156 (1985).

164. T. J. Chuang, Multiple photon excited SF_6 interaction with silicon surfaces, *J. Chem. Phys.* **74**(2), 1453–1460 (1981).

165. D. J. Ehrlich, R. M. Osgood, Jr., and T. J. Deutsch, Laser chemical technique for rapid writing of surface relief in silicon, *Appl. Phys. Lett.* **38**(12), 1018–1020 (1981).

166. D. J. Ehrlich and J. Y. Tsao, in: *Laser Diagnostics and Photochemical Processing for Semiconductor Devices* (R. M. Osgood, S. R. J. Brueck, and H. R. Schlossberg, eds.), North-Holland/Elsevier, Amsterdam (1983).

167. S. S. Cohen and G. H. Chapman, in: *VLSI Electronics Microstructure Science*, Vol. 21, Academic Press, New York (1989); W. W. Duley, *Laser Processing and Analysis of Materials*, Plenum Press, New York (1983).

168. I. P. Herman, R. A. Hyde, B. M. McWilliams, A. H. Weisberg, and L. L. Wood, in: *Laser Diagnostics and Photochemical Processing for Semiconductor Devices* (R. M. Osgood, S. R. J. Brueck, and H. R. Schlossberg, eds.), North-Holland/Elsevier, Amsterdam (1983); D. Bauerle, in: *Laser Diagnostics and Photochemical Processing for Semiconductor Devices* (R. M. Osgood, S. R. J. Brueck, and H. R. Schlossberg, eds.), North-Holland/Elsevier, Amsterdam (1983); D. J. Ehrlich and J. Y. Tsao, in: *Laser Diagnostics and Photochemical Processing for Semiconductor Devices* (R. M. Osgood, S. R. J. Brueck, and H. R. Schlossberg, eds.), North-Holland/Elsevier, Amsterdam (1983).

3

Plasmas

Physics and Chemistry

3.1. GENERAL BACKGROUND

3.1.1. Introduction

Plasmas are used extensively in microfabrication processes, including

- Deposition of thin films by sputtering, ion plating, plasma-enhanced chemical vapor deposition, and plasma polymerization
- Removal of surface layers and films by sputter etching, reactive ion etching, and resist stripping
- Sources of ions for ion implantation, and focused ion beam patterning

The purpose of this chapter is to provide a background for an understanding of the physics involved in these processes.

3.1.2. Plasmas

A plasma may be considered as a region of a gas discharge which contains essentially equal quantities of positive and negative charge associated with various species of charge carrier. The types of plasma used in microfabrication processes are only partially ionized and most of the plasma consists of neutral molecules. The positive charge carriers are generally molecules or molecular fragments which have lost electrons and the negative charge carriers are electrons or molecules to which are attached one or more electrons. The energies of the charge carriers and neutrals are randomized by various collision processes and interactions. For a plasma to remain stable, the charge carriers lost from the plasma by recombination and diffusion must be replaced by some charge-generating mechanism. The resulting dynamic equilibrium generally gives rise to different energy distributions for the various plasma components, as well as the continuous emission of photons. It is often a useful

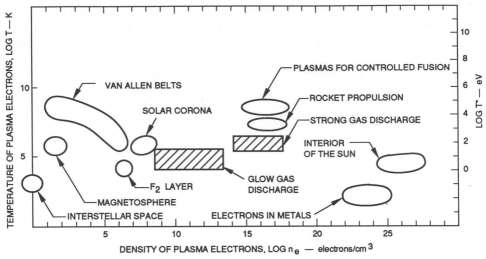

FIGURE 3.1.   Characteristics of various plasmas.

fiction to assume that the kinetic energies have a Maxwell–Boltzmann distribution corresponding to an equivalent temperature.

Plasmas may be characterized by the number density of the electrons, ions, and neutrals they contain (n_e, n_i, and n) and their corresponding temperatures (T_e, T_i, and T_n). Figure 3.1 gives the regions in the n_e versus T_e plot corresponding to various natural and laboratory plasmas.[1] Note that T_e is often expressed in terms of a mean electron energy T^* given in electron volts ($T^* \sim kT_e/e \sim T_e/10^4$). This enables a more physical "feel" for the energies involved to be obtained. We will be concerned mainly with glow discharges which occur at gas pressures of a few mTorr, involving T^*'s up to a few electron volts, and n_e's between 10^8 and 10^{14} electrons per cm^3.

When an energetic electron collides with a large molecule, many particle species can be formed including electrons, photons, molecular ions, multiply charged ions, and excited metastables.[2] Atoms or molecules are said to be metastable if they are excited to an electronic state for which direct transition to lower energy states is forbidden. Molecules and molecular ions may be excited into rotational and vibrational as well as electronic states. To simplify the discussion, we shall usually assume a simple monoatomic gas such as helium or argon.

3.1.3. Voltage Breakdown

In its simplest form, a gas discharge tube consists of two plane-parallel electrodes of area A and separation d, filled with a gas at pressure p. Initially, as the potential difference between the electrodes V is increased, only very small currents, usually in bursts, are observed. These are due to ionization of the gas by cosmic rays, the emanations from local radioactive materials or ambient ultraviolet light. However, when the voltage reaches a critical value V_s, a breakdown occurs. Breakdowns may be self-extinguishing or self-sustaining. The random nature of the breakdown-initiating phenomena gives rise to an uncertainty in the value of V_s and a variable time lag between the application of V_s and the appearance of the initiating electrons, called the statistical time lag. The value of V_s has been found to depend on pd and the particular gas being tested, as shown in Fig. 3.2. This is called a Paschen curve.[3] For each gas there is a minimum value of V_s that is usually in the range of 200 to 500 V. The Paschen curve also depends to some extent on the electrode configuration, the electrode materials and surface conditions, and the purity of the gas. For breakdown to occur, there must be some electrons present, and the energy gathered by these electrons when they collide with a gas molecule must be great enough to produce an electron–ion pair for avalanching to take place. To maintain the discharge, a continuous source of electrons is required. These usually arise from the residual electrons generated by the avalanching process mentioned above, but can also arise from photons and positive ions from the plasma striking the cathode surface which then emits electrons by secondary processes. The photons are generated by relaxation of the excited molecules formed by electron collision as well as by recombination of positive ions with electrons. Controlled electron-emitting cathode surfaces can also be provided by using thermionic or photoelectric emission thus eliminating the statistical time lag and the necessity for secondary processes to maintain the discharge. Field electron emission from whiskers on the cathode surface occurs when V/d is greater than about 10^5 V/cm. This source of electrons initiates the breakdown when there are very few gas molecules between the electrodes (i.e., at low pd) so that charges generated in the gas from cosmic rays and other sources are not necessary.

If an alternating voltage[4] is applied across the electrodes, breakdown becomes much more controlled since the field rapidly decreases immediately after breakdown occurs. If, after the first breakdown cycle, not all of the electrons and ions have had

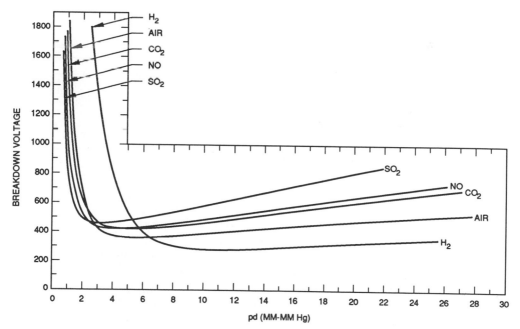

FIGURE 3.2. Paschen curves for various gases.

sufficient time to recombine or diffuse away from the field region, they are immediately available to initiate current flow in the following half-cycle. Obviously the higher the frequency, the more marked is the effect due to the shorter time available for loss of electrons. Use of an alternating electric field also enables the electrodes to be placed external to the chamber containing the gas (provided the chamber walls are insulating). In this case the field may be coupled capacitatively by plate electrodes, or inductively by a coil around the chamber.

3.1.4. Gas Discharges

After voltage breakdown (or sparking) has occurred, several types of continuous discharge can develop depending on the external circuit and the physical condition inside the discharge tube. These discharges are categorized[5] as follows:

- *The Townsend discharge*—These are discharges in uniform electric fields characterized by very small current densities (10^{-15}–10^6 A/cm^2) in which the effects of space charge are not dominant. The discharge is usually maintained by photoelectric or thermionic electron emission from the cathode.
- *The corona discharge*—Coronas are essentially Townsend discharges in inhomogeneous electric fields such as obtained from coaxial cylinders or point-to-plane electrode geometries. Space-charge effects may be dominant close to the small electrode, giving rise to oscillating current flow even with dc voltages across the electrodes. The energy gathered per mean free path for charges originating at the small electrode and traveling to the large electrode is, within a short distance, reduced to a value insufficient for avalanching, and in addition the current density in the discharge becomes progressively smaller. This type of discharge has found practical application for the charging of electrophotographic (xerographic) photoreceptors, and in Geiger counters. The physics of corona discharges is quite complicated and has been treated exhaustively by Loeb.[6]
- *The glow discharge*—These are brightly glowing discharges occurring at pressures in the range of 10^{-3} to 1 torr with current densities in the range of 10^{-4} to 10^{-1} A/cm^2 where space-charge effects are important. They are characterized by a brightly glowing positive column, within which the electric field is essentially zero, with dark spaces separating it from the cathode and anode. The positive column is a plasma sustained in dynamic equilibrium mainly by energetic electrons injected from the cathode end. These are the type of plasma used for both deposition and etching processes in microfabrication.
- *The arc discharge*—Arc discharges are characterized by very high current densities (> 10 A/cm^2), low applied voltages and the discharge current is emitted only from one or a few small spots on the cathode. The high currents usually originate thermionically at the cathode. The emitting spots may be heated directly by the discharge current or indirectly from an external heater. Liquid metals such as mercury are also used as the cathode producing their high currents by complex mechanisms involving thermal-field emission from unstable peaks moving rapidly over the liquid surface, similar to the more controlled liquid metal sources discussed in Section 2.2.3.2.

Comprehensive descriptions of these and other classes of discharges are given in the now-classical literature.[3,5–8]

3.1.5. Collision Cross Section

The concept of the collision cross section[2] derives from the classic kinetic theory model of a gas as consisting of mobile noninteracting hard spheres. The probability, dp, that a test particle (such as an electron), when traveling through a gas containing a random dispersion of identical molecules of average number n per unit volume, will collide within a distance dx is proportional to n and dx. The constant of proportionality, σ, has the dimensions of area and is called the collision cross section. Thus,

$$\frac{dp}{dx} = \sigma n \tag{3.1}$$

From a quantum mechanical viewpoint, the probability of collision depends on the energy states of the two particles. A large number of collision processes can occur in a given plasma each with its own collision cross section. The most important are

- Elastic interactions involving only exchange of kinetic energy
- Excitation of electronic states: these states may be radiative or metastable
- Ionization or the removal of one or more electrons from the impacted particles
- Excitation of vibrational states of polyatomic molecules
- Excitation of rotational states of polyatomic molecules
- Dissociation of polyatomic molecules
- Superelastic collision with an excited metastable molecule
- Attachment of a charged particle to a neutral particle
- Recombination of oppositely charged particles
- Photoionization (between a photon and a neutral molecule)
- Charge transfer (between a charged particle and a neutral)

The cross sections for many specific processes have been measured, and computational methods have enabled many to be calculated based on appropriate atomic and molecular models.[2] The total collision cross section is the sum of the cross section for each of the processes that can take place, given the types and energies of the colliding particles. Thus, for an electron of kinetic energy ε, striking a molecule

$$\sigma(\varepsilon) = \sum_i \sigma_i(\varepsilon) \tag{3.2}$$

Figure 3.3 shows total ionization cross sections for the noble gases. When a collision takes place the incident particle will be scattered into a solid angle $d\theta = \sin\theta \, d\omega \, d\phi$ making an angle θ with its original trajectory with probability $p(\theta) \, d\omega$. The term $\sigma p(\theta) \, d\omega = \sigma p(\theta) \sin\theta \, d\theta \, d\phi$ is called the differential cross section.

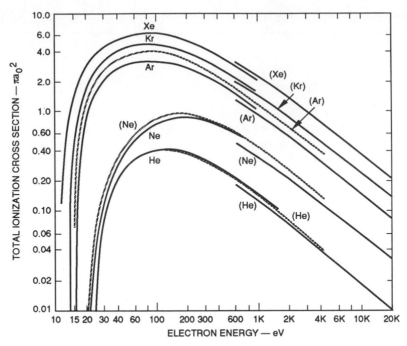

FIGURE 3.3.   Total ionization cross sections of the noble gases. $\pi a_0^2 = 8.82 \times 10^{-17}$ cm². From Ref. 4.

If the mean free path of the test particle traveling through the gas is λ, then the volume $\sigma\lambda$ must contain one molecule, i.e.,

$$\lambda\sigma = \frac{1}{n} \tag{3.3}$$

or

$$\lambda = \frac{1}{n\sigma} \tag{3.4}$$

Instead of the collision cross section, the collision probability is often used denoted by P_c. P_c is defined as the average number of collisions that occur when an electron travels a distance of 1 cm in a gas at a pressure of 1 Torr at 0°C, where the number of molecules per cm³, $n = 3.535 \times 10^{16}$ (Loschmidt number/atmospheric pressure in mm Hg), thus,

$$P_c = \frac{1}{\lambda} = \sigma n \tag{3.5}$$

or

$$\sigma = 0.283 \times 10^{-16} P_c \quad \text{cm}^2 \tag{3.6}$$

If $\bar{c}$ is the mean velocity of the test particle, then the mean free time between collisions is given by

$$\tau = \frac{\lambda}{\bar{c}} = \frac{1}{n\sigma\bar{c}} \qquad (3.7)$$

The number of collisions suffered in a unit time by the test particle, or the collision frequency is given by

$$\omega = \frac{1}{\tau} = n\sigma\bar{c} \qquad (3.8)$$

3.1.6. Energy States for Single and Polyatomic Molecules

The total energy of a molecule is given by the sum of the external and internal energies. The external (kinetic) energy is not quantized but all of the internal states are quantized. Electrons in the outer (valence) shell of the atom or molecule may be excited to various energy levels from which they rapidly ($\sim 10^{-9}$ s) make a transition to a lower level, usually the ground state, with the emission of a characteristic photon. Metastable states also exist in which the transition of the electron directly to a lower level is forbidden. A metastable molecule stores its internal energy but this energy is available for release when it collides with another particle. For example, when a metastable collides with a neutral molecule, it may enable the neutral to dissociate into its component parts. Valence electronic transitions usually give rise to photons of a few electron volts of energy.

3.2. COLLISION PROCESSES

Collisions involving electrons dominate the macroscopic behavior of the gas discharge. This subject has been reviewed extensively and additional material can be obtained in Refs. 2, 4, 9–18.

3.2.1. Elastic Collisions

In elastic collision processes, only the kinetic energies of the colliding particles change; the internal or potential energies of the atom or molecule remain unchanged. Consider the collision between two particles of mass m_i and m_t. Assume that m_t is initially stationary and that m_i collides with velocity v_i at an angle θ to the line joining the centers of m_i and m_t at the moment of collision (see Fig. 3.4a). By conservation of linear momentum

$$m_i v_i \cos \theta = m_i u_i + m_t u_t \qquad (3.9)$$

By conservation of energy

$$1/2 m_i v_i^2 = 1/2 m_i (u_i^2 + v_i^2 \sin^2 \theta) 1/2 m_t u_t^2 \qquad (3.10)$$

Eliminating u_i in Eq. (3.10) from (3.9) gives

$$m_i v_i^2 \cos^2 \theta = \frac{m_i}{m_t^2} (m_i v_i \cos \theta - m_t u_t)^2 + m_t u_t^2 \qquad (3.11)$$

The fractional energy transferred from mass m_i to mass m_t or the *energy transfer function* (ETF) is therefore given by

$$\text{ETF} = \frac{E_t}{E_i} = \frac{1/2 m_t u_t^2}{1/2 m_i v_i^2} = \frac{m_t}{m_i v_i^2} \left[\frac{2 m_i v_i}{m_t + m_i} \cos \theta \right]^2 \qquad (3.12)$$

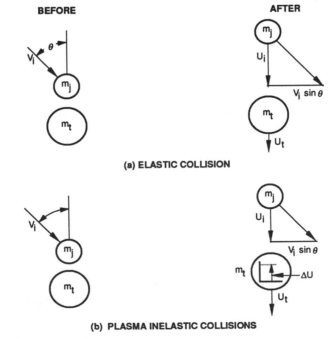

(a) ELASTIC COLLISION

(b) PLASMA INELASTIC COLLISIONS

FIGURE 3.4.   Plasma collisions. (a) Elastic collisions; (b) Inelastic collisions.

i.e.,

$$ETF = \frac{4m_i m_t}{(m_i m_t)^2} \cos^2 \theta \tag{3.13}$$

and this has a maximum of $\cos^2 \theta$, when $m_i = m_t$.
In the case of collisions where $\theta = 0$

$$ETF = 1 \qquad \text{when } m_i = m_t$$

$$ETF = \frac{4m_i}{m_t} \qquad \text{when } m_i \ll m_t \tag{3.14}$$

$$ETF = \frac{4m_t}{m_i} \qquad \text{when } m_i \gg m_t$$

For plasma interactions, this implies that there is a very efficient transfer in collisions between atoms and ions of the same gas, but little energy ($10^{-4} E_0$) can be transferred by elastic collision from the electron to the gas atom (ion) [Eq. (3.14)]. This inefficient energy coupling implies that in a glow discharge, most of the external electrical energy supplies heat to the electrons while the gas molecules and ions remain cool.

3.2.2. Inelastic Collisions

In inelastic collision processes, the internal energy of the molecules and ions ΔU can change in addition to their kinetic energies. Thus, the momentum and energy conservation equations can be written as (see Fig. 3.4b)

$$m_i v_i \cos \theta = m_i u_i + m_t u_t \tag{3.15}$$

$$1/2 m_i v_i^2 = 1/2 m_i (u_i^2 + v_i^2 \sin^2 \theta) + 1/2 m_t u_t^2 + \Delta U \tag{3.16}$$

Eliminating u_i as before gives

$$m_i v_i^2 = \frac{m_i}{m_t^2} (m_i v_i \cos \theta - m_t u_t)^2 + m_t u_t^2 + 2\Delta U \tag{3.17}$$

which simplifies to

$$2 m_t u_t v_i \cos \theta = \frac{m_t}{m_i} (m_t + m_u) u_t^2 + 2\Delta U \tag{3.18}$$

where ΔU and u_t are the only variables. To maximize ΔU,

$$2\frac{d}{du_t}(\Delta U) = 2m_t v_i \cos\theta - \frac{m_t}{m_i}(m_t + m_i)2u_t = 0 \tag{3.19}$$

Substituting into Eq. (3.18) we obtain

$$2\Delta U = \left[\frac{m_t m_i}{m_t m_i}\right]v_i^2 \cos^2\theta \tag{3.20}$$

Hence, the fraction of the kinetic energy of the first particle that can be transferred to the internal energy of the second has a maximum value

$$\text{ETF} = \frac{\Delta U}{1/2m_i v_i^2} = \left[\frac{m_t}{m_t + m_i}\right]\cos^2\theta \tag{3.21}$$

so the inelastic energy transfer function tends to 1 when $m_t \gg m_i$. From Eq. (3.18) it is seen that the fractional kinetic energy transfer is

$$\text{ETF} = \frac{1/2m_t u_t^2}{1/2m_i v_i^2} = \frac{m_t u_t^2}{m_i}\left[\frac{2m_t u_t}{\dfrac{m_t}{m_i}(m_t + m_i)u_t^2 + 2\Delta U}\right]^2 \tag{3.22}$$

Excluding the cases when ΔU can be negative, this function has a maximum value when the denominator has a minimum value given by $\Delta U = 0$; then the fractional energy transfer reverts to the $4m_i m_t/(m_i + m_t)^2$ value, as expected. Accordingly, even though a large amount of internal energy can be transferred when $m_i \ll m_t$, it is still impossible to transfer any significant amount in the form of kinetic energy.

3.2.3. Recombination

Recombination is the inverse process to ionization in which an electron combines with a positive ion to form a neutral atom. However, it has been shown that a simple two-body recombination ($e^- + A^+ \rightarrow A$) is unlikely because the momentum and energy conservation laws cannot be satisfied simultaneously. Thus, the dominant recombination process is that which occurs when three particles encounter each other simultaneously. This is called a three-body process, e.g.,

$$e^- + A^+ + M \rightarrow A + M \tag{3.23}$$

If there are n_e electrons and n_+ ions per unit volume randomly distributed, and there is no source of ionization, then the number of ion pairs lost by neutralizing encounters dn in time dt is proportional to n_e, n_+, and dt, i.e.,

$$dn = \alpha n_+ n_e\, dt \tag{3.24}$$

where the constant of proportionality α is termed the *recombination coefficient*. In a plasma $n_+ = n_e = n$, thus

$$\frac{dn}{dt} = -\alpha n^2 \tag{3.25}$$

Integrating gives the number of electrons/ions per unit volume remaining at time t as

$$n = \frac{n_0}{1 + n_0 \alpha t} \tag{3.26}$$

where n_0 is the number of electrons/ions per unit volume at $t = 0$. The recombination coefficient is related to the recombination cross section $\sigma_R(v)$ by the relation

$$\alpha = \int_0^D v \sigma_R(v) f(v)\, dv \tag{3.27}$$

where $f(v)\, dv$ is the number of encounters in which the relative velocity between the electron and ion lies between v and $v + dv$. For most purposes

$$\alpha \approx \bar{v} \sigma_R(\bar{v}) \tag{3.28}$$

At atmospheric pressure, the total value of α, due to all processes, is in the range 10^{-10} to 10^{-8} cm^3/sec.

3.2.4. Attachment and Dissociation

Attachment of an electron to an atom or molecule can occur when the outer electronic shells are nearly filled. The electron affinity is defined as the difference between the potential energy of the normal atom or molecule and that of the ion formed by electron attachment. The electron affinity varies from a maximum of about 4 V for gases like O_2 and F, for which strong electron attachment, to negative values for gases like H_2 and CO, which exhibit no electron attachment, is highly unfavorable. If, for n electrons per unit volume, dn attachments occur in time dt, then

$$dn = -hn\nu\, dt \tag{3.29}$$

where ν is the collision frequency and h is a constant of proportionality termed the attachment probability. Integrating gives the number of electrons per unit volume remaining at time t as $n = n_0\, e^{-hvt}$ where n_0 is the number of electrons per unit volume at $t = 0$.

The attachment probability is related to the attachment cross section σ_a by

$$\sigma_a \lambda_a = \frac{1}{n} \tag{3.30}$$

where λ_a is the mean distance traveled by an electron before attachment, but

$$\lambda_a = \frac{\lambda}{h} \tag{3.31}$$

hence

$$\sigma_a = \frac{h}{n\lambda} \tag{3.32}$$

Dissociation is the breaking apart of a molecule. In glow discharges, electron impact dissociation may be used to generate chemically active species, e.g.,

$$e^- + O_2 \rightarrow O + O + e^- \tag{3.33}$$

$$e^- + CF_4 \rightarrow CF_3 + F + e^- \tag{3.34}$$

O and F are more reactive than O_2 and CF_4. Dissociation may also be acompanied by ionization:

$$e^- + CF_4 \rightarrow CF_3^+ + F + 2e^- \tag{3.35}$$

A molecule may dissociate into a number of possible fragments charged and uncharged, depending on its original atomic components and the energy of the impinging electron. For complex organic molecules, the charged fragments form a "cracking pattern" of charged molecular ions whose distribution may be measured in a mass spectrometer.

3.2.5. Secondary Processes

An electrode associated with a plasma may be bombarded by primary elecrons, ions, photons, and neutral atoms and molecules originating in the plasma. The interaction of these particles with the electrode may produce secondary electrons, ions, photons, and neutral species originating at the electrode surface, the yields depending on the energy of the incident particles and the electrode surface material. Electron interactions are discussed in detail in Section 2.6, ion interactions in Section 2.7, and photon interactions in Section 2.9.

For ion-assisted processing, the most important secondaries from the target area are:

- Target atoms (or ions) for sputter deposition or etching
- The "reflected ions" (mostly neutralized) for sputter deposition
- Secondary electrons. These sustain the discharge by creating electron (ion pairs in the glow, and additionally can break up or ionize component molecules of a gas mixture for chemical processing.

3.2.6. Cluster Ion Formation

If polar or polarizable molecules are present in the plasma, then it becomes possible to form charged species that consist of an ionized atom or molecule to which is attached a number of neutral molecules held by forces due to the interaction of the field of the ion with the dipole moment of the neutrals.[19] The binding energy is usually less than that for chemical bonds, but if it is greater than the random kinetic energy, then the cluster can be stable. The dipole moment of a molecule in an electric field, E, is given by

$$\mu = \mu_0 + \alpha E \qquad (3.36)$$

where μ_0 is the permanent dipole moment and αE is the induced dipole moment where α is termed the polarizability. The field due to the dipole acting on the ion is proportional to μ/r^3, and that due to the ion acting on the dipole is proportional to e/r^2. Thus, if μ_0 is negligible, the force holding the two particles together

$$F \propto \frac{1}{r^5} \qquad (3.37)$$

but if μ_0 is large

$$F \propto \frac{1}{r^3} \qquad (3.38)$$

To attach the first molecule to a monatomic ion, a three-body collision is required to take up the excess energy. However, for more attachments the excess energy can be taken up in vibrational or rotational states of the cluster, and this combined with the larger cross section for collisions means that a complete monolayer cluster, with about six attached neutrals, is likely to be produced once the first neutral has attached. Because of the $1/r^5$ dependence of the binding force, second layers are unlikely to form with polarizable molecules. Large organic molecules often have large permanent dipole moments but this tends to make r large for the first monolayer so that even with a $1/r^3$ dependency for the field a second layer is unlikely to form. Cluster irons of this type would appear to be important for plasmas containing organic reagents, but no definitive studies have been reported.

Cluster ions have also been generated in supersaturated vapors[20,21] (see Section 2.3.7). In this case the supersaturated gas condenses into tiny droplets that nucleate on an ion. By evaporating a vapor through a small nozzle into a high vacuum, adabiatic cooling reduces the vapor temperature below saturation. At high nozzle speeds a sufficient number of atoms are ionized by frictional contact to nucleate cluster ions close to the nozzle. Subsequent to formation, the clusters do not grow since they have been injected into a high vacuum. The velocity of the clusters is the gas nozzle velocity which can be substantially greater than the corresponding thermal velocity and tends to an upper limit given by

$$v_\infty = \left[\frac{2\gamma}{\gamma - 1} \cdot \frac{kT_0}{m} \right]^{1/2} \tag{3.39}$$

where T_0 is the source temperature, γ is the ratio of the specific heats, and m is the mass of the cluster.

3.3. ELECTRON MOTION IN A GAS DISCHARGE

3.3.1. Electron Motion in Electric and Magnetic Fields

Between collisions, an electron moves under the action of electric (E) and magnetic (B) fields in trajectories given by solution of the Lorentz equation

$$- \mathbf{F} = m\ddot{\mathbf{r}} = e\mathbf{E} + e\mathbf{B} \times \dot{\mathbf{r}} \tag{3.40}$$

where r is the vector coordinate of the electron. In a gas discharge, four cases can be distinguished:

3.3.1.1. No Fields Present, i.e., E = B = 0. This is the condition inside a simple plasma. In this case the energy (velocity) distribution of the electrons becomes randomized by collisions. For a plasma to remain stable, charge pairs must be generated at the same rate at which they are being lost by recombination and diffusion so that that a state of dynamic equilibrium exists. If the forward and reverse rates for all electron energy exchange processes are equal, then the principle of detailed balancing[22,23] leads to a Maxwell–Boltzmann velocity distribution corresponding to an electron temperature T_e. However, such a state rarely exists in a glow discharge. The deviation from the Maxwell distribution and its importance is discussed in Section 3.3.2.

3.3.1.2. Magnetic Fields Only, i.e., E = 0. In this case the solution to equation (3.40) shows that the electrons gain no energy from the magnetic field but move in spiral paths around the magnetic field line. Between collisions, the motion in the plane perpendicular to the field line is a circle of radius

$$r_g = \frac{mv_1}{eB} \tag{3.41}$$

where v_1 is the random component of the electron velocity perpendicular to the field line and the motion along the field line is simply v_2, the random component of velocity parallel to the field line. If $\varepsilon_1 = 1/2mv_1^2$ is expressed in electron volts and B in gauss, then

$$r_{\rm g} = \frac{3.37}{B}(\varepsilon_1)^{1/2} \text{ cm} \tag{3.42}$$

Typically, $r_{\rm g}$ for electrons is small compared with the dimensions of the volume occupied by the plasma. On the other hand, the effect of a magnetic field on the motion of ions ($m_{\rm i} \gg m$) is negligible, because $r_{\rm g}$ for ions is usually much larger than the dimensions of the plasma. Because of the tight spiral path taken by the electrons in a magnetic field, they take longer to diffuse out of a plasma. This in turn leads to a higher electron density for the same rate of production of charge pairs. Since the ions are held by increased electrostatic forces and can only leave by ambipolar diffusion (equal quantities of positive and negative charge), a magnetic field leads to an increase in the plasma density.

The frequency of rotation, or cyclotron frequency for B in gauss is given by

$$\omega_{\rm c} = \frac{eB}{m} = 1.76 \times 10^7 B \quad \text{rad/s} \tag{3.43}$$

If S is the number of turns made between collisions, then

$$S = \frac{\lambda}{2\pi r_{\rm g}} = \frac{1}{2\pi r_{\rm g} n \sigma} \tag{3.44}$$

The energy exchange that takes place upon collision of an electron with a gas molecule alters the electron motion and may cause it to rotate about an adjacent magnetic field line. This allows collisional diffusion across magnetic field lines to take place.

Plasmas can also be confined within a "magnetic bottle." In this arrangement a magnetic field is increased rapidly at the boundary of the region within which it is desired to confine the plasma. Electrons moving in a circle form a magnetic dipole with moment $\mu_{\rm M}$. The force on a dipole due to a nonuniform magnetic field is given by

$$F_{\rm r} = \mu_{\rm M} \frac{dB_{\rm r}}{dr} \tag{3.45}$$

When the field is increasing, the force acts to repel the dipole and if $dB_{\rm r}/dr$ is large enough it can give rise to electron reflection.

3.3.1.3. Electric Fields Only, i.e., $B = 0$. In this case the electron gains on average an energy ε from the electric field between collisions of an amount

$$\varepsilon = Ee\lambda \tag{3.46}$$

where λ is the mean free path. The value of ε is crucial to the behavior of electron flow in a gas discharge.

Because the mass of the electron is small compared with the gas molecules, it loses, on elastic collision, only a small portion of the energy gained from the electric field. But to reach an equilibrium drift velocity requires eventual loss of essentially all of its energy by an inelastic collision. The electron gains energy from the field during a time in which a number of elastic collisions may have occurred. Until it has sufficient energy for an inelastic collision at which point its velocity in the field direction is reduced substantively. Thus, there is a net electron drift in a direction following electric field lines characterized by a drift velocity superimposed on a random velocity distribution function.

In the case of ions of mass comparable to that of the gas molecules, essentially all of the energy gathered from the electric field is dissipated on collision whether elastic or not. Equation (3.40) with $B = 0$ gives

$$m_i \ddot{x} = Ee \tag{3.47}$$

integrating twice and assuming $x = 0$ at $t = 0$, and $x = s$ at $t = \tau$:

$$s = \frac{Ee_i}{2m_i} \tau^2 \tag{3.48}$$

where τ is the time between collisions
   The drift velocity is given by

$$v = \frac{\bar{s}}{\bar{\tau}} = \frac{Ee_i}{2m_i} \frac{\bar{\tau}^2}{\bar{\tau}} = \frac{Ee_i}{m_i} \frac{\lambda_i}{\bar{c}_i} \tag{3.49}$$

where c_i is the mean ion thermal velocity. The constant μ by which the electric field is multiplied to obtain the drift velocity is called the mobility, and this simple theory gives for ions

$$\mu_i = \frac{e_i}{m_i} \frac{\lambda_i}{\bar{c}_i} \tag{3.50}$$

For electrons, the same simple approach gives

$$\mu_e = \frac{e}{m} \frac{n\lambda_e}{\bar{c}_e} \tag{3.51}$$

where n is the number of collisions before the electron has sufficient energy to have an inelastic collision.

In an alternating field, an electron oscillates with the equation of motion

$$m\ddot{x} = eE_0 \sin \omega t \tag{3.52}$$

where ω is the angular frequency. Hence,

$$\dot{x} = \left[\frac{eE_0}{m\omega}\right]\cos \omega t \tag{3.53}$$

and

$$x = \left[\frac{eE_0}{\omega^2 m}\right]\sin \omega t \tag{3.54}$$

The maximum energy gained from the field

$$U_{max} = \tfrac{1}{2}m(\dot{x}_{max})^2 = \frac{1}{2m}\left[\frac{eE_0}{\omega}\right]^2 \tag{3.55}$$

If the amplitude is less than the electron free path λ_e and the interelectrode spacing d, the electron is trapped in the interelectrode region until an ionizing collision occurs. Thus, the condition for efficient transfer of rf energy to ionization

$$\frac{eE_0}{\omega^2 m} = \frac{eV_0}{d\omega^2 m} < 1 \tag{3.56}$$

or

$$\omega^2 > \frac{eV_0}{d^2 m} \tag{3.57}$$

Because ions are 10^3–10^5 times as heavy as electrons, their amplitude of motion and maximum energy due to an ac field are small compared with the electrons. The common frequency used for generating rf plasmas is 13.56 MIIz, where $\omega^2 = 7.3 \times 10^{15}$. For typical values of $d = 0.02$ m, $V_0 = 2000$ V, we note

$$\frac{eV_0}{d^2 m} = \frac{1.6 \times 10^{-19} \times 2000}{9.11 \times 10^{-31} \times 4 \times 10^{-2}} = 8.8 \times 10^{15} \tag{3.58}$$

so that the condition for trapping the electrons is almost met.

    3.3.1.4. Both Electric and Magnetic Fields Present. The mode of operation of a gas discharge with both electric and magnetic fields present is termed the magnetron mode. While detailed analysis of this mode of operation is complex,[18] the main advantage comes from the increased times the electrons are confined in the discharge before being collected. During this time they are constantly gathering energy from the electric field and using it to create electron–ion pairs and energetic photons. Thus, the density of the plasma formed by the discharge can be substantially increased and the power dissipated by the electrons at the electrodes may be substantially decreased.

3.3.2 Electron Energy Distribution

It was noted above that a critical parameter for describing a gas discharge is the energy gathered per mean free path $eE\lambda$. Since λ is inversely proportional to the gas density, this parameter is generally expressed in the form E/p_0, where p_0 is the gas pressure at a reduced temperature of $0°C$.

If $\rho(\varepsilon)d\varepsilon$ is the number of electrons per unit volume with energies between ε and $\varepsilon + d\varepsilon$, then $\rho(\varepsilon)$ is termed the energy distribution function. The mean energy

$$\bar{\varepsilon} = \frac{\int_{\varepsilon=\infty}^{\infty} \varepsilon\rho(\varepsilon)\,d\varepsilon}{\int_0^\infty \rho(\varepsilon)\,d\varepsilon} \tag{3.59}$$

We distinguish three cases[23]:

- $E/p_0 = 0$ leads to the Maxwell–Boltzmann distribution

$$\rho(\varepsilon) = C\varepsilon^{1/2} \exp\left[-1.5\frac{\varepsilon}{\bar{\varepsilon}}\right] \tag{3.60}$$

- E/p_0 sufficiently small that only elastic collisions occur leads to the Dryvesteyn distribution

$$\rho(\varepsilon) = C\varepsilon^{1/2} \exp\left[-0.55\frac{\varepsilon^2}{\bar{\varepsilon}^2}\right] \tag{3.61}$$

An assumption in producing this equation is that the electron collision cross section is independent of electron energy. Since many gases show maxima in collision cross section around a few electron volts, this distorts the Dryvesteyn distribution in this range.
- E/p_0 sufficiently large that inelastic collisions occur. The distribution function is then given approximately by the solution to the differential equation[20]

$$\frac{\partial^2\rho(\varepsilon)}{\partial\varepsilon^2} = \frac{3k}{(eE\lambda)^2}\rho(\varepsilon) \tag{3.62}$$

where k is the ratio of inelastic to elastic collision cross sections. If λ is assumed constant and k assumed proportional to ε, the solution is given by the sum of two Bessel functions.

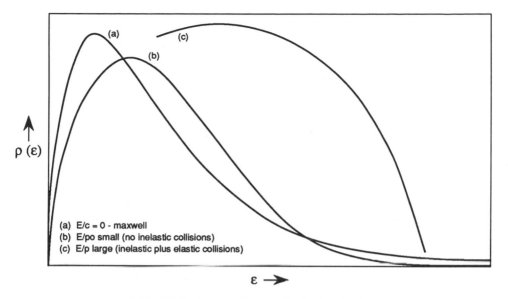

FIGURE 3.5. Shapes of energy distribution functions.

The shapes of the three distributions are given in Fig. 3.5. An important aspect of any energy distribution function is the number of very high-energy electrons predicted in the "tail" of the distribution. Since these electrons have energies sufficient for inelastic collisions, it seems likely that in practice the tails will be foreshortened compared with the predictions of the zero or low E/p_0 theories.

3.3.3. The Townsend Coefficients

Consider an electron moving in a gas under the influence of a uniform electric field. The first Townsend coefficient (named after a pioneer worker in gas discharge), α is defined as the number of electron–ion pairs formed in a unit distance per electron. Thus, the increase in electrons dn produced in moving distance dx

$$dn = \alpha \, dx \tag{3.63}$$

hence

$$n = n_0 \, e^{\alpha x} \tag{3.64}$$

where n_0 is the initial concentration. It is usual to present data in the form of a plot of α/p versus E/p (Fig. 3.6). Equation (3.64) fairly describes the current growth with distance from the cathode using photoelectric emission and in low electric fields.

If it is assumed that the electrons have attained an equilibrium energy distribution in the discharge, then the number of ionizing collisions per unit volume, per second per electron of energy ε is

$$\left[\frac{2\varepsilon}{m}\right]^{1/2} \cdot \frac{P_i(\varepsilon)}{\lambda} \tag{3.65}$$

where $P_i(\varepsilon)$ is the probability of ionization for an electron of energy ε on collision and λ is the mean free path.

The total number of ionizing collisions per unit volume per second is therefore given by

$$N_i = \left[\frac{2}{m}\right]^{1/2} \int_{\varepsilon_i}^{\infty} W(\varepsilon)\varepsilon^{1/2}\frac{P_i(\varepsilon)}{\lambda}\,d\varepsilon \tag{3.66}$$

where $W(\varepsilon)d\varepsilon$ is the number of electrons per unit volume with energy in the range ε to $\varepsilon + d\varepsilon$, i.e., $W(\varepsilon)$ is the energy distribution function. The number of electrons

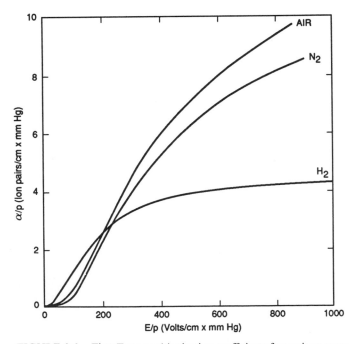

FIGURE 3.6.   First Townsend ionization coefficients for various gases.

crossing a unit area of a plane perpendicular to E per second is given by

$$N = v \int_0^\infty W(\varepsilon) \, d\varepsilon \tag{3.67}$$

where $v = \mu E$ is the average drift velocity. Since $N_i dx = \alpha N dx$ by definition, we have $\alpha = N_i/N$, i.e.,

$$\alpha = \left[\frac{2}{m}\right]^{1/2} \frac{\displaystyle\int_{\varepsilon_i}^\infty W(\varepsilon) \varepsilon^{1/2} \frac{P_i(\varepsilon)}{\lambda} \, d\varepsilon}{\displaystyle\mu E \int_0^\infty W(\varepsilon) \, d\varepsilon} \tag{3.68}$$

To solve this equation the functions $W(\varepsilon)$ and $P_i(\varepsilon)$ must be known, and also in general λ is a function of ε, and μ a function of E/p. Thus, α can only be calculated by numerical methods.

However, there is an approximate method for estimating α based on considerations of energy balance. If i is the current passing through the discharge and dV the voltage dropped in distance dx, then in equilibrium the energy given to the electrons by the electric field in dx is equal to $i\,dV$. This energy is dissipated in ionizing (requiring an average energy ε_i per electron formed), exciting the gas molecules (requiring energy $r\varepsilon_i$ per electron formed), and in heating the new electrons generated to the average electron energy ($\bar{\varepsilon}$ per electron formed), i.e.,

$$i\,dV = \varepsilon_i \frac{di}{e} + r\varepsilon_i \frac{di}{e} + \bar{\varepsilon} \frac{di}{e} \tag{3.69}$$

however,

$$\frac{di}{e} = \alpha \frac{i}{e} \, dx = \frac{\alpha}{E} e \, dV \tag{3.70}$$

Substituting above gives

$$\frac{\alpha}{E} = \frac{1}{[1 + r + \bar{\varepsilon}/\varepsilon_i]} \tag{3.71}$$

where $V_i = \varepsilon_i/e$ is the ionization potential. The term $\alpha/E = \eta$ is called the *ionization coefficient*. To obtain values for α by this approach requires a knowledge of r and $\bar{\varepsilon}$ as a function of E/p_0.

At large values of the interelectrode spacing d, the simple form of current growth predicted by Eq. (3.64) is not followed. Townsend found empirically that

$$n = \frac{n_0 \, e^{\alpha d}}{1 - \gamma[e^{\alpha d - 1}]} \tag{3.72}$$

where γ is termed the second Townsend coefficient. Note that $n \to \infty$, i.e., a complete breakdown occurs, when

$$1 = \gamma[e^{ad-1}] \tag{3.73}$$

The existence of this effect is due to electrons being produced at the cathode by secondary processes (Section 3.2.5). Let n_0 be the number of electrons emitted from the cathode by the primary process, and n_s be the number of electrons formed by secondary processes. We assume n_s is proportional to n, i.e., $n_s = \gamma n$.

Since

$$n = (n_0 + n_s)\, e^{ad} + n_s \tag{3.74}$$

i.e.,

$$n = (n_0 + \gamma n)\, e^{ad} + \gamma n \tag{3.75}$$

hence

$$n = \frac{n_0 e^{ad}}{1 - \gamma(e^{ad} - 1)} \tag{3.76}$$

in agreement with the empirical result. The vaue of γ depends on the nature of the photons, metastables and ions formed in the interelectrode region, and the cathode surface properties. After breakdown, the current increases until it is limited either by the external circuit or by the physical limitations in the discharge itself.

3.4. COLLECTIVE PHENOMENA IN PLASMAS

3.4.1. Debye Screening

Consider a positive ion at a point P in a plasma. In the region immediately around P, electrons will be attracted and positive ions repelled giving rise to an average spherically symmetric potential distribution of the form $\phi = \phi(r)$ about P. Using Boltzmann statistics, the number of positive ions within a spherical shell radius r thickness dr where the potential energy is $\phi(r)$ given by

$$dn_+ = \exp\left\{\frac{\phi(r)e}{kT_i}\right\} 4\pi r^2\, dr \tag{3.77}$$

and similarly for electrons

$$dn_e = \exp\left\{\frac{\phi(r)e}{kT_i}\right\} 4\pi r^2\, dr \tag{3.78}$$

where n_+ and n_e are the average number densities in the plasma of positive ions assumed singly charged, and electrons, respectively, and T_i and T_e their respective temperatures.

Thus, the net charge density in the shell

$$\rho = e\left[\frac{dn_e - dn_+}{4\pi r^2 \, dr}\right] \tag{3.79}$$

for $n_+ = n_e = n$, and $\phi(r)e \ll kT_e$ and kT_i

$$\rho = en\left[\left(1 + \frac{\phi(r)e}{kT_e}\right) - \left(1 - \frac{\phi(r)e}{kT_i}\right)\right] \tag{3.80}$$

i.e.,

$$\rho = \frac{ne^2\phi(r)}{k}\left[\frac{1}{T_e} + \frac{1}{T_i}\right] \tag{3.81}$$

For glow discharges, the gas temperature $T \cong T_i \ll T_e$, thus

$$\rho \approx \frac{ne^2\phi(r)}{kT} \tag{3.82}$$

Poisson's equation in spherical coordinates states

$$\frac{1}{r^2}\frac{d}{dr}\left(r^2\frac{d\phi}{dr}\right) = -\frac{\rho}{\varepsilon_0} = \frac{n^2\phi(r)}{\varepsilon_0 kT} \quad \text{(SI units)} \tag{3.83}$$

where ε_0 is the permittivity of free space.

The solution to this differential equation is

$$\phi(r) = A\frac{1}{r}e^{-r/\lambda_D} + A'\frac{1}{r}e^{r/\lambda_D} \tag{3.84}$$

where

$$\lambda_D^2 = \frac{\varepsilon_0 kT}{ne^2} \tag{3.85}$$

At $R = \infty$, the potential energy is zero, hence $A' = 0$. Thus,

$$\phi(r) = A\frac{1}{r}e^{-r/\lambda_D} \tag{3.86}$$

The quantity $\lambda_D = (\varepsilon_0 kT/ne^2)^{1/2}$ is termed the Debye length, and may be considered as the distance beyond which the ion under consideration is screened from local changes in the plasma potential. Thus, in order to treat the spatial charge distribution as uniform, the volume containing the plasma must have linear dimensions L much greater than the Debye length, i.e.,

$$L \gg \lambda_D = 6.96(T/n)^{1/2} \quad \text{for } T \text{ in K and } n \text{ in charges per cm}^3$$
$$= 743\,(T^*/n)^{1/2} \quad \text{for } T^* \text{ in electron volts}$$

$$(3.87)$$

For the glow discharge $T \cong 300$ K and n lies in the range 10^8 to 10^{14} per cm^3, giving λ_D in the range 10^{-2} to 10^{-5} cm.

In order for Debye shielding to be valid, the number of particles in the Debye sphere N_D must be large enough for Boltzmann statistics to apply, i.e.,

$$N_D = 4/3\pi\lambda_D^3 n = 1412T^{3/2}n^{-1/2} \gg 1 \qquad (3.88)$$

For the glow discharge N_D lies in the range 1000, at $n \cong 10^8$ cm^3, to 1, at $n \cong 10^{14}/$ cm^3.

3.4.2. Ambipolar Diffusion

A plasma has a tendency to remain neutral; thus, in the case where the charge species are being lost by diffusion, the rates of diffusion of ions and electrons must adjust themselves so that the two species leave at essentially the same rate. The mechanism is as follows. The electrons, being lighter, have higher thermal velocities and tend to leave the plasma first. A positive charge is left behind, and an electric field (E) is created so as to retard the loss of electrons and accelerate the loss of ions. In equilibrium the number of ion pairs crossing a unit area per second is given by

$$\Gamma = n\mu_i E - D_i\nabla n = -n\mu_e E - D_e\nabla n \qquad (3.89)$$

where μ_i and μ_e are the mobility and D_i and D_e are the thermal diffusion coefficients for ions and electrons, respectively. Thus,

$$E = \left(\frac{D_i - D_e}{\mu_i + \mu_e}\right)\frac{\nabla n}{n} \qquad (3.90)$$

The common flux Γ is then given by

$$\Gamma = \mu_i\left(\frac{D_i - D_e}{\mu_i + \mu_e}\right)\nabla n - D_i\nabla n \qquad (3.91)$$

or

$$\Gamma = -\left(\frac{\mu_i D_e + \mu_e D_i}{\mu_i + \mu_e}\right)\nabla n \tag{3.92}$$

This is simply Fick's law for the diffusion of gases with a new diffusion coefficient

$$D_a \equiv \frac{\mu_i D_e + \mu_e D_i}{\mu_i + \mu_e} \tag{3.93}$$

called the *ambipolar diffusion coefficient*.

Since the electron mobility μ_e is much larger than the ion mobility μ_i, then

$$D_a \approx D_i + \frac{\mu_i}{\mu_e} D_e \tag{3.94}$$

The Einstein relation between μ and D states

$$\mu = |q|\frac{D}{kT} \tag{3.95}$$

where q is the charge on the particle. Thus,

$$\frac{\mu_i}{\mu_e} D_e = \frac{T_e}{T_i} D_i \tag{3.96}$$

leading to

$$D_a = D_i\left(1 + \frac{T_e}{T_i}\right) \tag{3.97}$$

For $T_e = T_i$, the net effect of the ambipolar electric field is to enhance diffusion of ions by a factor of two. The important point to notice is that the diffusion rate is primarily controlled by the slower species.

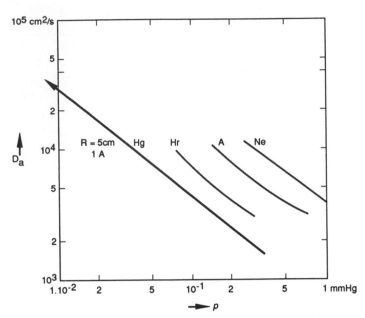

FIGURE 3.7.   Ambipolar diffusion coefficient D_a as a function of the gas pressure p ($D_a = 10^5$ cm^2/s calculated at $p = 2\mu$ in Hg). From Ref. 5.

Figure 3.7 shows how D_a varies with gas pressure for several gases for which $T_e = T_i = 300$ K.

3.4.3. Plasma Oscillations

Consider the propagation of an electromagnetic wave through a plasma under conditions where the amplitude of the electric field is not large enough to give sufficient energy to an electron for it to have an inelastic collision, and the electron collision frequency is small compared to the wave frequency. For a time-varying electric field of the form $\mathbf{E} = \mathbf{E}_0 \, e^{i\omega t}$, the motion of an electron is given by

$$m \frac{d^2 \mathbf{r}}{dt^2} = -e\mathbf{E}_0 \, e^{i\omega t} \tag{3.98}$$

Integrating gives

$$\frac{d\mathbf{r}}{dt} = \frac{-e}{m} \mathbf{E}_0 \int e^{i\omega t} \, dt = -\frac{e}{m} \mathbf{E}_0 \frac{e^{i\omega t}}{i\omega} \tag{3.99}$$

but since

$$\frac{\partial \mathbf{E}}{\partial t} = \mathbf{E}_0 i\omega\, e^{i\omega t} \tag{3.100}$$

$$\frac{\partial \mathbf{r}}{\partial t} = \frac{e}{m} \cdot \frac{\partial \mathbf{E}}{\partial t} \cdot \frac{1}{\omega^2} \tag{3.101}$$

Now from Maxwell's equations for a partial conductor we have

$$\nabla \times \mathbf{H} = \rho\mathbf{u} + K\varepsilon_0 \frac{\partial \mathbf{E}}{\partial t} \tag{3.102}$$

where ρ is the charge density, u its velocity, and K the dielectric constant of the medium, leading in this case to

$$\nabla \times \mathbf{H} = -ne \frac{d\mathbf{r}}{dt} + \varepsilon_0 \frac{\partial \mathbf{E}}{\partial t} \tag{3.103}$$

since $K = 1$ for the space (vacuum) in which the electron is moving. Substituting for dr/dt from Eq. (3.101) gives

$$\nabla \times \mathbf{H} = \left(1 - \frac{ne^2}{\varepsilon_0 \omega^2 m}\right) \varepsilon_0 \frac{\partial \mathbf{E}}{\partial t} \tag{3.104}$$

This is the same as the equation for wave propagation in a dielectric medium but with the dielectric constant given by

$$K = \left(1 - \frac{ne^2}{\varepsilon_0 \omega^2 m}\right) \tag{3.105}$$

Resonance occurs at the *angular plasma frequency* ω_p where $K = 0$, i.e.,

$$\omega_p^2 = \frac{ne^2}{\varepsilon_0 m} \tag{3.106}$$

The plasma frequency $f_p = \omega_p/2\pi = 8968n^{1/2}$ Hz for n in electrons per cm^3. For glow discharges, plasma frequencies are in the range 10^8–10^{11} Hz.

If the motion of the ions is similarly taken into account, we obtain

$$K = 1 - \frac{ne^2}{\varepsilon_0 \omega^2}\left(\frac{1}{m} + \frac{1}{m_i}\right) \tag{3.107}$$

Since the mass of the ions $m_i \gg m$, the plasma resonant frequency is essentially unchanged with ions present. Thus, measurement of the effective dielectric constant of a plasma at frequencies below the plasma frequency enables the electron concentration in a plasma to be estimated.[5] Since K for a plasma is always less than unity, this implies that the phase velocity of electromagnetic waves through a plasma is greater than the speed of light.

3.4.4. Plasma Sheaths in Glow Discharges

Consider a gas discharge taking place between two plane-parallel electrodes spaced a distance L apart with a potential difference $(V_A - V_C)$ applied between them. For $L \gg \lambda_D$ the major portion of the space is filled with a plasma which takes a uniform potential V_P (termed the plasma potential), but dark regions (called sheaths) exist between the plasma and the electrodes. Across the cathode sheath thickness d_c, the potential varies from V_P to V_C, and similarly across the anode sheath thickness d_a, the potential varies from V_P to V_A. The potential difference $(V_P - V_C)$ is usually much greater than the ionization potential of the gas.

From kinetic theory (Section 3.7.2), the flux of particles crossing a unit area of a plasma in a direction perpendicular to the plane of the electrodes is given by

$$J = \tfrac{1}{4} n\bar{c} = \frac{n}{4}\left(\frac{8kT}{\pi M}\right)^{1/2} \tag{3.108}$$

Thus, assuming a sharp delineation of the plasma boundary, the largest possible value for the positive ion current density reaching an electrode is

$$j+ = \frac{ne}{4}\left(\frac{8kT_i}{\pi M_+}\right)^{1/2} \tag{3.109}$$

Electrons leaving the cathode (density j_c usually generated by secondary processes) are accelerated across the cathode sheath but can only commence ionizing the gas after have traveled about a mean free path. Thus, the thickness of the cathode sheath, d_c, is approximately λ_e (assuming the potential drop across the sheath is much greater than the ionization potential of the gas).

The electron current density moving in a direction from the plasma toward the boundary of the sheath is given by

$$j- = \left(\frac{8kT_e}{\pi M_e}\right)^{1/2} \gg j+ \quad \text{since } M_e \ll M_+ \tag{3.110}$$

There is a large retarding field between the plasma and the cathode which effectively prevents any electrons reaching the cathode. There must also be a retarding field for electrons between the boundary of the sheath and the anode, but this must be small enough to allow a sufficient number of electrons with energy greater than $e(V_A - V_P)$

to escape in order to maintain charge neutrality in the plasma. Assuming a Maxwell–Boltzmann distribution, we have

$$j_- = \tfrac{1}{4}ne\bar{c}\, e^{-e(v_A - v_P)/kT_e} \tag{3.111}$$

This means that the plasma potential must be more positive than the anode potential. Thus, positive ions can also flow from the plasma to the anode and secondary electrons (density j_a) from the anode to the plasma. The thickness of the anode sheath d_a will adjust itself so that the flow of ions across the sheath is space charge limited. If $d_a < \lambda_i$ and the number of electrons in the sheath at any time is very much less than the number of positive ions, then the space charge flow equation (2.24) may be used, giving

$$j_+ = \tfrac{1}{4}ne\left(\frac{8kT_i}{M_i\pi}\right)^{1/2} = \frac{4\varepsilon_0}{9}\left(\frac{2e}{m_i}\right)^{1/4}\frac{(V_A - V_P)^{3/2}}{d_a^2} \tag{3.112}$$

Assuming $(j_c + j_a) \ll j_+$, then

$$2j_+ = \tfrac{1}{4}ne\left(\frac{8\pi kT_e}{\pi M_i}\right)^{1/2} e^{e(v_A - v_P)/kT_e}$$

$$= \tfrac{1}{2}ne\left(\frac{8\pi kT_i}{\pi M_i}\right) \tag{3.113}$$

$$= \tfrac{8}{9}\varepsilon_0\left(\frac{2e}{M_i}\right)^{1/4}\frac{(V_A - V_P)^{3/2}}{d_a^2}$$

relating d_a and V_P to the other plasma parameters.

The concept that there is a sharp plasma boundary between the plasma and the sheath is, of course, very simplistic. As Bohm has suggested, the sheath must be composed of several regions as shown in Fig. 3.8. Comprehensive theories for the plasma sheaths are given in Ref. 4.

If an electrically isolated metal electrode is inserted in the plasma, it takes a floating potential V_f relative to the plasma potential V_P, such that the ion and electron currents leaving it are equal, i.e.,

$$\frac{en\bar{c}_i}{4} = \frac{en\bar{c}_e}{4}\exp - (eV_f/kT_e) \tag{3.114}$$

thus

$$V_f = \frac{-kT_e}{2e}\log\left(\frac{T_i}{T_e}\frac{M_e}{M_i}\right) \tag{3.115}$$

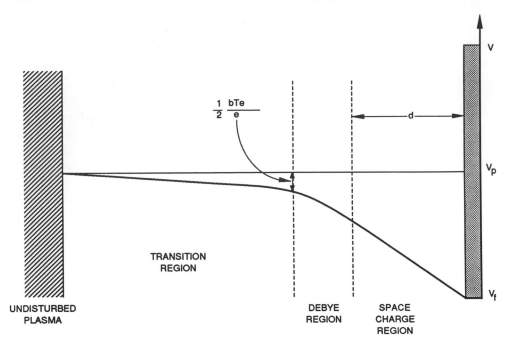

FIGURE 3.8.   A more realistic model for a cathode sheath.

If a potential V is applied to the electrode, the current voltage characteristic takes the form shown in Fig. 3.9. A small electrode used in this way is called a Langmuir probe.

The net current density to the probe is always the sum of electron and ion current densities. For $V < V_P$, this sum can be written as

$$j = \frac{en_i\bar{c}_i}{4} - \frac{en_e\bar{c}_e}{4}\exp -\left[\frac{e(V_P - V)}{kT_e}\right] \qquad (3.116)$$

For large negative V,

$$j \rightarrow \frac{en_i\bar{c}_i}{4} \qquad (3.117)$$

For $V > V_P$, the sum of fluxes can be expressed as

$$j_e = \frac{en_i\bar{c}_i}{4}\exp -\left[\frac{e(V - V_P)}{kT_i}\right] - \frac{en_e\bar{c}_e}{4} \qquad (3.118)$$

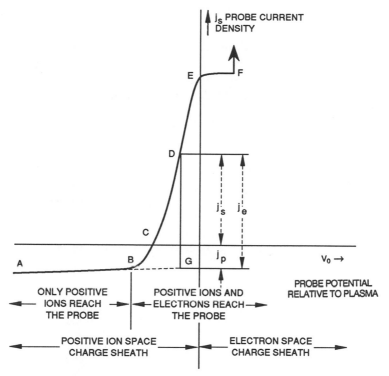

FIGURE 3.9. Characteristics of the Langmuir probe. From Ref. 7.

Because $T_i \ll T_e$, the ion current density term rapidly goes to zero with $V > V_P$. Hence,

$$j \rightarrow \frac{e n_c \bar{c}_e}{4} \qquad (3.119)$$

i.e., only the electron current density remains. Thus, a fairly well-defined V_P can be obtained from the j–V characteristic. In practical measurement the "saturation currents" are not independent of V, because the effective area of the probe changes with voltage; in addition, secondary electron emission is voltage dependent. In the intermediate range of V, the plot of $\log j$ versus V is linear and the slope may be used to estimate T_e. The use of a lock-in amplifier to measure the slope provides a fast, accurate method. To gain an insight into the complexity of interpreting Langmuir probe data, the specialized literature should be consulted.[8]

Clearly the properties of the electrodes strongly influence the nature of the sheath. The cathode is bombarded by photons and ions from the plasma causing it to emit secondary electrons, sputtered neutral atoms, and possibly negative ions with a wide range of energies up to the potential dropped across the cathode sheath. The anode is also bombarded by photons and ions from the plasma. Since the potential dropped across the anode sheath is usually much smaller than that across the cathode, the anode sputter yield is negligible.

To determine secondary particle yields from an electrode, we must know the energy distribution and densities of the particles that strike the target. At low pressure (10^{-2} Torr), most of the particles are ions with energies close to the voltage across the target dark space. At higher pressures, however, charge exchange can play an important role in reducing the mean energy per particle.

The particles that strike an electrode can be broken down into four groups:

- Ions accelerated through the dark space without suffering collisions or charge exchange
- Ions formed somewhere in the dark space and then accelerated the remaining distance without suffering charge exchange
- Energetic neutral species produced by charge exchange with ions in the plasma
- Energetic neutral species produced by secondary charge exchange with ions in the dark space

The four groups are shown in Fig. 3.10. Theoretical energy distribution relations have been made.[24]

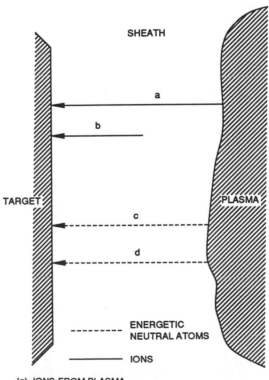

FIGURE 3.10.   Particles striking the electrode. (a) Ions from plasma; (b) ions formed in dark space; (c) energetic neutrals from the plasma; (d) energetic neutrals formed in the sheath.

3.5. RADIO FREQUENCY GLOW DISCHARGES

An ac discharge is considered "rf" when the frequency of operating is high enough that the amplitude of the motion of electrons approaches that of the interelectrode spacing. The rf discharge has proved to be the most popular for producing plasmas that are to be used for processing microdevices. This is because of their simplicity in assembling apparatus, their stability, versatility, reproducibility, the uniformity and high density of the plasmas generated. Furthermore, plasmas can be generated at lower pressures using rf than using dc discharges.

3.5.1. Planar Conducting Electrodes

For metal electrodes directly coupled to the rf source the discharge is symmetrical, the plasma takes a high positive dc potential V_P with respect to the electrodes whose average potential is at the center of the rf voltage (usually ground). Between the plasma and the electrodes are dark sheaths, thickness d, across which positive ions flow, space charge limited, as if it were a dc discharge, i.e.,

$$j_+ = \tfrac{1}{4} n_i c_i e = \tfrac{8}{9} \varepsilon_0 \left[\frac{2e}{M_i} \right]^{1/4} \frac{V_p^{3/2}}{d^2} \tag{3.120}$$

The plasma itself behaves as a conductor held at dc potential V_p, hence the instantaneous value of the rf voltage impressed on an electrode is dropped entirely across the sheath. Electrons originating at an electrode are instantaneously acted upon by a field of value

$$E = \frac{1}{d}(V_0 \sin 2\pi ft - V_p) \tag{3.121}$$

where f is the alternating frequency. For the plasma to remain neutral, the net number of electrons reaching an electrode from the plasma during a cycle must equal the number of positive ions arriving plus the number of electrons leaving the electrode during the same time period. If $V_0 > V_p$, this means that during a portion of the cycle the electrode will be positive with respect to the plasma so that some electrons can flow from the plasma to the electrode. If the time per cycle allowed for flow is t', then

$$\tfrac{1}{4} n_e \bar{c} \cdot t'f = \tfrac{1}{4} n_i \bar{c}_i + \gamma \tfrac{1}{4} n_i \bar{c}_i \tag{3.122}$$

where γ is the secondary electron emission for ions and photons striking the electrode (assumes the photon flux is proportional to the ion current). Hence, for $n_e = n_i$,

$$t' = \frac{\bar{c}_i}{\bar{c}_e f}(1 + \gamma) \tag{3.123}$$

Since $c_i \ll c_e$ and γ is on the order of unity, the ratio of t' to the cycle time $1/f$ is always small so that $V_p \to V_0$, i.e., the plasma potential is close to the rf amplitude.

Rf discharges of this type are useful for plasma chemistry where the main function of the discharge is to create reactive ion species and heat up the reagents. A detailed account of the ionization processes in an rf discharge is given in the literature.[25] In summary, the rf glow discharge has

- Sufficient ionization reactions within the plasma already that there is no need for secondary electrons from the electrodes to maintain the discharge.
- Enhanced efficiency due to the trapping of electrons in the sheath which allow the discharges to be sustained at lower pressures (1 mTorr).
- Does not require conductive electrodes and can operate in a wide frequency range.

3.5.2. One Electrode Capacitatively Coupled to the rf Source

By placing a capacitor in only one arm of the rf source, an asymmetric discharge is created.[26] Electrons arriving at the capacitatively coupled electrode during the first positive half cycle charge it up negatively to the full amplitude of the rf, V_0, from which it can only be discharged by the flow of ions (assuming no bypass resistance). The time to discharge a capacitor C with ion current i_i is

$$\delta t = \frac{C V_0}{i_i} \tag{3.124}$$

Clearly if $1/\delta t \ll$ the frequency of the rf, then the dc potential of the capacitatively coupled electrode will be close to $-V_0$. Thus, the dc potential difference between the plasma and the capacitatively coupled electrode

$$(V_p + V_0) \to 2V_0 \tag{3.125}$$

That is, the capacitatively coupled electrode takes a potential of $-2V_0$ with respect to the plasma, whereas the directly coupled electrode remains at $-V_0$ with respect to the plasma. The FCC regulated frequency of 13.56 MHz is sufficiently high that this occurs in most practical situations. An electrode may be capacitatively coupled simply by covering it with an insulator, an aspect that has proved extremely useful for the sputter coating of refractory insulators (see Section 4.7).

3.5.3. The rf Magnetron Glow Discharge

A problem with rf discharges as described above, is that the uncoupled electrode is intensely bombarded by electrons.[27] The power dissipated at this electrode tends to limit the intensity of the plasma that can be generated. The effect is particularly severe when the capacitatively coupled electrode is covered with an insulator. This is due to secondary electrons which are efficiently generated by ions (and photons) striking the insulated electrode. These secondaries are accelerated across the sheath, and depending on the mean free path, may strike the opposite electrode with a high energy. By placing magnets behind the capacitatively coupled electrode, these electrons tend to be contained in the sheath and plasma regions until they have lost their energy by ionizing and excitation collisions. This both reduces the counter-electrode bombardment and increases the plasma density.

3.5.4. Geometrically Asymmetric rf Glow Discharge

If the electrodes have different areas, as in the case of coaxial cylinders, the sheath dc voltages (V_1, V_2) can be related to the electrode areas (A_1, A_2).[28] The ion current density across the sheaths, thickness D_1 and D_2, are equal and are space charge limited, i.e.,

$$j_{i1} = \frac{KV_1^{3/2}}{D_1^2} = j_{i2} = \frac{KV_2^{3/2}}{D_2^2} \tag{3.126}$$

hence,

$$\frac{V_1^{3/2}}{D_1^2} = \frac{V_2^{3/2}}{D_2^2} \tag{3.127}$$

The sheath behaves as a conductor; therefore, the capacitances between other electrodes and the plasma are

$$C_1 = \frac{k'A}{D_1} \quad \text{and} \quad C_2 = \frac{k'A_2}{D_2} \tag{3.128}$$

Since the total voltage is divided across these capacitances

$$\frac{V_1}{V_2} = \frac{C_2}{C_1} = \frac{A_2}{D_2} \frac{D_1}{A_1} = \frac{A^2}{A_1} \left(\frac{V_1}{V_2}\right)^{3/4} \tag{3.129}$$

i.e.,

$$\frac{V_1}{V_2} = \left(\frac{A_2}{A_1}\right)^4 \tag{3.130}$$

This is known as the Koenig–Maissel relation. Experimentally, the exponent is found to be in the range of 1–1.5 rather than 4.

3.5.5. Microwave Discharges

Electromagnetic waves in the frequency range 1–100 GHz (corresponding to wavelengths from 30 to 0.3 cm) are termed *microwaves*. Such waves are conveniently transmitted by the use of waveguides which are usually tubes of conducting material

of rectangular or circular cross section.[29] A cavity with metallic conducting walls
resonates at a specific frequency or harmonics thereof. The energy dissipated within
the cavity determines the frequency range about the resonant frequency that the
cavity will accept and is characterized by

$$Q = f_0/2\Delta f \tag{3.131}$$

where $\pm \Delta f$ are the frequencies at which the amplitude of the wave is one-half that
at the resonant frequency for the same input power. For an empty cavity, the dissipation is determined by the skin depth δ of the electric field into the metal walls which
is given by

$$\delta = \sqrt{\frac{\rho}{\pi \mu f}} \text{ meters} \tag{3.132}$$

where ρ is the resistivity in ohm meters, μ is the permittivity of free space $= 4\pi \times 10^{-7}$
henry per meter, and f is the frequency in hertz. For the generation of microwave
plasmas for microfabrication purposes, a cavity in the form of a right circular cylinder
radius a, length $2h$ is generally used operating in the $TM_{0,1}$ mode, i.e., with the electric
field parallel to the axis of the cylinder. For this case

$$\lambda_0 = 2.61a \tag{3.133}$$

$$Q = \frac{a}{\delta}\left(1 + \frac{a}{2h}\right)^{-1} \tag{3.134}$$

If a portion of the cavity enclosed by a quartz envelope is filled with a gas at a low
pressure p (in the range 0.1 to 10 Torr) and the microwave power input to the cavity
increased, breakdown occurs when free electrons within the quartz chamber can gain
sufficient energy to ionize the gas, usually at an amplitude of the electric field of a
few hundred volts per centimeter (see Section 3.3.1.3). The microwave discharge
remains stable at this amplitude or slightly less. The resonant frequency of the cavity
will be changed due to the change in dielectric constant of an ionized gas [Eq. (3.105),
Section 3.4.3], and of course the Q will be decreased due to the increased power
dissipated in generating the plasma.

　　　If a static magnetic field is also present, then the electrons also move in a circular
motion at the electron cyclotron resonant frequency for that magnetic field [Eq.
(3.43), Section 3.3.1.2]. In the electron cyclotron resonance (ECR) discharge, a
region or surface layer is identified within the discharge chamber where the electron
cyclotron frequency and the microwave frequency are equal and the electric field
component is perpendicular to the static magnetic field. Under these conditions, the
electrons within the ECR region are continuously accelerated until they lose energy
by an ionization or excitation collision with a neutral gas molecule. In this way the
density of a plasma may be simply controlled by varying the input microwave power
and the gas pressure. The maximum plasma density occurs if the input power is such
that the plasma frequency [Eq. (3.106)] is equal to the microwave frequency which

is itself equal to the ECR frequency. Typically, a 2.45-GHz microwave power source is used with gases in a pressure range from 0.5 to 10 Torr.[30] Thus, the maximum density of electrons (and ions) that can be formed within the quartz chamber at 2.45 GHz is given by (see Section 3.4.3)

$$n = \left(\frac{2.45 \times 10^9}{8968}\right)^2 = 7.5 \times 10^{10} \text{ electrons per cm}^3 \qquad (3.135)$$

3.6. PLASMA CHEMISTRY

From the chemical viewpoint, the main advantage of an electrical discharge is its ability to controllably heat gases to temperatures not otherwise easily attainable.[31, 32] The discharge is in dynamic equilibrium; i.e., the rate of generation of electrons, excited and ionic species is equal to their rate of loss. Electrons and ions extract energy from the electric fields and impart some of this energy to the neutral gas. Typically, the energy distribution of the electrons, T_e, and that of the neutral gas and ions, T_g, are different with $T_e > T_g$. Because of the similarity of masses of the ions and gas neutrals, the energy of these particles tend to equilibrate at high collision frequencies. The electrodes and walls of the containing vessel can affect both the generation and loss of the charged species and hence contribute indirectly to T_e and T_g.

As the temperature of a molecular gas is increased, it first dissociates and then ionizes. All excited species that can exist between the extremes of unexcited neutral molecules to fully ionized atoms are produced. The number density of each species is determined by the appropriate equilibrium constants, which in turn depend on the temperature and the activation energy for the reaction. Activation energies for molecular dissociation are typically in the range 1–10 eV, while ionization energies are typically in the range 10–20 eV. Equilibrium constants vary as $\exp[-\phi/kT]$, where ϕ is the appropriate activation energy and T is the absolute temperature.

Figure 3.11 shows graphically the behavior of matter as the temperature is increased. At temperatures above 10,000 K (or an average energy of 1 eV per particle), all molecules dissociate, and their constituent atoms are mostly ionized.[33] The upper practical limit of furnaces is about 3000 K, whereas average temperatures above 10,000 K can easily be attained in plasmas, depending on the discharge conditions. Under low-power discharge conditions, a "cold" plasma can be generated. In this case, although the electrons may have a high average energy (corresponding to a high temperature), very little of their energy is transferred to the gas molecules, which remain essentially unheated. Cold plasmas can exist in dc discharges at high pressures (atmospheric) if the specific power is low. Rf discharges at industrial band frequencies (13.56 MHz) are typically "hot." Thus, by controlling the nature of the electrical discharge, the temperature of the gas can be varied from room temperature to above 10,000 K, i.e., from an essentially neutral molecular form to a completely dissociated and ionized form, and all molecular fragments (ionized, excited, and

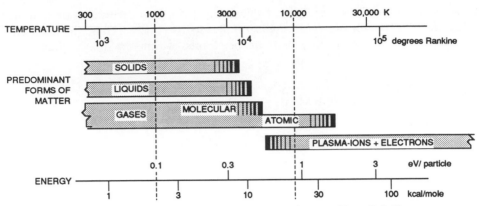

FIGURE 3.11.   High-temperature behavior of matter. From Ref. 33.

neutral forms) that can exist between these two extremes. Because equilibrium constants depend exponentially on the temperature, precise control of the plasma temperature is necessary, if a particular ratio of species is required.

The primary role of the plasma in plasma chemistry is to produce chemically active species, which can subsequently react via conventional pathways. Ions and free radicals are the principal active species. In conventional chemical vapor deposition[34] (CVD), gaseous components react at a hot surface to form a solid material, while the gaseous reaction products are removed. Plasma deposition[35] may be considered as an enhanced CVD process in which

- The gaseous components can be heated to high temperatures.
- New, short-lived gaseous components that do not exist outside a plasma, can be formed.
- Components from the gas phase condensing on the substrate can have their reaction rates enhanced by bombardment by energetic electrons, ions, or photons arising from the plasma.

The situation obtained in a plasma is often too complex for the deposition rates (or even the film composition and properties) to be predicted in any quantitative fashion. Thus, techniques developed to deposit specific films have been largely empirical, albeit based on the broad general concepts discussed above. It is to be hoped that quantitative theories may be developed in the future to enable these techniques to be put on a more scientific basis.

The reactions of plasma-produced species with gaseous or solid materials can give rise to either the deposition or etching of thin films. Several examples of elementary processes occurring in the plasma and at the interface between a plasma and a solid surface are listed below[36]:

1. Inelastic collisions between heavy particles

- Penning dissociation: $M^* + A_2 \rightarrow 2A + M$
- Penning ionization: $M^* + A_2 \rightarrow A_2^+ + M + e^-$
- Charge transfer: $M^+ + A_2 \rightarrow A_2^+ + M$; $M^- + A_2 \rightarrow A_2^- + M$

- Collisional detachment: $M + A_2^- \rightarrow A_2 + M + e^-$
- Association detachment: $A^- + A \rightarrow A_2 + e^-$
- Ion–ion recombination: $M^- + A_2^+ \rightarrow A_2 + M$; $M^- + A_2^- \rightarrow 2A + M$
- Electron–ion recombination: $e^- + A_2^+ \rightarrow 2A$; $e^- + A_2^+ + M \rightarrow A_2 + M$
- Atom recombination: $2A + M \rightarrow A_2 + M$
- Atom abstraction: $A + BC \rightarrow ABC + C$
- Atom addition: $A + BC + M \rightarrow ABC + M$

2. Electron-impact reactions

- Excitation (rotational, vibrational, and electronic): $e^- + A_2 \rightarrow A_2 + e-$
- Dissociative attachment: $e^- + A_2 \rightarrow A^- + A^+ + e^-$
- Dissociation: $e^- + A_2 \rightarrow 2A + e^-$
- Ionization: $e^- + A_2 \rightarrow A_2^+ + 2e^-$
- Dissociative ionization: $e^- + A_2 \rightarrow A^+ + A + 2e^-$

3. Heterogeneous reactions (S is a solid surface in contact with the plasma)

- Atom recombination: $SA + A \rightarrow S + A_2$
- Metastable deexcitation: $S + M^* \rightarrow S + M$
- Atom abstraction: $SB + A \rightarrow A + AB$
- Sputtering: $SB + M^+ \rightarrow S^+ + B + M$

3.7. VACUUM PHYSICS

3.7.1. Introduction

Vacuum physics deals with the production and measurement of low pressures and the flow of gases at these pressures. The term *vacuum* (Latin for "empty") is loosely applied to characterize gases held at pressures below that of the earth's atmosphere. The average barometric pressure at sea level corresponds to that of a column of mercury 760 mm in height. Initially low pressures were measured relative to atmospheric pressure (bars) or in millimeters of mercury (Torr). More rigorously, pressure is measured in force per unit area. In SI units the measure is newtons per square meter or pascals, and in cgs units dynes per square centimeter. For reference: 1 Torr = 1 mm Hg = 0.76 mbar = 133 PA = 1334 dyn/cm^2.

The following terms are conventionally used:

Rough vacuum (RV)	Atmosphere to 1 Torr
Medium vacuum (MV)	1 to 10^{-3} Torr
High vacuum (HV)	10^{-3} to 10^{-7} Torr
Ultrahigh vacuum (UHV)	Less than 10^{-7} Torr
Ultimate vacuum	The lowest pressure that can be attained in a given vacuum system
Perfect vacuum (unattainable)	A volume containing no gaseous molecules whatever

In microfabrication technology we are mainly concerned with gases at medium vacuum pressures and below, where the kinetic theory of gases, which assumes that molecules consist of perfectly elastic spheres in random motion, can be used to accurately predict the physical properties. In these (Knudsen flow) regions the mean free path of air molecules ranges from a few centimeters at 10^{-3} Torr to a few thousand kilometers at 10^{-9} Torr.

3.7.2. Review of the Kinetic Theory of Gases

Assume that a gas consists of n_0 molecules per unit volume of mass m in thermal equilibrium colliding as perfectly elastic spheres so that all of the energy in the system is stored in the kinetic energy of the molecules. Initially the energy stored in rotational, vibrational, and electronic states is ignored. At any instant, any given molecule is assumed to be moving with velocity c. Let $n(c)dc$ be the number of molecules per unit volume with velocity c in any direction; then those that have a component within a solid angular spread $d\omega = \sin \theta \, d\theta \, d\phi$ is given by

$$n(c) \, dc \, \frac{\sin \theta \, d\theta \, d\phi}{4\pi} \tag{3.136}$$

Each of these molecules has a component $c \cos \theta$ in a direction perpendicular to a small area element dS placed within the gas. Thus, the number crossing dS per second in one direction with velocity c is given by the number of molecules in a cylinder base area dS and height $c \cos \theta$

$$dN = n(c) \, dc \, \frac{\sin \theta \, d\theta \, d\phi}{4\pi} c \cos \theta \, dS \tag{3.137}$$

The number crossing dS per second with velocity c for all positive values of $\cos \theta$ is therefore

$$dN(c) = n(c) \, dc \, \frac{cdS}{4\pi} \int_{\phi=0}^{2\pi} d\phi \int_{\theta=0}^{\pi/2} \sin \theta c \cos \theta \, d\theta$$
$$= n(c) \, dc \, \frac{cdS}{4} \tag{3.138}$$

Hence, the number of molecules with velocity between c and $c + dc$ crossing a unit area in one direction perpendicular to any plane within the gas per second is

$$N(c) = \tfrac{1}{4} n(c) \, dc \cdot c \tag{3.139}$$

The number of molecules crossing per unit area per second of a plane in one direction within a gas for all values of c is therefore

$$N_s = \tfrac{1}{4} \int_0^\infty cn(c) \, dc \tag{3.140}$$

i.e.,

$$N_s = \tfrac{1}{4} n_0 \langle c \rangle \qquad (3.141)$$

since the average value of c is defined by

$$\langle c \rangle = \frac{\displaystyle\int_0^\infty cn(c)\, dc}{\displaystyle\int_0^\infty n(c)\, dc} = \frac{\displaystyle\int_0^\infty cn(c)\, dc}{n_0} \qquad (3.142)$$

Referring to Eq. (3.137), the component of momentum in a direction perpendicular to dS is

$$n(c)\, dc\, \frac{\sin\theta\, d\theta\, d\phi}{4\pi}\, c\cos\theta\, mc\cos\theta\, dS \qquad (3.143)$$

If dS is part of the containing wall of the gas chamber at which the molecules are perfectly elastically reflected, the rate of change in momentum or force on dS due to this class of molecules, is given by

$$2n(c)\, dc\, \frac{\sin\theta\, d\theta\, d\phi}{4\pi}\, mc^2\cos^2\theta\, dS \qquad (3.144)$$

Integrating over θ, ϕ, and c gives the total force on dS as

$$dF = \frac{m}{2\pi}\int_{c=0}^\infty n(c)c^2\, dc \int_0^\pi \sin\theta\cos^2\theta\, d\theta \int_0^{2\pi} d\phi\, dS \qquad (3.145)$$

Hence, the pressure

$$p = \frac{dF}{dS} = \frac{m}{3}\int_0^\infty n(c)c^2\, dc \qquad (3.146)$$

or

$$p = \tfrac{1}{3} mn_0 \langle c^2 \rangle \qquad (3.147)$$

since the average value of c^2 is defined by

$$\langle c^2 \rangle = \frac{\int_0^\infty n(c)c^2 \, dc}{\int_0^\infty n(c) \, dc} = \frac{\int_0^\infty n(c)c^2 \, dc}{n_0} \tag{3.148}$$

The temperature of a gas may be defined as a linear function of the mean kinetic energy of the molecules. That is, if $\varepsilon = 1/2mc^2$ is the kinetic energy of a given molecule, then the average value of ε

$$\langle \varepsilon \rangle = \tfrac{1}{2}m\langle c^2 \rangle \propto T \tag{3.149}$$

but from Eq. (3.147) p is proportional to $n_0 T$ i.e.,

$$p = n_0 kT = \tfrac{1}{3}mn\langle c^2 \rangle \tag{3.150}$$

The constant of proportionality k is called Boltzmann's constant and

$$\langle \varepsilon \rangle = \tfrac{1}{2}m\langle c^2 \rangle = \tfrac{3}{2}kT \tag{3.151}$$

T is measured in Kelvin if the unit of temperature is such that 100 units corresponds to the difference between the freezing and boiling points of water (at one standard atmosphere). Since Eq. (3.150) indicates that p is independent of m, gases at the same temperature and pressure contain the same number of molecules per unit volume. This is known as Avogadro's law. The gram molecular weight (mole) of gas is defined as that weight of gas which occupies 2.24146×10^{-2} m^3 at 0°C (273.16 K) and 1 atm pressure (10^5 PA). This is the volume measured to be occupied by exactly 2 g of hydrogen. There are two hydrogen atoms per hydrogen molecule, and the weight of the hydrogen atom was originally used to define the unit of atomic mass (atomic mass unit or amu).

The number of molecules in a gram-molecule N_A is known as Avogadro's number and has been measured to be 6.0228×10^{23}, giving Boltzmann's constant $k = 1.3803 \times 10^{-23}$ J/K, and the atomic mass unit as 1.66×10^{-27} kg.

If N is the total number of molecules in volume V, then Eq. (3.150) may be written as

$$p = \frac{N}{V}kT \tag{3.152}$$

i.e.,

$$\frac{pV}{T} = \text{constant} = Nk \tag{3.153}$$

For a volume containing 1 mole, R = gas constant = $N_A k$ = 8.313 in SI units. Equation (3.153) is called the perfect gas law. A gas thermometer is a device in which the pressure and volume of a fixed quantity of gas can be measured precisely to estimate the temperature.

If a gas comprises a mixture of n_1 molecules per unit volume of mass m_1, n_2 of mass m_2, etc., then since the total pressure is independent of molecular mass,

$$p = n_0 kT = \sum_i n_i kT \tag{3.154}$$

i.e.,

$$p = \sum_i p_i \tag{3.155}$$

where $p_i = n_i kT$ is the partial pressure of the ith molecule. That is, each molecular species behaves as if the other species were not present. Equation (3.155) is Dalton's law of partial pressure.

In thermal equilibrium, the invariants of the molecular motion are the total number of molecules per unit volume (n_0), the total kinetic energy stored per unit volume E, and the total momentum M stored per unit volume. If $n(c)\,dc$ is the number of molecules per unit volume with total velocity in any direction between c and $c + dc$, then

$$n_0 = \int_{c=0}^{\infty} n(c)\,dc \tag{3.156}$$

$$M = \int_{c=0}^{\infty} mcn(c)\,dc = n_0 m \langle c \rangle \tag{3.157}$$

$$E = \int_{c=0}^{\infty} \tfrac{1}{2} mc^2 n(c)\,dc = n_0 \tfrac{1}{2} m \langle c^2 \rangle \tag{3.158}$$

The simplest function for which all of these integrals are invariant is of the form

$$n(c) = ac^2 \exp -(\beta c^2) \tag{3.159}$$

giving

$$n_0 = \alpha \tfrac{1}{4} \left(\frac{\pi}{\beta^3} \right)^{1/2} \tag{3.160}$$

$$M = \alpha m \frac{1}{2} \frac{1}{\beta^2} \tag{3.161}$$

$$E = \frac{\alpha m}{2} \frac{3}{8} \left(\frac{\pi}{\beta^5} \right)^{1/2} \tag{3.162}$$

Since

$$\langle \varepsilon \rangle = \tfrac{1}{2} m \langle c^2 \rangle = \tfrac{3}{2} kT = \frac{E}{n_0} = \frac{3}{2} \frac{m}{2\beta} \tag{3.163}$$

$$\beta = \frac{m}{2kT} \tag{3.164}$$

and

$$\alpha = 4 \frac{n_0}{\pi^{1/2}} \left(\frac{2kT}{m} \right)^{3/2} \tag{3.165}$$

leading to Maxwell's law for the velocity distribution

$$n(c) \, dc = \frac{4n_0}{\pi^{1/2}} \left(\frac{2kT}{m} \right)^{3/2} c^2 \exp - \left(\frac{c^2 m}{2kT} \right) dc \tag{3.166}$$

From whence

$$\langle c \rangle = \left[\frac{8kT}{\pi m} \right]^{1/2} \tag{3.167}$$

and

$$\langle c^2 \rangle = \frac{3kT}{m} \tag{3.168}$$

Figure 3.12 shows the typical shape of $n(c)$ as a function of c, rising from zero at $c = 0$ to a maximum at c_m and asymptotically approaching zero as c tends to infinity.

If the case of identical spherical molecules of diameter d, the collision cross section

$$\sigma = \pi d^2 \tag{3.169}$$

If λ is the mean free path, then a volume $\sigma\lambda$ contains 1 molecule, hence

$$n_0 - \frac{1}{\sigma\lambda} \quad \text{or} \quad \lambda = \frac{1}{\sigma n_0} \tag{3.170}$$

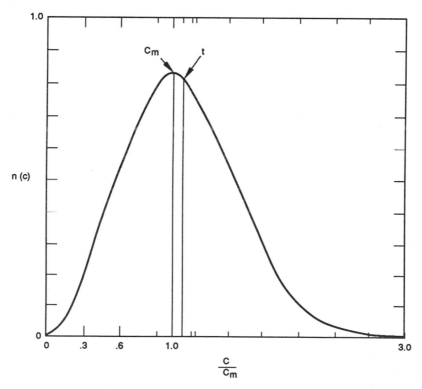

FIGURE 3.12. Maxwell velocity distribution.

The total path swept out by the molecules contained in a unit volume per second is $n_0 \langle c \rangle$. Thus, the collision frequency

$$v = \frac{2n_0 \langle c \rangle}{\lambda} = \frac{2n_0}{\lambda} \left[\frac{8kT}{\pi m} \right]^{1/2} \tag{3.171}$$

TABLE 3.1. Useful Equations Derived from Kinetic Theory

Quantity	Approximate value for air at $T = 300$ K, $P = 10^{-6}$ Torr $= 1.33 \times 10^{-4}$ Pa
1. $n_0 = \dfrac{p}{kT} =$ number of molecules per unit volume	3×10^{16} per m^3
2. $n_1 = \frac{1}{4} n_0 \langle c \rangle =$ number of molecules crossing unit area per second in one direction	4×10^{18} per m^2 per s
3. $\lambda = \dfrac{1}{n_0} \sigma =$ mean free path ($\sigma =$ collision cross section)	51 m
4. $n = \dfrac{2n_0 \langle c \rangle}{\lambda} =$ collision frequency	6×10^{17} per m^3 per s
5. $\langle c \rangle = \left[\dfrac{8kT}{\pi m} \right]^{1/2} =$ mean velocity	
6. $\langle c^2 \rangle^{1/2} = \left[\dfrac{3kT}{m} \right]^{1/2} =$ root mean square velocity	$\approx 5 \times 10^2$ m
7. $c_m = \left[\dfrac{2kT}{m} \right]^{1/2} =$ most probable velocity	
8. $\eta = \frac{1}{3} n_0 \langle c \rangle \lambda m =$ viscosity coefficient	670 SI units
9. $C_v = \dfrac{3k}{2m} =$ specific heat at constant volume (assuming no rotational energy stored)	0.8 J/kg
10. Thermal conductivity $K = \frac{1}{3} m n_0 \langle c \rangle \lambda C_v$ $= \frac{1}{4} k n_0 \langle c \rangle \lambda$	535 SI units

the factor of 2 arising because each collision involves two molecules. In chemical equilibrium the pressure of a saturated vapor over its own liquid is

$$p_s = \tfrac{1}{3}mn_0\langle c^2 \rangle \tag{3.172}$$

The number leaving the surface per unit area per second, N, equals the number arriving, i.e.,

$$
\begin{aligned}
N &= \tfrac{1}{4}n_0\langle c \rangle \\[4pt]
&= \frac{3p_s}{4\dot{m}}\frac{\langle c \rangle}{\langle c^2 \rangle} \\[4pt]
&= \frac{3p_s}{4\dot{m}}\left[\frac{8kT}{\pi m}\right]^{1/2}\frac{m}{3kT} \\[4pt]
&= p_s[2\pi kT]^{-1/2} \text{ molecules per unit area per second}
\end{aligned}
\tag{3.173}
$$

This relationship is useful for vacuum evaporation (Section 4.6).

Using Maxwell's law and the basic concepts of kinetic theory, a number of physical properties can be derived. Those important for vacuum physics are summarized in Table 3.1. The complete kinetic theory is of course quite sophisticated and the reader is referred to the classical texts.[37-39]

3.7.3. Pumping to Low Pressures

The discussion in this section is mainly qualitative since a detailed discourse on the operational principles and physics of vacuum pumps is beyond the scope of this book and is well treated in classical texts[40] and trade literature.[41] The type of pumps now most often used, their ranges and applications are illustrated in Fig. 3.13. From the viewpoint of microfabrication process technologies the next most important consideration after achievement of the desired pressure is to eliminate, insofar as possible, all unwanted contaminants in the vacuum chamber. These arise from:

1. The original air in the chamber before it was evacuated
2. Impurities in gases deliberately introduced for a given application
3. Backstreaming of vapors from the pumps
4. Outgassing from the interior walls and the contents of the vacuum chamber
5. Leaks in the walls and joints of the vacuum chamber

Even if a system is to be operated at medium vacuum for, say, plasma experiments, it is often desirable to first attain an ultrahigh vacuum in the chamber to ensure that all sources of contamination have been eliminated. There is not much that can be done about source (1) above in the very first pumpdown of a system; however, it is good practice when "letting down" a vacuum chamber to atmospheric pressure, to change or introduce new components, to do this in a controlled way so that only filtered dry nitrogen from a compressed gas cylinder is used to raise the pressure.

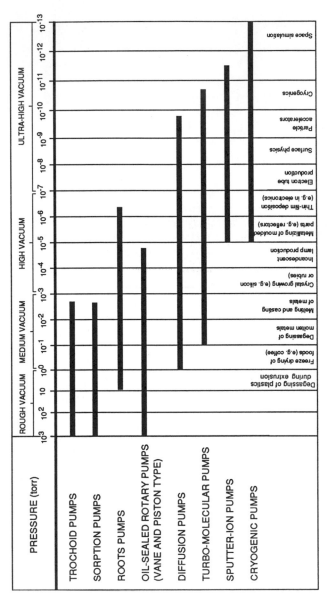

FIGURE 3.13. Types of vacuum pumps and typical applications.

Although in principle source (3) can be prevented by using foreline sorption traps for the roughing pumps and liquid nitrogen-cooled traps for the high-vacuum pumps, for the least contamination, it is best to avoid all pumps which require oil or other low-vapor-pressure liquids for their operation. The cleanest method for removing most of the air in a vacuum chamber is to use an adsorption pump connected to the chamber with a high-vacuum valve. These pumps consist of a vessel containing dry zeolite (a microporous alkali-alumino-silicate) which is cooled to liquid nitrogen temperature (77 K). When the valve is opened, most of the gases in the main chamber are adsorbed on the surfaces of the zeolite and pressures of 10^{-2} Torr may be obtained. Since the bulk of the gas originally in the chamber is now in the pump and the ultimate vacuum that can be obtained depends on the amount of gas adsorbed, it is recommended that it be valved off and a second adsorption pump be valved in to attain the lowest possible pressure by this means. The residual pressure is due to nonadsorbable gases, mainly hydrogen and the moble gases. This technique for roughly clearly introduces the least amount of contaminants.

For ultrahigh vacuum pumping, sublimation pumps, sputter-ion pumps, cryopumps, and turbomolecular pumps all have their place depending upon the application. Sublimation or getter pumps operate by evaporating a chemically active metal such as titanium or zirconium which reacts with the residual gases forming a solid compound on the walls of the pumping chamber where it is buried by the condensing metal vapor. The sputter-ion pump operates in a similar way except that the active metal is sputtered rather than evaporated from a heated pot. The magnetically confined Penning discharge allows ions to be formed in the pump down to very low pressures ($<10^{-12}$ Torr) once the discharge has been initiated. Furthermore, since the residual gas is ionized, it can create highly chemically active species and even unreactive gaseous ions can still be accelerated to the cell walls where they may be buried or implanted at sufficient depth to prevent their escape. Contamination can arise if sputter-ion pumps are allowed to operate at a sufficiently high pressure (about 10^{-3} Torr) that the cell walls become heated to the point of outgassing.

Cryopumps operate by simply condensing the residual gases on vanes cooled to very low temperatures (usually liquid helium temperature 4.2 K). Their pumping speed for condensable gases is very high but after a certain quantity of gas has been condensed they can become saturated, and must be valved off from the vacuum chamber while they are warmed up to remove the excess condensed gases. The Stirling cycle refrigerators used to produce the liquid helium produce strong low-frequency vibrations and an irritating squeaking noise.

Turbo-molecular pumps depend for their operation on the action of rapidly rotating vanes that strike the residual gas molecules so as to give them some momentum in the flow direction. These have been made practical by the use of jet engine technology. Rotational speeds of 25,000–60,000 rpm are used and ultimate pressures of 10^{-10} Torr may be obtained. The main source of contamination from these pumps arises from the bearing lubricants but for the cleanest requirements a magnetic levitation technique has been developed to suspend the rotor without the use of lubricants. Vibrations from the pumps are remarkably low and can be completely isolated from the vacuum chamber by flexible joints.

In certain plasma process technologies, such as reactive ion etching, it is necessary to have a pump that can handle a high throughput of gases (3×10 liters/s) in the medium vacuum range. This need has been met by the Roots blower, which consists of two approximately figure-eight cross-section rotors which rotate in opposite directions around parallel axes. They are synchronized such that they move past the walls of their housing and each other without touching. In this way a volume of gas entering from one side is trapped, compressed, and then expelled from the opposite side. There is a small gap (approximately a few thousandths of an inch) between the rotor and the walls so that no lubricant is required except on the bearings, yet the rotors can rotate at high speeds (3500 rpm) without wear. The high throughput is attained at the expense of a lower compression ratio, in the range of 10 to 100.

The walls of the vacuum chamber and all of the components within it contain gases adsorbed on their surface (mostly water) and adsorbed in the interior even after having been chemically cleaned. Much of this material can be removed by baking the whole system under high vacuum. A 150°C bake for 24 h is very effective for stainless steel systems, but bakes up to 600°C or higher have been employed to obtain the lowest possible residual pressures. To reduce this source of contamination still further, it is sometimes appropriate to use materials that have been previously melted in high vacuum to allow the dissolved gases to escape. Since the chamber walls may be pervious to some of the components of air at high temperatures, the bake is often carried out with the exterior at a medium vacuum. Since most materials are pervious to hydogen at high temperatures, this gas (or a nonexplosive mixture of 2% hydrogen in nitrogen) is also used as an exterior environment for the vacuum chamber during a bake. Hydrogen reacts with impurities within the wall material forming high-vapor-pressure compounds which can escape more easily. Furthermore, it ensures that subsequent outgassing will be mostly hydrogen which is in most cases an inoffensive contaminant.

The fifth source of contamination, leaks, is the bane of the vacuum technologist. However, the development of cleverly designed flanges, valves, feedthroughs, and a whole range of other components, using copper, polyimide, and other vacum-compatible gasket materials, together with helium arc welding techniques, have made it relatively straightforward to assemble an experimental leak-tight high-vacuum system from commercially available components. The most sensitive method to detect small leaks uses a simple mass spectrometer to detect helium which is introduced into the system when a probe in the form of a fine jet of helium gas is close to the external entrance of the leak. Helium has been selected as the optimum probe gas because of its unambiguous identification at mass 4 amu in a simple mass spectrometer (its closest neighbor is hydrogen at 2 amu), its fast effusion rate because of its low mass (second only to hydrogen), and its low occurrence in the atmosphere of about 1 part in 200,000 of air.

3.7.4. Pumping Speed and Conductance

In this section we consider pumping in the high-vacuum region where $\lambda >$ dimensions of the pumping elements. The pumping speed, S, is defined as the volume V passing through the pump entrance per unit time, and is usually given in liters per second, i.e.,

$$S = \frac{dV}{dt} \qquad (3.174)$$

If the pressure in the chamber is p, then the throughput of the pump, q, is defined as the volume pumped per second when measured at unit pressure, usually 1 Torr, i.e.,

$$q = pS \text{ torr liters per second} \qquad (3.175)$$

The number of molecules pumped per second

$$N = \frac{q}{kT} = \frac{pS}{kT} \text{ per second} \qquad (3.176)$$

The mass pumped per second

$$M = N_m = \frac{qm}{kT} = \frac{pSm}{kT} \text{ kg per second} \qquad (3.177)$$

where m is the molecular mass (kg).

The pump is usually connected to the chamber by a tube or at very least an aperture. If p is the pressure in the pump and p_0 the pressure in the chamber, then the conductance of the connecting element C is defined by

$$q = C(p - p_0) \qquad (3.178)$$

Conductance has the dimensions of liters per second. The conductance of the element connecting a pump to a chamber obviously should not be less than the pumping speed of the pump for optimum operation. For a circular aperture diameter d, the number of molecules arriving from the chamber per second

$$N = \frac{\pi d^2}{4} \tfrac{1}{4} n \langle c \rangle = \frac{\pi d^2}{4} \frac{1}{4} \frac{3p}{m} \qquad (3.179)$$

and from the pump

$$N_0 = \frac{\pi d^2}{4} \frac{1}{4} n_0 \langle c \rangle = \frac{\pi d^2}{4} \frac{1}{4} \frac{3p_0}{m} \qquad (3.180)$$

Therefore, the number of molecules pumped per second is

$$N - N_0 = \frac{3\pi d^2}{16m} (p - p_0) = \frac{q}{kT} \qquad (3.181)$$

hence,

$$q = \frac{3\pi d^2 kT}{16m}(p - p_0) \equiv C(p - p_0) \tag{3.182}$$

and

$$C = \frac{3pd^2 kT}{16m} \tag{3.183}$$

Putting appropriate values into Eq. (3.183), we find C equals approximately $9d^2$ liters/ s, for d in centimeters, pumping air at room temperature.

The theory for molecular flow down tubes is more complex[42, 43]; however, for cases when the tube is straight and its length l is greater than a few times the tube diameter d, then $C = 12d^3/l$ liters/s (for d and l in centimeters, pumping air at room temperature). From this we see that the effect of l on C is not substantial for $l \leq d$.

3.7.5. Measurement of Low Pressures

The useful pressure range for microfabrication processes extends from 10^{-2} Torr for plasma and low-pressure CVD to 10^{-10} Torr for ultraclean vacuum evaporation processes such as molecular-beam epitaxy. This range is comfortably covered by a variety of ionization gauges. An ionization gauge is a device that measures the number of positive ions generated by a stream of electrons as they travel from their source to a collector. The number of positive ions, or ion current, i_p produced is proportional to the electron emission current, i_e, and the pressure p, i.e.,

$$i_p = Si_e p \tag{3.184}$$

or

$$p = \frac{1}{S}\frac{i_p}{i_e} \tag{3.185}$$

where S is called the sensitivity of the gauge, and is often given in microamperes (of ioncurrent) per milliampere (of electron current) per Torr (pressure).

The comprehensive theory gives

$$S = \frac{Bp}{Vg}\left[\beta - (V_g - V_i + \beta)\exp-\left(\frac{V_g - V_i}{\beta}\right)\right] \tag{3.186}$$

where B is a constant determined by the geometry of the electrodes and the gas, β depends only on the gas, and V_i and V_g are the potentials relative to the cathode of the ion collector and grid, respectively. Typically for Bayard Alpert gauges (Ref. 38, p. 331), $V_g = +150$ V, $V_i = -45$ V, $i_e = 10^{-4}$ A, $S = 12~\mu$A per mA per Torr for nitrogen. At very low pressures, the following physical effects must be taken into account:

1. The long electron free path before a collision with a residual gas molecule
2. Secondary electron emission from the ion collector due to photons, generated by the impact of the electrons on the electron collector reaching the ion collector
3. Removal of gas from the region being measured by ionization and condensation on the walls (ion pumping) of the gauge
4. Removal and generation of gas by the hot electron (thermionic) source, usually used to generate large, stable electron currents

Errors in pressure measurement due to these problems, once recognized, have been reduced to acceptable levels by various inventions and refinements. The effects due to the last two items are reduced by installing the gauge directly in the vacuum chamber ("nude") rather than as an appendage connected to the chamber by a relatively narrow tube. The first item necessitates finding a method to make an electron oscillate or rotate about a fixed position with an amplitude smaller than the overall dimensions of the gauge, so that the electron can travel a long distance in a small volume before being collected. This may be accomplished by using a magnetic field (Penning gauge) or a grid of fine wires widely spaced held at a positive potential with respect to the electron source so that one electron traverses the grid many times before being collected by it. The photoelectric effect described in item 2 leads to an apparent positive current measured by any negatively charged electrode that serves as the ion collector in the gauge. This current is observed even if there are no gas molecules present. This limit to the lowest pressure that can be measured is called the X-ray limit. To reduce this effect to a negligible value, the ion collector is formed of fine wire which is strung along the axis of a cylindrical gauge. The grid is wound around this axis, usually a few centimeters away, and the hot wire cathodes are wound external to the grid. The solid angle subtended by the grid wires, where the energetic photons are formed by electron impact, to the ion collector is thus made extremely small, and the X-ray limit can be reduced to below 10^{-14} Torr. For the measurement of low pressures, a tungsten filament cathode, which can be cleaned by flashing at very high temperatures (2500 K), is usually used. For the measurement of high pressures, up to 1 Torr, a thoria-coated platinum ribbon which is insensitive to oxygen reactions is used as a stable electron source.

The Penning gauge is essentially a sputter-ion pump and the electrons are generated by what is essentially a magnetically confined Townsend discharge. However, such gauges have to be started by introducing some electrons by a high-voltage breakdown usually induced by a Tesla coil.

Higher pressures are measured using thermocouple or Pirani gauges. In these devices a wire is heated by passing a current through it. The gas cooling changes with decreasing pressure. The wire temperature may be measured with very fine thermocouple wires attached to the heated wire (thermocouple gauge) or by comparing the resistance of the wire to that of a similar wire in a sealed high-vacuum tube (Pirani gauge) in a Wheatstone's bridge arrangement.

In most process applications, a knowledge of the absolute value of the pressure is less important than the ability to measure the rate of change of pressure with time, and to precisely reproduce vacuum conditions. Ultimately, a knowledge of the gas constituents may be required for which a mass spectrometer is necessary. The most popular form is the quadrupole mass spectrometer, which does not require a magnetic field for operation. Spectrometers are available commercially from low-cost devices with mass ranges up to 150 amu, to more expensive devices with mass ranges up to 3000 amu. The quadrupole mass spectrometer[44] consists of four rigidly held parallel rods of hyperbolic cross section. Ions, generated by electron collision within an ionization chamber, are extracted and injected along the axis of the quadrupole. A dc voltage V is applied between one set of opposite pairs of electrodes, and an ac voltage of amplitude U and frequency f is applied between the other pair. Ions of mass M (singly charged) can travel along the axis provided

$$M = \frac{V}{7.2f^2r_0^2} \tag{3.187}$$

and

$$U/V = 0.167 \tag{3.188}$$

where r_0 is the distance from the axis to an electrode. Upon emerging from the quadrupole system the ions can be collected. The mass spectrum is obtained by plotting the ion current as a function of V (while maintaining U/V constant). The ability to resolve masses spaced ΔM apart remains constant in this type of mass spectrometer. The actual value of ΔM depends on the length of the rods and the precision with which they are made and assembled. By counting ions, partial pressures down to 10^{-14} Torr may be measured.

PROBLEMS

1. a. What types of discharge can occur in a gas chamber at low pressure when a dc voltage is applied between two metal electrodes inside the chamber?

   b. Discuss the collision ionization processes that can take place in a glow discharge.

2. a. What is the advantages of using an alternating voltage over a static voltage for energizing the plasma?

   b. What is the Townsend ionization coefficient?

   c. What effect does the collision process have on the energy distribution of the electrons?

   d. Sketch the potential distribution at the cathode. Indicate the Debye region, the space-charge region, and the transition region.

3. A 10-liter vacuum chamber is maintained at a pressure of 5 mTorr and room temperature. Assume the gas is nitrogen with molecular weight 28 g/mole, and molecular diameter 3.7 Å. (Avogadro's number = 6.02×10^{23} molecules/mole.)

   a. Calculate the number of gas molecules in the chamber using the ideal gas law.

b. Define the mean free path.

c. Calculate the mean free path.

d. Calculate the mean gas molecule velocity.

e. Calculate the wall flux.

f. Calculate the approximate residency time of a gas molecule in a chamber if the flow of gas into the chamber is $100\ cm^3/min$ (SCCM) when measured at atmospheric pressure.

REFERENCES

1. J. A. Thornton, Diagnostic methods for sputtering plasmas, *J. Vac. Sci. Technol.* **15**, 188 (1978).
2. H. S. W. Massey, E. H. S. Burhop, and H. B. Gisbody, *Electronic and Ionic Impact Phenomena*, 3 volumes, Oxford University Press, New York (1971).
3. J. J. Thomson and G. P. Thomson, *The Conduction of Electricity Through Gases*, Cambridge University Press, London (1933).
4. B. Chapman, *Glow Discharge Processes*, Wiley, New York (1980).
5. S. C. Brown, *Basic Data of Plasma Physics*, Wiley, New York (1959).
6. L. B. Loeb, *Electrical Coronas*, University of California Press, Berkeley (1965).
7. J. D. Cobine, *Gaseous Conductors*, McGraw-Hill (reprinted by Dover) (1941).
8. A. von Engle, *Ionized Gases*, Oxford University Press (Clarendon), London (1955).
9. V. N. Kondratiev, *Chemical Kinetics of Gas Reaction*, Addison–Wesley, Reading, Mass. (1964).
10. E. W. McDaniel, *Collision Phenomena in Ionized Gases*, Wiley, New York (1964).
11. E. H. Holt and R. H. Haskell, *Foundations of Plasma Dynamics*, Macmillan Co., New York (1965).
12. S. C. Brown, *Introduction to Electrical Discharges in Gases*, Wiley, New York (1966).
13. L. J. Kieffer, Bibliography of Low Energy Electron Collision Cross Section Data, NBS Publ. 289, AD-649 862, and Suppl. AD-865 520, Government Printing Office, Washington, D.C. (1967).
14. G. Carter and J. S. Colligan, *Ion Bombardment of Solids*, Elsevier, Amsterdam (1969).
15. E. W. McDaniel, V. Cermak, A. Dalgarno, E. E. Ferguson, and L. Friedman, *Ion–Molecule Reactions*, Wiley–Interscience, New York (1970).
16. A. T. Bell, in: *Techniques and Applications of Plasma Chemistry* (J. R. Hollohan and A. T. Bell, eds.), Wiley, New York (1974).
17. A. Rutscher, Progress in electron kinetics of low pressure discharges and related phenomena, *Proc. 13th International Congress on Phenomena in Ionized Gases*, Physical Society of the German Democratic Republic, Leipzig, pp.269–289 (1977).
18. J. Thornton, in: *Deposition Technologies for films and Coatings* (J. R. Bunshah, ed.), pp. 19–62, Noyes Press, Park Ridge, N.J. (1982).
19. T. Halicioglu and W. Bauschlicher, Jr., Physics of microclusters, *Rep. Prog. Phys.* **51**, 883–921 (1988).
20. T. Takagi, I. Yamada, and A.Sasaki, Ionized cluster beam technology, *Proc. Conf. Ion Plating and Allied Technologies*, Edinburgh, Scotland (1977).
21. I. Yamada, in: *Semiconductors and Semimetals* (J. I. Pankove, ed.), Academic Press, New York (1984).
22. J. R. Acton and I. D. Swift, *Cold Cathode Discharge Tubes*, Academic Press, New York (1963).
23. M. F. Druvesteyn and F. M. Penning, The mechanism of electrical discharges in gases of low pressure, *Rev. Mod. Phys.* **12**, 88 (1940).
24. J.H. Keller and R. G. Simmons, Sputtering process model of deposition rate, *IBM J. Res. Dev.* **23**(1), 24–32 (January, 1979).
25. W. B. Pennebaker, Influence of scattering and ionization on R.F. impedance of glow discharge sheaths, *IBM J. Res. Dev.* **23**(1), 16–23 (January, 1979).
26. H. S. Butler and G. S. Kino, Plasma sheath formation by radiofrequency fields, *Phys. Fluids* **6**(9), 1346–1355 (September, 1963).
27. I. Brodie, L. T. Lamont, Jr., and D. O. Myers, Substrate bombardment during RF sputtering, *J. Vac. Sci. Technol.* **5**(5), 175 (September, 1968).

28. J. H. Keller and W. B. Pennebaker, Electrical properties of R.F. sputtering systems, *IBM J. Res. Dev.* **23**(1), 3–15 (January, 1979).
29. R. E. Collins, *Foundations for Microwave Engineering*, McGraw–Hill, New York (1966).
30. A. D. MacDonald, *Microwave Breakdown in Gases*, Wiley, New York (1966).
31. F. K. McTaggart, *Plasma Chemistry in Electrical Discharges*, Elsevier, Amsterdam (1967).
32. A. T. Bell, Fundamentals of Plasma Chemistry, *J. Vac. Sci. Technol.* **16**(2), 418–419 (March/April, 1979).
33. T. B. Reed, in: *The Application of Plasmas to Chemical Processing* (R. F. Baddour and R. S. Timmins, eds.), pp. 26–34, MIT Press, Cambridge, Mass. (1967).
34. C. F. Powell, J. H. Oxley, and J. M. Blocker, Jr. (eds.), *Vapor Deposition*, Wiley, New York (1966).
35. M. J. Rand, Plasma-promoted deposition of thin inorganic films, *J. Vac. Sci. Technol.* **16**(2), 420–427 (March/April 1979).
36. A. von Engel, *Handbuch der Physik*, Vol. 21, p. 504, Springer-Verlag, Berlin (1956).
37. D. ter Haar, *Elements of Statistical Mechanics*, p. 381, Holt, Rinehart & Winston, New York (1960).
38. J. Jeans, *An Introduction to the Kinetic Theory of Gases*, Macmillan Co., New York (1940).
39. S. Chapman and T. G. Cowling, *The Mathematical Theory of Non-Uniform Gases*, Cambridge University Press, London (1939).
40. S. Dushman and J. M. Lafferty, *Scientific Foundations of Vacuum Technique*, 2nd ed., Wiley, New York (1962).
41. Product and Vacuum Technology Reference Book, Leybold-Heraus, Export, Pa. (1988).
42. G. L. Saksagaskii, *Molecular Flow in Complex Vacuum Systems*, Gordon & Breach, New York (1988).
43. W. Stechelmacher, Knudsen flow 75 years on, *Rep. Prog. Phys.* **49**, 1083–1107 (1986).
44. R. W. Kiser, *Introduction to Mass Spectrometry and Its Applications*, Prentice-Hall, Englewood Cliffs, N.J. (1965).

4

Layering Technologies

4.1. INTRODUCTION

Planar technology requires that thin layers of materials be formed and patterned sequentially, commencing with a flat rigid substrate. The key aspects of each layer are its

- Thickness
- Interface properties
- Structure, composition, and topography
- Electronic properties
- Optical properties
- Mechanical properties
- Chemical properties

These aspects relate to the mechanism of operation of the final device, adhesion of adjacent layers, the stability of the film, the etching properties when patterning the layer, and the elimination of unwanted defects in the layer such as pinholes.

Layers may be produced by:

- Modifying the substrate material at its surface by chemical or physical means
- Physically depositing a layer on a surface
- Chemically forming a layer at a surface

Because producing thin layers of diverse physical and chemical properties is extraordinarily varied and complex, the objective of this chapter is limited to providing the basic knowledge and terminology required to tackle the larger body of literature, rather than provide specific recipes. In general, the base substrate on which a microdevice is built may be single crystal, polycrystalline, or amorphous (glassy). It may be conducting, semiconducting, insulating, or semi-insulating, and it may or may not participate in the operation or formation of the device. However, the majority of microdevices today are built on single-crystal semiconducting wafers of silicon or gallium arsenide. For the preparation of substrate materials, as opposed to its surface modification, the reader is referred to the specialist literature.[1]

315

Thin deposited films, which have been of general interest since 1857 when Faraday[2] deposited gold films on glass in vacuum using an exploding wire technique, are especially important in fabricating microdevices. This is because by depositing a thin film, at least one dimension of the device is relatively straightforward to control with high precision. Single molecular layers (monolayers) a few angstroms thick to a layer several light wavelengths thick (microns) can be formed. Of equal consequence are the unique physical properties exhibited by thin films because of the influence of the surfaces, their special structure, the crystal arrangements, and the chemical compositions that may be deposited.

To form a coating on the surface of a solid substrate, particles of the coating material must travel through a carrier medium in intimate contact with the surface. Upon striking the surface a substantial fraction of the coating particles must adhere to it or by chemical reaction at the surface form a new compound that adheres. The particles may be atoms, molecules, atomic ions, ionized fragments of molecules, or grains of material, both charged and uncharged. The carrier medium may be a solid, liquid, gas, or vacuum.

In the case of a solid carrier, intimate contact between the substrate and carrier surfaces is often difficult to obtain, and only particles that can diffuse through the solid at a reasonable rate are useful. This restricts solid carrier usage to only a few specific combinations. A common example is the diffusion of oxygen through silicon dioxide, which is used in growing silicon dioxide layers on silicon (see Section 4.3.2). The oxygen reaching the surface of the silicon reacts, forming silicon dioxide, thereby increasing the thickness of the oxide layer from beneath. The diffusion rate decreases as the oxide layer grows thicker and essentially ceases when the oxide layer is about 1.3 μm thick.

A liquid carrier medium is much more versatile, since many compounds can be dissolved and intimate contact with the substrate surface is readily obtained. Microgranules of compounds that cannot be dissolved can be put in suspension. In the case of dissolved ions or charged grains, the transport rate to the substrate surface may be substantially increased by the application of an electric field.

We distinguish between a gaseous carrier medium and a vacuum by the mean free path of the transported particles. At high carrier gas pressures the particle has many collisions with the carrier gas molecules before arriving at the surface to be coated, but in a vacuum the particle moves from its source to the substrate surface with a very low probability of collision with the residual gas molecules. The coating particles may be premixed, suspended, or dissolved in the carrier medium, which is depleted as deposition proceeds. Often the coating material is continuously replaced by introduction into the carrier medium from some source such as an electrode in electrolytic deposition.

Thus, in order to characterize any given deposition process, we must identify the following:

- The carrier medium (solid, liquid, gas, vacuum)
- The nature of the coating particles (atom, molecule, ion, grain)
- The method of introducing coating material into the carrier medium (premixed or dissolved material, precipitation of premixed material in carrier, evaporation, electrochemical reaction at a supply electrode surface, bombardment of a supply surface by energetic particles)

- The mechanism by which the coating particles are transported from source to substrate (free flight, gaseous diffusion, liquid diffusion)
- The surface reaction involved (simply condensation, chemical reaction between transported components at the surface, evaporation of carrier liquid, electrochemical reaction at the surface, implantation)

Table 4.1 characterizes a number of common coating methods. The choice of the deposition process will be dictated by the composition, purity, stoichiometry, and morphology required of the coating and physical properties sought.

The condition of a surface prior to deposition of a film upon it is of prime importance for adhesion of the film and other interface properties. The major culprits are oil or greasy residues left from protective coatings or fingers. These are best removed by warm ionic detergents that are ultrasonically agitated to render them water-soluble, followed by rinsing in pure water, which is usually in plentiful supply through modern filtering and deionizing techniques.[3] Other sources of surface contamination are corrosion or the reaction products formed between the material of the surface and the ambient to which it has been exposed. These can be removed by using chemical or electrochemical processes which react at the surface with the corrosion or the underlying material-forming compounds that can be dissolved in a carrier fluid (if possible, water). Removal of the reactants and reaction products by rinsing with pure water is again important.

Even after such careful treatment, most surfaces remain contaminated at levels from one to several molecular layers with adsorbed oxygen, water, or other air contaminants. At worst, adherent layers of oxide many molecular layers thick may be found, such as form on aluminum or silicon surfaces. These layers can only be removed in high vacuum to prevent them from re-forming. Outgassing and the removal of weakly adsorbed layers can be accomplished by baking at moderate temperatures up to 600°C. However, heating to remove the last monolayer of oxygen is practical in only a few cases of highly refractory materials, such as tungsten, that can be heated to the very high temperature (3000 K) required to desorb the last monolayer of oxygen, or in materials that have low adsorption energies for oxygen, such as SiO_2. Sputtering away the surface with an inert gas such as argon is the most popular way of removing this last atomic layer of contamination, but low bombardment energies (<500 V) should be used to prevent surface damage and low current densities (100 $\mu A/cm^2$) should be used to prevent heating of the surface, which may give rise to thermal etching.[4] Electron bombardment can also be effective in removing certain residual monolayers.

The cleavage of single crystals in a high vacuum produces perhaps the most perfect surfaces from the point of view of cleanliness. However, many cleaved crystal surfaces exhibit cleavage steps. If atomically smooth surfaces are required, these can be obtained only by the cleavage of strongly layered crystals such as mica. Atomically smooth surfaces can also be obtained on glasses and metals by long annealing close to the melting point (see Ref. 7, p. 63).

We emphasize that there are no really general methods for creating a thin layer that can be used in all or even a majority of situations. Each device requires specific physical/chemical properties for the layer and must be treated individually. Research to develop the optimum technique for a given layer–substrate combination often

TABLE 4.1. Characterization of Common Coating Methods

Process	Example	Carrier medium	Coating particles	Method of injecting coating particles into medium	Surface reaction	Transport mechanism
Oxidation	Growth of SiO_2 on silicon	Solid	Atoms	Diffusion from substrate	Chemical reaction with gas at surface	Diffusion
Settling	Cathodoluminescent screens (for CRT)	Liquid	Grains	Premixed	—	Settling under gravity
Electrophoresis	Coating of insulation on heater wires (tungsten, alumina)	Liquid	Charged grains	Premixed	Particle neutralization	Electromigration
Electroplating	Copper on steel	Liquid	Ions	From copper electrode	Ion neutralization	Electromigration
Liquid pyrolysis	CdS on metal	Liquid	Molecules	Predissolved	Chemical reaction at surface	Diffusion and stirring
Electrostatic toning	Xerography	Liquid or air	Charged grains	Premixed	Charge neutralization	Electromigration
Spray pyrolysis	CdS on metal	Air	Droplet of liquid containing dissolved reactants	Premixed	Chemical reaction at surface	Carried by air flow
Chemical vapor deposition	Tungsten on metal	Gas	Molecules	Premixed	Chemical reaction at surface	Diffusion
Evaporation	Aluminum on glass	Vacuum	Atoms	Heating source material	Condensation	Diffusion vacuum
Sputtering	Gold on silicon	Gas	Atoms	Bombardment by positive ions of carrier gas	Condensation	Diffusion
SPIN coating	Resists on wafers	Liquid	Organic molecules	Dissolved	Evaporation of solvents	—
Liquid-phase epitaxy	Gallium arsenide	Liquid	Atoms/molecules	Acts as its own carrier medium	Crystal growth	Diffusion
Molecular beam epitaxy	Gallium arsenide	Vacuum	Atoms/molecules	Evaporation from heated source	Condensation	Free flight in vacuum
Flame spraying	Crankshaft buildup (steel)	Gas	Molten droplets	Injection from heated source	Condensation	Transfer from carrier gas
Ion implantation	Boron into silicon	Vacuum	Ions	Plasma or ion gun	Implantation	Free flight in vacuum

takes years. A number of excellent books and papers that review the current status are available.[5-9] The layers often have to be annealed. A number of transient annealing techniques have been explored using lasers, electron beams, and radiant heating techniques.[10]

Since the major thrust in the microdevice area for the past 30 years has been in silicon microelectronics, most work has been carried out on silicon layering technologies, which are more often discussed in this chapter to illustrate particular concepts.

4.2. EPITAXY

Epitaxy, which is essentially ordered growth on a crystal plane to form a thin layer of crystalline material with controlled orientation and impurity content, is an important method for producing semiconductor layers with specific physical properties. Epitaxial techniques[11,12] have been credited with increasing the yield of silicon ICs because they eliminate the long diffusion times required to produce uniformly doped layers (see Section 4.4). The focus of this section is on the epitaxial growth process and its simple physical modeling, illustrated by appropriate applications in silicon, gallium arsenide, or related technologies.

4.2.1. Direct Processes

Growth mechanisms can be divided into direct and indirect processes.[13] In the direct process, silicon is transferred in atomic form directly from the source to the substrate (Si wafer) surface. Under proper deposition conditions, the arriving Si

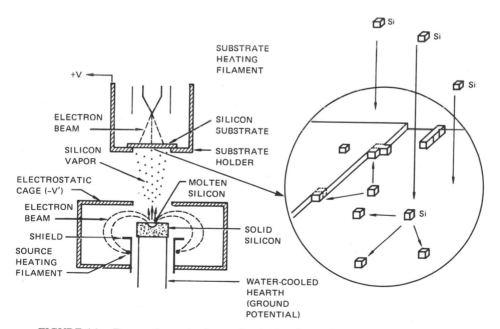

FIGURE 4.1. Evaporation and substrate heating by electron bombardment. From Ref. 6.

atoms are mobile on the heated substrate surface and align themselves according to the substrate crystal structure. Epitaxial growth occurs due to the formation of nucleation sites and subsequent lattice matching as two-dimensional islands grow across the surface (Fig. 4.1) (see Section 4.14).

The rate of formation R_n of nucleation sites depends on the concentration of the silicon in the vapor and on the free energy of formation of the site, i.e.,

$$R_n \approx n_0 \exp\left(-\frac{\Delta G}{kT}\right) \tag{4.1}$$

where n_0 is the silicon concentration in the vapor, ΔG is the free energy of formation of the nucleation sites, T is the temperature of deposition, and k is Boltzmann's constant.

In addition to evaporation, other examples of the direct deposition method include sublimation and sputtering (see Section 4.7). Although epitaxial films can be obtained by direct methods, the indirect techniques are better suited for the fabrication of controlled doped layers.

4.2.2. Indirect Processes

Indirect processes which may also be considered as chemical vapor deposition (Section 4.3) are those in which silicon atoms are obtained by the decomposition of a silicon compound at the heated substrate surface. Figure 4.2 shows the schematic of a rf-heated horizontal epitaxial reactor.[14]

Table 4.2 describes the various source gases and their functions in the reactor.[15]

Silane ($SiCl_4$) is the most common deposition chemical because it is readily available and its high temperature of deposition makes it less sensitive to oxidation, which (at a few ppm) can lead to surface defects. The minimum deposition temperature and the maximum growth rate are limited by the trapping of structural defects, which occurs if the surface diffusion rate is lower than the arrival rate of the silicon atoms.

At the silicon surface, silane decomposes according to

$$SiH_{4(g)} \xrightarrow[1000°C]{} Si_{(S)} + 2H_{2(g)} \tag{4.2}$$

and the silicon growth rate is proportional to the partial pressure of silane. For cases where the reaction is surface controlled, epitaxial growth proceeds according to the following steps[15]:

- Mass transfer of the reactant molecules (SiH_4) by diffusion from the turbulent layer across the boundary layer to the silicon surface
- Adsorption of the reactant atoms on the surface
- The reaction or series of reactions that occur on the surface
- Desorption of the by-product molecules
- Mass transfer of the by-product molecules by diffusion through the boundary layer to the main gas stream
- Lattice arrangement of the adsorbed silicon atoms

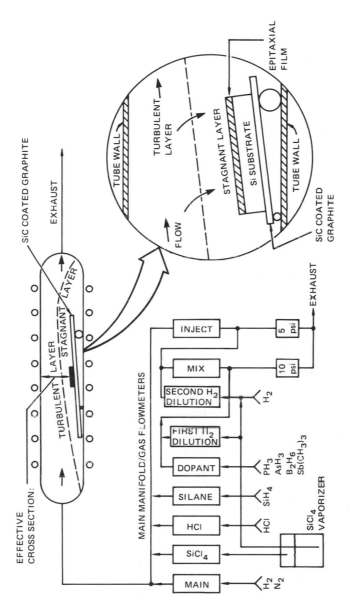

FIGURE 4.2.   rf-heated horizontal epitaxial reactor.

TABLE 4.2. Source Gases and Their Functions as Used in Epitaxial Deposition

Gas	Function	Comments
N_2	Main flow	Purges out explosive/poisonous gases prior to opening the reactor tube to air
H_2	Main flow	Most common ambient for the growth of epitaxial layers
$SiCl_4$	Si source	Common liquid Si source; vaporized in a H_2 bubbler; corrosive vapor; temperature range, 1150–1200°C; growth rate, 0.2–10 μm/min
SiH_4	Si source	Common gaseous Si source; pyrophoric gas; temperature range, 1000–1050°C; growth rate 0.2–1.0 μm/min
HCl	Si etchant	Most common Si etchant used for substrate preparation; corrosive poison gas
PH_3	Si dopant	Most common phosphorus source for doping epitaxial silicon; flammable poison gas
AsH_3	Si dopant	Behaves like PH_3
$Sb(CH_3)_3$	Si dopant	A liquid antimony source used as a vapor at a concentration of a few hundred ppm in H_2; used because SbH_3 is unstable; poisonous vapor
B_2H_6	Si dopant	Behaves like PH_3

The overall deposition rate is determined by the slowest process in the list above. Under steady-state conditions, all steps occur at the same rate and the epitaxial layer grows uniformly. One of the major objectives of silicon epitaxy is the controlled inclusion of dopants.[16-20] Doping is normally accomplished by the incorporation, in hydride form, of B, As, or P in the main gas flow. The amount of dopant incorporated is approximately proportional to its partial pressure. Unwanted sources of impurities are the reactor (walls, structure) and the wafer itself.

Autodoping is the term used to describe the dopant contribution from the wafer to the epitaxial layer. Autodoping can be divided into two categories:[14] macroautodoping, where unwanted dopants are transported within the reactor from one wafer to another, and microautodoping, where unwanted dopants are concentrated into a localized region of the same wafer. The trends in silicon epitaxy are toward reduced growth temperatures to provide better control of autodoping and growth rates, and also to avoid wafer distortion.

An interesting application of silicon epitaxy is the silicon-on-sapphire (SOS) technology, where a component is constructed within a silicon island that is totally electrically isolated from other such islands within the same IC.[21,22] The sapphire crystal lattice closely matches that of silicon. For this process a 0.5- to 1-μm silicon layer is grown epitaxially on the sapphire substrate in which these islands are etched, and p and n wells fabricated for further silicon processing. The result is a very densely packed IC without the parasitic effects obtained with bulk silicon processing.

4.2.3. Liquid-Phase Epitaxy

The fabrication of devices from the III–V compound semiconductors is based totally on epitaxial layers grown on bulk single-crystal substrates. For many materials (e.g., GaAs), epitaxial layers are grown from liquid solutions. In liquid-phase epitaxy

(LPE), the substrate is thermally equilibrated with the solution and growth takes place upon supersaturation of the solution brought about by a temperature decrease.

In current-controlled LPE[23,24] (electroepitaxy), an electric current is made to flow through the growth interface while the temperature of the system is maintained constant. Layers of semiconductors, including InSb, GaAs, InP, and GaAlAs as well as garnet layers, have been successfully grown in this way. The growth behavior is related to the Peltier cooling of the liquid–substrate interface, which leads to supersaturation and electromigration[25,26] (caused by electron–momentum exchange).

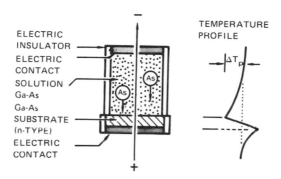

(a) GROWTH CELL USED FOR ELECTROEPITAXIAL GROWTH OF GaAs FROM Ga·As SOLUTION

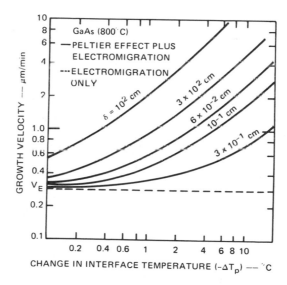

(b) GROWTH VELOCITY OF GaAs FROM Ga·As SOLUTION AS A FUNCTION OF ΔT_p

FIGURE 4.3. Electroepitaxial growth. (a) Growth cell used for electroepitaxial growth of GaAs from Ga·As solution; (b) growth velocity of GaAs from Ga·As solution as a function of ΔT_p. From Ref. 23.

Standard LPE reactors are easily modified to permit current flow through the solution–substrate interface (see Fig. 4.3a). Since the substrate and solution have different thermoelectric coefficients, a current flow across their interface is accompanied by Peltier cooling or heating, depending on the current direction. The magnitude of this heat, Q, is equal to the difference in Peltier coefficients π_p times the current density j. For III–V semiconductors, Q is on the order of 1 W/cm^2 and the current density is about 10 A/cm^2. ΔT_p is typically on the order of 0.1–1°C.

The growth rate is strongly influenced by the formation of a solute boundary layer of thickness δ, as shown in Fig. 4.3b, where the growth velocity is plotted as a function of the interface temperature difference $(-\Delta T_p)$. From Fig. 4.3b it is evident that the contribution of electromigration to growth is dominant for large boundary-layer thicknesses, but the Peltier effect dominates the growth for small δ.

4.2.4. Molecular-Beam Epitaxy (MBE)

MBE is essentially the coevaporation of elemental components onto a heated single-crystal substrate, but it is so named to emphasize the control of the crystal composition attained by this method. The technique consists of directing controlled "beams" of the required atoms from effusion-cell ovens (which can be shuttered to change from the growth of one type of crystal to another) toward the heated substrate in a vacuum of 10^{-10} Torr. The technology of this technique has been extensively reviewed.[27,28] An important feature of MBE is that its growth rate is relatively low (60–600 Å/min), which permits the precise control of the thickness of the epitaxial layers. In fact, it is possible to grow AlAs and GaAs in alternate monolayers, permitting the growth of structures with tailored transport properties.

Probably the most important advantage of the MBE growth technique is that it is possible to do several different types of analyses *in situ* during the growth of the epitaxial layers. Commercial MBE growth systems usually include reflection electron-diffraction equipment, a mass spectrometer, and an Auger spectrometer with ion-sputtering capability. Although an ultrahigh vacuum is required, most commercial systems also incorporate a sample-exchange interlock so that atmospheric contamination of the growth chamber is prevented and rapid throughput exchange can be accomplished. Most of the work on MBE has been directed toward GaAs and AlGaAs to fabricate novel single-crystal structures. Single-crystal multilayered structures or "superlattices" having component layers that differ in composition but are lattice-matched throughout form the basis for semiconductor devices in which both signs of charge carriers (holes and electrons) can be manipulated. When the layers are only a few atoms thick, electrons can move relatively long distances ballistically, i.e., without colliding with the lattice, thus providing new possibilities for device design. The fabrication of the double-heterostructure (DH) laser is an important example of this type of device.[27]

The GaAs–Al$_x$Ga$_{1-x}$As, DH laser in its simplest version is a small, rectangular single-crystal parallelepiped consisting of an n-type GaAs substrate with at least three layers (n-Al$_x$Ga$_{1-x}$As, p-GaAs, and p-Al$_x$Ga$_{1-x}$As) grown epitaxially onto it, as illustrated in Fig. 4.4a. The alignment of the conduction and valence bands of the composite structure when forward biased (n side negative) with a voltage of about the width of the GaAs energy gap is shown schematically in Fig. 4.4b. As a result of the forward bias, electrons are injected into the conduction band of the p-GaAs layer, where they recombine with the holes and emit radiation with approximately the energy of the GaAs energy gap, $E_{g\text{GaAs}}$.

Note in Fig. 4.4b that at the heterojunction there are potential barriers that prevent holes (e^+) and electrons (e^-) from diffusing beyond the GaAs region. The injected electrical carriers are thus confined to the GaAs layer. In addition, because GaAs has a higher refractive index than does $Al_xGa_{1-x}As$, the $Al_xGa_{1-x}As$–GaAs–$Al_xGa_{1-x}As$ three-layer sandwich is a waveguide so that the generated light tends to be confined to the GaAs layer. The cleaved ends of the parallelepiped act as partial mirrors. Thus, light of energy of approximately $E_{g\text{GaAs}}$ is generated by an electronic transition in a waveguide within a Fabry–Perot cavity formed by the mirrors. With a sufficiently high current through the device, stimulated emission and lasing result.

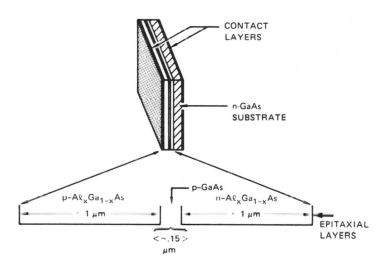

(a) THE GaAs-$Al_xGa_{1-x}As$ DOUBLE HETEROSTRUCTURE LASER

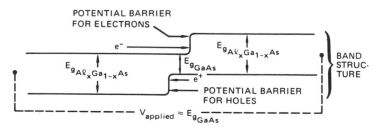

(b) SCHEMATIC OF THE ALIGNMENT OF THE CONDUCTION AND VALENCE BANDS OF THE COMPOSITE STRUCTURE

FIGURE 4.4. A double heterostructure laser. (a) The GaAs–$Al_xGa_{1-x}As$ double heterostructure laser; (b) schematic of the alignment of the conduction and valence bands of the composite structure. From Ref. 25.

The DH laser provides an excellent illustration of how heterostructures are used to manipulate light and electrical carriers in a single solid-state device. MBE techniques also can be used to fabricate monolayer structures and quantum wells, which consist of alternating layers of Ga, As, and Al, with each layer about 50–400 Å thick. The steps joining the conduction and valence band edges at the hetero-junctions form the boundaries of potential wells that tend to contain electrons in the conduction band and holes in the valence band. These quantum wells modify the conduction properties of the multilayer structures so that they differ from the bulk semiconductors. Since the wells act as one-dimensional containers for the electrons, the energy of the electrons is quantized, giving rise to a band structure for these confined carrier quantum states.[27]

MBE, unlike other crystal growth methods, also permits the growth of complex epitaxial structures with controlled lateral dimensions by using shadow masking. These developments have resulted in the fabrication of very thin (5–500 Å) structures and novel optical and microwave devices.

4.2.5. Artificial Epitaxy (Graphoepitaxy)

The growth of oriented polycrystalline or, preferably, single crystal layers on amorphous substrates, such as are used as insulating or encapsulation layers, could be of great importance for the fabrication of multilayers of devices. Techniques involving rapid thermal recrystallization using electron beam, laser beam, and infra-red heating are being explored. However, if the amorphous substrate is first patterned with a crystallographically symmetric microrelief structure control can be gained on the orientation of the depositing film. As an example, techniques have been developed at MIT Lincoln Laboratory[124] which produce a layer of crystalline silicon, about 0.5 μm thick, with a predetermined crystallographic orientation. To produce the crystalline layer, a layer of amorphous silicon is deposited onto a slab of fused silica that has a shallow grating etched into it by photolithography. The grating has a square-wave cross section with a 3.8-μm spatial period, a 1000-Å depth, and a corner radii of about 50 Å. The amorphous silicon is then fused by heating it with a laser beam. It then recrystallizes as a single crystal, with the $\langle 100 \rangle$ direction perpendicular to the plane of the substrate and the $\langle 001 \rangle$ direction along the grating.

This technique for growing an oriented semiconductor film on an amorphous substrate in turn opens new opportunities in the fabrication of microelectronic devices. One can have different crystal orientations on the same substrate or fabricate three-dimensional devices by growing a second semiconductor layer on the top of an amorphous overlayer, such as SiO_2. Such techniques are currently being extensively studied under the titles of artificial epitaxy or graphoepitaxy. This subject is treated comprehensively in Ref. 125 and is discussed further in Section 4.11.7.

4.3. CHEMICAL VAPOR DEPOSITION (CVD)

4.3.1. CVD Processes

In CVD, reagents in the vapor phase, often diluted with an inert carrier gas, react at a hot surface to form a solid film of the desired material; all other reaction products, being gaseous, are carried away.[29] Its importance lies in its versatility for depositing a large variety of elements and compounds at relatively low temperatures and a wide range of pressures including atmospheric. Amorphous, polycrystalline, epitaxial, and uniaxially oriented polycrystalline layers can be deposited with a high degree of purity and control.

Aspects of CVD include the chemical reactions involved, the thermodynamics and kinetics of the reactors, and the transport of material and energy to and from the reaction site. The following is a list with examples of some of the common types of chemical reactions used in CVD.

Pyrolysis. The simplest CVD process is pyrolysis, in which a single gaseous compound decomposes on a hot surface to deposit a stable residue. Examples are the following: deposition of pyrolytic graphite from methane (CH_4), which takes place at a substrate temperature of 2200°C; deposition of silicon from silane (SiH_4), which takes place in the range 800–1350°C; and deposition of nickel from the carbonyl [$Ni(CO)_4$], which takes place at about 100°C. Metal organic compounds such as gold or platinum resinates are oily liquids that decompose at a few hundred degrees to leave the noble metal residue (this process is referred to by the acronym MOCVD). This process was developed in the 19th century for depositing gold films on porcelain objects, and is currently used for producing superlattice structures.[30]

Reduction. Hydrogen is the most commonly used reducing agent. Examples are deposition of silicon by the hydrogen reduction of silicon tetrachloride, which takes place at about 1000°C, and deposition of tungsten by the hydrogen reduction of tungsten hexafluoride, which takes place at about 800°C. Hydrogen reduction is also used to accelerate the pyrolytic process by removal of the unwanted by-products as gaseous hydrogen compounds, for which less energy is required.

Oxidation. Silicon dioxide films can be deposited by the reaction of silane with oxygen.

Nitridation. Silicon nitride films can be deposited by reaction of silane with ammonia.

Carbidization. Titanium carbide films can be deposited by reaction to titanium tetrachloride with methane at a substrate temperature of 1850°C.

Chemical-transport reaction. For these processes the transport of the desired material from the source to the substrate on which it is to form a film depends on the difference in equilibrium constants between the reactant source and a carrier phase, and the substrate and the carrier phase, when each is held at a different temperature. For example, the deposition of gallium arsenide by the chloride process depends on the reversible reaction

$$6GaAs_{(g)} + 6HCl_{(g)} \underset{T_2}{\overset{T_1}{\rightleftharpoons}} As_{4(g)} + GaCl_{(g)} + 3H_{2(g)} \tag{4.3}$$

where T_1 is the temperature of the solid GaAs source, T_2 is the temperature of the solid GaAs substrate, and $T_1 > T_2$. This allows, in effect, indirect distillation of the gallium arsenide from the hot source at temperature T_1 to the cooler substrate at temperature T_2 through an intermediate gas phase of different chemical composition.

Spray pyrolysis. In this process the reagents are dissolved in a carrier liquid, which is sprayed onto a hot surface in the form of tiny droplets. On reaching the

TABLE 4.3.   Sprayable Materials[a]

Film	Starting materials	Solvents	References
SnO_2	$SnCl_4$	H_2O + HCl	1, 7, 78, 79
		H_2O + alcohol	9, 15, 80–82
	$SnCl_2$	Ethanol	5, 56, 83, 84
	$SnBr_4$	HBr	78
	$(NH_4)_2SnCl_6$	H_2O	85, 86
	$(CH_3COO)_2SnCl_2$	Ethyl acetate	86
In_2O_3	$InCl_3$ or $InCl_2$	Butyl acetate or butanol	19, 77
	Indium acetylacetonate	Acetylacetone	7
	$InCl_3$	Methanol–water	9, 56
ITO	$InCl_3$, $SnCl_4$	Methanol–water or ethanol–water	9, 15, 87, 88
Cu_2S	$Cu(C_2H_3O_2)_2$ + thiourea	Water	45
PbS	$Pb(CH_3COO)_2$, $PbCl_2$, or $Pb(NO_3)_2$ + thiourea	Water	56
PbO	$PbCl_2$	Water	56
Cr–Co oxides	Cr acetylacetonate + Co acetylacetonate	Toluene–methanol	90
Cr_2O_3	Cr acetylacetonate	Butanol	7
Fe_2O_3	Iron acetylacetonate	Butanol	7
V_2O_3	Vanadium acetylacetonate	Butanol	7
Pd	Palladium acetylacetonate	Butanol	7
Ru	Ruthenium acetylacetonate	Butanol	7
$CdSnO_4$, $CdSnO_3$	$CdCl_2$ + $SnCl_4$	HCl	69, 91, 92
$CuInSe_2$	$InCl_3$, CuCl, N,N-dimethylselenourea	Water	6, 74
$CuInS_2$	$InCl_3$, CuCl or $CuCl_2$, N,N-dimethylthiourea	Water	6, 74
$CuGaS_2$	CuCl or $CuCl_2$, $GaCl_2$, N,N-dimethylthiourea	Water	6, 73, 74
$CuGaSe_2$	CuCl or $CuCl_2$, $GaCl_2$, N,N-dimethylselenourea	H_2	6, 73, 74
ZnO	$ZnCl_2$	Water	56
B_2O_3	Boron chloride	Water	56
Al_2O_3	Al Cl_3	Water	56

[a]Source: Ref. 31.

hot surface the solvent evaporates and the remaining components react, forming the desired material. An example is the formation of cadmium sulfide films by spray pyrolysis of cadmium chloride and thiourea dissolved in water with the substrate at about 300°C.[32] Table 4.3 taken from a review article[31] lists many materials that can be deposited by this technique.

4.3.2. Oxidation of Silicon

The oxidation of single-crystal silicon to form a passive and protective SiO_2 layer is one of the most important processes in microelectronics, and for this reason is accorded more comprehensive treatment.

4.3.2.1. The Deal–Grove Model.   The thermal oxidation of silicon is commonly performed in a water vapor or oxygen atmosphere over the temperature range

700–1250°C. The oxidation process consists of the diffusive transport of oxygen through the growing amorphous SiO_2 layer, followed by the reaction with Si at the interface.

Oxidation proceeds much more rapidly in a wet-oxygen ambiance, which is used for the formation of thicker protective layers. The basic theory of thermal oxidation[33–35] can be illustrated with the help of the simple (Deal–Grove) model shown in Fig. 4.5.

It is assumed that oxidation takes place at the Si–SiO_2 interface, so the oxidizing species must diffuse through any previously formed oxide and then react with the silicon at this interface. According to Henry's law, the equilibrium solid concentration is proportional to the bulk gas partial pressure P. Thus,

$$C^* = HP \tag{4.4}$$

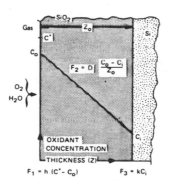

(a) FLUX CONDITIONS FOR THE SiO_2–Si SYSTEM

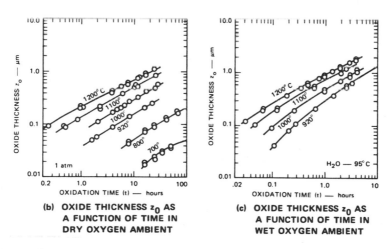

(b) OXIDE THICKNESS z_0 AS A FUNCTION OF TIME IN DRY OXYGEN AMBIENT

(c) OXIDE THICKNESS z_0 AS A FUNCTION OF TIME IN WET OXYGEN AMBIENT

FIGURE 4.5.   Thermal oxidation in wet and dry oxygen ambients. (a) Flux conditions for the SiO_2–Si system; (b) oxide thickness z_0 as a function of time in dry oxygen ambient; (c) oxide thickness z_0 as a function of time in wet oxygen ambient. From Ref. 33.

where C^* is the maximum oxidant concentration for a given P, and H is Henry's law coefficient. The oxidant concentration in this nonequilibrium case is less than C^* on the solid surface. The flux F_1 is determined by the difference in oxidant concentrations:

$$F_1 = h(C^* - C_0) \tag{4.5}$$

where C_0 is the oxidant surface concentration and h is the main transfer coefficient.

The oxidant concentration C_0 at the oxide surface is determined by the temperature, the gas flow rate, and the solid solubility in the oxide. To determine the oxide growth rate, we consider the fluxes of the oxidant in the oxide (F_2) and at the oxide–silicon interface (F_3). From Fick's law, the flux across the oxide is given by the gradient of the oxidant concentration:

$$F_2 = D \frac{dC}{dz} = \frac{D(C_0 - C_i)}{z_0} \tag{4.6}$$

where C_i is the oxidant concentration, in molecules/cm^3, at $z = z_0$; D is the diffusion constant at a given temperature; and z_0 is the oxide thickness.

The flux (F_3) at the oxide–silicon interface is determined by the surface reaction rate constant K and is given by

$$F_3 = KC_i \tag{4.7}$$

Under steady-state conditions, these fluxes must be equal so that $F_3 = F_1 = F_2 = F$. Therefore, we can equate Eqs. (4.6) and (4.7) to yield C_i and C_0 in terms of C^*:

$$C_i = \frac{C^*}{1 + K/h + Kz_0/D} \tag{4.8}$$

and

$$C_0 = \left(1 + \frac{Kz_0}{D}\right)C_i \tag{4.9}$$

To describe the rate of oxide growth, the flux equation at the SiO$_2$–Si interface is written in the form

$$N_i \frac{dz_0}{dt} = F_3 \frac{KC^*}{1 + K/h + Kz_0/D} \tag{4.10}$$

The oxide growth rate is determined by the flux (F_3) and the number of oxidant molecules (N_i) needed to form a unit volume of oxide. Since there are 2.2×10^{22} SiO$_2$ molecules cm^3 in the oxide, we need 2.2×10^{22} molecules/cm^3 of O$_2$, or 4.4×10^{22} molecules/cm^3 of H$_2$O.

The following integral defines the relationship of z_0 and t:

$$N_i \int_{z_1}^{z_0} \left(1 + \frac{K}{h} + \frac{Kz_0}{D}\right) dz = KC^* \int_0^t d\tau \tag{4.11}$$

Integrating gives

$$z_0^2 + Az_0 = B(t + \tau) \tag{4.12}$$

where

$$A \equiv 2D\left(\frac{1}{K} + \frac{1}{h}\right) \approx \frac{2D}{K} \quad \text{(since generally } h \gg K) \tag{4.13}$$

$$B \equiv \frac{2DC^*}{N_i} \tag{4.14}$$

$$\tau \equiv \frac{z_i^2 + Az_i}{B} \tag{4.15}$$

and z_i is the initial value of the oxide thickness at $t = 0$. Solving for the oxide thickness z_0,

$$z_0 = \frac{A}{2}\left[\left(1 + \frac{t + \tau}{A^2/4B}\right)^{1/2} - 1\right] \tag{4.16}$$

Equation (4.16) reduces to

$$z_0 \approx \frac{B}{A}(t + \tau) \approx \frac{KC^*}{N_i}(t + \tau) \quad \text{for } (t + \tau) \ll \frac{A^2}{4B} \tag{4.17}$$

and

$$z_0 \approx (Bt)^{1/2} \approx \left(\frac{2DC^*}{N_i}t\right)^{1/2} \quad \text{for } t \gg \frac{A^2}{4B} \tag{4.18}$$

B/A is called the "linear" rate constant and B is called the "parabolic" rate constant. These limiting forms indicate that the oxidation process in the parabolic domain is diffusion limited and in the linear region it is surface-reaction limited.

For short oxidation times [Eq. (4.17)], the oxide thickness is determined by the surface reaction rate constant (K) and is linearly proportional to the oxidation time. If the oxidation time is long [Eq. (4.18)], the oxide growth is determined by the diffusion constant D. The oxide thickness in this case is proportional to the square root of the oxidation time.

Using experimentally determined values of A, B, and τ, we may plot the oxide thickness as a function of time for several commonly used oxidation temperatures for both dry and wet oxidation [see Fig. 4.5b, c).

Oxide thicknesses of a few tenths of a micron are often used, with 1–2 μm being the upper practical limit for conventional thermal oxidation. A significant advance was the addition of chlorine-containing chemicals during oxidation. This resulted in improved device threshold voltage (V_T) stability, increased dielectric breakdown strength,[36,37] and an increased rate of silicon oxidation. The primary function of chlorine in SiO_2 films (10^{16}–10^{20} per cm^3) is to passivate or electrically inactivate impurity ions (such as sodium and potassium) that have inadvertently been incorporated into the SiO_2).

At the present time, B, B/A, and τ are known only for $\langle 111 \rangle$-oriented lightly doped conditions in which case,

$$B = C_1\, e^{-E_1/kT} = 2D_{\text{eff}}\,\frac{C^*}{N_i} \tag{4.19}$$

$$\frac{B}{A} = C_2\, e^{-E_2/kT} = \frac{C^*}{N_i(1/K + 1/h)} \tag{4.20}$$

and

$$\tau = (z_i^2 + Az_i)/B \tag{4.21}$$

where z_i is the effective value of the oxide thickness at $t = 0$. The parameters for dry oxidation[38] are the following:

$$N_i = 2.2 \times 10^{22} \text{ per cm}^3$$
$$C_1 = 7.72 \times 10^2 \ \mu\text{m}^2/\text{h}$$
$$C_2 = 6.23 \times 10^6 \ \mu\text{m}/\text{h}$$
$$E_1 = 1.23 \text{ eV/molecule}$$
$$E_2 = 2.0 \text{ eV/molecule}$$
$$z_i = 0 \text{ Å}$$

For the wet oxidation process the parameters are

$$N_i = 4.4 \times 10^{22} \text{ per cm}^3$$
$$C_1 = 2.24 \times 10^2 \ \mu\text{m}^2/\text{h}$$
$$C_2 = 8.95 \times 10^7 \ \mu\text{m}/\text{h}$$
$$E_1 = 0/71 \text{ eV/molecule}$$
$$E_2 = 1.97 \text{ eV/molecule}$$
$$z_i = 200 \text{ Å}$$

4.3.2.2. Crystal Orientation Effects.   Orientation differences in the oxidation rate appear only after the oxide thickness exceeds about 100 Å. The growth rate difference is largest (40%) at relatively low temperatures (700°C) and decreases gradually at higher temperatures (2% at 1200°C). Generally the $\langle 111 \rangle$ orientation oxidizes faster than the $\langle 100 \rangle$ orientation. The orientation dependence should appear[29] in the linear rate constant B/A, since it involves the reaction rate constants (K, h) via the constant A.

4.3.2.3. Impurity Doping.   The effects of impurity doping on thermal oxidation rates are intimately connected to impurity redistribution. As a thermal oxide is grown over a doped Si substrate, there results a redistribution of the impurity (as seen in Fig. 4.6). Phosphorus, arsenic, and antimony dopant atoms tend to pile up at the interface ($C_s > C_B$), but in boron a surface depletion ($C_s < C_B$) takes place.

It has been observed that the oxidation rates are generally faster for heavily doped ($C_B \cong 10^{19}$ cm^{-3}) p and n substrates. The two concentrations that are involved in this increased oxidation are the dopant concentration of the SiO$_2$–Si interface (C_s) and the impurity concentration in the oxide (C_{ox}). From recent experiments it has been concluded that the pileup of n-type dopants at the SiO$_2$–Si interface affects the oxidation kinetics far more than does the increased impurity concentration (C_{ox}) in the oxide. It is the linear rate constant B/A that is changed by a high impurity dopant concentration. Figure 4.7 shows the linear and parabolic rate constants as functions of the dopant concentration at 900°C.[14]

4.3.2.4. High-Pressure Oxidation.   The advantage of high-pressure (20-atm) oxidation is its ability to oxidize at lower temperatures, thereby decreasing the redistribution of dopants and decreasing the generation of silicon defects. The relationship between oxide thickness and steam pressure is shown in Fig. 4.8.

The relation for the parabolic rate constant B should be modified to include the effect of the steam pressure, since according to Henry's law the equilibrium solid

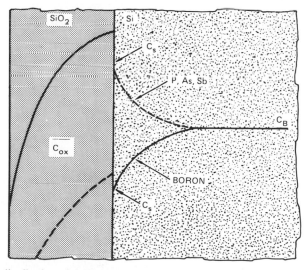

FIGURE 4.6.   Redistribution of the impurity doping profiles near the SiO$_2$–Si interface. From Ref. 15.

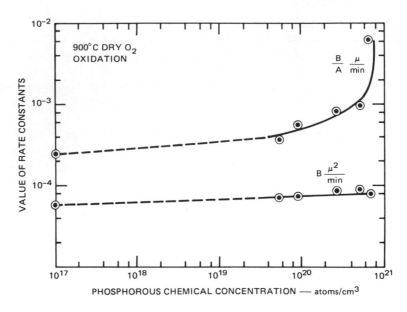

FIGURE 4.7.   Value of the rate constants A/B and B as a function of dopant concentration at 900°C dry O_2 oxidation. From Ref. 15.

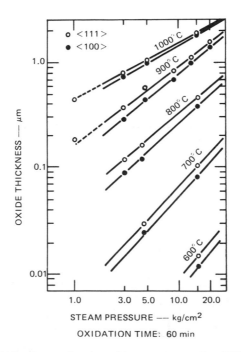

FIGURE 4.8.   Oxide thickness as a function of steam pressure for $\langle 111 \rangle$ and $\langle 100 \rangle$ silicon. From Ref. 38.

concentration is proportional to the bulk gas particle steam pressure P_g, and $C^* = HP_g$, but using Eq. (4.19), $B = 2D_{eff}C^*/N_1$. A is independent of the steam pressure and affected by the Si crystal orientation and the doping concentration. A rule of thumb is that for every increase of 1 atm, the temperature of oxidation may be reduced by approximately 30°C.[39] As a consequence of the increased pressure, processes operating at 1100°C may be reduced to about 800°C without incurring a time penalty. For a 25-atm system, 1 μm of oxide may be grown in about 15 min at 920°C, as opposed to approximately 10 h at 1 atm.

  4.3.2.5. Oxidation in HCl–O₂ Mixtures. A significant development in thermal oxidation has been the addition of chlorine during oxidation. Such additions result in an improved threshold stability for MOS devices and an increased dielectric strength. In addition, chlorine increases the rate of silicon oxidation. Experiment[36] shows that the monotonic increase in B with HCl concentration (Fig. 4.9) is related to the effective diffusion coefficient D_{eff}. The increased chlorine concentration may cause the SiO₂ lattice to be strained, thereby allowing the diffusion of the oxidant to occur more easily. Figure 4.9 shows the parabolic rate constant as a function of the percentage of HCl for different orientations and oxidation temperatures.

  4.3.2.6. Oxide Charges. The densities of the charges and states associated with the Si–SiO₂ system are shown in Figure 4.10.[40,41] These consist of charges trapped in the interface of density Q_{it} (C/cm²), charges fixed in the oxide close to the interface of density Q_f, mobile impurity ions in the oxide density Q_m, and space charge trapped in the bulk of the oxide density Q_{ot}. Each charge density has a corresponding number density given by $|Q|/q = N$ (number/cm²), where q is the electronic charge. The sign of Q is either positive or negative, depending on the sign of the majority charge. By definition, however, N is always positive.[41] The interface trap states are located at the oxide–silicon interface and are introduced into the forbidden gap near the

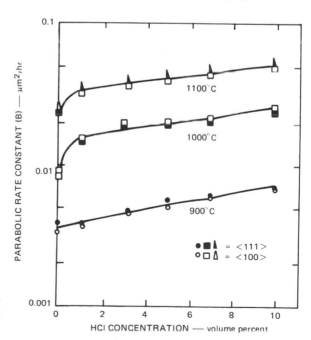

FIGURE 4.9. Parabolic constant (B) as a function of percent HCl for $\langle 111 \rangle$- and $\langle 100 \rangle$-oriented n-type silicon at 900, 1000, and 1100°C. From Ref. 33.

semiconductor surface because of the disruption of periodicity of the lattice. It is believed that there should be approximately one surface state for every surface atom, resulting in a number of approximately one surface state for every surface atom, i.e., $N_{it} \sim 10^{15}$ per cm^2. The charges Q_f are fixed in the oxide near the interface between the oxide and the silicon. Located within 200 Å of the SiO$_2$–Si interface, they cannot be discharged or moved, but the value of Q_f is a strong function of oxidation, annealing conditions, and the orientation of the Si crystal. The dependence of Q_f on the ambient temperature of the final heat treatment suggests excess ionic silicon as

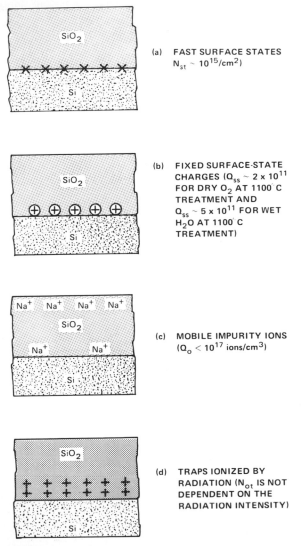

(a) FAST SURFACE STATES
$N_{st} \sim 10^{15}/cm^2$)

(b) FIXED SURFACE-STATE
CHARGES ($Q_{ss} \sim 2 \times 10^{11}$
FOR DRY O$_2$ AT 1100 C
TREATMENT AND
$Q_{ss} \sim 5 \times 10^{11}$ FOR WET
H$_2$O AT 1100 C
TREATMENT)

(c) MOBILE IMPURITY IONS
($Q_o < 10^{17}$ ions/cm^3)

(d) TRAPS IONIZED BY
RADIATION (N_{ot} IS NOT
DEPENDENT ON THE
RADIATION INTENSITY)

FIGURE 4.10.   Charges associated with the Si/SiO$_2$ system. (a) Fast surface states ($N_{st} \sim 10^{15}$ per cm^2); (b) fixed surface state charges ($Q_{ss} \sim 2 \times 10^{11}$ for dry O$_2$ at 1100°C treatment and $Q_{ss} \sim 5 \times 10^{11}$ for wet H$_2$O at 1100°C treatment); (c) mobile impurity ions ($Q_0 < 10^{17}$ ions/cm^3); (d) traps ionized by radiation (N_{ot} is not dependent on the radiation intensity). From Ref. 40.

a possible oxide in the origin of Q_f, although bond coordination defects have been shown to be a likely cause.[42] In addition, there may be mobile charges Q_m present within the oxide layer caused by impurity ions (Na⁺, Li⁺, K⁺). These charges Q_m are responsible for the voltage drift in MOS device characteristics. Finally, a positive space-charge buildup Q_{ot} can be generated in the oxide from external ionizing radiation (X ray, electron beams).[43]

Radiation-induced oxide space charge can be annealed out at relatively low temperatures ($>300°C$). Figure 4.11 shows the location of these oxide charges in thermally oxidized silicon structures.

So far we have assumed that the oxidizing species proceeding through the oxide are uncharged. However, experiments[44] studying the effect of an electric field applied across the oxide on the oxidation rate have indicated that the oxidizing agent (O_2) is negatively charged. It is possible that the molecular oxygen dissociates upon its entry into the oxide as

$$O_2 \Leftrightarrow 2O^- + 2 \text{ holes}$$

and both the oxygen ion and the hole diffuse through the oxide coupled by an electric field (ambipolar diffusion). Another theory[45] predicts that as long as the oxide thickness is smaller than some critical distance, oxidation will be rapid. At greater oxide thicknesses the growth rate will slow down. Estimates of these critical thicknesses yield 150 Å for silicon oxidation in O_2 and 5 Å for oxidation in H_2O. These values are very close to the z_i values used in the Deal–Grove theory of oxidation (Section 4.3.2.1).

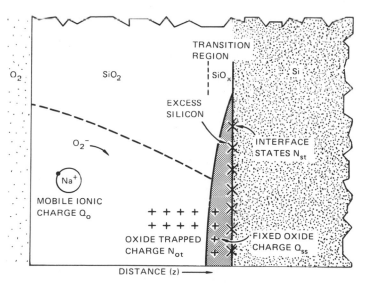

FIGURE 4.11. Location of oxide charges in thermally oxidized silicon structures. From Ref. 41.

4.3.3. *Plasma-Enhanced Chemical Vapor Deposition (PECVD)*

In conventional CVD the reagent gases are converted at the substrate surface which must be held at an appropriately high temperature for the surface film to be deposited. Forming a plasma in the gas mixture both heats up the reagent molecules and forms more highly reactive radicals or ionized radicals (see Section 3.6). This results in much lower substrate temperatures for the deposition process to take place, typically in the range of 200°C to 600°C. Deposition rates in the range of 1–100 Å per second are obtained.[9]

TABLE 4.4.   Inorganic Films Made by Plasma-Enhanced CVD[a]

Film	Reagent gases	References
Silicon oxide	$SiH_4 + N_2O$	Sterling and Swan (1965), Alexander *et al.* (1970)
	$SiCl_4 + O_2$	Kuppers *et al.* (1976)
	$SiH_4 + O_2(N_2)$	Kalnina *et al.* (1971)
	$Si(OC_2H_5)_4 + O_2$	Ing and Davern (1964, 1965)
Silicon nitride	$SiH_4 + N_2$	Kuwano (1969)
	$SiH_4 + NH_3$	Sterling and Swan (1965), Alexander *et al.* (1970)
	$SiBr_4 + N_2$	Androshuk *et al.* (1969)
	$SiI_4 + N_2$	Shiloh *et al.* (1977)
Silicon oxynitride	Combine above processes, e.g., $SiH_4 + N_2O + NH_3$	
Amorphous Si (*a*-Si)	SiH_4	Sterling and Swan (1965)
Doped *a*-Si	$SiH_4 + PH_3$ or B_2H_6	Spear and LeComber (1975)
Epitaxial Si	SiH_4	Townsend and Uddin (1973)
Amorphous carbon	C_4H_{10}	Holland and Ojha (1978)
	C_2H_2	Whitmell and Williamson (1976)
Silicon carbide	$SiH_4 + C_2H_4$ or CH_4	Sterling and Swan (1965)
Silicon carbonitride	$[Si(CH_3)_2NH]_3$	Wrobel and Kryszewski (1974)
Amorphous germanium	GeH_4	Chittick (1980)
Germanium oxide	$Ge(OC_2H_5)_4 + O_2$	Secrist and Mackenzie (1966)
Germanium carbide	$GeH_4 + C_2H_4$	Anderson and Spear (1977)
Amorphous arsenic	AsH_3	Knights and Mahan (1977)
Boron oxide	$B(OC_2H_5)_3 + O_2$	Secrist and Mackenzie (1966)
Boron nitride	$BBr_3 + NH_3(H_2)$	Alexander *et al.* (1970)
	$B_2H_6 + NH_3$	Dell'Oca *et al.* (1971)
Si-Ge-B oxides	Chlorides + O_2	Kuppers *et al.* (1976), Jaeger *et al.* (1978)
Aluminum oxide	$AlCl_3 + O_2$	Katto and Koga (1971)
Aluminum nitride	$AlCl_3 + N_2$	Bauer *et al.* (1977)
Titanium oxide	$TiCl_4 + O_2$	Alexander *et al.* (1970)
	Ti isopropylate + O_2	Secrist and Mackenzie (1966)
	$TiCl_4 + CO_2$	Sterling *et al.* (1966)
Tantalum oxide	Not given	Alexander *et al.* (1970)
Molybdenum, nickel	Carbonyls	Sterling *et al.* (1966)
Iron oxide	$Fe(CO)_5$	Secrist and Mackenzie (1966)
Tin oxide	Dibutyltin diacetate	Secrist and Mackenzie (1966)
Phosphorous nitride	$P + N_2$	Veprek and Roos (1976)
Gallium nitride	$Ga(CH_3)_3 + NH_3$	Wiemer (1973)

[a]Source: Rand (1979).

Table 4.4 lists the reagent gases and references to PECVD processes for the deposition of a number of useful films. Those of prime interest are:

- Silicon nitride—Applications include use as an oxidation mask, an alkali ion barrier, a dielectric in MNOS nonvolatile memories, a passivation layer, and as an antireflection coating on electron-optical devices.
- Silicon—Applications include use for IC layers, solar photovoltaic energy conversion, and electrophotography.
- Oxides of gallium arsenide and silicon—Used for MIS devices.
- Tungsten—Used for interconnections.

Chambers used for PECVD are usually radial flow (Fig. 4.12). Operating problems include:

- Air leaks into the gas preparation and vacuum lines. Small amounts of contaminants can have a major effect on the film properties and deposition rates.
- Effects of reactive species in exhaust lines, traps, and pumps. These give rise to unwanted powdery products, which must be regularly cleaned. The pump oil is continuously exposed to the exhaust gases and must be frequently changed to eliminate contaminants.

For experimental purposes (as opposed to production), tube reactors are often used, as shown in Fig. 4.13. For inductive coupling, an rf coil is placed outside a glass or quartz tube through which the reactive gases are passed. Alternatively, capacitor plates may be placed inside the tube.

The choice of chemical reactions and the design of the reaction chamber to deposit specific films by CVD is a highly specialized field. The reader who wishes to explore this field more thoroughly is recommended to read the excellent review article in Ref. 9, p. 258, before proceeding to the open literature to find out what processes may have been used previously for the specific film he or she has in mind.

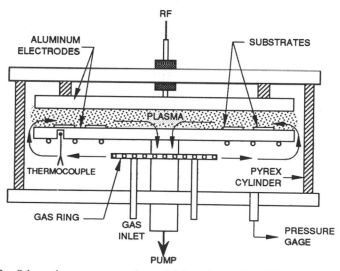

FIGURE 4.12.   Schematic arrangement of a radial-flow plasma deposition chamber. From Ref. 174.

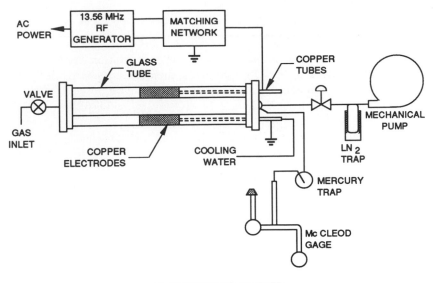

(a) CAPACITIVELY COUPLED

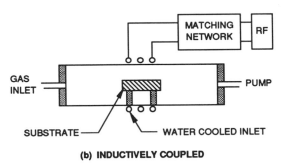

(b) INDUCTIVELY COUPLED

FIGURE 4.13.   Schematic arrangements for tubular flow reactors for plasma deposition. (a) Capacitively coupled; (b) inductively coupled.

4.4. DOPING BY THERMAL DIFFUSION

4.4.1. The Thermal Diffusion Process

The principal purpose of doping semiconductors is to control the type and concentration of impurities within a specific region of the crystal to modify its local electrical properties. Diffusion has been found to be a highly practical way of doing this and has been extensively studied.[35,46–48]

The two steps in the diffusion process, the predisposition and the drive-in, are described by the diffusion equation

$$\frac{\partial N(z, t)}{\partial t} = D \frac{\partial^2 N(z, t)}{\partial z^2} \tag{4.22}$$

corresponding to boundary conditions as shown in Fig. 4.14.

STEP	BOUNDARY CONDITIONS	SOLUTIONS	MODEL
PREDEPOSITION	$N = N_0$ at $z = 0$ $N = 0$ at $z = \infty$	$N(z,t) =$ $N_0 \operatorname{erfc} \dfrac{z}{2\sqrt{Dt}}$ $Q(t) = \displaystyle\int_0^\infty N(z,t)\,dz$ $= \dfrac{2}{\sqrt{\pi}}\sqrt{Dt}\,N_0$	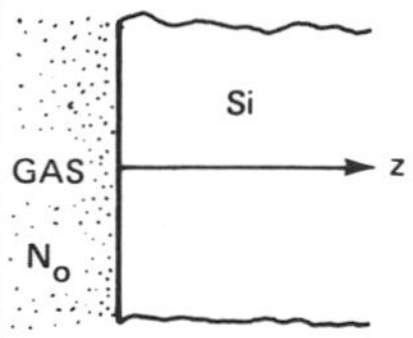
DRIVE-IN	$\left.\dfrac{\partial N}{\partial z}\right|_{(0,t)} = 0$ $N(\infty,t) = 0$	$N(z,t) =$ $\dfrac{Q}{\sqrt{\pi Dt}}\, e^{-\frac{z^2}{4Dt}}$	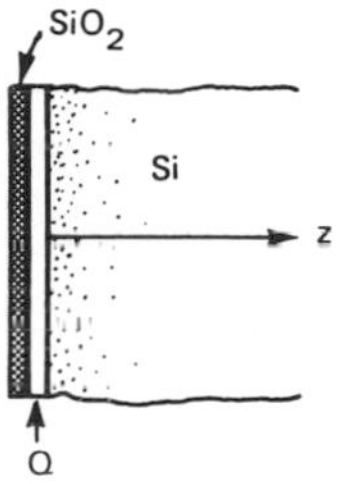

FIGURE 4.14. Diffusion process for impurity atoms.

The complementary error function, erfc, is tabulated in Appendix A, and the solid solubilities of several common dopants in silicon as functions of temperature are shown in Fig. 4.15. Figure 4.16 shows the diffusivities of the dopants as a function of temperature.

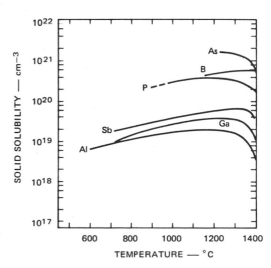

FIGURE 4.15. Solid solubilities of several common dopants in silicon, expressed as functions of temperature.

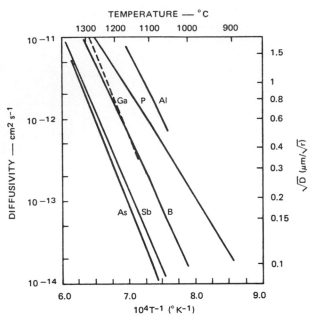

FIGURE 4.16. Diffusivities of several common dopants as functions of temperature.

A p–n junction can be formed when a p-type impurity is diffused into an n-type material. In this case the impurity distribution becomes

$$N(z, t) = N_0 \operatorname{erfc} \frac{z}{l\,(Dt)^{1/2}} - N_c \tag{4.23}$$

where N_c is the background doping density of the n-type silicon. The junction depth z_j from the surface can be obtained by setting $N = 0$:

$$z_j = 2(Dt)^{1/2} \operatorname{erfc}^{-1} \frac{N_c}{N_0} \tag{4.24}$$

This junction depth is an important device parameter.

Figure 4.17 shows the impurity profiles for a p–n junction and npn transistor fabricated by the diffusion process. The dopant impurity profile changes every time the wafer is reheated for another process step such as oxidation or diffusion. This means that during the planning of a fabrication process, each proposed heat treatment must be taken into account to achieve the desired overall junction depths and dopant profiles.

4.4.2. The Physics of Thermal Diffusion

Thermal movement of a diffusing atomic or ionic species through a crystal lattice takes place in jumps.[46–48] These jumps occur in all three dimensions, their net flux

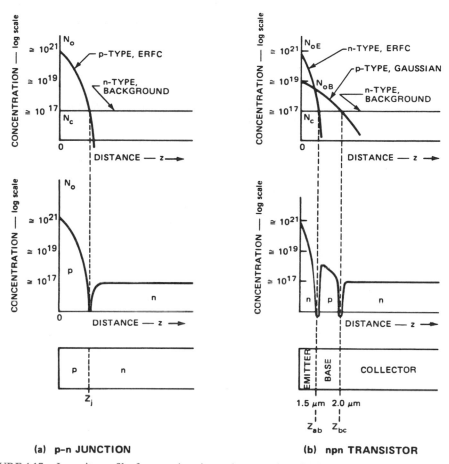

FIGURE 4.17. Impurity profiles for a *p–n* junction and *npn* transistor fabricated by the diffusion process. (a) *p–n* junction; (b) *npn* transistor.

being the statistical average over a period of time. The mechanism by which jumps take place is as follows: The atoms of the crystal form a series of potential hills (see Fig. 4.18a) that impede the motion of the impurities. The height of the potential barrier is typically on the order of 1 eV in most materials. The distance between successive potential barriers is on the order of the lattice spacing, which is 1–3 Å. If a constant electric field is applied, the potential distribution as a function of distance will be tilted, as shown in Fig. 4.18b.[46] This will make the passage of positively charged particles to the right easier and their passage to the left more difficult. Let us now calculate the flux F at position z. This flux will be the average of the fluxes at positions $z - a/2$ and $z + a/2$. In turn, these two fluxes (Fig. 4.18b) are given by $F_1 - F_2$ and $F_3 - F_4$, respectively.

Consider the component F_1. It will be given by the product of (1) the density per unit area of impurities in the plane of the potential valley at $z - a$, (2) the probability of a jump of any of these impurities to a valley at z, and (3) the frequency of attempted jumps v. Thus, we can write

$$F_1 = [aC(z - a)] \exp\left[-\frac{q}{kT}(W - \tfrac{1}{2}aE)\right]v \qquad (4.25)$$

where $aC(z - a)$ is the density per unit area of particles situated in a valley at $z - a$, and the exponential factor is the probability of a successful jump from the valley at $z - a$ to the valley at z. Note the lowering of the barrier due to the electric field E.

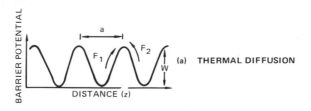

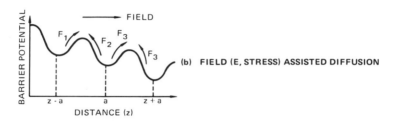

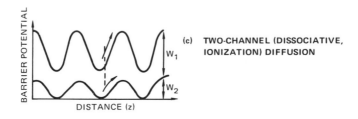

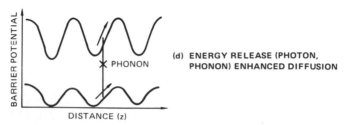

FIGURE 4.18. Model of diffusion mechanisms in crystals. (a) Thermal diffusion; (b) field (E, stress)-assisted diffusion; (c) two-channel (dissociative, ionization) diffusion; (d) energy release (photon, phonon)-enhanced diffusion.

Similar formulas can be written for F_2, F_3, and F_4. When these are combined to give a formula for the flux F at position z, with the concentrations $C(z \pm a)$ approximated by $C(z) \pm a(\partial C / \partial z)$, we obtain

$$F(z) = -(va^2 \, e^{-qW/kT}) \frac{\partial C}{\partial z} \cosh \frac{qaE}{2kT} + (2av \, e^{-qW/kT})C \sinh \frac{qkT}{2kT} \qquad (4.26)$$

An extremely important limiting form of this equation is obtained for the case when the electric field is relatively small, i.e., $E \ll kT/qa$. In this case we can expand the cosh and the sinh terms in Eq. (4.26). Noting that $\cosh z = 1$ and $\sinh z = z$ for $z \rightarrow 0$, this results in the limiting form of the flux equation for a positively charged species

$$F(z) = -D \frac{\partial C}{\partial z} + \mu E C \qquad (4.27)$$

where

$$D = va^2 \, e^{-qW/kT} \qquad (4.28)$$

and

$$\mu = \frac{va^2 \, e^{-qW/kT}}{kT/q} \qquad (4.29)$$

Note that the mobility μ and the diffusivity D are related by the well-known Einstein relationship:

$$D = \frac{kT}{q} \mu \qquad (4.30)$$

It is customary to identify the contribution to the flux, which is proportional to the concentration gradient, as the diffusion term, while the contribution that is proportional to the concentration itself is referred to as the drift term. The migration of the diffusing species follows Fick's law:

$$\frac{\partial C}{\partial t} = D \frac{\partial^2 C}{\partial z^2} - \mu_+ E \frac{\partial C}{\partial z} \qquad \text{for positive charge}$$

$$\frac{\partial C}{\partial t} = D \frac{\partial^2 C}{\partial z^2} + \mu_- E \frac{\partial C}{\partial z} \qquad \text{for negative charge}$$

$$(4.31)$$

Interstitial diffusion takes place when the impurity atoms move through the crystal lattice by jumping from one interstitial site to the next. The diffusion of *substitutional* impurities, i.e., impurities that occupy sites in the silicon lattice, usually proceeds by the impurities jumping into silicon vacancies in the lattice. Here the activation energy includes the energy required to form a Si vacancy as well as the energy required to move the impurity. The diffusion of interstitial species usually proceeds more rapidly than does the corresponding species located on a substitutional site.

In some systems, an atom, which is normally substitutional, has a very small energy difference between the substitutional and the interstitial site. In this case, another diffusion mechanism, called dissociative diffusion, can occur. In this mechanism, the substitutional atom is excited to become an interstitial atom, leaving behind a vacancy. The interstitial atom diffuses until it encounters another vacant site, at which point it becomes substitutional again. The one-dimensional diffusion equations now become

$$\frac{\partial}{\partial t} S = D_s \frac{\partial^2}{\partial z^2} S - R_s + Q_I$$

$$\frac{\partial}{\partial t} I = D_I \frac{\partial^2}{\partial z^2} I + R_s - Q_I$$

(4.32)

with S denoting the concentration of substitutional atoms and I the concentration of interstitials. That is, the diffusion equation is augmented by reaction terms, R_s denoting the rate of loss per unit time from the S channel and Q_I the rate of loss from the I channel. Unless $R_s = Q_I$ (i.e., the channels are in equilibrium), the diffusion is non-Fickian. There are several impurity–host systems that yield a non-Fickian diffusion.[36]

In addition to the normal thermally activated diffusion process, the diffusion mechanisms can be enhanced by particle bombardment. Particle bombardment creates both ionization and recoil atoms, both of which can enhance diffusion. If the energy of the recoil atom is high enough to create vacancies and interstitials, these defects can enhance diffusion (see Section 4.4). (Defect-enhanced diffusion is often referred to as radiation-enhanced diffusion.) Vacancies and interstitials can each create one or more diffusing species in the system, so that the proper description of the diffusion process requires a number of coupled equations such as those for dissociative diffusion. Moreover, the reaction terms in the equations are frequently not in thermal equilibrium, so that the diffusion is non-Fickian.

4.4.3. Field-Enhanced Diffusion

Field-enhanced diffusion occurs in the presence of a field, whether electric, stress, or other. As shown in Fig. 4.18b, the potential-energy barriers traversed by the diffusion species through the lattice now experience a bias, resulting in a net drift in the diffusion process. The diffusion equation for the field-enhanced case is given by

$$\frac{\partial C}{\partial t} = \frac{\partial}{\partial z} D(z, t) \frac{\partial}{\partial z} C - v \frac{\partial}{\partial z} C$$

(4.33)

where the last term describes the drift due to the field. Field-enhanced diffusion can readily happen at surfaces. For example, particle-induced electron emission at the surface of a low-conductivity material creates an electric field, while the evaporation of one constituent of an alloy could create a strain field at a surface. Recoil-enhanced diffusion occurs when a high-energy collision imparts a recoil energy to the diffusion species so that its net probability of diffusing is enhanced (see Section 4.4.4).

Clearly, if there is a directionality to the bombarding particle beam, then there will be a new direction for the drift of the recoil atoms. The diffusion equation for recoil-enhanced diffusion is essentially the same as that for field-enhanced diffusion with a different origin for the coefficients. Recoil-enhanced diffusion in fact occurs in the form of recoil implantation, e.g., the recoil implantation of oxygen atoms from a SiO_2 layer on top of a silicon lattice. This process is also responsible for the alteration of an alloy composition at a sputtered surface, since the sputtering collisions not only favor the "emission" of one constituent from the surface but also favor driving the other constituent(s) into the bulk.

4.4.4. Ionization-Enhanced Diffusion

Ionization-enhanced diffusion occurs when the diffusion species possesses charge states, one of which has a lower barrier to migration than does the other. Figure 4.18c shows the potential-energy curve as a function of distance through the lattice of two channels for the ionization-enhanced diffusion mechanism.

4.4.5. Energy Release-Enhanced Diffusion

The creation of an electron–hole pair in a material system with a forbidden gap results in the temporary storage of energy, since the energy of the electron is substantially higher than that of the hole. Recombination of the electron and the hole then provides energy that may enhance diffusion, particularly if that recombination occurs at the site of the diffusing species, e.g., at a recombination center as shown in Fig. 4.18d. Of course, if the energy is released in the form of a photon, then the photon can escape the vicinity of the diffusing species and will not contribute to an enhancement of the diffusion.

If, on the other hand, some or all of the energy is released locally in the form of phonons, these phonons can enhance the diffusion. This energy-release mechanism has been observed to occur frequently in III–V compounds.[49] If the energy is released in the form of heat in the vicinity of the defect, it may also enhance diffusion. The released energy contributes to the jump probability and can be incorporated into the diffusion coefficient of the simple (one-channel) Fick equation.

4.5. DOPING BY ION IMPLANTATION

4.5.1. Background

The use of ion beams in microelectronics has become increasingly important.[50–52] Ions are heavy—they have 10^3–10^5 times the mass of an electron. Thus, for a given energy, an ion carries approximately 10^2–10^4 times more momentum than an electron. As a result of this characteristic, ions produce quantitatively different effects in their interactions with crystal lattices (see Section 2.7 on ion interactions).

An ion retains most of the chemical properties of the original atom. After it has been injected into the lattice, it may recombine or remain ionized, but it usually behaves chemically just like an atom introduced into the lattice by diffusion, alloying, or during epitaxial growth. Thus, ion implantation can be used to "dope" a semiconductor. In addition, because an ion is electrically charged, it can be accelerated to any desired velocity and electrically deflected and focused. Thus, an ion beam can be positioned at will and caused to dope only a tiny volume of material.

In comparison to diffusion, ion implantation has a number of advantages, the more important of which follow:

- Doping levels can be controlled precisely, since the incident ion beam can be measured accurately as an electrical current.
- Doping uniformity across a surface can be accurately controlled.
- The depth profile can be regulated by choice of the incident ion energy.
- It is a low-temperature process. This feature is not necessarily important with silicon, but some compound semiconductors are unstable at high temperatures.
- Extreme purity of the dopant can be guaranteed by mass analysis of the ion beam.
- Particles enter the solid as a directed beam, and since there is little lateral spread of the beam, smaller and faster devices can be fabricated.
- Dopants can be introduced that are not soluble or diffusible in the base material.

Because of these features, circuits of identical characteristics can be made across a large wafer. In addition, because of the precision of doping levels, devices can be made that are either very difficult or even impossible to make by other means. Together with this process control, implantation provides improved yields. By far the greatest use of ion implantation to date has been in the fabrication of devices with a turn-on, or threshold, voltage less than 1.5 V. By diffusion techniques it is very difficult to get the turn-on voltage as low as 1.5 V in a routine and controlled manner, because the amount of dopant that has to be introduced is so small by normal diffusion standards. Using ion implantation, the process becomes simple. It is largely because of ion implantation that hand calculators and digital watches can be operated by standard 1.5-V batteries. The disadvantages of ion implantation in comparison to diffusion are that expensive and complicated equipment is required, the junctions are not automatically passivated, and the crystal structure may be damaged.

4.5.2. Basic Features of Ion Implantation

In the process of ion implantation,[53] atoms of the desired doping element are ionized and accelerated to high energies and then made to enter a substrate lattice by virtue of their kinetic energy (or momentum). After the energetic ion comes to rest and equilibration has occurred, the implanted atom may be in a position in which it serves to change the electronic properties of the substrate lattice; i.e., doping occurs. Lattice defects caused by the energy-loss process of the ions may also change the electronic properties of the substrate.[54]

The technique by which dopant atoms are ionized and then accelerated to high energies (30–350 keV) by an electric field is shown in Fig. 4.19. A mass-separating magnet eliminates unwanted ion species. After passing through the deflection and focusing control, the ion beam is aimed at the semiconductor target so that the high-energy ions can penetrate the semiconductor surface. The energetic ions lose their energy through collisions with the target nuclei and electrons and finally come to rest. The total distance that an ion travels before coming to rest is called its range (R); the projection of this distance onto the direction of incidence is called the projected range (R_p).

The doping profile is usually characterized by the projected range and its standard deviation ΔR_p. Figure 4.20 shows R_p and $R_p + \Delta R_p$ for boron, phosphorus, and arsenic ions implanted into silicon. Provided that the semiconductor is not aligned in a major crystallographic direction with the ion beam, profiles resulting from implants are nearly Gaussian in shape. The dopant concentration as a function of the distance from the silicon surface is given as

$$N(Z) = N_{\max} e^{\dfrac{-(z - R_p)^2}{\Delta R_p^2}} \tag{4.34}$$

where

$$N_{\max} = \frac{N_p}{2.5 \Delta R_p} \tag{4.35}$$

and

$$N_p = \frac{\text{number of implanted atoms}}{\text{cm}^2} \tag{4.36}$$

Some ions from a well-collimated beam are directed toward a channel (certain crystallographic directions in which open spaces exist among the rows of atoms) are able to penetrate deeply into the crystal lattice before coming to rest in interstitial or substitutional sites (Fig. 4.21). The directions that allow large amounts of channeling are limited. When the crystal is examined from directions other than along a channel or plane, the atoms appear more randomly oriented, roughly as in a dense atomic gas. However, even when ions are injected along a nonchanneling direction, it is difficult to completely prevent some channeling.

The penetration depth and final distribution of the ions in the crystal depend on the ion energy, the crystal and the ion species, and the angular alignment of the ion beam with the crystal axis. These characteristics do not depend strongly on the crystal temperature. One can distinguish two primary classes of ion–crystal interaction that yield quite different penetration depths and ion density distributions. Ions penetrating a target that is either amorphous or crystalline, but with the ion path

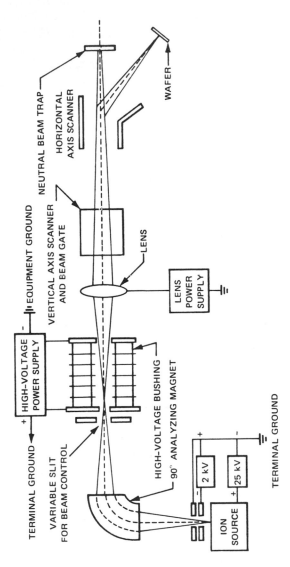

FIGURE 4.19.   Ion-implanted technique for selective introduction of dopants into the silicon lattice.

misaligned from any crystal axis, will suffer collisions that reduce their inward motion in such a way that the resulting density profile with depth is roughly Gaussian. This case is illustrated in Fig. 4.21, where it is shown as the "amorphous" distribution near the surface.

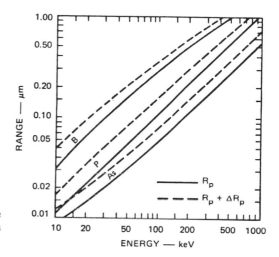

FIGURE 4.20. Projected range R_p and the standard deviation in projected range ΔR_p as a function of the implanted ion energy.

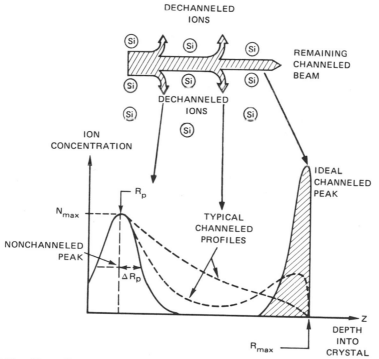

FIGURE 4.21. Channeling effect of an ion entering the Si crystal parallel to a major crystal axis or plane.

If the ion dose is low and the trajectories are directed precisely in an open crystallographic direction, the ions can penetrate deeply into the crystal and will stop rather abruptly at the end of their range. The resulting profile is also shown in Fig. 4.21, by the curve marked "channeled peak." Some fraction of the ions traveling along a channel leave the channel prematurely. Channeled profiles cannot be maintained at high dose levels, since energetic ions striking the surface of a crystal will displace the crystal atoms, resulting in a near-amorphous condition close to the surface. Thus, with high dose levels, the implanted distribution will always be near Gaussian.

Typical parameters for an ion-implanting machine are as follows:

- Dopants: P, As, Sb, B
- Dose: 10^{11}–10^{16} per cm^2
- Energy: 10–400 keV
- Depth of implant: typically 1000–5000 Å
- Reproducibility and uniformity: ±35
- Temperature: normally room temperature
- 50-μA current/3-inch wafer corresponds to a dose rate of 6×10^{12} atoms/ cm^2 s.

4.5.3. Range Theory—Lindhard, Scharff, and Schiott (LSS)[55]

It is customary to assume that there are two major forms of energy loss for an ion entering a target (see Section 2.7). These are interactions of the ion with the electrons in the solid and collisions of the ion with the nuclei of the target. Therefore, the total energy loss can be written as the sum [Eq. (2.395)]

$$-\frac{dE}{dz} = N[S_n(E) + S_e(E)] \tag{4.37}$$

where $-dE/dz$ (eV/cm) is the average rate of energy loss with distance along R, E (eV) is the energy of the ion at point z along R, $S_n(e)$ (eV $\cdot$ cm^2) is the nuclear stopping power, $S_e(E)$ (eV $\cdot$ cm^2) is the electronic stopping power, and N is the target atom density $= 5 \times 20^{22}$ per cm^3 for Si. The total distance that the ion travels before coming to rest is called its range (R); the projection of this distance onto the direction of incidence is called the projected range (R_p). The distribution of the stopping points for many ions in space is the range distribution.

If $S_n(E)$ and $S_e(E)$ are known, then the total range can be calculated [Eq. (2.396)]:

$$R = \int_0^R dz = \frac{1}{N} \int_0^{E_0} \frac{dE}{S_n(E) + S_e(E)} \tag{4.38}$$

R is the *average* total range and one should expect a distribution of R in the direction of incidence (ΔR_p) around R_p, the projected range, and a transverse distribution (ΔR_T) around R_T, the transverse range. Generally, the shape of the distribution depends on the ratio between the ion mass M_1 and that of the substrate atoms, M_2. The relative width $\Delta R_p/R_p$ of the distribution depends on the ratio of M_1 to M_2. For light ions such as B, $\Delta R_p/R_p$ is large, and for heavy ions such as As and Sb, $\Delta R_p/R_p$ is small.

4.5.4. Useful Approximations

The expression for the nuclear stopping power can be written[54] in the form [Eq. (2.385)]

$$S_n = 2.8 \times 10^{-15} \frac{Z_1 Z_2}{Z^{1/3}} \frac{M_1}{M_1 + M_2} \, \text{eV} \cdot \text{cm}^2 \tag{4.39}$$

where S_n is the nuclear stopping power (independent of E), Z_1 is the ion atomic number, M_1 is the ion atomic mass, Z_2 is the substrate atomic number (14 for Si), M_2 is the substrate atomic mass (28 for Si), and $Z^{1/3} = (Z_1^{2/3} + Z_2^{2/3})^{1/2}$.

The electronic stopping power can be approximated as [Eq. (2.390)]

$$S_e(E) = kE^{1/2} \tag{4.40}$$

Here E is the energy of the ion and k is a constant that depends on both the ion and the substrate and is given by

$$k = Z_1^{1/6} \frac{0.0793 Z_1^{1/2} Z_2^{1/2} (M_1 + M_2)^{3/2} C_R}{(Z_1^{2/3} + Z_2^{2/3})^{3/4} M_1^{3/2} M_2^{1/2} C_E^{1/2}} \tag{4.41}$$

where

$$C_R = \frac{4\pi a^2 M_1 M_2}{(M_1 + M_2)^2} \tag{4.42}$$

and

$$C_E = \frac{4\pi \varepsilon_0 a M_2}{Z_1 Z_2 q^2 (M_1 + M_2)^2} \tag{4.43}$$

where ε_0 is the permittivity of free space $= 8.85 \times 10^{-15}/\text{F} \cdot \text{cm}^2$, a is the Bohr radius $= 0.529 \times 10^{-8}$ cm, and $q = 1.602 \times 10^{-9}$ C. For an amorphous Si substrate, k is independent of the type of ion and Eq. (4.40) reduces to

$$S_n(E_c) = S_e(E_c) \tag{4.44}$$

and

$$E_c^{1/2} = \frac{S_n^0}{k}$$

(4.45)

hence

$$E_c^{1/2} = 14 \frac{14Z_1}{(14^{2/3} + Z_1^{2/3})^{1/2}} \frac{M_1}{M_1 + 28}$$

(4.46)

For amorphous Si the critical energy is given as

$$E_c \cong 10 \text{ keV} \quad \text{for B } (Z = 5, M = 10)$$
$$E_c \cong 200 \text{ keV} \quad \text{for P } (Z = 15, M = 30)$$
$$E_c > 500 \text{ keV} \quad \text{for As and Sb}$$

Thus, B tends to be stopped by electronic interactions; P, As, and Sb tend to be stopped by nuclear collisions. If $E \ll E_c$, then

$$\frac{dE}{dz} = kS_n$$

(4.47)

and from Eq. (4.39)

$$R = (0.7 \text{ Å}) \frac{Z^{1/3}}{Z_1 Z_2} \frac{M_1 + M_2}{M_1} E_0$$

(4.48)

Equation (4.48) is useful for heavy ions such as As, Sb, and sometimes P. If $E \gg E_c$, then

$$\frac{dE}{dz} = NkE^{1/2}$$

(4.49)

$$R = 20E_0^{1/2} \text{ Å} \quad \text{for } E_0 \text{ in eV}$$

Equation (4.49) is useful for boron if channeling does not take place. In Eqs. (4.48) and (4.49) R is the total range, not the projected range R_p. For the general case when the above limiting cases do not apply,

$$R = \frac{1}{N} \int_0^{E_0} \frac{dE}{S_n + kE^{1/2}}$$

(4.50)

TABLE 4.5.   Projected Range Corrections R_p/R (Si Substrate)

Ion	R_p/R values				Rule of thumb value $(1 + M_2 3 M_1)^{-1}$
	20 keV	40 keV	100 keV	500 keV	
Li	0.54	0.62	0.72	0.86	0.4
B	0.57	0.64	0.73	0.86	0.54
P	0.72	0.64	0.79	0.86	0.77
As	0.83	0.84	0.86	0.89	0.89
Sb	0.88	0.88	0.89	0.91	0.93

4.5.5. Projected Range

The previous equations have been in terms of R, the total range, but we really require R_p, the projected range. It can be shown that[53]

$$\frac{R}{R_p} = 1 + b\frac{M_2}{M_1} \tag{4.51}$$

where $b \cong 1/3$ for nuclear stopping and $M_1 > M_2$, i.e., for Sb and As. Although b is smaller for electronic stopping (B, P), 1/3 is still correct for the first order. For example, As: $M = 75$, therefore $R/R_p \cong 1.12$. For other mass and energy combinations, the projected range can be estimated using the R_p/R calculations in Table 4.5.

The dopant concentration distribution as a function of the distance in the target is shown in Fig. 4.22, where

$$R_p = \frac{R}{1 + bM_2/M_1} \tag{4.52}$$

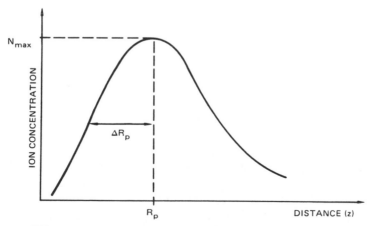

FIGURE 4.22.   Definition of Gaussian profile for implanted ions.

and the standard deviation

$$\Delta R_p \cong \frac{2}{3} \frac{M_1^{1/2} M_2}{M_1 + M_2} R_p = \frac{\text{half-width at } (1/2) \, N_{max}}{(2 \ln 2)^{1/2}}$$

$$N_{max} \cong \frac{N_D}{2.5 \Delta R_p}$$

(4.53)

and N_D is the number of implanted atoms per square centimeter. The ion concentration $N(z)$ will be normally distributed about R_p; i.e.,

$$N(z) = N_{max} \exp \frac{-(z - R_p)^2}{2 \Delta R_p^2}$$

(4.54)

dropping by one decade at

$$z = R_p \pm 2\Delta R_p \qquad N(z) = \frac{1}{10} N_{max}$$

(4.55)

two decades at

$$z = R_p \pm 3\Delta R_p \qquad N(z) = \frac{1}{100} N_{max}$$

(4.56)

and five decades at

$$z = R_p \pm 4.8\Delta R_p \qquad N(z) = 10^{-5} N_{max}$$

(4.57)

In practice, R_p and ΔR_p are tabulated for most common impurities as a function of the ion energy.

Thus, to find an implantation profile one needs to do the following:

- Look up R_p and ΔR_p in Table 4.6
- Assume that the distribution is Gaussian
- Calculate N_{max} from

$$N_{max} = \frac{N_p}{2.5 \Delta R_p}$$

(4.58)

The profile is then given by Eq. (4.54).

The doping profile is usually characterized by the projected range R_p and the standard deviation in the projected range, ΔR_p. Figure 4.23 shows R_p, ΔR_p, and the transverse distribution ΔR_T for boron, phosphorus, and arsenic ions implanted into silicon.[56] The penetration depth is roughly linear with energy and is shallower the more massive the ion.

TABLE 4.6. Projected Range (Å) and Standard Deviation in Projected Range for Various Ions in Silicon[a]

Ion		keV									
		20	40	60	80	100	120	140	160	180	200
B	R_p	714	1413	2074	2695	3275	3802	4289	4745	5177	5588
	(ΔR_p)	276	443	562	653	726	793	855	910	959	1004
N	R_p	491	961	1414	1847	2260	2655	3034	3391	3728	4046
	(ΔR_p)	191	312	406	479	540	590	633	672	710	745
Al	R_p	289	564	849	1141	1438	1737	2036	2335	2633	2929
	(ΔR_p)	107	192	271	344	412	476	535	591	644	693
P	R_p	255	488	729	976	1228	1483	1740	1998	2256	2514
	(ΔR_p)	90	161	228	291	350	405	495	509	557	603
Ga	R_p	155	272	383	492	602	712	823	936	1049	1163
	(ΔR_p)	37	64	88	111	133	155	176	197	218	238
As	R_p	151	263	368	471	574	677	781	885	991	1097
	(ΔR_p)	34	59	81	101	122	141	161	180	198	217
In	R_p	133	223	304	381	456	529	601	673	744	815
	(ΔR_p)		23	38	51	63	75	86	97	108	119
Sb	R_p	132	221	300	376	448	519	590	659	728	797
	(ΔR_p)	22	36	49	60	71	82	92	102	112	122

[a]Source: Ref. 53.

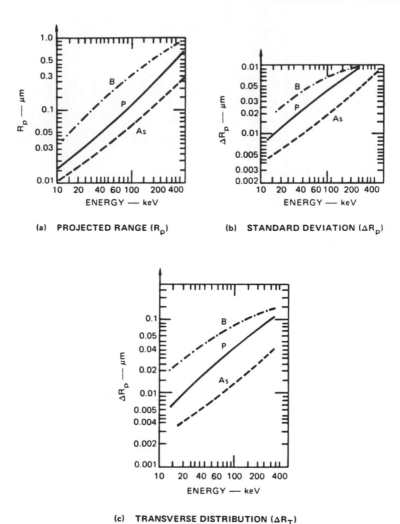

FIGURE 4.23.   Projected range (R_p), standard deviation (ΔR_p), and transverse distribution (ΔR_T) for boron, phosphorus, and arsenic as functions of the implanted ion energy. (a) Projected range (R_p); (b) standard deviation (ΔR_p); (c) transverse distribution (ΔR_T). From Ref. 53.

4.5.6. Masking

Spatially selective doping is achieved by means of "masking." Being a mechanical process, ions only require a physical barrier of a sufficient thickness in order to be stopped. There is actually a considerably wider choice of masking materials usable with ion implantation than with diffusion. Figure 4.24 shows some of them and offers a comparison of different masking layer thicknesses required to prevent all but about 0.1% of an implant from penetrating. SiO_2 makes a good mask, Si_3N_4 is even better, Al is good, and photoresist (KTFR) in thicknesses commonly applied is adequate for all but high-energy boron implants.

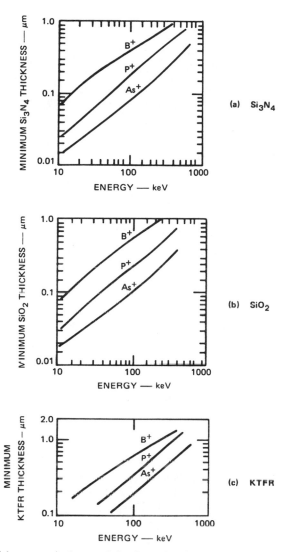

FIGURE 4.24.   Thickness required to mask implants. (a) Si_3N_4; (b) SiO_2; (c) KTFR. From Ref. 52.

4.5.7. Channeling

The range distributions in single-crystal targets are different if the ion beam is aligned along a crystal axis (see Section 2.7.2.4). In this case, because projectiles can channel along open directions, a deeper penetration can occur.[57] The channeling angle ψ is given as (see Fig. 4.25)

$$\psi \cong \left(\frac{a}{d}\,\psi_1\right)^{1/2} \tag{4.59}$$

where $\psi_1 = [(2Z_1 Z_2 q^2)/Ed]^{1/2}$, $a \cong 0.5$ Å, and d is the atomic spacing (2.5 Å for silicon).

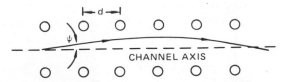

(a) SCHEMATIC TRAJECTORY OF A CHANNELED PARTICLE

Ion	Energy (keV)	Channel Direction		
		<110>	<111>	<100>
Boron	30	4.2	3.5	3.3
	50	3.7	3.2	2.9
Nitrogen	30	4.5	3.8	3.5
	50	4.0	3.4	3.0
Phosphorus	30	5.2	4.3	4.0
	50	4.5	3.8	3.5
Arsenic	30	5.9	5.0	4.5
	50	5.2	4.4	4.0

(b) CRITICAL ANGLES FOR CHANNELING OF SELECTED IONS IN SILICON

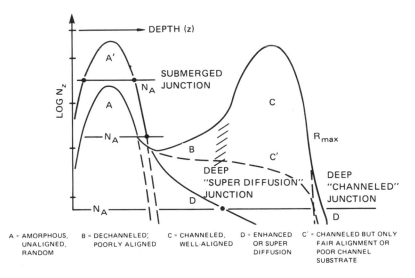

A = AMORPHOUS, UNALIGNED, RANDOM B = DECHANNELED; POORLY ALIGNED C = CHANNELED, WELL-ALIGNED D = ENHANCED OR SUPER DIFFUSION C' = CHANNELED BUT ONLY FAIR ALIGNMENT OR POOR CHANNEL SUBSTRATE

(c) DEPTH PROFILE

FIGURE 4.25.   Channeling effect. (a) Schematic trajectory of a channeled particle; (b) critical angles for channeling of selected ions in silicon; (c) depth profile. From Ref. 54.

Critical angles as a function of energy for several different projectiles are also given in Fig. 4.25 for the three major orientations of a Si target. The channeling effect can result in any of the several types of profiles shown in Fig. 4.25. The amorphous peak A or A′ is generated by ions A′ entering a crystal in random directions or by ions A in well-aligned trajectories which impact substrate atoms at the surface end of the lattice rows. This amorphous distribution exhibits a Gaussian shape characterized by a mean projected range R_p and a standard deviation ΔR_p. The channeling peak C comprises ions whose trajectories are well within the critical angle of the crystal direction. These ions lose energy by electronic collisions and channel to approximately the maximum range. Ions which are subject to higher values of electronic stopping produce more-pronounced (higher) channeling peaks. A channeling peak is not exhibited for all ions incident at room temperature. Channeling is purposely avoided in most devices because it is difficult to control accurately. Usually, implantation is done with the wafer tilted to get an "amorphous" profile. For example, a $7°$ tilt off the $\langle 100 \rangle$ axis results in greater than 99% of the ions being stopped as if the silicon were amorphous.

4.5.8. Implantation

To summarize the previous results,

- For light ions (B) and high energies, the electronic stopping is the major energy-loss mechanism. For heavy ions (As, Sb) and low energies, the dominant energy loss is due to nuclear collisions.
- The ion distribution in the target is Gaussian in both the z and x directions (see Fig. 4.26), such that

$$N(z, x) = N_{\max} \exp\left(\frac{-(z - R_p)}{2\Delta R_p^2}\right) \exp\left(\frac{-x^2}{2\Delta R_T^2}\right) \tag{4.60}$$

Lindhard–Scharff–Schiott (LSS) theory allows the prediction of R_p, ΔR_p, and ΔR_T for an arbitrary ion and substrate.[55] Channeling can be a problem if the beam is aligned along a crystal axis, but this is normally avoided in devices by tilting the

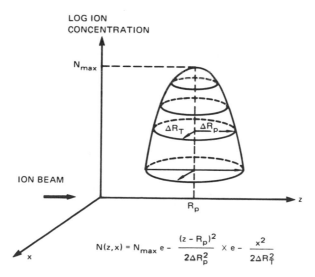

FIGURE 4.26.   Implanted longitudinal (z) and transverse (x) ion distribution.

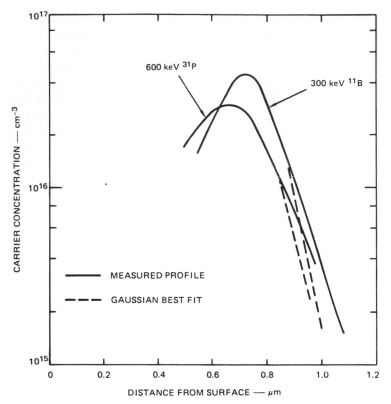

FIGURE 4.27.   Measured profiles and Gaussian fit. From Ref. 53.

wafer with respect to the beam. Deviations from the Gaussian shape are usually caused by the unwanted channeling effect or by the enhanced diffusion which occurs during annealing of the damaged substrate (see Section 4.4). Figure 4.27 shows two actual measured profiles along with a Gaussian fit.[56]

4.5.9. Damage

For range determination, both electronic and nuclear interactions are important. However, when considering atomic displacements in the substrate lattice, generally only the atomic or nuclear collisions should be considered because only these interactions can directly transfer enough energy to the target atoms.[54]

For crystalline substrates, atoms are bound together with energy E_d. The necessary condition to cause damage is that $E_{transf} \geq E_d$. Heavy ions (Sb, As) are stopped primarily by nuclear collisions; therefore, they cause more atomic displacements than B or P, which are stopped mainly by electronic interactions. Nuclear stopping dominates at low energies; therefore, one should expect that most of the damage for B occurs near where the ions stop.

For heavy ions such as Sb and As, stopping is primarily due to nuclear collisions ($E_c > 500$ keV for As and Sb and $E_c \cong 200$ keV for P), therefore one would expect a great deal of damage.

The typical situation is understood by observing that the nuclear stopping power S_n of a 100-keV Sb projectile in a silicon target is about 0.2 keV/Å averaged over the entire trajectory.[54] Since the spacing between the lattice planes in silicon is about 2.5 Å, this means that the Sb projectile loses about 500 eV per lattice plane on the average. One can expect the majority of this energy to be given to one primary silicon recoil atom, as the average recoil will have an energy approaching 500 eV; it would then have a range of roughly 25 Å. For comparison, the range of the 100-keV Sb ions is 500 Å, which is not substantially larger than the possible range of the average primary recoil atoms it creates.

The collision cascade of each primary recoil will contain roughly 15 displaced target atoms, since the number of recoils $\cong E/2E_d$ for heavy ions, where E_d is the bonding energy of the target atom, $\cong 15$ eV for Si. The damage volume $V_D \cong \pi(25 \text{ Å})^2 (500 \text{ Å})$ (see Fig. 4.28). Within this volume there are roughly 15 displacements per lattice plane, or 3000 total displacements. The average vacancy density is about $3000/V_D \sim 10^{22}$ per cm^3, or about 20% of the total number of atoms in V_D. As a consequence of the implantation, the material has turned essentially amorphous.

For light ions such as B, much of the energy loss is a result of electronic interactions, which do not cause atomic displacement. There is more atomic displacement near the end where nuclear stopping dominates. For boron $E_c \cong 10$ keV. As a consequence, most of the damage is near the final ion position. For example, 100-keV B has $R_p \cong 3400$ Å, which equals 500 recoils or vacancies created in the last 1700 Å, or about 200 over the first 1700 Å. In the damage volume $(V_D \sim 1.6 \times 10^{-18} \text{ cm}^3)$ there are 500 displacements, and with this the average vacancy density is $500/V_D \sim 3 \times 10^{20}$ per cm^3, which is less than 1% of the atoms; higher doses are needed to create amorphous material with light ions.

To create amorphous material, an energy dissipation of $\cong 10^{21}$ keV/cm^3 is required, the same as needed for melting. To estimate the dose for converting a crystalline material to an amorphous form by heavy ion bombardment, consider a 100-keV Sb ion that has a range $R_p \cong 500$ Å (5×10^{-6} cm). For this ion beam the dose to make amorphous Si is

$$D = \frac{(10^{21} \text{ keV/cm}^3)R_p}{E} = 5 \times 10^{13} \text{ per cm}^2 \qquad (4.61)$$

For 100-keV B, $R_p \cong 3300$ Å $= 3.3 \times 10^{-5}$ cm. The dose to make amorphous Si $D \cong 10^{21}(3.3 \times 10^{-5})/100 = 3.3 \times 10^{14}$ per cm^2.

In practice, higher doses are required for boron because the damage is not uniformly distributed along the path of the ion.

To the first order the damage profile may be assumed to be Gaussian. It is generally shallower than the ion profile (see Fig. 4.29). Based on LSS theory, it is possible to calculate the range values for the damage profile. For example, for boron

$$\frac{M_{Si}}{M_B} = \frac{12}{5} = 2.4 \qquad (4.62)$$

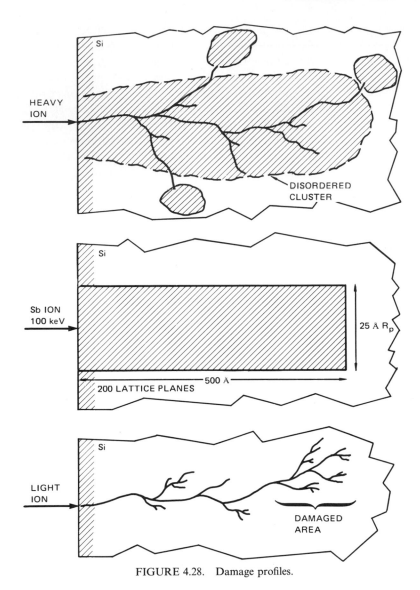

FIGURE 4.28.   Damage profiles.

Using the Sigmund–Saunders tables,[58] one can estimate the damage distribution by noting that

$$\langle z \rangle_D = R_D \tag{4.63}$$

and

$$\langle \Delta z \rangle_D = \Delta R_D \tag{4.64}$$

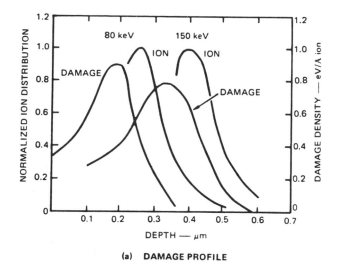

(a) DAMAGE PROFILE

$\dfrac{M_{target}}{M_{ion}}$	$\dfrac{(z)}{E/NC_1}$	$\dfrac{(\Delta z^2)}{(z)^2}$	$\dfrac{(x^2)}{(z)^2}$	$\dfrac{(zx^2)}{(z)\,(x^2)}$	$\dfrac{(\Delta z^3)}{(z^3)}$
1/10	0.842	0.058	0.018	1.07	0.007
1/4	0.577	0.125	0.044	1.16	0.021
1/2	0.453	0.195	0.089	1.20	0.043
1	0.369	0.275	0.176	1.20	0.079
2	0.297	0.409	0..343	1.16	0.135
4	0.229	0.710	0.674	1.12	0.221
10	0.153	1.684	1.671	1.07	0.345

(b) RANGE DISTRIBUTION PARAMETERS

$\dfrac{M_{target}}{M_{ion}}$	$\dfrac{(z)_D}{E/NC_1}$	$\dfrac{(\Delta z^2)_D}{(z)^2_D}$	$\dfrac{(x^2)}{(z)^2_D}$	$\dfrac{(zx^2)_D}{(z)_D(x^2)_D}$	$\dfrac{(\Delta z^3)_D}{(z)^3_D}$
1/10	0.692	0.434	0.192	1.80	0.580
1/4	0.489	0.437	0.181	1.71	0.433
1/2	0.376	0.386	0.152	1.47	0.218
1	0.295	0.380	0.157	1.41	0.172
2	0.241	0.457	0.257	1.42	0.272
4	0.198	0.623	0.485	1.31	0.391
10	0.143	1.215	1.153	1.16	0.567

(c) DAMAGE DISTRIBUTION PARAMETERS

FIGURE 4.29.   Damage profile (a) and associated range and damage distribution parameters (b and c, respectively).

The damage range can be calculated by taking ratios from the second column of the tables in Fig. 4.29. Thus,

$$\langle z \rangle_D = \frac{0.22}{0.26} R_p \approx 0.8 R_p \qquad (4.65)$$

The standard deviation in the damage distribution is given from ratios in the third column of the table as

$$\langle \Delta z \rangle_D = 0.75 (\Delta R_p) \qquad (4.66)$$

which can be used together with Eq. (4.65) to construct an actual damage distribution.

For low doses, many of the ions end up on lattice sites; they are substitutional and also electrically active. For high doses, most of the ions end up as interstitials and are not electrically active. In general, boron shows a higher percentage of ions in interstitial or nonelectrically active sites.[49,59]

4.5.10. Radiation-Enhanced Diffusion

All of the impurities used in fabricating silicon devices diffuse by either a substitutional or an interstitial mechanism. As a consequence, the diffusion process is

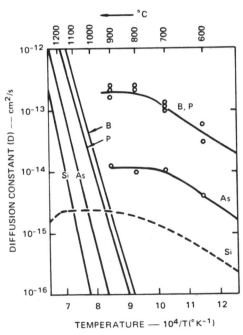

FIGURE 4.30. Temperature dependence of enhanced diffusion (5×10^{11} protons/cm$^2 \cdot$ s). From Ref. 56.

sensitive to the vacancy concentration and can be readily influenced by the supply of excess vacancies. One way of generating the vacancies is by means of displacement reactions caused by nuclear collisions in the ion-implantation process. This scheme is known as radiation-enhanced diffusion.[60] Figure 4.30 shows the experimental results for B, P, As, and Si under the intrinsic condition $N_A < n_i$. From the theory at low temperatures and high fluxes, the enhanced self-diffusion coefficient in Si depends on the flux by a square-root law

$$D_s \approx k(\text{flux})^{1/2} \tag{4.67}$$

and at low fluxes

$$D_s \sim k'(\text{flux}) \tag{4.68}$$

depends linearly on the flux.

4.5.11. Recoil Phenomena

Nuclear collisions are responsible not only for the atomic displacement in the target materials, which leads to the radiation-enhanced diffusion of impurity profiles, but also for the modification of impurity distributions by the introduction of additional impurity species by recoil from surface layers (see Fig. 4.31). This effect occurs when dielectric films are applied to the surface of a silicon wafer for selective doping. The total number of recoil atoms from the surface film that reach the silicon substrate depends on the type of bombarding ions, their energy (E_0), the type of atoms in the film, and the film thickness (W).

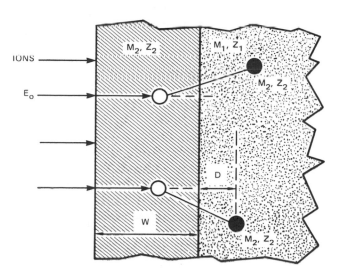

FIGURE 4.31. Diagram of recoil collisions.

The depth distribution for the dielectric–silicon interface is shown in Fig. 4.31 and is given by.[61]

$$N(D) = BN\left[\left(\frac{E_0}{W}\right)^{1/3} - \left(\frac{E_0}{D}\right)^{1/3}\right]\left[1 + \left(\frac{W}{W_0}\right)^{2/3}\right] \quad \text{atoms/volume} \quad (4.69)$$

where B is a coefficient that contains the atomic parameters of the film substrate and the projectile and W_0 is an empirical factor that characterizes the formation of the collision cascade. Figure 4.32 shows the oxygen profile obtained from the 50-keV recoil collisions of 10^{13} ions/cm^2 traversing an oxide layer of 10 Å.

Recoil, range, and damage distributions following ion implantation are of considerable importance in device design. Calculations based on the LSS theory (see Section 4.5.3) have been developed for range and damage profiles for ions implanted in semiconductors and they give a satisfactory accounting of the primary-ion range distributions in semi-infinite substrates.[60] However, practical device processing involves implantation into targets with one or more thin films. In these cases we are interested in the spatial distributions of not only the primary ions but also the recoil-implanted atoms in each layer.

Numerical integration of the extended Boltzmann equation can be used to obtain estimates of the primary-ion and recoil range distributions, the energy deposition profiles, and the energy and angular distributions.[62]

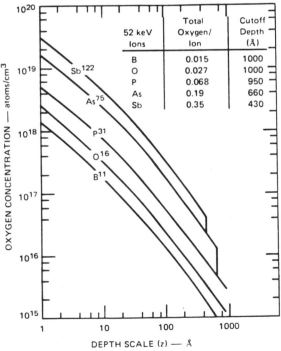

FIGURE 4.32.   Oxygen profile obtained from 50-keV recoil collision at 10^{13} ions/cm^2. From Ref. 56.

4.5.12. High-Dose Implantation Limits

One of the major advantages of ion implantation is the external control of the number of implanted ions. However, in the implantation process, ion sputtering transfers energy from the incident ion to the target material, resulting in the ejection of atoms from the surface. Thus, in high-dose implantation, control is lost and an upper limit to the concentration of an implanted species is set by sputtering erosion of the implanted surface.[63]

For a dose of 10^{17} ions/cm^2 (about 100 monolayers), it is possible to remove 100–1000 layers of target material, which corresponds to a thickness change of 500–5000 Å. Then during implantation the surface profile is a result of erosion and sputtering of both the target and the implanted ions. The simplest estimate gives the concentration of implanted species to be proportional to $1/S$, where S is the sputtering yield of the target.

Recent experiments on the sputtering of compounds[63] indicate that preferential sputtering effects can influence the maximum achievable ion concentration. The influence of preferential sputtering (r) is generally to reduce the concentration of higher-mass atoms at the surface of the target. According to this model, the maximum concentration is proportional to r/S, where S is the total sputtering yield and r is the preferential sputtering factor $12 < r < 2$. Since lower-mass elements tend to be preferentially sputtered, one can achieve a higher concentration of heavy elements than of lighter elements in the substrate.

For GaAs and other III–V compound materials, ion energies up to several megaelectron volts are required for deeper penetration. For a variety of ions, direct current and also pulsed machines have been built[64] recently for medium- and high-energy applications.

4.5.13. Neutron Transmutation Doping

Irradiating semiconducting materials in a nuclear reactor was initiated in 1950 at the Oak Ridge National Laboratory.[65] The primary interest at that time was the study of radiation damage effects, although the transmutation of silicon was used to determine isotropic cross sections. This work led to the direct use of neutron transmutation doping (NTD) to convert pure single-crystal ingots of high-resistivity silicon into homogeneously and precisely phosphorus-doped n-type material.[66,67] NTD, however, cannot be used for spatially selective doping as can diffusion and ion implantation.

The resistivity variation of NTD material can be as low at 1%, corresponding to doping variations of less than 1% across the wafer.[67]

Nuclear reactors provide the source of thermal ($E_n = 0.025$ eV) neutrons for NTD. However, the thermal neutron flux is always accompanied by a fast neutron component that is not useful in providing transmutations but which produces unwanted displacements of atoms that must be repaired by annealing after the irradiation.

Thermal neutrons interact only weakly with the atomic electrons through their magnetic moments, but, being neutral, they can impact the target nuclei and be captured by it. This interaction is described in terms of a capture cross section σ_c.

The number of captures per unit volume for a fluence (flux exposure time) of neutrons ϕ is given by

$$N = N_T \sigma_c \phi \qquad (4.70)$$

where N_T is the number of target nuclei per unit volume. At low energies

$$\sigma_c \approx E^{-1/2} \approx \frac{1}{v} \qquad (4.71)$$

where E is the neutron energy and v the corresponding velocity. For a given nuclear radius, $1/v$ is proportional to the interaction time, so that σ_c may be considered as representing the probability of interaction between the nucleus and the thermal neutron.

After neutron capture the target nucleus differs from the initial nucleus by the addition of one nucleon forming a new isotope in an excited state. The excited state must relax by the emission of energy, which is usually in the form of high-energy gamma rays. The time for the decay of this excess energy by gamma radiation can be very short (prompt gammas) or can take an appreciable time, in which case a half-life (the time for the number of particles emitted per second to decrease by a factor of 2) can be measured. The gamma emission spectrum is used to characterize the nuclear energy levels of the transmuted target nuclei.

The absorption of a neutron and the emission of gamma rays is represented by the notation $A_x(n, \gamma)(A + 1)_x$ where (n, γ) represents (absorption, emission), A is the initial number of nucleons in the target element x before neutron absorption, while $A + 1$ is the number after absorption. It is possible for the product isotope $(A + 1)_x$ to be naturally occurring and stable. In many cases, however, the product isotope is unstable. Unstable isotopes decay further by various modes involving the emission of electrons (β decay), protons, alpha particles, and K-shell electron capture or internal conversion until a stable isotopic form is reached. These decay processes produce various radioactive elements, characterized by their half-lives $T_{1/2}$.

In the case of silicon, absorbing the thermal neutron, three stable target isotopes are formed by (n, γ) reactions as in Table 4.7.[68]

The first two reactions produce no dopants and only slightly redistribute the relative abundances. The third reaction produces ^{31}P, the desired donor dopant, at a rate of about 3.355 ppb per $\phi = 10^{18} \ n_{th}/cm^2$. This production rate is calculated

TABLE 4.7. Nuclear Reactions for Silicon

Reaction[a]	Cross section σ_c (barns) $(1 \ b = 10^{-24} \ cm^2)$	Isotope abundance
1. ^{28}Si$(n, \gamma)^{29}$Si	0.08	92.3% ^{28}Si
2. ^{29}Si$(n, \gamma)^{30}$Si	0.28	4.7% ^{29}Si
3. ^{30}Si$(n, \gamma)^{31}$Si $\rightarrow$ ^{31}Si P $+ e^-$		

[a]The half-life of ^{31}Si is 2.62 h.

from the rate equation for ^{30}Si neutron capture ($\sigma_c = 0.11 \times 10^{-24}$, $N_T = 5 \times 10^{22}$ Si/cm$^2 \times 0.031$).

In addition to the desired phosphorus production reaction and its relatively short half-life for β decay, the reaction

$$^{31}P(n, \gamma)^{32}P \rightarrow {}^{32}S + e^-(\sigma_c + 0.19b, T_{1/2} = 14.3 \text{ days})$$

occurs as a secondary undesirable effect. The decay of ^{32}P is the main source of residual radioactivity in NTD Si.

Once the dopant phosphorus has been added to the silicon ingot by transmutation of the ^{30}Si isotope, this radiation-damaged and highly disordered material has to be annealed to make it useful for electronic device manufacture. Several radiation-damage mechanisms contribute to the displacement of the silicon atoms from their normal lattice positions. They are the following:

- Fast neutron knock-on displacements
- Fission gamma-induced damage
- Gamma recoil damage
- Beta recoil damage
- Charged particle knock-ons (n, p), (n, α), etc. reactions.

Estimates can be made of the rate at which Si atom displacements are produced by these various mechanisms, once a detailed neutron energy spectrum is known.[67]

NTD is now quite well established as a method of introducing a uniform distribution of phosphorus dopant into high-purity silicon in order to obtain a uniform n-type starting material. This is particularly interesting for power devices where n-type starting material is required for high-power device design and performance, including a more precise control of avalanche breakdown voltages, a more uniform avalanche breakdown, i.e., a greater capacity to withstand overvoltages and a more uniform current flow in the forward direction. In addition to the area and spatial uniformity of the dopant distribution, the NTD silicon offers three clearly identifiable advantages over silicon doped by more traditional techniques.

- Precise control of the doping level
- Elimination of dopant segregation at grain boundaries in polycrystalline silicon
- Superior control of heavy-atom contaminants

The precise control of doping levels is important in any application such as avalanche or infrared detectors, which require a high-resistivity material. The transmutation doping of epitaxial layers deposited on low-resistivity n-type or p-type silicon substrates has also been demonstrated.[69]

4.5.14. Application of Ion Implantation to Bipolar Device Fabrication

For bipolar applications, high-dose implantations are used for buried layers, emitters, and base contact doping, while low- or moderate-dose implants are employed to dope active base regions and to make resistors.

Figure 4.33 shows an example of how a low-resistivity As buried layer can be formed under a high-quality epitaxial layer. In this application a high-dose As implant is made through an oxide cut. A high-temperature drive-in and oxidation step serves as the anneal. The oxide grown is sufficient to consume the silicon in which the implanted layer originally resided. This eliminates residual damage typical of very-high-dose implants and allows the growth of high-quality epitaxial silicon over the buried layer. Sheet resistivities of less than 10 Ω per square can be achieved by this process.

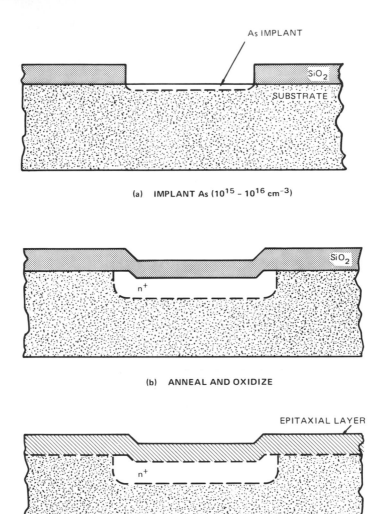

(a)   IMPLANT As $(10^{15} - 10^{16}$ cm$^{-3})$

(b)   ANNEAL AND OXIDIZE

(c)   STRIP OXIDE AND GROW EPITAXIAL LAYER

FIGURE 4.33.   Buried layer formation. (a) Implant As $(10^{15}-10^{16}$ cm$^{-3})$; (b) anneal and oxidize; (c) strip oxide and grow epitaxial layer.

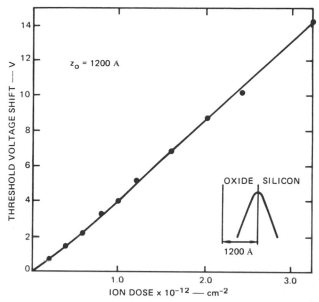

FIGURE 4.34. *p*-Channel MOS with 30-keV boron implant, fully annealed. From Ref. 53.

4.5.15. Application of Ion Implantation to MOS Device Fabrication

Ion implantation was first applied on a large scale in MOS devices, and its use continues to grow. It has provided a reproducible means of introducing small amounts of impurities near the surface and has led to many applications. Most MOS applications of ion implantation involve threshold voltage control. Adjustment of the active device threshold voltage by placing a controlled low dose close to the gate oxide–substrate interface was the first of these applications and remains the most widely used. Figure 4.34 shows some data for a *p*-channel MOS in which a boron implant is made through the gate oxide with the peak of the distribution at the interface so that approximately 50% of the dose goes into the silicon. For the ideal case of the sheet charge placed right at the interface, the threshold voltage will shift linearly with the implanted dose.

4.5.16. Application of Ion Implantation to Resistor Fabrication

Ion-implanted resistors can be used by any IC technology that has a need for them as well as for monolithic precision resistor networks. Dose control is also a key factor, as the higher sheet resistivities are very difficult to obtain reproducibly with diffusion. A wide variety of conditions are possible; some of these are illustrated in Table 4.8 for the case of boron-implanted resistors. Sheet resistivities of from less than 100 Ω per square to many thousands are achievable, although surface conditions make doping level control difficult for the very high resistivities. Of great interest is the temperature coefficient of resistance (TCR). For a fully annealed implant, represented by the last column of Table 4.8, the TCR is positive and higher the lower

TABLE 4.8. Boron-Implanted Resistors

Ion dose (cm⁻²)	Parameter range for 400–950°C anneals		Anneal temperature for TCR = 0 (°C)	TCR for 950°C anneal (%/°C)
	R (Ω/°C)	TCR (%/°C)		
10^{12}	1,000,000–25,000	~ +0.7	None	~ +0.7
10^{13}	13,000–2500	−0.1–+0.45	475	+0.45
10^{14}	3,000–400	−0.15–+0.3	475	+0.15
10^{15}	2,000–50	−0.2–+0.2	540	+0.10

the dose. At lower annealing temperatures, however, combined damage and doping effects can cause a negative TCR for all but very low doses. There exists an annealing condition that yields a zero TCR, representing partial annealing of the damage. This is shown in the fourth column of Table 4.8. This partial annealing has found use in precision resistor networks, which are stable over some desired operating temperature range. The second column of Table 4.8 represents the range of observed TCR values in the annealing range 400–950°C. For all but the very low doses, the TCR starts out negative, crosses to zero, is maximum at about 600–700°C, and falls back to its final value shown for 950°C. This final value represents a fully annealed implant with no residual damage effects.[56]

4.5.17. Characterization of Ion Implantation

Characterization of ion implantation processing (carrier profiles, depth distribution and contour maps, film thickness) is an important task in IC microfabrication. In MOS processing, implants as low as 10^{10} ions/cm² are used for threshold voltage adjustments, but for other applications doses can be as high as 10^{17} ions/cm².

There are several classes of instrumental techniques used for ion implantation characterization. These can be categorized as

$$\text{Electrical}^{\text{resistive}}_{\text{capacitive}}$$

Thermal wave analysis

$$\text{Optical}^{\text{dosimetry}}_{\text{ellipsometry}}$$

$$\text{Surface analysis}^{\text{Rutherford backscattering}}_{\text{Secondary ion mass spectroscopy}}$$

The electrical evaluation methods measure resistance, and resistance change due to crystal damage during implantation and capacitance.[70] These techniques measure electrically active ions in the dose range of 10^{10}–10^{17} ions/cm².

The four-point probing technique requires that the ion-implanted wafers be electrically activated by a postimplant anneal before measurements can be made. Since the four-point probe is a contact method and since it has a spatial resolution on the order of a centimeter or more, this method is limited to test wafers. Furthermore, the four-point probe method does not ordinarily have the sensitivity to measure the low-dose implants (10^{11}–10^{12} ions/cm²) that are used for the critical adjustment of threshold voltages in MOS devices. To obtain any data at the low-dose level, it is necessary either to use high-resistivity test wafers that have undergone a special

TABLE 4.9. Instrumental Techniques for Ion Implant Evaluation[a]

Measurement class	Electrical				Thermal-wave	Optical		Surface analysis	
Evaluation technique	Four-point probing	Spreading resistance probing	Four-point probing double	Capacitance voltage measurement	Thermal-wave analysis	Optical dosimetry	Ellipsometry	Rutherford backscatter analysis	Secondary-ion mass spectroscopy
Property measured	Sheet resistance	Spreading resistance	Crystal damage	Depletion capacitance	Crystal damage	Photoresist darkening	Crystal damage	Crystal damage	Sputtered ions
Ion species measured	Active	Active	Active and inactive	Active	Active and inactive	Active and inactive	Active and inactive	Active and inactive	Active and inactive
Lateral resolution[b]	0.25 mm	2.5 μm	0.25 mm	0.10 mm	1 μm	3 mm	1.5 mm	1 mm	5 μm
Sensitivity[b]	0.7–0.8	0.7–0.8	0.5–1	0.7–9.8	0.2–1	0.5–1	0.5	~1	~1
Dose range ($10^?$ ions/cm)	12–16	10–17	10–14	10–12	10–15	11–15	11–14	13–18	11–18
Tested material	Si and GaAs	Si	Si	Si	Si and GaAs	Photoresist	Si and GaAs	Si	Si
Measurement made on	Test and some product wafers	Product[c] or test wafers	Test wafers	Product or test wafers	Product or test wafers	Special glass wafers	Product or test wafers	Product[c] or test wafers	Product[c] or test wafers
Preprocessing of test wafer before implant	Screen optional	None	Implant, anneal, and measure	Oxide	None	Measure before	Measure before	None	None
Postprocessing required before measurement	Anneal and strip oxide	Anneal cut and lap	None	Anneal and metal dots	None	None	None	None	None

[a]Source: Ref. 62.
[b]Percent change of measured parameter/percent change of dose.
[c]Product wafer destroyed in preparation for test.

cleaning procedure for passivating the wafer surface, or to employ a double-implant technique whereby the sheet resistance of an activated high-dose implant is increased by the lattice damage introduced by a succeeding preanneal low-dose implant. Both of these special four-point probe techniques require considerable sample preparation and can provide low-dose data with only moderate precision.[71] (See Table 4.9.)

An optical technique is based on optical dosimetry whereby a transparent glass substrate is first coated with positive photoresist, the photoresist decolored by UV annealing to render it colorless, and then the photoresist-coated glass substrate is exposed to the ion beam. This method utilizes the principle that the photoresist darkens in the UV wavelength region due to the effect of the incident ion beam. The UV transparence of the photoresist is then recorded by measuring the transmission of a 3-mm spot of UV light through the photoresist. This method requires several processing steps: spinning on the photoresist, annealing it with UV radiation, ion implantation, a fairly lengthy postimplant stabilization waiting period, and then the optical dosimetry measurements.

4.5.17.1. Thermal Wave Analysis for Implants. Thermal wave analysis measures the level of crystalline disorder created by ion implantation. The measured periodic signal amplitude is proportional to the dose (damage) by the implanted active and inactive ions. Thermal wave physics is playing an ever-increasing role in the nondestructive analysis and testing of materials. It has been employed for several years in optical investigations of solids, liquids, and gases by means of photoacoustic[72] and thermal-lens spectroscopy.[73] Thermal waves are present whenever there is periodic heat generation by an energy beam (electron, ion, or photon beams) and heat flow in a medium.

In the modulated optical reflectance technique, thermal waves are generated by an Ar^+ ion laser beam which is focused to a 1-μm-diameter spot at the sample surface.[74] This pump beam is acousto-optically modulated at frequencies in the 1–10 MHz range and has an incident peak power of $\sim$10 mW at the sample surface. The thermal waves are detected with an unmodulated He–Ne probe laser that is collinear with the pump beam and is also focused to a 1-μm spot size. The probe beam that is retroreflected from the sample surface is directed onto a photodiode.

The existence of a modulated optical relfectance signal is a result of the fact that the optical reflectance of a sample surface depends to some extent on its temperature.[75] Thus, if we denote the sample reflectivity by R,

$$R = R_0 + (\partial R/\partial T)\, \Delta T$$
$$= R_0 + \Delta R \tag{4.72}$$

Then

$$\Delta R/R_0 = (1/R_0)(\partial R/\partial T)\, \Delta T \tag{4.73}$$

where R_0 is the sample reflectivity at temperature T_0, $(1/R_0)(\partial R/\partial T)$ is the temperature coefficient of reflectivity, and ΔT is the variation of the temperature of the sample surface from T_0. For most materials, $(1/R_0)(\partial R/\partial T)$ is on the order of 10^{-5} to $10^{-4}/°C$, and since the laser-induced surface temperature oscillations, ΔT,

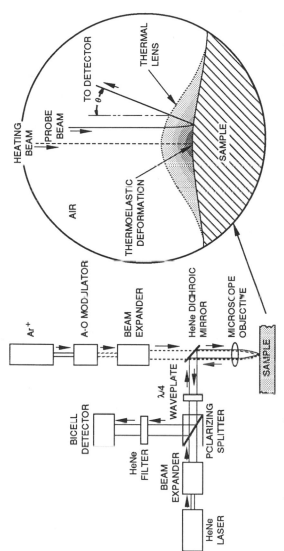

FIGURE 4.35. Schematic depiction of the laser beam deflection technique and the physical process affecting the laser probe. From Ref. 72.

produced in materials like Si are usually <10°C, the thermo-reflectance signal $\Delta R/R_0$ is on the order of 10^{-4} to 10^{-3}. Although this is a small signal, it is readily measured in a low-noise modulation detection system. Thus, this modulated thermoreflectance technique is an excellent noncontact means for generating and detecting thermal waves in most materials.

Figure 4.35 shows the block diagram of the laser beam deflection technique used to monitor the local thermoelastic deformations depicted in detail in the insert.[72] In this system a 488-nm beam of a 100-mW Ar^+ ion laser is intensity-modulated with an acousto-optic modulator, directed through a beam expander, and then focused to a ~1-μm-diameter spot on the sample. This is the heating beam, and it has a sample incident power of ~30 mW. The 633-nm beam of a 5-mW He–Ne laser, the probe beam, is directed through a beam expander, a polarizing beam splitter and quarterwave plate, reflected off a dichroic mirror, and then focused onto a ~1-μm-diameter spot on the sample with an incident power of ~2 mW. The two laser spots are displaced ~1 μm from each other at the sample surface. The 488-nm heating beam reflects back on itself, while the 633-nm probe beam undergoes a periodic deflection arising from the periodic change in the local slope of the thermoelastically deformed sample surface. The reflected probe beam passes through the quarterwave plate again, and since it is now 90° out of phase from the beam leaving the He-Ne laser, it is directed by the polarizing beam splitter to the bicell photodetector, which measures the periodic deflections of the probe beam. Unlike the laser interferometric technique, which measures local surface displacements in the vertical direction, the laser probe method measures changes in the local slope of the thermoelastically deformed surface as depicted in Fig. 4.35.

In the case of implantation analysis, the damage introduced into the crystal lattice by the ion implantation process extends only a short distance beneath the crystal surface. The damage in this layer increases as the implant dose increases. Damaged silicon will have a lower thermal conductivity than nondamaged silicon because of the increased phonon scattering from the various point defects and dislocations in the crystal. The damaged layer thus represents a subsurface thermal feature and is therefore detectable with the thermal waves. The magnitude of the thermoelastic displacement or deformation at the surface would thus be expected to increase monotonically with increasing lattice damage and thus with increasing implant dose.

The thermal wave analysis is a highly accurate (1–2%) noncontact measurement technique for implants in the range 10^{10}–10^{15} ions/cm^2 (see Table 4.9). Furthermore, the measurement spot is on the order of 1 μm, thus providing truly high spatial resolution on the order of actual device dimensions.

Using thermal wave analysis, the implant dose measurements can be made immediately after implantation and require no intervening annealing or other process step and thus represent a true monitor of the ion implantation process itself.

Two-dimensional contour mapping of implant dose is an established procedure for the detection and diagnosis of implanter problems. This technique because of its high spatial resolution (~1 μm) is useful for two-dimensional implantation dose maps.

4.5.17.2. Ellipsometry. Ellipsometry is a very sensitive optical technique for characterizing solid surfaces. In fact, it is widely used as a thin film thickness monitor in semiconductor process technologies. However, ellipsometric characterization has usually been restricted to rather simple applications such as the measurement of optical constants of bulk semiconductors or thin film thicknesses on Si substrates.

Ellipsometry also has been successfully applied to characterize ion-implanted layers of Si.[75] Damage profiles have been determined based on the three-layer model (SiO_2 surface layer, amorphous Si layer, transition layer). Isothermal annealing processes of damaged layers have been studied and microdefects found, and a new method of measurement of carrier depth profiles in annealed samples is described.

Other applications are also shown: analysis of the interface between Si and its thermally grown oxide by spectroscopic ellipsometry, *in situ* measurements of $Ga_{1-x}Al_xAs$ epitaxial growth on GaAs substrates, and characterization of the chemisorption layer in an ultrahigh vacuum chamber.

In ellipsometric measurement, linearly polarized monochromatic light is incident on a sample surface with angle of incidence θ, and the reflected light is generally elliptically polarized. The polarization of the incident (reflected) light can be described by the ratio of perpendicular components (termed the p and s components) of the electric vectors, E_p/E_s (E'_p/E'_s).

If the sample material is optically isotropic, then the p and s components are not mixed after reflection, i.e., $E'_q = R_q \exp(i\phi_q)E_q$ ($q = p, s$). The parameters $R_p(R_s)$ and $\phi_p(\phi_s)$ respresent the changes in the amplitude and phase for the p (s) component after reflection. Ellipsometry determines Δ and ψ defined as

$$\Delta = \phi_p - \phi_s \quad \text{and} \quad \tan\psi = R_p/R_s \tag{4.74}$$

The ellipsometric parameters Δ and ψ, or R_q and ϕ_q ($q = p, s$) are determined by solving the Maxwell equations

$$\nabla \times E = i(\omega/c)H \quad \text{and} \quad \nabla \times H = -i(\omega/c)\varepsilon(\omega)E \tag{4.75}$$

where ω is the angular frequency of the incident light, c is the velocity of light in vacuum, and $\varepsilon(\omega)$ is the dielectric function of the sample and is used to describe its electrical and optical properties. In this way, the properties of the sample can be investigated by ellipsometry.

The optical constants n (refractive index) and k (extinction coefficient) are also used to describe the optical properties. They are defined by

$$\varepsilon^{1/2} = n + ik \tag{4.76}$$

4.5.17.3. Rutherford Backscattering. Rutherford backscatter analysis (RBS) (see Section 6.3.2), a broadly applied surface analysis technique, is often used to analyze ion-implanted layers, particularly in the range of 10^{13} to 10^{18} ions/cm². RBS measures the energy spectrum of backscattered high-energy helium nuclei created when a sample is bombarded with helium ions.

In RBS experiments a monoenergetic high-energy beam of ions (e.g., H^+ or He^+) impinges on a target from which some are backscattered: part of these backscattered ions are energy analyzed and counted, and the data stored.[50,76] Figure 4.36 is a schematic drawing of an ion backscattering experiment divided into its essential components.

The accelerator consists of an ionization chamber for creation of the ions, followed by a column where the ions are accelerated. After passing through a short drift tube, the ions enter a magnet analyzer, where the various species of ions are separated and only those energies are selected which are of use in the experiment. After passing through the magnet, the analyzed beam enters a drift tube and is collimated and directed onto the target. A few of the ions elastically backscatter from the target, but those which do scatter in a particular direction are detected by a

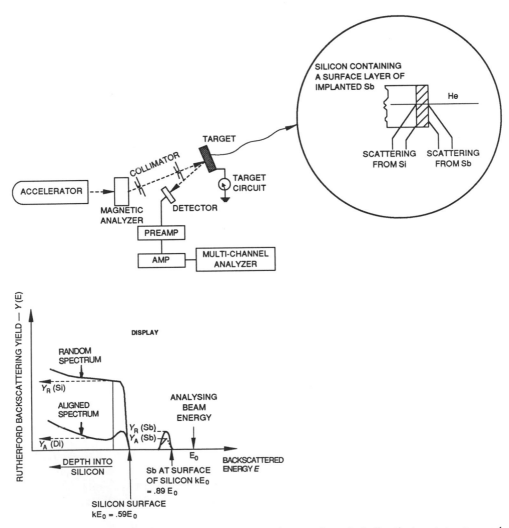

FIGURE 4.36. Schematic of a Rutherford backscattering experiment including the target structure and the energy spectrum of the scattered particles. From Ref. 50.

particle detector; their energy is analyzed and stored in an appropriate data storage system.

In the case shown in Fig. 4.36, He ions are scattered on silicon and implanted antimony. The structure of the target and the display of the energy spectrum are also depicted in Fig. 4.36. In this example the scattering from an impurity heavier than silicon results in less energy loss than scattering from the silicon. Thus, helium scattering from antimony located at the surface of the silicon lattice appears at a higher energy ($0.89E_0$). Provided the Sb is located in sufficient quantities ($>10^{-4}$ atom fraction) and is located within a shallow region, a well-defined yield is obtained. With the energies usually employed in ion implantation, the ranges of heavy ions such as Sb result in depth distributions close to the sample surface.[50]

The introduction of a heavy ion into a crystal lattice will usually result in radiation damage. The location of the Sb within the lattice can also be investigated by similar methods. Exposure of the sample to helium beams in both aligned and random directions will produce both types of spectra for the substrate and for the heavy implant.

Another important surface analyzing tool, *secondary ion mass spectroscopy* (SIMS) (see Section 6.3.2), is also used for implant analysis across the whole dose range (10^{10} to 10^{18} ions/cm^2). In this destructive technique the wafer is bombarded by a beam of atoms, often of oxygen or cesium, and the sputtered secondary ions mass analyzed.

4.6. EVAPORATION IN ULTRAHIGH VACUUM (UHV)

Evaporation in UHV[5, 9] is particularly useful when it is necessary for the film material to remain uncontaminated by residual gases and where the interface between the substrate and the film must have specific properties that could be influenced by contaminants. For this latter case, the substrate surface once cleaned must remain uncontaminated until deposition is started.

Table 4.10 gives the relevant data for air at various pressures, illustrating that pressures of less than 10^{-8} Torr are necessary to maintain the substrate uncontaminated for a reasonable time after cleaning. In vacuum evaporation, the following

TABLE 4.10. Kinetic Data for Air

Pressure (Torr)	Mean free path (cm)	Number impingement rate ($s^{-1}\,cm^{-2}$)	Monolayer impingement rate (s^{-1})
10^{-2}	0.5	3.8×10^{18}	4400
10^{-4}	51	3.8×10^{16}	44
10^{-5}	510	3.8×10^{15}	4.4
10^{-7}	5.1×10^4	3.8×10^{13}	4.4×10^{-2}
10^{-9}	5.1×10^6	3.8×10^{11}	4.4×10^{-4}

physical factors must be considered:

- The "cleanliness" of the vacuum
- The method of heating the evaporant
- The method for outgassing the evaporant
- The method for cleaning the substrate surface
- The method for ensuring film uniformity

Pumping is discussed in Section 3.7.3. While oil diffusion pumps have high pumping speeds and produce adequate pressure for many applications, they are also recognized as being quite "dirty," since even with careful trapping there is a good chance that the high-boiling-point silicone oils used in these pumps will eventually find their way into the evaporation chamber and contaminate the evaporant and substrate surface. Hence, for coatings with stringent purity specifications, oil-containing systems are avoided. In pumping down a system, the major portion of the air to be removed is from atmospheric pressure to, say, 10^{-3} Torr. This is preferentially done by sorption pumping in which a material with a large surface-to-volume ratio (Xeolite) is cooled to liquid-nitrogen temperatures. The air is adsorbed on the Xeolite surfaces, and with this method very fast pumping speeds and pressures as low at 10^{-4} Torr may be obtained without contaminating the system. The pressure may be further reduced cleanly by the use of sputter ion pumps, titanium getter pumps, or cryopumps (operating at liquid-helium temperatures). Turbomolecular pumps are also increasing in popularity despite the minor possibility of contamination by bearing lubricants and backing pumps. The walls of the vacuum chamber and all of the components within it contain gases adsorbed on their surfaces and dissolved inside. Much of this material can be removed by baking the whole system under high vacuum. A 150°C bake for 24 h is extremely effective for stainless steel systems,[77] but bakes up to 600°C or greater have been employed to obtain the lowest possible residual pressures.

If a solid or liquid is heated to a temperature T (K) in vacuum, the number of molecules leaving a unit area per second is given from thermodynamic considerations by the relation

$$N = N_0 \exp - \left(\frac{\phi_e}{kT} \right) \tag{4.77}$$

where N_0 may be a slowly varying function of T and ϕ_e is the energy required to remove one molecule of the material from the bound state in the surface to the vapor state above the surface: ϕ_e is called the activation energy for evaporation and is usually given in electron volts. It is related to the latent heat of evaporation of a given molecule, Q, by $Q = e\phi_e \times$ Avogadro's number (kJ/mol). Data are usually available in the form of plots of the equilibrium vapor pressure p as a function of temperature. Figure 4.37 shows the data for a few elements. Data on more elements are available in Ref. 78. The vapor pressure may be converted to evaporation rates

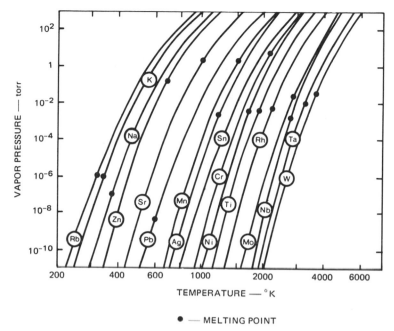

● — MELTING POINT

FIGURE 4.37. Temperature dependence of vapor pressure for several metal elements.

from kinetic theory considerations (see Section 3.7.2) by the relation

$$N = \frac{p}{(2\pi MkT)^{1/2}} = \frac{3.513 \times 10^{22}}{(MT)^{1/2}} \quad \text{molecules} \cdot \text{cm}^2/\text{s} \qquad (4.78)$$

where M is the molecular weight of the vapor molecule in atomic mass units and p is in Torr.

For the deposition of thin films it is often useful to think of the deposition process in terms of the number of atomic layers of the film arriving per second at the substrate surface. The number of atoms per unit area corresponding to a monolayer can be calculated from the lattice constants and is about 10^{15} atoms/cm^2.

Typically, vapor pressures at the source on the order of 10^{-2} Torr or greater are required for useful rates of film buildup. Heating the evaporant can present difficulties if high temperatures are required to obtain a fast evaporation rate or if the evaporant is highly chemically reactive at the operating temperature. Methods of heating that are commonly employed include resistive, eddy-current, electron bombardment, and laser heating.

Boats or holders are made of a highly refractory metal that can be heated by passing a current through it. Tungsten, tantalum, platinum, and graphite are preferred boat materials. Of course, if the evaporant is a conductor and remains solid at temperatures at which the vapor pressure is high, it can be made in the form of a wire or other resistive element and evaporated directly. Electron-beam heating offers

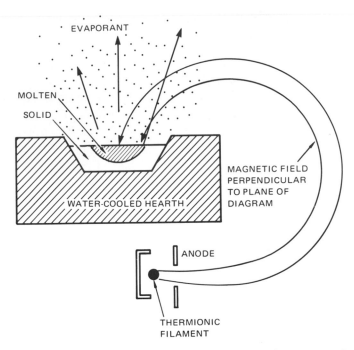

FIGURE 4.38.   Diagram of a 270° magnetized deflective electron-beam evaporation system.

the greatest flexibility (see Fig. 4.38). In the preferred technology the material to be evaporated is placed in a recess in a water-cooled copper hearth. An electron current of about 100–5000 mA is generated by a tungsten filament, hidden from direct viewing of the evaporant, and accelerated to a high voltage of 3–20 kV. The electron beam is magnetically directed onto a small spot on the evaporant material, which melts locally. Thus, the material forms its own crucible and the contact with the hearth is too cool for chemical reaction. The main disadvantage of this type of source is that the substrate and coating film will be bombarded by X rays and possibly energetic ions as well as the neutral thermal evaporant. This may be avoided by using a focused high-power laser beam instead of the electron beam.

Due to its high resistivity and ease of fabrication, graphite crucibles are particularly useful for eddy-current heating, since large amounts of rf power may be coupled into the crucible for heating evaporants that do not react with graphite.

Some compounds dissociate before vaporization and hence the higher-vapor-pressure component is preferentially distilled from the evaporation source. To overcome this the various components are evaporated from separate sources at a rate appropriate to the molecular composition required in the condensate. This requires very precise temperature control, since evaporation rates depend exponentially on the temperature. In such circumstances it can be advantageous to elevate the temperature of the substrate during deposition to a temperature at which the compound condenses but the components do not.

Many materials contain absorbed gases and volatile impurities. For this reason a shutter is usually placed between the evaporation source and the substrate to protect the substrate surface from contamination during preheating of the source. The shutter is swung open when the experimenter is satisfied that the impurities in the vapor stream have been reduced to a tolerable level.

Having the substrate surface atomically clean before deposition can be of vital importance for adhesion and determining other physical properties that are dependent on the interface between the substrate and the film, such as electrical contact. Thermal removal of the contaminant materials can be used only with highly refractory materials, such as tungsten or graphite. Atomically clean single-crystal faces can be exposed by cleavage in a high vacuum. Bombardment by energetic inert gas ions (sputtering) is a good general way of removing surface layers but can result in rough surfaces due to preferential erosion of the different crystal faces of polycrystalline substrates and in unwanted ion implantation.

The geometrical relationship between the source and the substrate determines the uniformity of deposition and the ratio of the source evaporation rate to the substrate arrival rate. Most evaporation sources can be treated as point or line sources, so that the uniformity of deposition on the substrate can be estimated by simply using the Knudsen relation

$$\text{Arrival rate} \propto \frac{\cos \theta}{r^2} \tag{4.79}$$

where r is the distance between the source and the substrate and θ is the angle between the normal to the substrate and the radial vector joining the source to the point being considered. A compromise must be effected between obtaining a film of uniform thickness over the substrate while maintaining a reasonable condensation rate without excessive waste of the evaporant. This topic is treated comprehensively in the literature.[5]

On the other hand, film uniformity can be maintained even for nonuniform sources by the simple expedient of moving the substrate so that each part of the sample integrates material from different source directions. Various combinations of rotary motions are typically used.

In summary, an evaporation source in which the desired material is transported in an ultrahigh vacuum to condense on a clean substrate is a preferred technique in cases where freedom from contamination both in the film and at the interface between film and substrate is of prime importance.

4.7. CATHODIC SPUTTERING

In Section 2.7 we discussed in some detail the physical processes involved when energetic ions strike the surface of a target material. One important effect is sputtering, namely, the ejection of atoms from the target surface. If the surface of a target material is bombarded by a uniform flux of energetic ions, the target may be utilized as a source of the material of which it is composed. In this section we discuss the properties of a sputter source for the deposition of thin films on a substrate and how these properties depend on the method used for producing the bombarding ions. We are concerned with the rate at which sputtered material leaves the target surface, the angular distribution of the ejected atoms, and the energy distribution of the ejected atoms. In addition, the effects of secondary particles produced at the target and the effects of the pressure of the transport medium on the purity and morphology of the film are of interest.

4.7.1. Sputter Rate, Angular and Energy Distributions

The number of atoms N per unit area per second leaving the target is given by

$$N = \frac{J_+}{ge} S(V, A, B) \tag{4.80}$$

where J_+ is the current density of the bombarding ions, g is the number of electronic charges per ion, and S is the sputter yield in atoms per incident ion, which is a function of the ion energy V, the ion species A, and the target material B.

The features of a typical sputter yield characteristic as a function of ion energy are shown in Fig. 4.39. In the low-energy region the yield increases rapidly from a threshold energy, which is essentially independent of the ion species used, with a value about four times the activation energy for evaporation, and usually in the range 10–130 eV. The threshold energy decreases with increasing atomic number within each group of the periodic system. Above a few hundred volts there is a changeover region whose value depends on the ion species and the target used, after which the sputter yield increases much more slowly with increasing ion energy up to tens of kilovolts. In this region the sputter yields are typically in the range of 0.1 to 20 atoms per ion. At ion bombardment energies in the range 10 keV–1 MeV, the

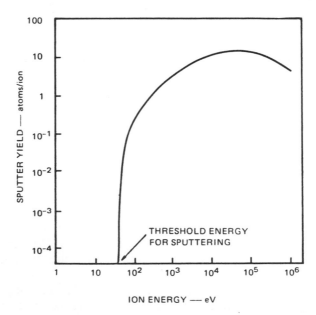

FIGURE 4.39. Typical sputter yield characteristic.

sputter yield reaches a maximum, after which it gradually declines in value as a result of deep implantation of the incident ion.

When a target is bombarded by an ion stream, most of the energy is dissipated in the target. The rear face of the target is usually put in good thermal contact with a water-cooled copper holder; nevertheless, an important limitation to the sputtering rate is set by the maximum temperature to which the target surface may be heated. To be most efficient, ion energies in the range in which the sputter yield changes more slowly are utilized, namely, 0.5–3 kV, so as to minimize the energy required to eject an atom from the target. Even so, the energy dissipated at the target surface to liberate one atom can be 100–1000 times the activation energy for evaporation. Considerations such as this appear to limit useful sputtering rates to about one atomic layer per second. This is considerably less than the rates of about 1000 atomic layers per second available from a typical evaporation source operating at a vapor pressure of 10^{-1} Torr.

If a target is composed of several atomic species, the ratio of the species in the sputtered material may initially differ greatly from their ratio in the target because of their different probabilities of being ejected by the impinging ion. However, this condition will not persist for long, since the surface will be more rapidly denuded of the species with the higher yields until an equilibrium condition is reached in which the atoms are removed in the same ratio as they are present in the bulk. This holds whether the target is in the form of a compound, alloy, or mixture.

The process of sputtering appears to have two underlying mechanisms that contribute to the ejection of the sputtered atoms, namely, thermal spikes and collision sequences.[79] Thermal spikes arise from hot spots, where the energy of an incident ion heats up a small volume as it is slowed down in the lattice; evaporation occurs from this region for a short time until the heat has diffused away. The contribution to the sputter yield from thermal spikes is negligible at ion energies below 10 keV, but correlation between the energy distribution of the sputtered atoms at low energies with the heat of vaporization of volatile target materials (such as potassium) has demonstrated that the effect exists. Collision sequences, which are the major source of ejected atoms, are essentially cascades of atomic collisions resulting from the primary recoil of the lattice with the incident ion (see Section 2.7.2). The cascade ends with transfer of momentum to an atom at another part of the surface, resulting in its ejection.

In the case of an amorphous target material, the cascades take a random path, but for crystalline materials it has been shown that there is a focusing action related to the crystal planes.[80] Thus, polycrystalline or amorphous target materials should give a cosine angular distribution to the ejected atoms. However, since many polycrystalline materials have some preferential orientation to the grains of which they are composed, measurements of the angular distribution often show deviations from the cosine law, being "under cosine" or "over cosine."[81] For the purposes of calculating the uniformity of a thin film deposited by sputtering, the cosine distribution gives sufficiently accurate results so the results generated for evaporation may be used directly.[5] Single-crystal targets should be avoided for sputter sources because of their preferential directions of ejected atoms.

The theories predict and experiment confirms that the energy spectrum of the ejected atoms between the surface-binding energy of the target material and the primary recoil energy of the incident ion takes an inverse-square form.[79] Average

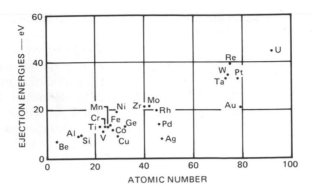

FIGURE 4.40.   Average ejection energies under 1200-eV Kr ion bombardment.

ejection energies, however, are much higher than thermal energies, typically being between 10 and 100 eV (see Fig. 4.40). This results in better adhesion of the sputtered atoms when condensing on the substrate and greater disorder in the condensed film and interface.

4.7.2. Implementation of Sputtering for Thin-Film Deposition

The major advantage of sputtering is its generality. In principle, any vacuum-compatible material can be coated by this method. Conceptually, the use of an ion gun focusing a beam of ions of controlled voltage, current density, and area on a target electrode might appear to be the simplest way to produce a source of sputtered atoms. However, this approach has practical difficulties, and the utilization of electrical discharges in an inert gas with the target and substrate as electrodes has proved more useful. Both dc and rf gas discharges in a variety of configurations have been used. Power is required to generate the plasma from which the useful ions are to be extracted and accelerated to the target. In the case where the target material is insulating, means must be found to neutralize the surface charge built up on it by the impinging ions. The substrate is bombarded by inert gas molecules, electrons, photons, and negative ions from the plasma as well as the desired atoms from the target. Means must be found for avoiding deleterious effects from these extraneous sources.

The physics of electrical discharges in gases is highly complex,[82] and are discussed in more detail in Chapter 3. The following simplistic discussions are added to enable the nonspecialist reader to grasp some basic principles. In its simplest form the electrical discharge is initiated between two parallel-plate electrodes, one composed of the target material and the other of the substrate that is to be coated. In the case of dc discharge by applying a potential of several kilovolts between planar electrodes in an inert gas atmosphere (argon or xenon) at pressures between 10^{-1} and 10^{-2} Torr, a glowing plasma is formed separated from each electrode by a dark space. The plasma is essentially at a constant positive potential with respect to the cathode. Electrons are drawn out of the boundary to the plasma close to the anode and accelerated across the anode dark space. These electrons then strike the anode, where the energy they carry is dissipated. Ions are drawn out of the other boundary of the plasma, accelerated across the cathode dark space, and strike the target cathode, causing sputtering. The plasma is replenished by electron–ion pairs formed by the collision of neutral molecules with the secondary electrons generated at the cathode surface, accelerated across the cathode dark space, and injected into the plasma region. The cathode dark space has a thickness of about that for an electron free path for an electron–ion collision. The anode dark space has a thickness determined by space-charge-limited flow of electrons from the plasma to the anode (see Section 3.4.4).

Direct current discharges have difficulties with initiating the discharge and with the sputtering of insulating materials. These disadvantages are eliminated by using high-frequency ac discharges (5 to 30 MHz). If one of the electrodes is capacitatively coupled to a rf generator, it develops a negative dc bias with respect to the other electrode. This is because electrons are much more mobile than ions, which hardly respond to the applied rf field at all. When the rf discharge is initiated, the capacitatively coupled electrode becomes negatively charged as electrons are deposited on it. However, this charge is only partially neutralized by positive ions arriving during the negative half-cycle. Since no charge can be transferred through the capacitor, the electrode surface retains a negative bias such that only a few electrons arrive at the surface during the next positive half-cycle, and the net current averaged over the next full cycle is zero.[83] (see Section 33.5).

Ions diffusing from the plasma are not strongly affected by the rf fields and behave as if only the negative dc bias voltage existed and sputter the negatively charged electrode. The conducting anode, being directly connected to the power supply, receives pulses of energetic electrons accelerated essentially to the full amplitude of the rf field. If the target is an insulator, the target itself provides the capacitive coupling.

In the case of gas-discharge sputtering, where the substrate is an electrode of the system, the energy dissipated in the sample by electrons from the plasma can be substantial.[84] A small amount of heating can be advantageous, since the coating is continuously annealed, resulting in a stress-free film. However, excess heating can cause distortion of the film by differential expansion with the substrate, recrystallization of the film material, and eventually evaporation of the film material. Heating of the deposited film from electron bombardment limits the rate of film growth that can be utilized in sputtering, even before the target heating effect. To reduce electron bombardment of the film, magnetic fields are used to deflect the electrons from striking the substrate. The presence of the magnetic fields has the additional benefit of increasing the efficiency of generating ions. The evolution of the various magnetically confined discharge configurations is given in Ref. 9 (pp. 76–170). (See Section 3.5.)

Originally the coaxial magnetron configuration was used (Fig. 4.41). It consists of electrodes made in the form of concentric cylinders, the target cathode being the innermost electrode and the substrates placed on the inside surface of the outer cylinder. The magnetic field is axial. Electrons from the cathode surface are trapped by the crossed electric and magnetic fields in the cathode dark space and can escape only by losing energy in ionizing collisions. The loss of energy results in the electrons moving in a tighter spiral path as they drift toward the anode. As this process

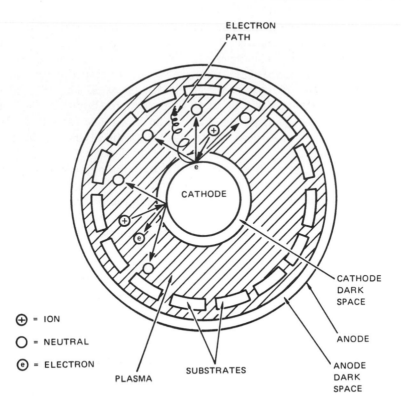

FIGURE 4.41. Coaxial magnetron. In axial-magnetic radial-electric fields, an electron originating at the cathode and accelerated across the cathode dark space makes spiral orbits of decreasing radii as it loses energy to the plasma and drifts toward the anode.

continues, the energy carried by the electrons becomes smaller and the spiral path becomes tighter. This process results in a greater efficiency of ionization and a substantial lowering of the power dissipated by the electrons at the anode. The configuration can be used for dc or rf discharges.

The planar magnetron (Fig. 4.42) is a simple extension of the parallel-plate discharge in which magnets (permanent or electromagnets) are placed behind the cathode. The magnetic field lines enter and leave normal to the cathode surface. The transverse component of the magnetic field in front of the target is arranged to be in the range 200–500 G. Electrons are trapped in the crossed electric and magnetic fields, both enhancing the efficiency of the discharge and reducing the energy dissipated by electrons arriving at the anode.

Circular magnetrons (sputter guns or "S" guns) in which the cathode is in the form of a ring that surrounds a planar disk-shaped anode are also popular (Fig. 4.43). Magnets behind the cathode enhance the discharge and the electrons are collected by the anode. A large fraction of the sputtered material from the target cathode is ejected in a forward direction and deposited on a substrate that no longer needs to be an electrode of the system. In this way, very high power densities (50 W/cm^2) in the erosion zone of the target can be dissipated.

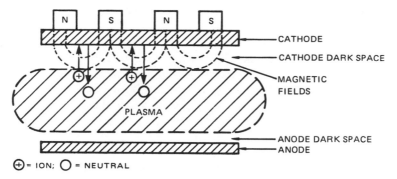

FIGURE 4.42. Planar magnetron. Electrons originating at the cathode surface are confined by the fields of permanent magnets placed behind the cathode.

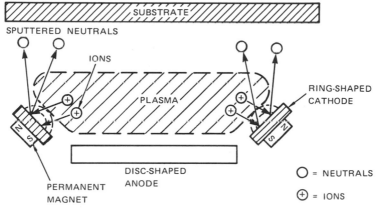

FIGURE 4.43. Circular magnetron. Electrons originating at the cathode are confined by fields from permanent magnets placed behind the cathode. They lose energy in spiral paths in the plasma and are collected by the disk anode. The substrate, not being an electrode of the system, collects only sputtered neutrals.

The main disadvantage of planar and circular magnetrons is the fact that the deposition rate is not uniform across the region where the substrates are placed. This, as in evaporation, is overcome by an appropriate mechanical motion of the substrate table so that each substrate is exposed to the same average flux.

If an active gas such as oxygen, water, or hydrogen sulfide is intentionally added to the inert working gas, it can react with the target material either at the target surface or in the gas phase, resulting in the deposition of a new compound. By varying the pressure of the active gas, control can be gained on the stoichiometry of the film. This technique is called *reactive sputtering*.

Another variant of the sputtering process is *ion plating*. This essentially involves purposeful bombardment of the substrate with positive ions from a glow discharge so that the film is being sputter eroded at a slow rate at the same time as it is receiving a film material from an appropriate source, usually a simple evaporation source. This technique supposedly improves adhesion and film uniformity.

4.8. ORGANIC LAYER FORMATION

Organic compounds are used extensively in microfabrication, principally as photon- or electron-beam resists. These are materials whose structure and solubility change when exposed to radiation. Materials that are rendered insoluble in a developer solution by the action of radiation are called negative resists, while those that are rendered soluble are called positive resists. The material left behind after pattern generation must be chemically inert, or treated to become inert, so that removal of the unprotected substrate may proceed without affecting the resist or the substrate under the resist.

Because of its importance to pattern generation, resist technology is discussed in more detail in Section 5.3.

The most interesting class of organic films to be recently applied has been the polyimides[85] which are now available commercially as photoresists. These materials are exceedingly inert and can be heated up to 450 or 500°C without decomposing, which enables them to survive subsequent high-temperature processing in the fabrication of microdevices.

4.8.1. Spin Coating

In the spin coating process a polymer is dissolved in an appropriate solvent forming a viscous solution usually with a viscosity coefficient in the range of 20–500 cP. Particulate impurities are removed by centrifuging. A specified volume of material is dispersed onto the substrate (a silicon wafer in the case of spin coating a resist in planar technology) which can rotate in a horizontal plane about a central vertical axis. After dispensation, which usually takes place at slow rotational speed, the substrate is accelerated up quickly to a high angular velocity, usually in the range of 3000 to 8000 rpm, held at this speed for between 10 and 30 s, and then rapidly brought to rest. During the high-speed spin, centrifugal forces drive the liquid radially outward. Excess material is driven off the rim of the wafer but the remainder is retained on the substrate as a thin surface layer by surface tension and viscous forces. Also, while spinning, some solvent may evaporate leaving a highly viscous or solid film on the surface. Photoresists are then usually prebaked at about 80°C for 20 min or so to remove all residual solvent. Films produced in this way are found to be remarkably uniform in thickness, except at the periphery. The thickness of organic films may precisely set in a range of 2000 Å to a few micrometers by controlling the angular speed, the initial viscosity, and the spinning time.

The theoretical model for the process[86] assumes that the presence of the boundary at the periphery of the substrate has no effect on the spreading process. The velocity of the liquid in the radial direction, r, is given by equating the viscous drag in the radial direction r, with the radial accelerating force per unit volume at a point, z, above the substrate in the liquid. Assuming Newton's law for viscous drag, we have

$$\eta \frac{\partial^2 v}{\partial z^2} = \rho \omega^2 r \tag{4.81}$$

where η is the viscosity coefficient, ρ the liquid density, and ω the angular velocity.

Integrating gives

$$v = \frac{\rho\omega^2 r}{\eta}\left[-\frac{z^2}{2} + hz\right] \tag{4.82}$$

Assuming $\partial v/\partial z = 0$ at the liquid surface, i.e., at $z = h$, where h is the film thickness, and $v = 0$ at $z = 0$, i.e., at the substrate/liquid interface. Thus, the radial mass flow per unit length across a circle radius r is

$$q = \int_0^h v\, dz = \frac{\rho\omega^2 rh^3}{3\eta} \tag{4.83}$$

The equation of continuity requires

$$r\frac{\partial h}{\partial t} = \frac{\partial}{\partial r}(rq) \tag{4.84}$$

Thus, substituting for q from Eq. (4.83) we have

$$\frac{\partial h}{\partial t} = \frac{\rho\omega^2}{3\eta}\frac{1}{r}\frac{\partial}{\partial r}(r^2 h^3) \tag{4.85}$$

Accepting the experimental evidence that h is independent of r, this reduces to

$$\frac{\partial h}{\partial t} = \frac{\rho\omega^2 h^3}{3\eta} \tag{4.86}$$

giving

$$h = \frac{h_0}{\left[1 + \dfrac{4\rho\omega^2}{\eta}h_0^2 t\right]^{1/2}} \tag{4.87}$$

where h_0 is the value of h at $t = 0$.
     For large t, where

$$\frac{4\rho\omega^2 h_0^2}{\eta} t \gg 1 \tag{4.88}$$

Eq. (4.87) reduces to

$$h = \left[\frac{\eta}{4\pi\rho\omega^2}\right]^{1/2} t^{-1/2} \tag{4.89}$$

That is, the film thickness is independent of the amount of liquid initially placed on the substrate.

The full description of the spinning process must take into account the time increase in viscosity and the viscosity profile as the solvent evaporates from the surface, the non-Newtonian behavior of the viscous forces (i.e., a nonlinear dependence of the viscous force on $\partial^2 v/\partial z^2$), and the effect of surface tension, particularly at the rim. An attempt has been made to take these effects into consideration using computer methods.[87]

4.8.2. Electron-Beam Polymerization

Pinhole-free polymer dielectric films down to 100 Å thick have been made by radiating monomer layers with ultraviolet light or energetic electrons.[88] Materials that have been polymerized by electron bombardment include dimethyl polysilane, butadiene, styrene, and methacrylate. Direct deposition by electrical discharges in gases containing organic vapors can also be utilized to make polymeric films. A large number of substances have been explored and the mechanism has been identified as

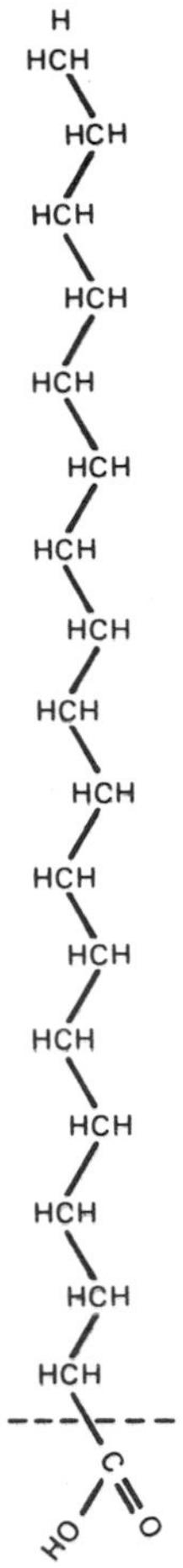

FIGURE 4.44.   Structure of stearic acid [$CH_3(CH_2)_{16}COOH$].

the formation of cross-linked, high-molecular-weight polymers by activation of the adsorbed monomer by ionic bombardment of the surface. The organic vapor simply dispenses the monomer to the surface.

4.8.3. Langmuir–Blodgett Films

Organic insulator films from 1 to 10 monolayers thick have also been formed by the classical Langmuir–Blodgett process.[89,90] This process consists of spreading a monolayer of film-forming molecules (such as stearic acid) on an aqueous surface, compressing the monolayer into a compact floating film, and transferring it to a solid substrate by passing the substrate through the water surface.[91] Film-forming molecules have a polar- or water-soluble group attached to a long hydrocarbon structure that is sufficiently large to make the complete molecule insoluble in water. The archetypical molecule used is stearic acid, whose molecular structure is shown in Fig. 4.44. Monolayer films are studied using a film balance (Fig. 4.45). This consists of a long, shallow trough filled with pure water. A mica strip floats on the water and is attached to a stirrup by an unspun silk fiber and to the sides of the trough by thin flexible platinum foils. The stirrup is fixed to a calibrated torsion wire that indicates the surface pressure, usually measured in dynes per centimeter. A compressing barrier of brass can also be moved along the trough as shown. A small amount of the film-forming material in a volatile solvent is spread between the float and the main barrier. The barrier is then moved toward the mica float, thus compressing the film. Compression results in the molecules forming a carpetlike structure, with the soluble group attached to the water surface and the long chains perpendicular to the surface, as shown in Fig. 4.46a. The balance is used to measure the pressure–area isotherms that characterize the film material. The pressure against the mica float increases with decreasing area until it reaches a constant value or even begins to fall. This is called the collapse point and is usually in the range 10–100 dyn/cm. Electron microscopy has suggested that collapse occurs as the monolayer distorts and forms more layers. Figure 4.46 shows the probable mechanism of collapse of a monolayer

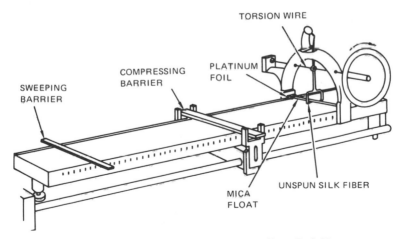

FIGURE 4.45.   Film balance apparatus. From Ref. 91.

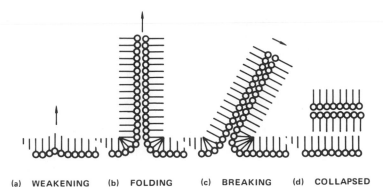

(a) WEAKENING   (b) FOLDING   (c) BREAKING   (d) COLLAPSED

FIGURE 4.46. Mechanism of monolayer collapse. (a) Weakening; (b) folding; (c) breaking; (d) collapsed. From Ref. 91.

of molecules consisting of a polar group (small circle) attached to the water surface and a long-chain hydrocarbon (line).

If a cylinder is rotated in a monolayer held at constant surface pressure, the monolayer is transferred quantitatively from the water surface to the cylinder and films of precisely known monolayer thicknesses can be built up. Langmuir films have been utilized for rust prevention, as very thin insulating films for MOS devices,[91] and for ion-selective membranes for sensors. It seems likely that other interesting uses for such films should be forthcoming as more microdevices requiring them are conceived.[90]

4.8.4. Plasma-Enhanced Polymerization

Organic films can be formed by plasma-enhanced polymerization of organic vapors at a surface. The first practical use made of such films was for forming polystyrene films as the dielectric for a nuclear battery.[92] A recent review of the subject[93] is available. The vast number of organic monomers available indicates that films with a wide range of properties should be producible in this way. Use of metal organic species increases the scope of these plasma materials even further[86].

The mechanisms of glow discharge polymerization by a competitive ablation and polymerization scheme[94] may be characterized, as shown in Fig. 4.47. Two mechanisms can account for polymer formation on a solid substrate. The first, *plasma-induced polymerization*, refers to conventional polymerization triggered by a reactive species created in the discharge; plasma-induced polymerization does not produce a gas-phase by-product and hence polymerization can only proceed via the utilization of a polymerizable structure from the gas phase. The second formation mechanism, *plasma state polymerization*, proceeds by first forming an intermediate radical that subsequently condenses and polymerizes on the surface.

The surface polymer may be simultaneously removed by reaction with the residual gases (ablation); the balance between ablation and polymerization may be controlled by adding gases that promote one or the other effect. For example, CF_4 is an effective plasma etch material under normal discharge conditions; however, if a small amount of hydrogen is added to the discharge, a polymer is deposited.

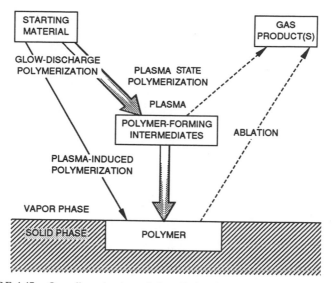

FIGURE 4.47.   Overall mechanism of glow-discharge polymerization. From Ref. 94.

Although the use of such deposition techniques in microelectronics is recognized to be of potential value, none is as yet used as a generally accepted technique.

4.8.5. Liquid-Crystal Layers

Liquid crystals[95] consist of rigid organic polymeric molecules in the form of rods or disks which remain ordered in the liquid phase. This order usually exists only over a limited temperature range between the melting point and a temperature at which the molecules become randomly oriented. For example, p-[N-(p-methoxybenz-ylidene)]-p'-n-butylaniline (MBBA) shows the following transitions:

$$\text{Solid} \underset{11^\circ C}{\overset{21^\circ C}{\rightleftarrows}} \text{nematic liquid crystal} \underset{41^\circ C}{\overset{41^\circ C}{\rightleftarrows}} \text{isotropic liquid} \tag{4.90}$$

Because of the ordering, the physical properties of the liquid including the refractive index and dielectric constant vary with direction. Three types of ordering are distinguished:

- The nematic phase, in which all of the long axes of the molecule are parallel.
- The cholesteric phase, in which molecular layers are formed in which within each layer the ordering is nematic but successive layers are aligned at a fixed angle to each other forming a helical (or spiral staircase) structure.
- The smectic phase, in which each layer is nematic and successive layers are aligned in the same direction.

Initial alignment is determined by the surface which bounds the material. Molecular alignment to the surface may be homeotropic (long axis perpendicular to the surface) or homogeneous (long axis parallel to the surface) and depends on the surface preparation. If the surfaces are polished so the microscratches are parallel, then the surface alignment is homogeneous but parallel to the scratches.[96] The most important application of liquid crystals in microstructures has been in displays (LCD) which depend for their operation on a change in optical properties of the liquid crystal with electron field.[97] The most common LCD cell[98–100] uses a twisted nematic liquid crystal. This type of liquid crystal is obtained when two plane-parallel plates spaced about 10 μm apart are prepared with homogeneous surfaces polished with their directions at 90° to each other and a nematic liquid-crystal layer formed between them. Light incident on the surface polarized parallel or perpendicular to the direction of the microscratching emerges from the other side with its plane of polarization rotated 90°. The mechanism of electromagnetic wave propagation down the twisted liquid crystal is quite complex and the interested reader is referred to the specialist literature.[97] When an electric field of about 10^4 V/cm is applied across the sandwich, the axis of the molecules tends to align with the electric field so that the twist structure in the nematic liquid crystal is modified sufficiently to alter the rotation of the plane of polarization of the emerging light. The electrooptic effect is observed by viewing the light transmitted between parallel Polaroid screens placed on either side on the cell. A typical result is shown in Fig. 4.48. The effect is not strongly dependent on waveform or frequency up to the kilo hertz range and the results shown in Fig. 4.48 were for a few hundred hertz. Note that there is a well-defined threshold voltage (V_{th}) and a fairly sharp saturation. Figure 4.49 shows a typical response characteristic when the field is applied. The shape of the response is characterized by three time constants: the delay time τ_d, the rise time τ_r, and the fall time τ_f. The delay time τ_d is defined as the time from the start of the pulse train to the instant

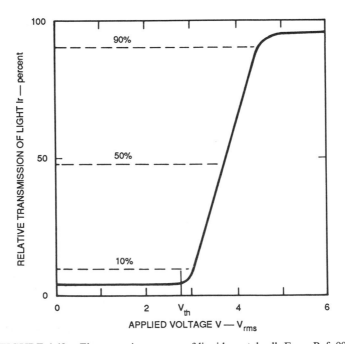

FIGURE 4.48.   Electro-optic response of liquid-crystal cell. From Ref. 99.

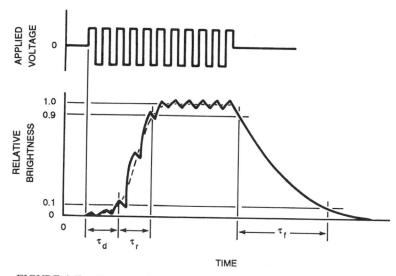

FIGURE 4.49. Response characteristics of liquid-crystal cell. From Ref. 99.

when the transmission of the light reached 10% of its final value. The rise time τ_r is defined as the time duration for the transmission to rise from 10% to 90% of its final value. Similarly, the fall time τ_f is the time required for the cell transmission to fall from 90% to 10% of its maximum value. For typical twisted-nematic cells at room temperature, the delay time is usually several milliseconds, the rise time is between 10 and 100 msec, and the fall time is between 20 and 200 msec. The rise time varies inversely as the square of the applied voltage for a cell of given thickness. The fall time depends on the liquid-crystal material, but for a given material it is proportional to the square of the cell thickness. The rise and fall times decrease as the ambient temperature increases.

4.9. SELECTIVE LAYER REMOVAL TECHNOLOGIES

4.9.1. Background

Removing a film from a substrate, usually in a pattern defined by a protective resist is called etching. Etching can be achieved by chemical methods, whereby an easily removable compound is formed with the film material, or physical methods, whereby the film is physically removed, as by abrasion or sputtering, or by a combination of both. Certain highly chemically active molecules or ions can only be formed in a glow discharge (or plasma), and the usage of such species is called reactive plasma etching.

4.9.2. Wet Chemical Etching

For wet chemical etching a solution is sought that dissolves the film material but not the resist or the substrate. The reaction products must dissolve in the carrier fluid or otherwise be carried away. When etching very small channels or cavities, the solution may quickly become saturated and must be removed by ultrasonic or other means of agitation so that fresh solvent may continuously be brought to the surface to be etched. If one of the reaction products is gaseous, small bubbles will form at the interaction surface. Ultrasonic removal of such bubbles also helps to keep the solution stirred. Making the film an electrode of an electrolytic cell allows the chemical reaction to be electrically driven, resulting in the deplating of the film layer and the removal of the reaction products by ionic flow. The choice of reagents for particular film removal properties is extremely vast. For application to specific insulators, dielectrics, semiconductors and metals, excellent tables are compiled in Ref. 9, p. 433.

Some etchants attack a given crystal face of a material much faster than others, giving rise to anisotropic etching.[101] Anisotropic etchants for silicon have been

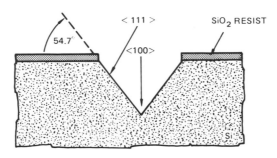

(a) THROUGH A PINHOLE <100> ORIENTED Si

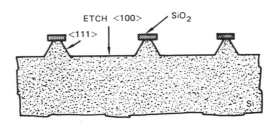

(b) THROUGH A SQUARE WINDOW FRAME PATTERN <100> ORIENTED Si

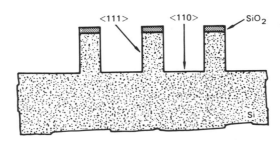

(c) THROUGH A SQUARE WINDOW FRAME PATTERN <111> ORIENTED Si

FIGURE 4.50.   Anisotropic etching of silicon. (a) Through a pinhole $\langle 100 \rangle$-oriented Si; (b) through a square window frame pattern $\langle 100 \rangle$-oriented Si; (c) through a square window frame pattern $\langle 111 \rangle$-oriented Si. From Ref. 101.

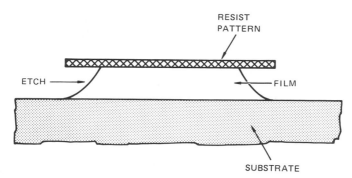

FIGURE 4.51. Undercutting of the resist film through chemical etching.

studied extensively. For example, in the case of the water–ethylene diamine pyrocate-chol (EDA) etchant the etch rates of $\langle 100 \rangle$, $\langle 110 \rangle$, and $\langle 111 \rangle$ oriented faces of silicon are in the ratio of approximately $50 : 30 : 3$ μm/h, respectively, at 100°C. Aniso-tropic etching of $\langle 100 \rangle$ Si through a patterned SiO_2 mask creates precise v-shaped grooves, the edges being $\langle 111 \rangle$ planes at an angle of 54.7° from the $\langle 100 \rangle$ surface, as shown in Fig. 4.50.[102]

If there is a pinhole in the oxide, the etch front through the hole will propagate into $\langle 100 \rangle$ silicon to form a perfect square-topped pit in the form of an inverted pyramid. On the other hand, if a square window frame pattern is opened in the oxide mask, a square-based pyramid structure can be etched, as shown in Fig. 4.50b. If $\langle 110 \rangle$ oriented silicon is used, essentially straight-walled grooves with sides of $\langle 111 \rangle$ orientation can be formed, as shown in Fig. 4.50c. In this way, 0.6-μm-wide openings on 1.2-μm centers and 600 μm deep have been made.[103] Scanning electron micro-graphs of the intersection of anisotropically etched planes show them to be sharp to the resolution of such instruments (about 50 Å) and likely to intersect in a single row of atoms.

An important disadvantage of chemical etching is the undercutting of the resist film that takes place, resulting in a loss of resolution in the etched pattern (see Fig. 4.51). In practice, for isotropic etching the film thickness needs to be about one-third or less of the resolution required. If patterns are required with resolutions much smaller than the film thickness, anisotropic or other special etching techniques must be sought.

Gas-phase etching methods using vapors that react with the substrate and form-ing gaseous reaction products have been used. For example, anhydrous hydrogen fluoride between 150 and 190°C at a pressure between 0.1 and 3.0 Torr reacts spe-cifically with silicon dioxide, but has no effect on silicon, silicon nitride, aluminum, and other common materials.[104] At high pressures, gas-phase reactions tend to be isotropic, but directing the gas flow in jets could conceivably help to reduce undercutting.

4.9.3. Sputter Etching

Directed ion bombardment of the surface (or sputter etching) is used to obtain good aspect ratios. Sources of ions for sputter etching are generally based on the

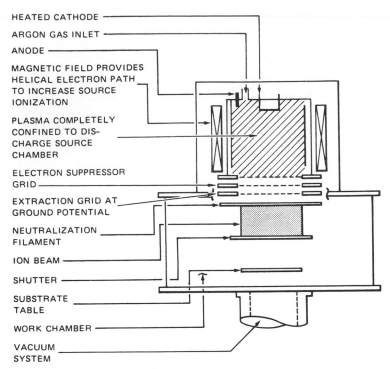

HEATED CATHODE

ARGON GAS INLET

ANODE

MAGNETIC FIELD PROVIDES
HELICAL ELECTRON PATH
TO INCREASE SOURCE
IONIZATION

PLASMA COMPLETELY
CONFINED TO DIS-
CHARGE SOURCE
CHAMBER

ELECTRON SUPPRESSOR
GRID

EXTRACTION GRID AT
GROUND POTENTIAL

NEUTRALIZATION
FILAMENT

ION BEAM

SHUTTER

SUBSTRATE
TABLE

WORK CHAMBER

VACUUM
SYSTEM

FIGURE 4.52.   Ion milling device (sputter etching). From Ref. 106.

Kaufman ion thruster developed by NASA[105] (Fig. 4.52). A magnetically confined gas discharge is obtained between a thermionic cathode and a concentric-anode cylinder. To extract the ion beam from the discharge region, a potential difference is applied between a pair of grids with aligned holes. The ions are injected into the working space as a well-collimated energetic beam. The beam is made neutral by extracting low-energy electrons from an auxiliary thermionic cathode so that it can be used for sputtering insulators as well as conductors.

Ion optical systems utilizing liquid-ion sources are now being built that are capable of forming intense energetic beams focused into spots of very small diameter (see Section 2.3.5). The spot can be positioned on a sample surface by deflector electrodes to directly generate and sputter-etch a pattern on the surface. This technology holds promise for the controlled fabrication of submicron structures without the use of resists (see Section 4.9.2). The selection of substrates and resists for sputtering etching is based simply on relative sputter coefficients. However, this offers limited opportunity for control, since the sputter yields of most materials do not differ by large factors.

4.9.4. Reactive Plasma Etching

Interest in reactive plasma etching is increasing because, in contrast to sputtering, large differences in etch rates for different materials can be obtained. In this technique the substrate is exposed to a plasma of a reactive gas, but there is no high-voltage acceleration of ions toward the substrate. Since the ions are not highly energetic, sputtering is not in itself an important surface-removal mechanism. Rather, the reactive gas components appear to be first adsorbed on the surface, where they are dissociated by electron or ion bombardment from the plasma, then they react with the film material, forming a gaseous product molecule. The reaction products are desorbed as gases and pumped away. Anisotropic etching can occur because the impingement rate on the horizontal surfaces is much greater than at the sidewalls at the low pressures used, where the mean free path of the molecules is typically much larger than the depth to be etched. Because of the chemical nature, a high degree of control over the relative etch rates for the film, resist, and substrate can be obtained by a choice of suitable materials.

The first important application of the technique was the stripping of photoresist.[106] Hardened photoresists are required to be extremely inert polymers so as to resist the etchants that attack the substrate they cover. This same property makes them difficult to remove by wet chemical processes or even organic solvents. However, when oxidized in a glow discharge, the oxides of carbon, being gaseous, are easily pumped away, whereas the oxides of many substrate materials (silicon, aluminum) rapidly form a protective film. This technique is generally used to remove carbonaceous contaminants.

Glow discharges in low-pressure fluorocarbon (CF_4) vapors or mixtures of fluorocarbons with various additive gases (usually oxygen) are currently being used for patterning silicon, silicon dioxide, silicon nitride (Si_3N_4), and other materials used in IC fabrication.[107] The chemically active radicals formed by the glow discharge react with silicon and its compounds to form gaseous SiF_4. CF_4 is assumed to decompose into CF_2 and atomic fluorine:

$$CF_4 \rightarrow CF_2 + 2F \tag{4.91}$$

The reactions are then assumed to proceed as follows:

$$Si + 4F \rightarrow SiF_4$$
$$SiO_2 + 4F \rightarrow SiF_4 + O_2 \tag{4.92}$$
$$Si_3N_4 + 12F \rightarrow 3SiF_4 + 2N_2$$

However, the detailed mechanisms are apparently much more complex.[108] Additives to fluorocarbon such as O_2, H_2, N_2, H_2O, and CF_4 are used to enhance or suppress various chemical reactions. Table 4.11 indicates that photoresists can be used for patterning Si and Si_3N_4. However, for the patterning of SiO_2 the addition of hydrogen to the fluorocarbon instead of O_2 is required to increase the etch rate of SiO_2 over silicon.

Reactive plasma etching is currently being intensively studied, and it is anticipated that both applications and an understanding of the detailed mechanisms involved will be expanded.[109]

TABLE 4.11. Relative Etch Rates in Fluorocarbon
(95% CF_4 + 4%O_2), Glow Discharge for Common
Microelectronic Materials[a]

Material	Relative etch rate
Si_3N_4	100
Si (111)	690
Si (poly)	990
SiO_2 (Thermal)	40
SiO_3 (CVD)	120
Al	0
W	100
Mo	100
Ti	100
Ta	100
Waycoat I (photoresist)	20
A21350J (photoresist)	40

[a]Source: Ref. 100.

4.10. THICKNESS AND RATE MEASUREMENT

While there are a very large number of physical phenomena that are functions of film thickness, there are only a few that are conventionally used for deposition control purposes because of their simplicity and convenience. These are usually based on classical measurements of mechanical, electrical, or optical properties. For ultrathin films less than a few tens of monolayers, the classical concept of thickness may not be meaningful owing to the possible nonuniform structure of the films. The weight per unit area of a film, however, does provide an unambiguous measure of the quantity of material deposited, and an equivalent thickness may be inferred from an assumed density; it also has the advantage of complete generality; that is, it may be used with any film molecule. Weights on the order of 10^{-9} g/cm^2 or less than 1/100th of a monolayer may be detected by gravimetric method in vacuums.

The simple *microbalance* (Fig. 4.53) consists essentially of a horizontal quartz torsion wire (120 mm long and 40 μm in diameter) fixed to a quartz frame and held taut by a quartz spring. The balance beam attached to the center of the torsion wire is also of quartz. At one end is a plate on which the material is evaporated, and on the other is a small magnet. Above the magnet is a small solenoid. Displacement, brought about by the deposition of the film on the plate, is magnetically compensated for by an appropriate current through the solenoid. A mirror on the torsion wire enables precise optical determination of the null position to be made. A copper cylinder surrounds the magnet to enable eddy-current damping of the balance beam oscillation to be effective. Such balances used in high-vacuum systems can be outgassed by baking to 500°C and are sensitive to 10^{-8} g. They are, however, very delicate and have largely been replaced by *quartz crystal oscillator* monitors. These devices operate on the principle that the resonant frequency of a piezoelectric crystal is changed by added weight. A quartz crystal plate is accurately ground to a thickness t, and metal electrodes evaporated on each side. When the electrodes are attached

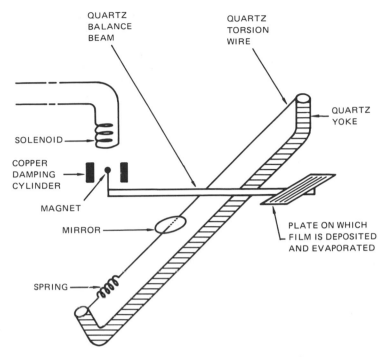

FIGURE 4.53. Simple microbalance. From Ref. 8.

to an oscillating circuit, resonance occurs at a frequency f, given by

$$f = \frac{v_T}{2t} = \frac{A}{t} \text{ Hz} \tag{4.93}$$

where v_T is the velocity of sound across the quartz wafer and A is the frequency constant. For the usual orientation along which quartz crystals are cut (called the AT cut), $A = 1670$ kHz/mm. If an amount of material Δm is deposited on the crystal area S, the effective increase in film thickness is given by

$$\Delta t = \frac{\Delta m}{\rho S} \tag{4.94}$$

where ρ is the quartz density. Using Eqs. (4.93) and (4.94) we get

$$\Delta f = \frac{-f^2 \Delta m}{A \rho S} \tag{4.95}$$

or

$$\frac{\Delta m}{S} = \frac{-\Delta f}{f^2} A\rho \text{ g/cm}^2 \tag{4.96}$$

Since the resonant frequency can be monitored continuously, the instrument can be used to measure the rate of deposition. Instruments based on this principle are commercially available and ruggedly built, yet they can measure routinely to 10^{-9} g/cm^2 (about one tenth of a monolayer) and, in circumstances in which errors due to extraneous effects are reduced as much as possible, measured to 10^{-12} g/cm^2.

Electrical and optical methods are useful for films above the thickness at which the films may be considered coherent. The resistance of a metallic film on an insulating substrate may be measured by bridge methods, and the thickness inferred, provided that a thickness-resistance calibration has been previously performed. The thickness of an insulating film on a conducting substrate may be inferred from its

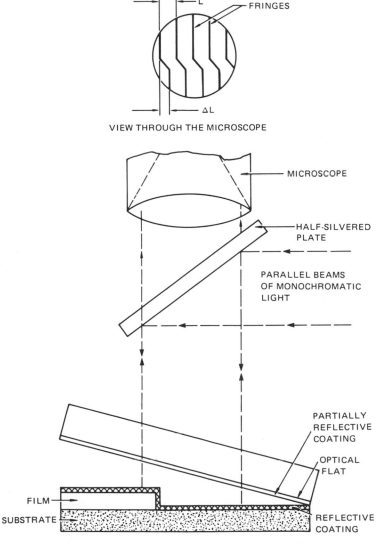

FIGURE 4.54.   Optical method of measuring film thickness and step height. From Ref. 8.

capacitance per unit area. This may be obtained by depositing a metallic electrode of known area on its surface and again using ac bridge methods.

A portion of the vapor from the deposition stream may be allowed to pass through a slit and be ionized by energetic electrons in an ionization gauge arrangement. The *positive ion current* measured in the gauge will be proportional to the deposition rate. Such devices have to be calibrated but are useful when repetitive work has to be done.

Electron-beam evaporation sources (see Section 4.6) generate their own ion current, since the beam of bombardment-heating electrons also passes through the vapor. By arranging to collect a portion of the ion current, the source can be made self-monitoring. If the substrate being coated has a portion of its surface appropriately shadowed, a sharp step may be produced whose height is of the thickness of the film. The step height can be measured with a *diamond stylus* arrangement. In commercially available instruments, the vertical movement of a tip of radius about 1 μm and weight 0.1 g is converted by a transducer to an electrical signal, which is amplified and recorded. Film thicknesses and surface irregularities as small as 25 Å can be detected by this method.

Another method of measuring the step height uses *optical interference* methods. The film and step are first coated with aluminum to make them highly reflecting. Then an optical flat is placed up against the surface at a small angle, forming a wedge-shaped air gap (Fig. 4.54). As is well known, illumination with parallel monochromatic light produces evenly spaced interference fringes at intervals where the air-gap thickness differs by $\lambda/2$. The displacement of the fringes at the step (Fig. 4.54) gives a direct measure of the height of the step in terms of λ. If L is the fringe spacing and ΔL is the displacement of the fringes at the step, then thickness t is given by $t = \Delta L/L(\lambda/2)$. Film thicknesses of less than 100 Å can be estimated by such methods.

4.11. LASER-ASSISTED PROCESSES

The use of lasers in semiconductor processing can be divided into two major areas: general processing and direct, on-chip/wafer processing. General processing refers to the use of lasers for:

- Wafer positioning and stepping
- Mask and wafer inspection and repair
- Packaging
- Marking

On-chip/wafer processing applications include:

- Laser annealing
- Laser-assisted diffusion, doping
- Laser-assisted crystalline conversion, graphoepitaxy
- Laser-based lithography
- Laser-assisted deposition and etching
- Laser pantography

4.11.1. Wafer Positioning and Stepping

The position of a wafer during processing is becoming increasingly important as linewidths decrease and wafer size increases. The accuracy and precision required are on the order of one-tenth of the wafer's smallest feature. Two major sources of registration inaccuracies are variations in mask dimension due to thermal expansion of the glass substrates and nonuniformities in the optics over the field of exposure.

A helium–neon laser interferometer has proved to be the most accurate method of positioning a step-and-repeat aligner. This technique is capable of stepping in increments of 1/32nd of a wavelength (~200 Å). An alignment accuracy of ±0.15 μm and a machine-to-machine overlay accuracy of ±0.25 μm can be readily obtained at the two-sigma level (95% probability).

4.11.2. Mask and Wafer Inspection and Repair

Lasers are also used to detect defects on masks and reticles. In the past, it was possible to adequately inspect masks and reticles by visual inspection. As linewidths have decreased, it is estimated that 50% of defects below 2 μm in size would be missed by a visual inspection.

Reticles must be defect free, as any defect would be repeated on each die image, whether on a mask or on a stepped wafer. Even on directly generated electron-beam masks, defect density must be kept to the barest minimum.

The monitoring of particulates on the surface of a mask of wafer employs a surface scan detection system after the contaminant has condensed on the surface. A focused He-Ne laser beam scans the surface in a raster pattern. On a contaminated surface, the laser beam is scattered and collected by a photomultiplier tube. The detection limit for particle size is around 0.3 μm.

In addition to particulate contamination, the optical inspection devices detect mask/reticle defects. Mask defects can be either opaque or clear as illustrated in Fig. 4.55. Lasers have been used for the removal of both types of defects.

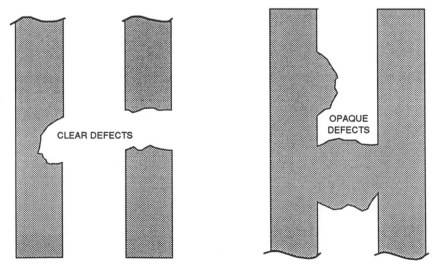

FIGURE 4.55. Comparison of clear and opaque defects on photomasks.

Clear defects may be removed by laser photolysis or pyrolysis. In laser photolysis (see Section 2.9.2) a laser is used to decompose an organometallic molecule such as $Cd(CH_3)_2$ in the vapor phase above the clear defect. This is shown in Fig. 4.56. With laser pyrolysis, the chemical reaction occurs at the surface of the mask, while the volatile components remain in the vapor phase. Using this technique, clear defects

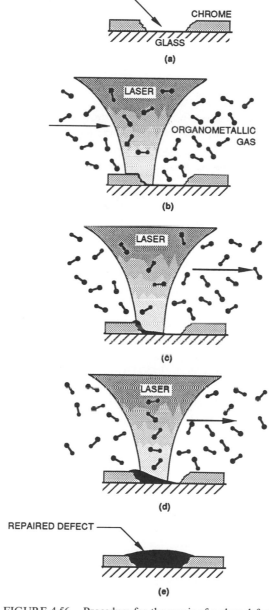

FIGURE 4.56. Procedure for the repair of a clear defect.

have been repaired approaching a lower limit of 1.6 μm in defect size. However, features of 1 μm have been repaired by using a combination of laser pyrolysis and laser trimming. Nevertheless, a practical limit of about 1 μm is expected (due to diffraction).

Opaque defects are removed by superheating them with a focused burst of energy from the laser beam. This process is difficult to control, and generally damages the underlying substrate or adjacent features. The target is first melted by the intense heat and then vaporized. During melting, the chromium tends to flow and build up around the edges of the repaired area, distorting them. The greatest limitation is removing a defect between two patterns, particularly as linewidths decrease below 1 μm. Focused ion beams are proving a viable alternative to laser repair (see Section 4.13).

4.11.3. Dicing and Bonding

After a wafer is processed, it is cut into individual dice to be subsequently packaged. Laser IC separation was first used in the early 1970s as a method to replace diamond saw dicing. However, residues deposited on the front and back side of the wafer resulted in circuit damage and wire bonding failures. The advent of the high-speed, rotating blade diamond saw that eliminated these problems has resulted in a decreased use of laser dicing for chips that are neither rectangular nor square.

In the bonding process, laser soldering systems can produce perfect joints every time by adjusting the heat they deliver to individual joints. An IR detector can monitor the formation of each joint as the Nd:YAG laser beam heats the solder to its melting point. This is an improvement over traditional laser soldering systems that apply the same amount of heat to each joint regardless of variations in the amount of solder at each joint or differences in surface dimensions.

4.11.4. Identification Marking

Laser marking is used for inventory control during processing. A problem that exists on the fabrication line is that all wafers look exactly alike. Mistaken identity can result in misprocessing. Laser marking involves writing a multidigit number on each wafer as it starts into the manufacturing line. The mark, placed in the surface layer by a medium-powered laser, is indelible so that it survives all of the processes the wafer undergoes during fabrication.

Once the wafer is marked, it can also be automatically read by a laser. There are two approaches: bar code lines and ASCII characters. Bar code systems are more reliable and less prone to reading errors, but are not readable by human operators. ASCII characters are read by an optical character recognition system; these characters can be read by operators.

Lasers can also be used to mark packaged wafers. Printed codes on the package tell the user product type, country of origin, and other essential information about the part that they are using. They are also used by the manufacturer for inventory control. The traditional method uses ink, deposited either by direct printing, offset printing, gravure printing, or ink jet printing. Laser marking, on the other hand, utilizes either a TEA (transversely excited atmospheric) CO_2 or Nd:YAG laser. The

TEA laser is permanent and fast, and, unlike ink, can mark moving devices and cannot smear. However, it does not work with metal parts as will the Nd:YAG laser.

4.11.5. Laser Annealing

The distribution of implanted ions in the solid substrate depends on both the acceleration voltage (ion energy) and mass relations between the implanted ion and the atoms of the target. Stopping of an energetic ion occurs as a result of collisions between the ion and the electrons and atoms of the target. If these collisions are sufficiently energetic, those atoms will be displaced from lattice sites, leaving vacancies behind. These displaced atoms may in turn displace other atoms in the lattice, forming as a result a disorder cluster around the ion track (see Section 4.5.9). Furthermore, since the process just described is far from thermal equilibrium, only some of the implanted impurities will be located on substitutional sites after the implantation. In general, a large fraction will be found in interstitial positions and still other impurity atoms will form complexes with simple lattice defects (lattice vacancies and interstitials). Moreover, the doping characteristics of the impurity may be overwhelmed by the damage. As a result, the as-implanted substrate will usually exhibit an electrical carrier concentration far below that which would be predicted simply from the implanted-impurity concentration profile. In other words, only a small percentage of the implanted impurities contribute to electrical carriers. Hence, an annealing process is required to restore crystal perfection.

In conventional furnace-annealing procedures, the entire wafer is heated to temperatures of 1000°C or higher for periods of about half an hour. This process is not always able to restore crystalline perfection in the implanted region and often leads to deleterious effects on both the crystalline perfection and the chemical purity of the silicon in the bulk of the wafer. Moreover, such thermal treatments often result in a high concentration of electrically inactive dopant atoms in the form of precipitates close to the surface of the wafer, leading in turn to short carrier lifetimes within this region.

One experimental method for obtaining high-quality anneals of near-surface displacement damage uses the heat generated by the absorption of radiation from high-powered lasers.[109-117] Several advantages are claimed for the laser annealing technique[110]:

- Fine spatial control of the area that is recrystallized on the wafer is possible.
- Control of the dopant diffusion depth is possible by varying the conditions of pulse length and pulse intensity.
- Because of extremely high recrystallization velocities (10^2–10^3 cm/s) of the melted surface layer, substitutional solid solutions are obtained with dopant impurity levels well in excess of conventionally observed values.
- Only the surface region of the wafer is exposed to elevated temperatures, so that bulk crystallinity is not degraded.
- Since laser annealing is an extremely rapid event, there is no need for vacuum conditions or special inert atmospheres to prevent oxidation or contamination during the annealing procedure.

In annealing ion-implanted damage in semiconductors by means of pulsed lasers, the defect layer, generally amorphous, must be melted by the laser; the melt depth must extend slightly into the single-crystal layer to achieve optimum annealing and substitution of the implanted ions in the lattice.

For semiconductor materials such as Si and Ge implanted with As, B, or P, and using a Q-switched Nd:glass, Nd:YAG, or ruby laser, the laser pulse is in the duration range 10–100 ns. The energy density of irradiation on the material is in the range 0.5–10 J/cm^2. The resulting duration of the melt is generally a few hundred nanoseconds. (Because the absorptive properties of host materials vary with wavelength, these parameters will vary also.) In addition, the absorptive properties of the liquid differ from those of the solid.

These conditions can create a problem in coupling adequate energy into the material during the short pulse. A way around this problem was found by using a Q-switched Nd:glass laser at both its basic wavelength of 1060 nm and a frequency-doubled wavelength of 530 nm.[118] The shorter-wavelength light couples well into the solid silicon and initiates the melt. The 1060-nm-wavelength light then couples well into the melted material and completes the process of annealing.

The laser annealing of ion-implanted damage in semiconductors basically completely restores the crystalline structure by epitaxial regrowth, with the implanted ions taking up substitutional positions in the lattice structure.[110] Pulsed-laser annealing results in none of the defects found in thermal annealing, i.e., dislocations, stacking faults, and loops in the crystal. In application, the laser beam is rastered, with the scan rate coupled with the spot size and interpulse delays so that the irradiated spots overlap.

Another approach to laser annealing is the use of a continuous-wave (cw) or CO$_2$ laser. In this technique the laser beam is scanned continuously over the surface of the semiconductor material at a rate that allows the surface to heat to a temperature just below the melting point. As in the case of the pulsed laser, the reordering of the crystal structure is essentially complete, with the implanted ions taking up substitutional positions in the lattice structure. In this case, however, there is no melting and there is a solid-phase epitaxial regrowth. With no melting, there is no redistribution of the dopant ions, so the crystal retains its high dopant density.[119]

The scan rate of the cw beam is usually about 0.5–10 cm/s, giving a dwell time of about 10–100 ms. The typical flux is about 200 J/cm^2. In application, the scanning can be done by moving either the laser beam or the table on which the semiconductor material is mounted. An experimental arrangement[120] of the former type is shown in Fig. 4.57. In cw laser annealing it has generally been necessary to preheat the

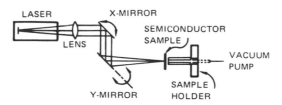

FIGURE 4.57.   Laser annealing apparatus. From Ref. 120.

substrate to a temperature in the 200–400°C range to reduce lateral thermal stress. This also allows the use of a lower-powered laser than would otherwise be necessary. Each mode of laser operation (pulsed, cw) has its advantages and disadvantages, depending in part on the materials and dopants used and on the size of the wafer to be processed.

In pulsed laser annealing, recrystallization follows the formation of a thin (0.1–1.0 μm) molten layer on the surface of the silicon wafer. Since the duration of the annealing pulse is $\leq 10^{-7}$ s, and from experiments we know that the migration of dopant atoms is several thousand angstroms, we can calculate the diffusion coefficient required to cause a diffusion length of l cm in a time τ by using the relationship

$$l = (D\tau)^{1/2} \tag{4.97}$$

or

$$D - \frac{l^2}{\tau} \tag{4.98}$$

With typical experimental values, $l = 10^{-5}$ cm and $\tau = 10^{-7}$ s, this formula gives $D \cong 10^{-3}$ cm^2/s. This value is more typical of diffusion constants in liquid materials than in solid-state materials, where $D_s = 10^{-12}$–10^{-14} cm^2/s. Therefore, both experimental (recrystallization) and theoretical evidence support the melted-layer model.

In typical experimental configurations (the diameter of the incident laser beam is $\cong 1$ cm and the pulse duration is <1 μs), the diffusion length for heat is

$$l \approx (k\tau)^{1/2} \approx 1 \ \mu\text{m} \tag{4.99}$$

where k is the thermal diffusivity and τ is the duration of the diffusion. Since l is much less than the beam diameter, the radial (x, y) diffusion can be neglected. The temperature distribution $T(z, t)$ can be calculated from the one-dimensional diffusion equation[103]

$$\frac{\partial T}{\partial t} = \frac{1}{\rho C} \left[\frac{\partial}{\partial z} \left(K(T) \frac{\partial T}{\partial z} \right) + A(z, t) \right] \tag{4.100}$$

where ρ is the density and C is the specific heat, which are effectively temperature independent. However, this equation is nonlinear, since the normal conductivity K is temperature dependent (see Fig. 4.58).

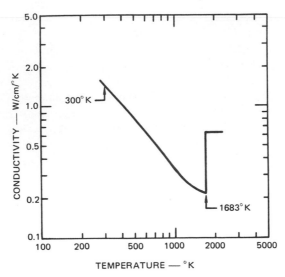

FIGURE 4.58.   Thermal conductivity of silicon as a function of temperature. From Ref. 109.

$A(z, t)$ represents the energy density absorbed from the incident laser pulse. The heat losses (radiation, conduction) are very small (1 W/cm^2) compared to the input pulse energies (10^6 W/cm^2) and they can be neglected. Equation (4.100) can be solved for the following limiting cases.

The distribution of the laser light intensity on the wafer is

$$I = (1 - R)I_0 \, e^{-\alpha z} \quad \text{W/cm}^2 \tag{4.101}$$

where $R(T)$ is the reflectivity of the silicon surface. The generation of heat can be calculated as

$$A(z) = \frac{dI}{dz} = \alpha(1 - R)I_0 \, e^{-\alpha z} \quad \text{W/cm}^3 \tag{4.102}$$

If diffusion is negligible during the pulse (adiabatic limit), the temperature distribution is given by

$$T(z) = \frac{A(z)\tau}{\rho C} \tag{4.103}$$

In this limit, $l = (k\tau)^{1/2} \ll \alpha^{-1}$, the temperature distribution in the wafer is determined by the initial distribution of the energy from the pulsed laser beam. Here one has to remember that the reflectivity is a function of the temperature.[109]

If $l = (k\tau)^{1/2} \gg \alpha^{-1}$, the temperature distribution in the silicon is determined by the heat diffusion into the bulk. In this case the temperature distribution is given as a solution of Eq. (4.101) in the following form

$$T(z, t) = \frac{2(1 - R)I_0}{K} \left[\left(\frac{kt}{\pi}\right)^{1/2} \exp\left(\frac{-z^2}{4kt}\right) - \frac{z}{2} \operatorname{erfc}\left(\frac{z}{2(kt)^{1/2}}\right) \right] \qquad (4.104)$$

and the temperature rise at the surface, $z = 0$, is

$$T(0, \tau) = \frac{2(1 - R)I_0}{K} \left(\frac{kt}{\pi}\right)^{1/2} \qquad (4.105)$$

In this diffusion limit the surface temperature rise depends on the absorbed flux $(1 - R)I_0$ and is independent of the absorption coefficient α.

Figure 4.59 shows the measured dependence of the threshold ($T = 1683°$K at the surface) incident power density I_0^T on the absorption coefficient α and pulse duration τ. The figure shows that in the adiabatic region.

$$I_0^T \approx \frac{1}{\alpha} \qquad (4.106)$$

Using Eq. (4.102), the energy required to elevate a surface layer with a thickness δz to the melting point T_M during the pulse duration τ can be obtained from

$$A\tau = \tau\alpha(1 - R)I_0^T \delta z = C\rho \delta z T_M \qquad (4.107)$$

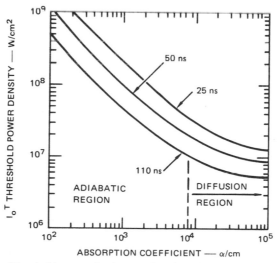

FIGURE 4.59.   Threshold power density as a function of coefficient $\alpha(\lambda)$. From Ref. 109.

so the threshold incident power density is given as

$$I_0^T = \frac{C\rho T_M}{(1 - R)\tau} \frac{1}{\alpha} \tag{4.108}$$

and in the diffusion-limited case

$$I_0^T = \frac{KT_M}{2(1 - R)} \left(\frac{\pi}{k\tau}\right)^{1/2} \tag{4.109}$$

Equation (4.109) is independent of α, as can be seen from Fig. 4.59 if $\alpha \geq 3 \times 10^4$. The threshold energy also can be calculated for the two limits, using Eqs. (4.108) and (4.109), as

$$E^T = I_0^T = \frac{C\rho T_M}{(1 - R)} \frac{1}{\alpha} \tag{4.110}$$

for the adiabatic limit and

$$E^T = I_0^T \tau = \frac{KT_M}{2(1 - R)} \left(\frac{\pi}{k}\right)^{1/2} (\tau)^{1/2} \tag{4.111}$$

for the diffusion limit. Here less energy is required for shorter pulses because the adsorbed energy does not diffuse far into the silicon and the surface can reach the melting point with less material being heated. The results of this one-dimensional thermal diffusion model[109] are in reasonable agreement with experiments over a wide range of conditions for λ, α, and τ.

4.11.6. Laser-Assisted Diffusion and Doping

Both n-type and p-type junctions can be obtained by the laser-assisted diffusion of dopants, with potentially higher activation densities and sharper profiles than those created by normal solubility techniques. In application, the dopant (such as As) may be spun onto the surface of the base material (such as Si) and irradiated by a laser at an energy density of about 10 J/cm^2 at a 1.06-μm wavelength, or 0.7 J/cm^2 at a 0.53-μm wavelength, providing sufficient mobility to allow the dopant ions to migrate to substitutional lattice sites at a desired depth. The resulting profile is sharp and an essentially uniform distribution of dopant is achieved in the melt.[121]

In UV-laser doping an XeCl excimer laser is used to form controllable diffusion junctions in substrates 50 to 120 nm deep.[122] The laser operates in a 1.3-μm-wide beam of wavelength 308 nm and forms a shallow molten layer on the substrate surface contained in a chamber filled with a dopant such as pyrolytic diborane. The laser dissociates the gas causing boron to diffuse into the molten silicon. The melting and recrystallization occurs in about 200 ns.

Gas immersion laser doping (GILDing) is a spatially selective doping technology. In this process[123] the dopant gas is introduced into the chamber and the wafer is exposed to energy from an excimer laser. Because of the short wavelength, the adsorption of the energy in silicon is confined to a depth of about 0.1 μm. The silicon surface is melted by the adsorbed energy and the dopant gas rapidly diffuses into the liquid. When the laser pulse terminates, the silicon regrows as a single crystal at a rate exceeding 1 m/s. By this means, almost perfect junctions can be made with ideal doping profiles.

Typically an XeCl excimer laser is used for boron gilding, and with this short wavelength (308 nm) a shallow junction is formed due to the increased absorption for short-wavelength photons.

If the energy density is high enough (on the order of 2 J/cm^2), the surface layer melts to the approximated depth of the electromagnetic radiation absorption during the flash. While the surface layer of the wafer is melting, some of the diborane at the surface is dissociating. The resulting boron dopant is incorporated into the melted layer forming the p-junction.

As the heat is conducted out of the melted layer, the melted boron-doped silicon recrystallizes. The process is completed in a few microseconds. This rapid quenching of the melted region occurs long before the melted material has an opportunity to agglomerate on the surface of the water. The resulting layer is now a p-doped junction.

The advantages of gilding are:

- The formed layer is fully annealed.
- The formed layer has very few crystallographic defects.
- The resulting junction is very abrupt.
- Although gilding is not a cold process, the wafer surface is heated for such a short period that many of the features of a cold process are also achieved.

4.11.7. Laser-Assisted Crystalline Conversion and Graphoepitaxy

The conversion of amorphous or polycrystalline materials to single-crystal or large-grain structures can be accomplished by laser processing without affecting the underlying characteristics of the substrates. In polycrystalline materials, grain boundaries act as barriers to the movement of charge carriers across junctions. Impurities may diffuse more rapidly along grain boundaries than they do in single crystals, catastrophically shorting out a junction. After conversion the resistivity is reduced by a factor of about 3. Applications of this crystalline conversion are found in the formation of interconnect structures on sophisticated IC devices and deposited polycrystalline Si on recrystallized metallurgical substrates.[119]

Laser radiation also can be used advantageously to alter the surface characteristics of a thin layer of polysilicon (5000 Å) by localized transient melting of the surface without producing deleterious heating effects at either interface of an underlying silicon dioxide layer on a silicon crystal substrate. The growth of high-quality oxides over polysilicon has long been a major problem in double-layer polysilicon structures, such as floating-gate memory devices (FAMOS) and charge-coupled devices (CCD).

Graphoepitaxy[124–127] (see Section 4.2.5) of silicon offers the possibility of forming single-crystal films over amorphous artificial microstructures. The film-formation method yields polycrystalline films on smooth amorphous substrates, and oriented films if the film formation is over a surface relief structure. It has been reported[64] that laser crystallization of amorphous silicon over a surface grating in fused silicon also yields a polycrystalline film with grains several tens of micrometers in diameter. These new techniques could lead to new combinations of substrates and films and new three-dimensionally integrated devices.

4.11.8. Laser-Based Lithography

Short-wavelength radiation has improved the resolution of optical photolithography to the submicron range and promises to make further improvements with electron-beam and X-ray techniques. The advantage of using mid-UV wavelengths (280–350 nm) is that improved resolution proportional to the square root of the wavelength, can be obtained without changing the source lamp, resist, or mask materials. Resolution as low as 0.4 μm has been obtained. The conventional source for the deep-UV range (240–300 nm) has been the Hg-Xe lamp. Cd-Xe lamps have extended the maximum wavelength to 230 nm giving a resolution of 0.2 μm.

Conventional coherent lasers are not used for lithography as they result in speckle (interference patterns that arise from varying optical path lengths) that makes it impossible to produce large-field light uniformity in the submicron range. Excimer lasers are different and deplete their pumped states faster than the time required for multiple reflections between the laser mirrors. Thus, the coherence length is negligible and these lasers operate more like a light bulb with a highly collimated light.

As the wavelength approaches 200 nm, it is becoming evident that quartz optical elements must be used. The high cost for the optics is one of the negative features of excimer lasers. However, excimer lasers have the advantages of:

- Higher intensity and thus shorter exposure times.
- Flash times of microseconds eliminate the need for maintaining the stage stationary for several hundred milliseconds or to wait for the alignment system to stabilize.
- Conventional chrome-on-glass mask-making can be used as opposed to X-ray lithography requiring difficult-to-make boron nitride masks. However, if an ArF excimer laser (193 nm) is used, quartz masks rather than soda lime or borosilicate glass are required.
- Since the contrast is very high ($\gamma > 10$), new inorganic resist and resist processes can be used.

Exposure with an XeCl excimer laser has been found to obey the law of reciprocity, i.e., the product of the light intensity and the required exposure time is intensity independent.[128] Thus, the higher flux available for excimer lasers can be used to reduce the required exposure times.

4.11.9. Laser-Assisted Deposition and Etching

The major application of the laser-assisted deposition technology is in maskless fuse linking of application-specific ICs (ASICs). Large semicustom circuits can be routed by the deposition of conductive boron-doped polysilicon links on chips using Ar lasers.[129] An array-programming system automatically calculates routing paths. Currently, ten links per second can be formed.

Numerous studies have been made for the selective deposition of various materials for IC applications. A large increase in the steady-state rate relative to conventional counterparts has been measured.[130] As an example, pyrolitic silicon chemical vapor deposition (CVD) in a standard furnace has a maximum useful rate $<10^{-2}\,\mu m/s$ but produces excellent electronic-grade polysilicon at rates $>10^2\,\mu m/s$ in laser direct writing. Tungsten on silicon have been selectively deposited by the reduction of WF_6.[131] In conventional CVD, the reaction occurs on the silicon surface at a temperature of 350°C. Using an argon laser, the reaction takes place between milliseconds and several seconds, depending on the scanning speed, spot size, and power of the laser beam.

High-quality gallium arsenide film has been grown with a CO_2 gas laser by adjusting the AsH_3 partial pressure, total pressure, and gas flow. By exciting PH_3 above the substrate, superlattices of $GaAs_xP_{1-x}/GaAs_yP_{1-y}$ have been grown.

Researchers at Sandia have used a cw Ar laser to repair circuits by depositing polysilicon conductors.[132] The wafer is placed in a chamber filled with a mixture of diborane and silane gases. The 0.4-W laser traces and heats a path over the insulating layer. When the path reaches 700–800°C, the two gases decompose and conducting polysilicon is deposited on the path. This process has been used to deposit thin films of tungsten on laser-written polysilicon interconnects.[133] Tungsten hexafluoride gas is exposed to the laser-written lines in a furnace at 400°C for 5–6 min to deposit films from 100 to 150 nm thick. Films are written at a rate of 2.5 mm/s.[134]

There are two distinct processes in laser etching, material vaporization and laser-assisted chemical etching. Material vaporization is presently used for the custom tailoring of ASICs.

A new process has been developed for fabricating CMOS gate arrays using a YAG laser that can cut interconnections directly on arrays of 1400 to 3200 gates with several hundred thousand to a million laser flashes per die. The die is already packaged in the uncovered ceramic package with nitrogen flowing across the wafer.

Laser-assisted chemical etching is accomplished in either the wet or dry mode. The wet etching process has been demonstrated[129] for Al that involves localized laser heating (<200°C) of a surface bathed in a capillary liquid etchant composed of 0.15% potassium dichromate in a 10% phosphoric acid/90% nitric acid solution that is trapped between the Si wafer and a thin cover glass. Laser-assisted dry etching is used to cut aluminum lines. The wafer is contained in a vacuum chamber filled with chlorine gas. A KrF excimer laser is aimed through a quartz window and pulsed between 20 and 30 times per second onto the aluminum line. This heats the aluminum, cracking the 2-nm oxide film so that chlorine and aluminum can spontaneously react to form volatile aluminum chloride.

4.11.10. Laser Pantography

In laser pantography[135] the patterning sequence of lithography, deposition, and etching is carried out entirely by laser-assisted processes.

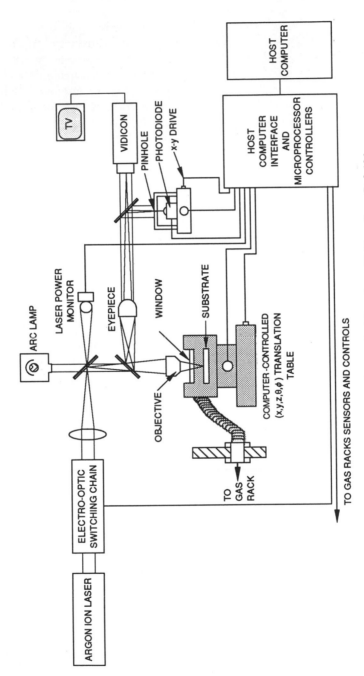

FIGURE 4.60.	The experimental set-up for Direct Laser Writing. From Ref. 136.

The lithography step is generally a direct-write process that avoids the use of photoresist. There are two methods for the lithography step:

- Direct laser writing. In this case a focused beam is used to induce direct deposition or etching as illustrated in Fig. 4.60. With this technique, dimensions as small as 0.2 μm[136] have been obtained.
- Laser projection. This method is similar to a conventional reduction stepper. In this case the laser beam is passed through a reticle projected onto the wafer housed in a vapor cell[137] and the reticle pattern is replicated in parallel.

4.12. ELECTRON-BEAM-ASSISTED PROCESSES

4.12.1. Electron-Beam Annealing

Pulsed electron-beam processing for annealing ion-implantation damage offers advantages over laser annealing in that it is independent of surface optical properties and it can irradiate large areas. These properties have been utilized for the low-cost automated manufacture of silicon solar cells[138] and the large-scale annealing of ICs. The electron-beam pulse-annealed wafers also show better electrical recovery than the furnace-annealed wafers. The furnace-annealed implant regions typically exhibit high residual defect densities, while the electron-beam pulsed-annealed implanted layer is found to be free of dislocations.[139]

Pulsed electron beams can also anneal through narrow (5–50 μm) windows in the oxide films. The insulating coating does not affect electron-beam parameters at moderate doses. However, at the very high dose rates (10^{15} rads/s), even the best

TABLE 4.12. Comparison of Electron- and Laser-Beam Processing

Characteristic	Electron beams	Lasers
Pulse width	10–200 ns	10–130 ns
Absorption and reflection dependence	Material density, electron-beam energy	Wavelength, temperature, crystal structure, doping level, dopant, surface preparation
Beam control	Electric and magnetic fields	Optics
Diameter of anneal area	To 76 mm	30 μm–20 mm
Uniformity		
Macroscopic	±5%	Gaussian beam profile
Microscopic	Improves by self-interaction	Hot spots, diffraction patterns
Energy flux	≥1 J/cm^2	1–10 J/cm^2
Processing depth	Controllable through electron energy spectrum	Maximum melt depth ≥1 μm because of damage mecanisms
Environment	Vacuum	Air or vacuum
Possible limitations	Residual charge (depends upon subsequent processing); charge carried across surface of insulators through radiation-induced conductivity	Patterned oxide surface films cannot be pulsed (interface pattern); perturbations in beam grow in terms of heating; and surface is not flat after pulsing

insulators become conducting. The trapped charges are dissipated with a subsequent low-temperature (500°C) processing step.

Finally, focused electron beams also can be used for local annealing.[140] The principle of this new experimental technique is that areas of the ion-implanted silicon chip are locally annealed to form conducting regions while the unannealed regions remain of high resistivity and form the isolation between adjacent conducting areas. Using this localized annealing technique, diode arrays can be fabricated to make integrated devices such as vidicon targets and optical sensors. Clearly, laser- and electron-beam-annealing techniques complement ion implantation. Table 4.12 shows a comparison[139] of pulsed electron-beam and pulsed laser processing.

4.12.2. Thermal Analysis of Electron-Beam Annealing

Electron-beam annealing models are similar in mathematical structure to laser models. But in the case of electron-beam heating the source is a volume source[140] and not a surface source, as it is for laser heating. This is because the electron penetration is typically much greater than the thickness of the layer to be annealed. The temperature distribution as the function of depth (z) is given as a solution of the diffusion equation in the following form [see Eq. (5.40)]:

$$T(z, t) = \frac{I_0}{K} \left\{ \left(\frac{kt}{\pi}\right)^{1/2} \left[\exp\left(\frac{-(z - z_0)^2}{4kt}\right) + \exp\left(-\frac{(z + z_0)^2}{4kt}\right) \right] \right.$$

$$\left. - \frac{|z - z_0|}{2} \operatorname{erfc} \frac{|z - z_0|}{2(kt)^{1/2}} - \frac{|z + z_0|}{2} \operatorname{erfc} \frac{|z + z_0|}{2(kt)^{1/2}} \right\} \tag{4.112}$$

where the plane source is at a depth z_0 and the decrease by a factor of 2 takes place because the heat is now conducted away in both z directions. Because it is assumed that no heat is lost from the boundary ($z = 0$) through convection and radiation, an image source of equal sign and strength I_0 is added, located at $z = -z_0$. Using this image method, temperature profiles can be calculated for stationary and moving sources.[141]

4.12.3. Radiation Chemistry

Radiation chemistry begins with the production of ions and excited molecules by incident fast electrons. These ions and excited molecules and the negative ions, secondary positive ions, and the free radicals that they produce can enter into chemical reactions. The pathways in radiation chemistry are graphically illustrated in Fig. 4.61[142]. At every stage, energy is either lost to collisions as heat or escapes as low-energy light photons in the form of luminescences, fluorescences, and Cerenkov radiation. This latter radiation, first observed in 1934 by P. A. Cerenkov, is the weak, bluish-white glow emitted when water and other transparent substances such as glass and mica are exposed to radiation. Such radiation rarely accounts for more than a few percent of the total energy loss, even at relativistic electron energies. Transition radiation is also an energy-loss mechanism,[143] but it is insignificant for chemical processing and is not shown in Fig. 4.61.

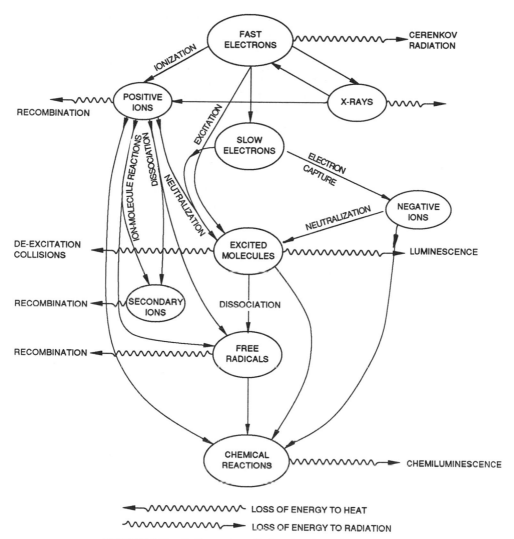

FIGURE 4.61. Pathways in radiation chemistry. From Ref. 142.

In recent years, much effort has been focused on electron–surface interaction for the purpose of materials modification. Electron bombardment can induce chemical reactions, break chemical bonds, and cause substrate heating, ionization, and electronic excitation. Localization of electron irradiation can allow for spatially selective thin-film deposition and etching. Electron-beam energies used in work done so far have ranged from 2 to 100 keV, to take advantage of the fine spatial resolution many instruments can provide at higher energy. Far less work has been done investigating beam-induced processes in the very low (<1 keV) electron energy regime.

Electron-assisted selective deposition and etching. Selective deposition can be carried out in a small area of a substrate (wafer) by using an electron-beam-induced surface reaction.[144–146] Thermal CVD and plasma CVD have been applied in the fabrication of microelectrical devices. A conventional deposition technology, such as thermal or plasma CVD, requires multiple steps to generate patterned thin films. Generally, photolithographic pattern definition, film deposition, and lift-off or etching are used. Therefore, in recent years, laser CVD has been studied for maskless deposition. The laser CVD resolution is around 1 μm, because it is limited by the laser wavelength. Even using an excimer laser with a shorter wavelength, it is hard to apply it as a deposition technology with which to fabricate devices with submicron dimensions. On the other hand, nanometer structures can be fabricated using electron-beam-induced selective CVD, because beam diameters as small as 0.5 nm can be formed with conventional electron-optical equipment. Computer-controlled direct writing also is possible for electron-beam CVD. Metals, semiconductors, and inorganic materials can be deposited by this means. Thus, electron-beam CVD has some excellent advantages over laser CVD, namely, processing methods in microelectronic devices can be simplified and new structure devices with nanometer dimensions realized.

The experimental apparatus for an electron-beam-assisted CVD system comprises a high-vacuum chamber that includes a focused electron-beam-assisted gun and the sources for the deposited elements. Electron-beam induced tungsten and chromium patterns can be deposited by using WF_6, WCl_6, and $Cr(C_6H_6)_2$ as sources, respectively. Figure 4.62 shows a scanning-electron-microscope-based electron CVD system for fine metal line deposition. In this system, the sample is cleaned by sputtering with argon ions. The electron beam breaks up the gas molecules on the surface and deposits the metal ions on the clean substrate. It has been reported[144,150] that

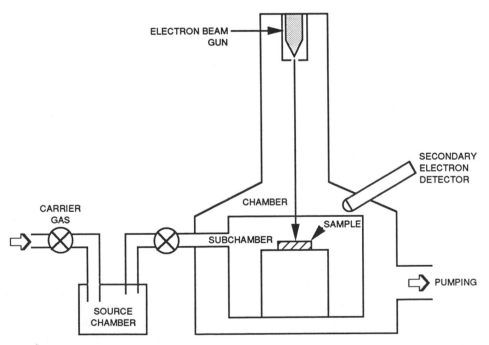

FIGURE 4.62. A scanning electron microscope (SEM)-based electron-assisted CVD system. From Ref. 136.

TABLE 4.13. Status of Electron-Beam Processing[a]

Process	1988	1998
Microlithography (20 keV)	Prod	Prod
Nanolithography (<100 keV)	Feas	Prod
Nanolithography (<2 keV)	Feas	Prod
Material modification	Lab	Prod
Oxidation (selective)	Feas	Prod
Etching (e-beam assisted)	Feas	Prod
Doping (using surface layers)	Feas	Lab
Annealing (selective)	Lab	Prod
Deposition (selective)	Lab	Prod
Voltage testing	Prod	Prod
Metrology	Prod	Prod
Defect analysis	Prod	Prod

[a]Prod, production; Lab, Laboratory; Feas, feasibility.

deposition took place at below $\cong 50°C$, but etching occurred at above $\cong 50°C$ substrate temperature, in a tungsten-deposition experiment using WF_6. These experimental results indicate that electron-beam irradiation causes not only dissociation of molecules adsorbed on the sample surface, but also a chemical reaction between the substrate and the molecules adsorbed on the sample surface.

Low-energy electron-induced chemical reactions. When a surface is bombarded with electrons from 5 to 100 keV, secondary electrons are emitted (see Section 2.6.3.3). The yield over this energy range for most materials is about unity. At low incident energy, however, the secondary electron yield increases and reaches a maximum, usually between 0.2 and 2 keV incident energy. This maximum can be as high as 24, as for MgO, but typical values are from 1 to 3. These secondary electrons enhance the chemical reaction rate at the gas–substrate interface. These effects have been investigated[144] by studying the electron-induced oxidation of silicon and the electron-enhanced etching of silicon by XeF_2.

Future of electron-beam processing. Based on the level of present development, we can infer that most fabrication processes can be implemented using electron or electron-assisted process steps. Table 4.13 lists the status of electron-beam technologies now and in the future. For complete IC processing, a few ion-beam process steps will still be needed for some processes, such as doping, materials modification, and micromachining. In some cases, the electron-beam process should be substituted to avoid radiation damage (changing, generation of interface status) in microelectronic devices.

4.13. ION-BEAM-ASSISTED PROCESSES

In the following sections we will review the status of ion-beam processes and their applications in microscience and materials research.[147–150]

Ion-beam technologies can be divided into two groups: focused ion beam (FIB) and showered ion beam (SIB). Ion implantation and dry etching techniques fall into the latter category.

An appreciation of the advantages of FIB and SIB for fabricating VLSI semiconductor devices must start with a basic understanding of ion-beam physics. Both focused and showered ion beams, which are made up of atoms or molecules that have lost one or more electrons, enable the interaction of ions with a solid target and thus change the chemical or physical characteristics of the target. FIB and SIB involve ion-to-substrate atom–momentum transfer and surface chemical reactions. Because of its narrow beam, FIB is capable of submicron spatial selectivity (or resolution). SIB requires a mask to achieve selectivity.

The most important parameters of ion beams as used in microfabrication are:

- Ion energy, which can be controlled precisely, sets the penetration depth (range) of the ions in the substrate. It is also influenced by the charge state of the ions.
- Ion current density, which can also be controlled precisely, controls the density of atoms implanted in the substrate if the irradiation time is set. It also controls the kinetic energy flux of the ions.
- Angle of ion-beam incidence, which affects the momentum transfer between the incident ion and the substrate in etching and deposition processes.
- Mass of the ions influences mechanical interaction processes.

Ion-beam lithography has developed in three directions: focused ion-beam lithography (FIBL), masked ion-beam lithography (MIBL), and ion projection lithography (IPL).

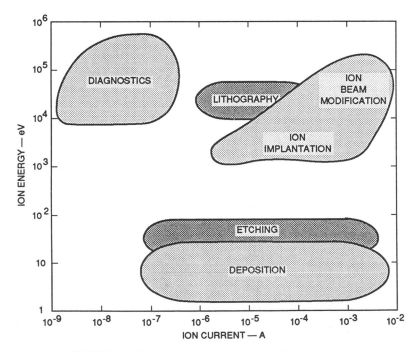

FIGURE 4.63.   Ion energy/current for ion-beam processes.

TABLE 4.14. Ion-Beam Characteristics for Processes

Beam parameters	Process lithography	Ion implantation	Surface modification	Deposition	Etching	Diagnostics
Size (μm)	Small (0.1–0.3)	Large $(1-1.5 \times 10^4)$	Very large $(1-2.0 \times 10^4)$	Large $(1-10 \times 10^4)$	Large $(1-10 \times 10^4)$	Very small (<0.1)
Types of ions	H, He, O, Ga	Dopant materials	All ions	Metals, semiconductors, insulators	Gases (F, Cl)	All ions
Energy (keV)	1 to 50	1 to 450	1 to 1000	10^{-3} to 10	10^{-3} to 10	1 to 30
Current (Å)	10^{-9} to 10^{-6}	10^0 to 10^{-3}	10^0 to 10^{-3}	Large (max)	Large (max)	10^{-11} to 10^{-6}
Purity	Low	High	Medium	High	Medium	High
Incident angle (deg)	90	Variable	90	Variable	Variable (0 to 180)	Variable
Typical dose (ions/cm²)	10^9-10^{16}	$10^{12}-10^{19}$	$10^{13}-10^{19}$	Max	$10^{14}-10^{18}$	Min

FIBL is a serial technique; a beam of ions from an ion source is focused on a target and is scanned just as in a scanning electron-beam system. Even with the high current densities achieved (1 A/cm^2), FIBL writing speeds are too slow for high volume production. The statistical nature of an ion-beam exposure, which requires at least 1000 ions to reach the smallest pixel, prohibits the use of more sensitive resist layers to improve writing speeds. The probe size for FIBL is currently limited to >10 nm.

MIBL is 1:1 shadow printing with ions. For submicron resolution, submicron masks are required, which are subjected to the same power density as the target. Mask heating problems restrict this technique to the exposure of sensitive organic resist layers.

IPL used ions as information carriers in a demagnifying step-and-repeat exposure system; submicrometer and even nanometer resolution can be obtained combined with a very high depth of focus (>100 μm). As a parallel exposure system, IPL permits significantly lower exposure times per chip than scanning systems through the use of masks that contain and store the complex design patterns. Step-and-repeat exposure is mandatory for the necessary chip-by-chip alignment to cope with wafer distortions induced by technological production steps. Demagnifying projection avoids the problems connected with the generation and maintenance of submicron masks, but the optics for IBL are of limited quality.

The high power densities possible with IPL not only permit pattern transfer in conventional organic resists but extend lithography to new processes using resistless ion-beam modification of materials. Because of statistical limitations, insensitive resist layers are mandatory for obtaining nanometer resolution.

Ion beams are also used for milling (micromachining, etching, and deposition) and materials modification.

Figure 4.63 shows the location of the ion-beam processes on the "ion energy and current plane," and Table 4.14 summarizes the ion-beam characteristics needed for different applications.

4.13.1. Ion Milling

Broad-beam ion milling, where ions (usually Ar$^+$) of a few hundred electron volts energy are incident on a surface, is a widely used method of material removal.[151] Usually one to three atoms of the substrate are removed per incident ion. This is called the sputtering yield (see Table 4.15). For focused ion beams, where the energy is about two orders of magnitude higher, the situation is expected to be similar.

With focused ion beams, the characteristics of the milling (such as yield) are geometry dependent. When an ion beam is scanned in a line on a surface, a trench is produced which initially has the shape of an inverse Gaussian as expected from the beam profile. However, when the dose is increased, the trench becomes sharp, narrow, V-shaped, and unexpectedly deep.[152-154]

It is important to note that surface topography will change under ion milling unless the surface is perfectly planar, because the sputtering yield $S(\theta)$ varies with the angle of incidence θ (see Fig. 4.64).

TABLE 4.15.   Sputter Yields[a]

Substrate	Ion	Energy (kev)	Yield (atoms/ion)
Cr	Ga^+	30–70	2.3
Glass	Ga^+	70	0.68[b]
Cr	Au^+	50–70	4.5
Si	Ga^+	30	2.6
Al	Ga^+	68	4.2
SiO_2	Ga^+	68	2.0[c]

[a]Source: Ref. 105.
[b]Corresponding to sputtering rate of 0.064 μm^3/nC.
[c]Molecules per ion.

The topography of the bombarded surface can be calculated for many surfaces (Fig. 4.65) using Monte Carlo simulation of the sputtering and redeposition processes.[151]

The redeposition process can be used to join two metal layers separated by an insulating layer[156] (see Fig. 4.66). In this fabrication process, a straight-sided hole is first milled through the top conductor and the insulator down to the lower conductor. The milling area is then reduced and the material redeposited on the sidewalls produces a short circuit between the two conductors. In addition to circuit reconfiguration using sputter removal and ion-beam redeposition, mask repair is the most important application for FIB processing. Figure 4.67 shows a removal process for chrome and two methods of repair for clear defects.[154,157]

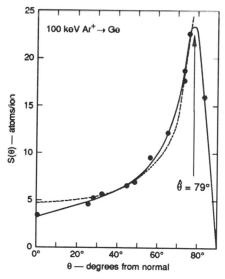

FIGURE 4.64.   Experimental $S(\theta)$ curve; sputtering yield as a function of angle of ion incidence. From Ref. 151.

To repair opaque defects, the unwanted chromium is simply milled off. In this process, the bombarding ions (Au$^+$) implanted in the glass plate cause staining due to light scattering.

Clear defects can be repaired by milling a light-scattering structure into the area to be rendered opaque.[143] A grating or a prism, for example, acts to scatter the

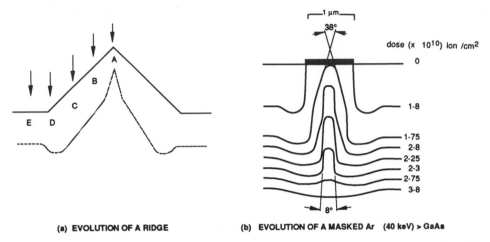

(a) EVOLUTION OF A RIDGE   (b) EVOLUTION OF A MASKED Ar (40 keV) > GaAs

FIGURE 4.65. Evolution of surfaces under ion-beam bombardment. (a) Evolution of a ridge; (b) evolution of a masked Ar (40 keV) > GaAs. From Ref. 151.

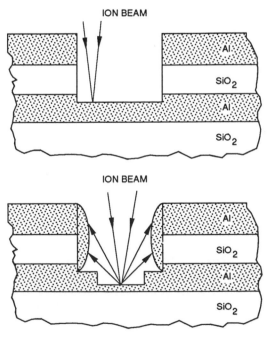

FIGURE 4.66. The joining of conductors using ion redeposition of the lower layer of metal. From Ref. 156.

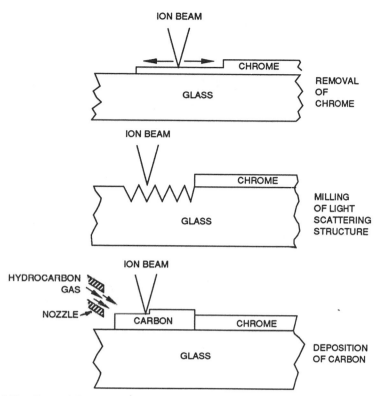

FIGURE 4.67. Focused ion-beam repair of photomasks, removal of unwanted chrome, and two methods of repair for clear defects. From Ref. 157.

light, and the area appears opaque when viewed in transmission, particularly when the incident light is collimated. This is usually the case in lithography applications.

4.13.2. Etching and Deposition

Energetic ions can induce chemical reactions near the surface of a target if an ambient gas is present. Using focused ion beams, spatially selective etching and deposition can be induced similar to the laser-induced surface reactions (see Section 4.11.10). The resolution of these processes are limited by the beam spots for ion and also for laser beams. For light beams the limit is set by the wavelength and for ion beams by the beam spot which can be in the nanometer range.[157]

The apparatus used in the FIB-assisted etching is shown in Fig. 4.68. The sample is located in a separate chamber which has a small hole 0.5 to 2 mm in diameter in a 0.2-mm-thick plate. The ion beam is incident on the sample through this hole and can be deflected to produce patterns. A gas such as Cl_2 is introduced into the chamber up to a pressure of 30 mTorr. The vacuum outside the chamber where the focused ion beam is generated is maintained near 10^{-6} Torr. This type of system has been used for etching of GaAs, InP, and Si. Chlorine or other etching gas is introduced into the chamber at pressures up to 30 mTorr while the vacuum chamber is maintained at

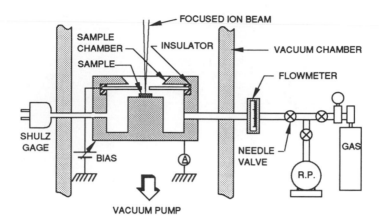

FIGURE 4.68.   Chamber used for ion-beam-assisted etching (and deposition). The focused ion beam enters through a small-bore hole. The gas, Cl_2, for etching is fed into the chamber. The top of the chamber can be biased to extract the secondary electrons due to the ion beam, thus imaging of the sample can be done

10^{-6} Torr. The etch rate obtained for GaAs is about ten times larger than the milling rate that would be obtained in the absence of the Cl_2 gas.

The same experimental arrangement can be used for deposition. In this case a metal-containing gas such as trimethyl aluminum (TMA) or tungsten hexafluoride (WF_6) is used. The energy of the ion beam induces decomposition of the gas at the

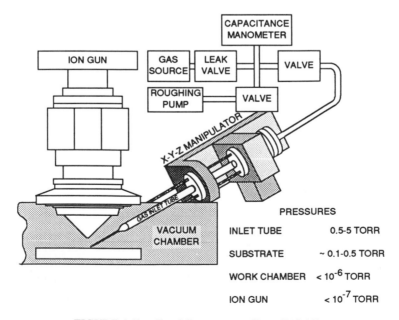

FIGURE 4.69.   Gas delivery system. From Ref. 149.

TABLE 4.16. Inorganic Film (Resist) Patterning

Material	Implant	Etch	Remarks
Cr	50- or 100-keV S_2^+ 3×10^{15} ions/cm	CCl_4 plasma (implant not etched)	0.3- to 0.4-nm-wide, 0.5-μm-thick, Cr lines formed
Si	50-keV In^+ ~ 5×10^{13} ions/cm^2 (50-keV Sn^+ and 50-keV Ga^+ also used)	H_3PO_4 at 195°C (implant etched)	50 nm-wide lines formed
GaAs	50-keV Ga^+ 1.2×10^{14} ions/cm^2	HCl at 50°C (implant etched)	30-nm-wide lines formed
Al	50-keV H_3^+, Sb^+ (N_2^+ & Ar^+ implanted) 5×10^{15} ions/cm^2	H_3PO_4 (irradiated area not etched)	Carbon contamination resist formed
Al/O Cermet	30- & 75-keV Ar^+, 70-keV Ga^+	H_3PO_4/HNO_3 (irradiated area not etched)	Al_2O_3 enhanced on surface
GaAs, InP, InGaAs, InGaAsP, InGaAsP (n-type)	20-keV Ga^+ > 5×10^{10} ions/cm^2	Photoelectrochemical, 2 M H_2SO_4 electrolyte	Very low dose needed

target. The metal deposits thus produced typically are contaminated by large amounts of carbon and oxygen and have high resistivity. For deposition, at high dose rates the physical sputtering rate can exceed the gas arrival rate and result in a net removal of surface material. Depositions will occur only when the net rates of ion arrival, gas arrival, and sputtering ejection favor material buildup.[146,148] The high local gas pressures (1 Torr) are accomplished by a gas jet such as depicted in Fig. 4.69. The needle orifice is placed 100 μm from the target surface.[149] Under these conditions, the vacuum pressure away from the target can still be maintained around 10^{-6} Torr.

4.13.3. Resists

Ion-beam exposure (doping) is a candidate for resistless processing still in most of the patterning applications (lithography, thin-film patterning). An organic or inorganic resist is used. The sensitivity of most resist materials such as PMMA is much higher for ions than for electrons. Thus, a dose of 2×10^{-5} C/cm^2 is needed to expose PMMA with 20-keV electrons while for argon ions at 120 or 150 keV a dose of 0.3 to 1×10^{-7} C/cm^2 (or 2 to 6×10^{11} ions/cm^2) is sufficient. However, this comparison is incomplete since electrons penetrate much deeper into the resist than do ions.[155,158,159] An important limitation of FIB exposure of resists is shot noise. The numbers for the dose needed for exposure quoted above are for large features. If one wants to expose features of, say, 0.1 μm, then at 10^{-7} C/cm^2, there would be only 70 ions incident per pixel area (see Section 7.3.3). The statistical fluctuation in this number would create variations in exposure level and, in addition, would create edge roughness of exposed features. Monte Carlo calculations of the energy deposited in the resist by incident ions have been used to simulate these fluctuations. The conclusion is that for an area of 0.1×0.1 μm, about 1200 ions are needed for complete exposure.[159]

Organic resists which are modified for etch resistance by ion beams are also used to pattern thin films. The implanted region can become either soluble or insoluble (i.e., the "resist" is either positive or negative). A number of materials have been patterned in this way as summarized[160] in Table 4.16.

4.13.4. Materials Modification

A number of very important materials properties, such as friction wear, hardness, corrosion resistance, surface chemistry, catalytic properties, bonding, and adhesion are markedly affected by the composition and structural changes in the surface layer due to ion bombardment[161].

Figure 4.70 depicts the direct ion implantation and its related techniques for preparing the surface-layer modification of metals. Table 4.17 shows the known materials transformation processes in the surface layer.

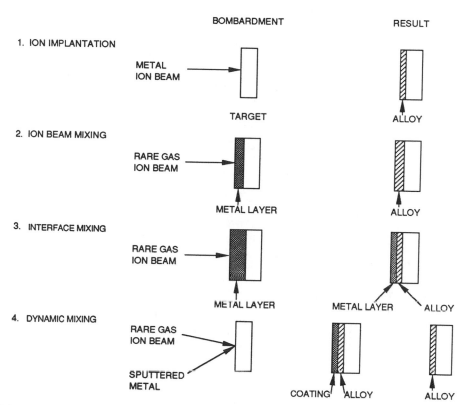

FIGURE 4.70. Different ion base beam technologies for surface layer modification and the resulting structures.

TABLE 4.17. Materials Modification

Ions		Substrate	New surface	Applications
Type	Dose			
O, N	10^{14}–10^{18}	II–VI, III–V	SiO_2, SiN_4	Ceramic layer formation
N, Ti, Ni, Al, Cr, Cu	10^{17}–10^{18}	Steel		Hardening (for drills, bearings, nozzles); extrusion dies, nozzle decoration; increase corrosion resistance
Cr, O, Ni	10^{17}	Low-steel carbon	Oxide layers	Surface ceramic formation
Cu, Ni	10^{17}–10^{18}			Increase wear resistance
Ni, Al, Cr, O	10^{17}	Steel	Al_2O_3, Cr_2O_3, SiO_2	Corrosion inhibition

4.13.5. Ion Cluster Beam (ICB) Deposition

The advantages of using cluster ions (see Sections 2.3.7 and 3.2.6) for depositing layers, as compared to vacuum evaporation, MBE, sputtering, and other ion-assisted

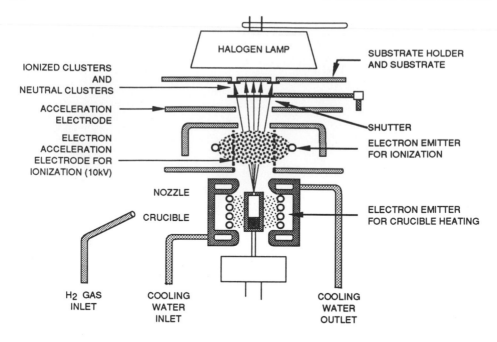

FIGURE 4.71.   Schematic diagram of the apparatus for ionized cluster beam. From Ref. 162.

processes are

- Control of particle energy in the useful range for film deposition leading to excellent surface coverage and adhesion.
- Enables very high deposition rates to be obtained with a large loading capacity.
- Can be operated at any pressure up to that where collision with residual gas molecules becomes dominant (i.e., up to about 10^{-4} Torr).

A typical cluster ion deposition system is shown schematically in Fig. 4.71. Typically singly charged clusters of 500 to 2000 loosely coupled atoms are accelerated to about 5 kV giving a particle kinetic energy in the range 2.5 to 10 eV, substantially higher than for evaporation sources (0.1 eV) but substantially lower than sputter or ion beam sources (1 keV–10 keV). Table 4.18 lists the film properties and parameters for the cluster ion deposition of various metals, semiconductors, glasses, and other materials.[162]

4.14. NUCLEATION AND GROWTH IN LAYER FORMATION

4.14.1. Overview

Thin films often have physical properties that differ substantially from bulk samples of the same material. This is due both to the influence of the surface, the interface, and to the structure formed by the initial condensation process. The case

TABLE 4.18. Deposition Parameters Using Cluster Ions[a]

Films/substrate	Accelerating voltage (kV)[b]	Substrate temperature (°C)	Advantages (applications)
Metals			
Au, Cu/glass, Kapton	1–10	RT	Strong adhesion, high packing density, good electrical conduction in a very thin film (high-resolution flexible circuits, optical coating)
Pb/glass	5	RT	Controllable crystal structure, strong adhesion, smooth surface, improved stability on thermal cycling (Superconductive device)
Ag/Si (n-type)	5	RT	Ohmic contact without alloying
Ag/Si (p-type)		400[c]	Ohmic contact at low temperature (semiconductor metallization)
Ag-Sb/GaP	2	400	Ohmic contact, strong adhesion (semiconductor metallization
Al/SiO_2, Si	0[d]–5	RT–200	Controllable crystal structure, strong adhesion, ohmic contact at low temperature, electromigration resistant (semiconductor metallization)
Intermetallic compounds			
C-axis preferentially oriented MnBi/glass	0[d]	300	Spatially uniform magnetic domain, high-density optical memory (magneto-optical memory)
GdFe/glass	0[d]	200	Thermally stable and uniform, amorphous (magneto-optical memory)
C-axis preferentially oriented PbTe/glass	0[d]–3	200	High Seebeck coefficient, low thermal conductivity (high-efficiency thermoelectrical converter)
Preferentially oriented ZnSb/glass	1	140	Thermally stable amorphous, p- or n-type controllable, high Seebeck coefficient (high-efficiency thermoelectrical converter)
Semiconductors			
Si/(111)Si	8	820	Low-temperature epitaxy in a high
Si/(100)Si			vacuum (10^{-7}–10^{-8} Torr), shallow
Si/(1102) sapphire			and sharp p–n junction (semiconductor devices)
Amorphous Si/glass	2	200	Thermally stable (solar cell, thin-film transistor)
Ge/(100)Si	0.5–1	300–500	Low-temperature epitaxy in a high vacuum (10^{-7}–10^{-6} Torr), shallow and sharp p–n junction (semiconductor devices)
Ge/(111)Si	6	550	Epitaxy in a high vacuum (10^{-7}–
GaAs/Cr-doped GaAs			10^{-6} Torr) (semiconductor devices)
GaP/GaP	4	550	Low-temperature epitaxy in high vacuum
GaP/Si		450	(10^{-7}–10^{-6} Torr) (low-cost LED)
ZnS/NaCl	1	200	Single-crystal formation (optical coating, ac–dc electroluminescent cells of low impedance)

(*continued*)

TABLE 4.18. (*Continued*)

Films/substrate	Accelerating voltage (kV)[b]	Substrate temperature (°C)	Advantages (applications)
ZnS:Mn/glass	1	200	Controllable crystal structure (magnetic sensor)
InSb/sapphire	3	250	Improved monocrystalline domains (infrared detector)
CdTeGe/glass	2	250	Very thin and multilayered films, low-temperature growth (optoelectronic devices, superlattice)
CdTe-PbTe/Si InSb	0[d]–5	200–300	Controllable optical bandgap and Faraday rotation angle (solar cell, magneto-optical devices)
Oxides, nitrides, and carbides			
FeO$_x$/Si, glass	3	250	Controllable optical bandgap (photovoltaic cell)
Li-doped ZnO/(1102) sapphire	−0.5–1	230	Single-crystal formation (optical waveguide)
C-axis preferentially oriented ZnO/glass	0[d]	150	Controllable crystal transmission (optical waveguide, SAW)
BeO/(0001) sapphire	0[d]	400	Single-crystal formation, transparent film (semiconductor device, heat sink, electrical insulator)
C-axis preferentially oriented BeO/glass	0[d]	400	High-electrical-resistivity and high-thermal-conductivity coating, low-temperature growth (semiconductor device, heat sink, electrically insulating and thermal conduction)
PbO/glass	3	RT	Low-temperature growth, smooth surface, accurately controllable thickness, good adhesion (superconductor)
SnO$_2$/glass	0[d]	400	Transparent low-resistivity film, strong adhesion (NESA glass)
SiO$_2$/Si	0[d]–2	200	Extremely low-temperature growth (Si devices)
GaN/ZnO/glass	0[d]	450	Low-temperature growth of GaN film on amorphous substrate (low-cost LED, photocathode electrode)
SiC/Si, glass	0[d]–8	600	Controllable crystal state (energy converter, surface protective coating)
Organic materials			
Anthracene/glass	0[d]–2	−10	Controllable crystal structure (detectors)
Cu-phthalocyanine/glass	0[d]–2	200	Controllable crystal structure (photovoltaic devices)
Polyethylene/glass	0[d]–1	0–110	Controllable crystal structure (photovoltaic devices)

[a] Source: Ref. 145.
[b] Since a cluster consists of 500–2000 atoms, one atom has an incident energy corresponding to 1/500 to 1/2000 of the acceleration voltage. For example, in a case of 5-kV acceleration voltage, one constituent atom has an energy of 2.5–10 eV, which is coincident to the optimum energy for film formation.
[c] Annealing temperature.
[d] Ejection velocity only.

where the arriving species consists of single identical atoms or clusters of identical atoms has been studied extensively.[163]

The cases where the arriving species are molecular, or are formed by chemical reaction at the surface are more difficult to study, but it may be assumed that the sequence of events in condensation will be similar after the condensing species has been formed. Film growth has been observed by electron microscopy (see Chapter 6) to proceed according to the following stages:

- Nucleation: Small nuclei appear (5-Å diameter) statistically distributed over the surface.
- Growth of nuclei: Nuclei grow into larger, three-dimensional islands.
- Coalescence of islands: The interconnected islands form a network containing empty channels.
- Filling of the channels to create a uniform layer.

If the arriving species is more strongly bonded to itself than the substrate, islands, many atomic layers high, grow before they coalesce. This growth mode is called the island or Volmer–Weber mode, and is the case for most of the situations met in microfabrication. On the other hand, if the arriving species is more strongly bonded to the substrate, the islands tend to be a single atomic layer thick, and completely coalesce before a second atomic layer is formed. As the atomic layers build up, the influence of the substrate is reduced and the condensate finally condenses upon itself in a normal bulk mode. This is called the layer or Frank–van der Merwe growth mode and is found in only a few cases which are of use to microdevices. Finally, in the layer plus island or Stranski–Krastanov growth mode, where the bonding of the species to itself and the substrate are comparable, the first layer is monatomic upon which multiatomic layer islands are formed.

Until many atomic layers are built up and all of the channels filled, substantial differences in such physical properties as the electrical conductivity or film density may be expected. Completion of this stage takes place in the range of a hundred to a few thousand angstroms apparent thickness, depending on the substrate, its temperature, and the evaporant. Beyond the last stage the film may become amorphous (short-range order), randomly oriented polycrystalline, uniaxially oriented (in direction of growth) polycrystalline, or identically oriented (epitaxial), depending upon the conditions of growth. When attempting to produce very thin films with specific properties, the initial growth phase is obviously of crucial importance.

4.14.2. The Island Mode of Nucleation and Initial Growth

When a particle impinges upon a surface, a number of interactions between them may take place, depending upon the chemical binding energy between the surface atoms and the impinging particles, the surface electric fields due to the asymmetry of the lattices at the surface, the polarizability of the impinging particle, the lattice constants of the substrate and film materials, the temperature of the substrate, and the energy of the impinging particle.[164]

In general, we may suppose that the impinging particle, after striking the surface, will be held to it by polar forces resulting from the surface electric fields. Its excess energy will enable it to move a certain distance along the surface until either it loses energy to the lattice and becomes chemically bound to a permanent site, or it gains sufficient energy from the lattice to escape the surface electric fields. On its travels along the surface the particle may be temporarily bound to sites with weak bonds. The time τ' it spends on the weakly bonded sites depends upon the activation energy for migration, ϕ_M, and the surface temperature T, and typically follows the relationship

$$\tau' = \tau'_0 \exp\left(\frac{\phi_M}{kT}\right) \tag{4.113}$$

The total time τ the particle spends on the surface is characterized by the activation energy for desorption, ϕ_D, i.e.,

$$\tau = \tau_0 \exp\left(\frac{\phi_D}{kT}\right) \tag{4.114}$$

Table 4.19 gives some typical values. At high substrate temperatures, τ' can be sufficiently short that the adsorbed particles essentially behave as a two-dimensional gas; at intermediate temperatures, where τ' is finite, the adsorbed particles hop from site to site in a random walk; whereas at low substrate temperatures, where τ' is very long, the particles soon lose their excess energy to become essentially permanently bonded. These processes have been well illustrated in the case of barium adsorbed on tungsten, where the low work function of barium on tungsten enables the movement of the barium to be observed in thermionic-emission microscopes (see Section 6.2.4).

Impinging particles are usually trapped on a cold surface, with kinetic energies corresponding to a higher temperature than that of the substrate, depending on their source, and move over the surface as a two-dimensional gas until it is possible for them to give up this excess energy. Condensation into a solid can only occur if sites for nucleation exist where the adsorbed particles can give up their excess energy. If the supply of the impinging particles exceeds their desorption rate, the film becomes supersaturated and condensation becomes more likely.

The conditions under which condensation starts depend on the ratio of the desorption energy ϕ_D to the heat of sublimation of the condensed film ϕ_S. If $\phi_D \ll \phi_S$, condensation occurs without supersaturation. If $\phi_D > \phi_S$, a high degree of supersaturation is required to start condensation. Obviously, after the film has started to form, we soon reach the condition $\phi_D = \phi_S$.

TABLE 4.19. Comparison of Activation Energies for
Migration (ϕ_M) and Desorption (ϕ_D)

System	ϕ_M (eV)	ϕ_D (eV)
Barium on tungsten	0.65	3.8
Cesium on tungsten	0.61	2.8
Tungsten on tungsten	1.21	5.83

Theoretical models for the condensation process usually assume that when the adsorbed atoms collide, there is a certain probability of their forming a bonded pair. Further collisions of atoms with the pair results in the formation of greater aggregates. The reverse process of aggregate disintegration proceeds simultaneously. The capillary theory (Ref. 8, p. 767), valid for aggregates containing a large number of atoms (>100), uses thermodynamic methods to show that an aggregate will be stable if its radius exceeds a value $r*$ given by

$$r* = \frac{2\sigma_{cv}V}{kT \ln p/p_e} \tag{4.115}$$

where σ_{cv} is the codensate-vapor interfacial free energy for a molecular volume V condensed from the unsaturated vapor at pressure p to its adsorbed state at equilibrium pressure p_e at surface temperature T.

The theory has been modified to take into account the nonspherical shape taken by the aggregate as a result of surface tension forces at the substrate. When the critical nucleus contains only a few atoms (1–10), it is necessary to consider the effects of individual bonding of the atoms to themselves and the substrate lattice. One approach has been to modify the capillary theory to take these into account,[165] whereas another has been to attempt to consider in detail the statistics of aggregate formation.[166] In the range appropriate to the theories, they yield results qualitatively in accord with observations.[167] The possibility for future quantitative comparison of theory and experiment exists, as the assumptions made in the theories can be made to correspond better to reality, and computational methods become more sophisticated.

The process of nucleation and growth has been confirmed by electron microscopy. As evaporation proceeds, the number of resolvable islands (>5Å in diameter) increases with a random position distribution until a saturation island density in the range 10^{10}–10^{12} atoms/cm^2 (for metal vapors condensing on insulating substrates at room temperature) is obtained, i.e., the islands are 100–10000 Å apart. As evaporation proceeds further, the islands grow and the island density decreases by mutual coalescence. The density of nucleation sites can be increased by external agents, such as by electron bombardment of the substrate. Nucleation also occurs preferentially on crystalline surfaces at the dislocations, impurity sites, and other irregularities in the lattice. This effect is utilized for the "decoration" of dislocations in single crystals to render them visible.

Impinging particles originating from sputter, plasma or cluster ion sources have much higher effective temperatures than from evaporation sources and hence can be used to obtain epitaxial films.[9] The nucleation site density is also higher at lower substrate temperatures as a result of the electron and ion bombardment of the substrate surface that takes place in such systems.

Nucleation processes for films deposited by reactive and electrochemical methods appear to have been studied less than the physical deposition methods and are obviously much more complex. More experimental techniques are being developed[163] but it must be concluded that although at this time the phenomenon of nucleation is understood in general terms, a more detailed understanding is necessary before quantitative predictions can be made for specific cases.

4.14.3. Postnucleation Growth Processes

When the nucleation sites on the substrate are filled, three-dimensional islands begin to grow from these sites. Depending on the substrate temperature, the islands may be liquid droplets or single crystals of the condensing materials. The melting point of the islands is lower than that of the bulk material T_m, and has been observed to be $2T_m/3$. Coalescence is assumed to take place only when the islands touch. Heat is liberated by coalescence, so that solid islands of many materials melt on contact. After coalescence they cool down, forming a new solid crystal, usually with the orientation of the larger of the two original island crystallites.

The occurrence of epitaxy, or the oriented growth of single-crystal films on monocrystalline substrates, suggests that in this case the orientation of most nucleation islands are similar and recrystallization with a preferred orientation takes place on coalescence. This is confirmed by the observation that the orientation of an epitaxial film is strongly dependent on the crystal structure of the substrate (see Sections 1.3.2 and 4.2). On the other hand, it is possible to grow single-crystal oriented films on amorphous or even liquid subtrates as well as single-crystal surfaces.

For every pair of condensate and substrate, there exists, for a given deposition rate, a critical substrate temperature above which a single-crystal oriented film grows, regardless of the degree of lattice mismatch. Condensation at temperatures below the critical temperature produces increasing disorder in the film, until at sufficiently low temperatures ($T_m/3$) the incident atoms condense close to the point of impingement, producing an amorphous (highly disordered) film. A laterally applied dc field on the order of 100 V/cm can lower the average film thickness at which coalescence occurs for certain combinations, such as silver on sodium chloride.[168] The applied field appears to flatten the islands, thus increasing their diameter and forcing them to touch earlier. Irradiating the surface with electrons or ions also promotes coalescence, although it is not immediately clear why this occurs. Films deposited from a sputter source are generally accompanied by electron and ion bombardment of the substrate, depending on the configuration being used. The earlier coalescence induced may produce thin films with properties different from those of evaporated films.

4.14.4. Structure and Composition

In this section the terms used to describe structure and composition are defined and the appropriate measurement techniques are indicated.

Morphology. Morphology describes whether the deposit consists of three-dimensional islands, a channeled layer film, or a uniformly dense film. This is best observed by high-resolution electron microscopy, but optical diffraction, electrical conductivity, and density measurements can offer clues.

Crystallinity. The film may be amorphous, randomly oriented polycrystals, uniaxially oriented polycrystals, or single crystals (epitaxial). This condition is measured by electron or X-ray diffraction. Amorphous films give no well-defined diffraction rings; randomly oriented polycrystals give well-defined diffraction rings. From the width of the diffraction lines it is possible to infer the average size, a, of the crystallite from the relationship

$$a = \frac{\lambda}{D \cos \theta}$$

where λ is the X-ray wavelength, D is the angular width of the diffraction lines, and θ is the Bragg angle.

Single-crystal films or identically oriented polycrystals give rise to round diffraction spots (Laue patterns). As the orientation of the polycrystal deviates from perfection, the spots become distorted to take elliptical shapes, and new spots appear. In the extreme case of complete randomization, the elliptical spots join to form the diffraction rings. If one axis of the crystal remains the same but the orientation of the other axes are random, this gives rise to missing or low-intensity rings, compared with the completely random case. If the electron beam is made small enough, the crystallinity of individual islands in the postnucleation stage can be ascertained by electron diffraction.

The lattice constants for the film material can, of course, be deduced from the angular positions of the diffraction patterns. The lattice structure of very thin films is affected by the presence of the substrate and surface.

Analysis of films is carried out by several methods (see also Section 6.3):

- Conventional chemical analytical techniques whereby the film is dissolved off the substrate and subjected to analysis. In this category we include volumetric, gravimetric, and optical spectrographic methods.
- X-ray fluorescence, a procedure in which the film is bombarded by electrons sufficiently energetic to excite inner-shell radiation, which is detected by solid-state counters and an energy spectrum of the emitted X rays displayed. By using a focused electron beam, very small areas can be analyzed (see Section 6.3.1).
- X-ray and (high-energy) electron-beam diffraction allow a comparison of the diffraction patterns obtained with those of known materials.
- Auger spectroscopy is the analysis of the energy spectrum of the secondary electrons emitted from the surface upon bombardment by primary electrons and enables identification of the surface atomic layer to be made. By sputtering the film away in steps, followed by Auger spectroscopy, a profile of the composition from the surface to the substrate may be obtained (see Section 6.3).
- Mass spectroscopy may be used if the surface is sputtered away by a beam of energetic ions. The sputtered material may be subsequently ionized and injected into a mass spectrometer for analysis.

X-ray fluorescence, Auger spectroscopy, and mass spectroscopy have been adapted for scanning systems so that variations in the composition of a film over an area can be obtained. With modern instrumentation, submicron resolutions may be obtained.

However, impurity levels below a few percent are difficult to detect by other than the conventional techniques and mass spectrometry.

4.14.5. Surfaces and Interfaces

The topography of surfaces can be visualized by scanning electron microscopy, with resolutions down to 30 Å (Section 6.3.1). A significant advantage of this device is its extraordinary depth of focus, which enables structures with large aspect ratios to be viewed clearly. Higher resolution can be obtained with transmission electron microscopy, but this technique requires very thin (electron transparent) films and substrates. Surface structures can be seen by reflection electron microscopy, using shadowing and replication techniques (Section 6.2.2).

The arrangements of the atoms in the surface can be detected by low-energy electron diffraction (LEED), and the atomic constituents of the surface atomic layer can be unambiguously identified from the energy spectrum of the secondary electrons (Auger spectroscopy). Small movements of the peaks in Auger spectroscopy can indicate how the atoms are bound together. These techniques, which view only the surface atomic layer due to the small penetration of low-energy electrons (less than a few kiloelectron volts), have revolutionized our understanding of surfaces since the instruments have become commercially available. This availability itself turned on the development of UHV technology, which enables experiments to be carried out in vacuums of less than 10^{-10} Torr so that the surface is not seriously contaminated by its environment during observation (see Section 3.7).

The substrate surface can be studied before deposition. After deposition the interface can be reached by sputtering away a portion of the film, step by step, with inert gas ions until the substrate is penetrated. However, this sputter "scalpel" is not precise enough to indicate exactly when the interfacial atomic layers are reached.

4.14.6. Rapid Thermal Annealing (RTA)

The use of tungsten halogen lamps in quartz envelopes has enabled very-high-power visible light photon fluxes (over 100 W/cm^2) to be obtained continuously over large-area wafers.[169] The photons with energy greater than the band gap of the semiconductor are absorbed close to the surface making the process ideal for annealing defects, particularly those caused by radiation damage after ion implantation. Typically the surface of a silicon wafer is heated from room temperature to the annealing temperature (1200°C) in 5 s, held at this temperature for a further 4 or 5 s, and cooled down to room temperature using a nitrogen purge in a further 15 to 20 s. Relatively inexpensive high-throughput annealing equipment has been developed for IC production using this technique. The same heating concept has also been developed for rapid thermal oxidation[170] and nitridation of silicon.[171] Unless heating is from both sides, a large temperature gradient will exist through the wafer. A temperature gradient of about 50°C/cm at about 1200°C is required for rapid thermomigration of aluminum[172] through silicon, and this concept has been used to pattern lamellar devices in silicon.[173]

PROBLEMS

1. Discuss the following terms:
   - Autodoping
   - Gettering
   - Frenkel defect

2. a. Express Fick's first law in words.

   b. Describe the mechanism of diffusion by interstitial motion. Will this be affected by vacancies in the lattice?

   c. A diffusion furnace operating at 1000°C has a ±1°C tolerance. What is the corresponding tolerance on the diffusion depth, assuming a Gaussian diffusion?

3. a. A resistor is made with a boron implant dose of 4×10^{12} per cm² into silicon at 100 eV. Calculate: (i) the peak concentration, (ii) the projected range and standard deviation in projected range, and (iii) the sheet resistivity of the layer. Assume $\mu = 250$ cm²/V · s.

   b. What are the advantages of using a rapid thermal anneal for this sample rather than a furnace anneal?

   c. Describe some of the techniques which are available for examination of an ion-implanted region of silicon. In particular, identify which techniques are destructive and which are nondestructive.

4. Consider the oxidation of silicon. Discuss the following:
   - Oxide charges
   - Impurity segregation at the Si/SiO_2 interface
   - The Deal and Grove model

5. Consider the formation of a metal film by rf sputter deposition on a silicon wafer. Describe the process of film growth on the microscopic scale. How does the process of film growth depend upon gas pressure and substrate temperature? What are the advantages of a magnetron source?

6. a. Discuss the differences and similarities between sputter etching, ion milling, reactive ion etching, and plasma etching.

   b. Given a source concentration of 10^{15} phosphorus atoms/cm², calculate the junction depth for a 2-h diffusion at 1100°C given that the substrate boron doping level is 5×10^{16} per cm³.

REFERENCES

1. D. T. J. Hurle, *Current Growth Techniques in Crystal Growth, a Tutorial Approach*, North-Holland, Amsterdam (1979).
2. M. Faraday, Experimental relations of gold (and other metals) to light, *Philos. Trans. R. Soc. London* **147**, 145–181 (1857).
3. M. K. Bolazs (ed.), *Proc. Semiconductor Pure Water Conference*, Santa Clara, Calif. (1982–1990).
4. H. E. Farnsworth, in: *The Surface Chemistry of Metals and Semiconductors* (H. Gates, ed.), p. 21, Wiley, New York, (1959).
5. L. Holland, *Vacuum Deposition of Thin Films*, Chapman & Hall, London (1956).

6. L. I. Maissel and R. Glang (eds.), *Handbook of Thin Film Technology*, McGraw–Hill, New York (1970).
7. K. L. Chopra, *Thin Film Phenomena*, McGraw–Hill, New York (1969).
8. L. Eckertova, *Physics of Thin Films*, Plenum Press, New York (1977).
9. J. C. Vossen and W. Kern (eds.), *Thin Film Processes*, Academic Press, New York (1978).
10. A. G. Cullis, Transient annealing of semiconductors by laser, electron beam and radiant heating techniques, *Rep. Prog. Phys.* **48**, 1155–1233 (1985).
11. D. W. Sah, in: *Crystal Growth Theory and Techniques* (C. H. Goodman, ed.), Vol. 1, Plenum Press, New York (1974).
12. R. M. Burger and R. P. Donovan (eds.), *Fundamentals of Silicon Integrated Device Technology*, Vol. 1, Prentice–Hall, Englewood Cliffs, N.J. (1967).
13. S. Nielsen and G. J. Rich, Preparation of epitaxial layers of silicon: I. Direct and indirect processes, *Microelectron. Reliab.* **1964**, 165–170 (1964).
14. M. L. Hammond, Silicon epitaxy, *Solid State Technol.* **21**(11), 68 (November, 1978).
15. R. W. Dutton, D. A. Antoniadis, J. D. Meindl, T. I. Kamins, K. C. Saraswat, B. E. Deal, and J. D. Plummer, Oxidation and epitaxy, Technical Report No. 5021–1, Integrated Circuit Laboratory, Stanford University, Stanford, Calif. (May, 1977).
16. T. I. Kamins, R. Reif, and K. C. Saraswat, in: *Electrochemical Society Fall Meeting, Las Vegas, October 17–22, 1976*, pp. 601–603, Electrochemical Society, Princeton, N.J. (1976).
17. R. Reif, T. I. Kamins, and K. C. Saraswat, Transient and steady-state response of the dopant system of an epitaxial reactor: Growth rate dependence, *Electrochem. Soc. J.* **124**(8), 912–923 (1977).
18. J. D. Meindl, K. C. Saraswat, and J.D. Plummer, *Semiconductor Silicon*, pp. 894–909, Electrochemical Society, Princeton, N.J. (1977).
19. J. D. Meindl, K. C. Saraswat, R. W. Dutton, J. F. Gibbons, W. Tiller, J. D. Plummer, B. E. Deal, and T. I. Kamins, Final report on computer-aided semiconductor process modeling, Stanford Electronics Laboratories, Report TR-4969-73-F, Stanford University, Stanford, Calif. (October, 1976).
20. R. Reif, T. I. Kamins, and K. C. Saraswat, A model for dopant incorporation into growing silicon epitaxial films, *J. Electrochem. Soc.* **126**(4), 644 (April, 1979).
21. B. V. Vanderschmitt, Silicon-on-sapphire: An LSI/VLSI technology, *RCA Eng.* **24**(1) (June–July, 1978).
22. S. Cristoloeanu, Silicon films on sapphire, *Rep. Prog. Phys.* **50**, 327–371 (1987).
23. L. Jastrzebski, Y. Imamura, and H. C. Gatos, Thickness uniformity of GaAs layers grown by electroepitaxy, *J. Electrochem. Soc.* **125**(7), 1140 (July, 1978).
24. L. Jastrzebski, J. Lagowski, H. C. Gatos, and A. F. Witt, Liquid-phase electroepitaxy: Growth kinetics, *J. Appl. Phys.* **49**(12), 5909 (December, 1978).
25. F. M. d'Heurle and P. Ho, in: *Thin Films: Interdiffusion and Reactions* (J. M. Poate, J. Mayer, and K. N. Tu, eds.), pp. 243–303, Wiley, New York, (1978).
26. F. M. d'Heurle and R. Rosenberg, in: *Physics of Thin Films*, Vol. 7, pp. 257–310, Academic Press, New York (1973).
27. A. Y. Cho, Recent developments in molecular beam epitaxy (MBE), *J. Vac. Sci. Technol.* **16**(2), 275 (March–April, 1979); M. B. Parish, Molecular beam epitaxy, *Science* **208**, 916 (May, 1980).
28. B. A. Joyce, Molecular beam epitaxy, *Rep. Prog. Phys.* **48**, 1637 (1985).
29. C. F. Powell, J. H. Oxley, and J. M. Blocher, Jr. (eds.), *Vapor Deposition*, Wiley, New York (1966).
30. C. E. Coates, *Organic Metallic Compounds*, Wiley, New York (1960).
31. J. B. Mooney and S. B. Radding, Spray pyrolysis processing, *Rev. Mater. Sci.* **12**, 81–101 (1982).
32. R. R. Chamberlain and J. S. Skarman, Chemical spray deposition process for inorganic films, *J. Electrochem. Soc.* **113**(1), 86–89 (January, 1966).
33. B. E. Deal and A. S. Grove, General relationship for the thermal oxidation of silicon, *J. Appl. Phys.* **36**(12), 3770–3778 (December, 1965).
34. J. Blanc, A revised model for the oxidation of Si by oxygen, *Appl. Phys. Lett.* **33**(5), 424 (September, 1978).
35. A. S. Grove, *Physics and Technology of Semiconductor Devices*, Wiley, New York (1967).
36. R. S. Ronen and P. H. Robinson, Hydrogen chloride and chlorine gathering. Effective technique for improving performance of silicon devices, *J. Electrochem. Soc.* **119**(6), 747–752 (1972).
37. C. M. Osburn, Dielectric breakdown properties of silicon dioxide films grown in halogen and hydrogen-containing environments, *J. Electrochem. Soc.* **121**(6), 809–815 (1974).

38. J. D. Meindl, K. C. Saraswat, R. W. Dutton, J. F. Gibbons, W. Tiller, J. D. Plummer, B. E. Deal, and T. I. Kamins, Computer aided engineering of semiconductor integrated circuits, Stanford University Integrated Circuit Laboratory, Report TR 4969-3, SEL-78-011, Stanford University, Stanford, Calif. (February, 1978).

39. M. Maeda, H. Kamioka, and M. Takagi, in: *Proceedings of the 13th Symposium on Semiconductors and IC Technology,* Tokyo (November, 1977).

40. B. E. Deal, The current understanding of charges in the thermally oxidized silicon structure, *J. Elecrochem. Soc.* **121**(6), 198C (June, 1974).

41. B. E. Deal, Standardized terminology for oxide charges associated with thermally oxidized silicon, *IEEE Trans. Electron Devices* **ED-27**(3), 606 (March, 1980).

42. G. Lucovsky and D. J. Chadi, in: *Physics of MOS Insulators*, pp. 301–305, Pergamon Press, New York (1980).

43. A. S. Grove and E. H. Snow, A model for radiation damage in metal-oxide semiconductor structures, *Proc. IEEE* **54**, 894 (June, 1966).

44. P. J. Jorgensen, Effect of an electric field on silicon oxidation, *J. Chem. Phys.* **37**(4), 874 (August, 1962).

45. N. Cabrera and N. F. Mott, Theory of the oxidation of metals, *Rep. Prog. Phys.* **12**, 163 (1948).

46. S. K. Ghandhi, *The Theory and Practice of Microelectronics*, Wiley, New York (1968).

47. B. I. Boltaks, *Diffusion in Semiconductors*, Academic Press, New York (1963).

48. A. R. Allnot, Statistical theories of atomic transport in crystalline solids, *Rep. Prog. Phys.* **50**, 373–472 (1987).

49. L. C. Kimerling and D. V. Lang, in: *Lattice Defects in Semiconductors* (J. E. Whitehouse, ed.), Institute of Physics, London (1974).

50. G. Carter and W. A. Grant, *Ion Implantation of Semiconductors*, Arnold, London (1976).

51. B. L. Crowder (ed.), *Ion Implantation in Semiconductors and Other Materials*, Plenum Press, New York (1972).

52. J. W. Mayer, L. Eriksson, and J. A. Davies, *Ion Implantation in Semiconductors, Silicon and Germanium*, Academic Press, New York (1970).

53. J. F. Gibbons, Ion implantation in semiconductors—Part I: Range distribution theory and experiments, *Proc. IEEE* **56**(3), 295–319 (March, 1968).

54. J. F. Gibbons, Ion implantation in semiconductors—Part II: Damage production and annealing, *Proc. IEEE* **60**(9), 1062 (September, 1972).

55. J. Lindhard, M. Scharff, and H. Schiott, Atomic collisions II. Range concepts and heavy ion ranges, *K. Dan. Vidensk. Selsk. Mat. Fys. Medd.* **33**(14), 1 (1963).

56. J. Sansbury, Applications of ion implantation in semiconductor processing, *Solid State Technol.* **19**(11), 31 (November, 1976).

57. J. Lindhard, Influence of crystal lattice on motion of energetic charged particles, *K. Dan. Vidensk. Selsk. Mat. Fys. Medd.* **34**(14) (1965).

58. P. Sigmund and J. B. Saunders, in: *Proceedings of the International Conference on Applications of Ion Beams to Semiconductor Technology*, Grenoble, France (P. Glotin, ed.), p. 215 Editions Ophrys (1967).

59. J. H. Crawford, Jr., and L. M. Slifkin (eds.), *Point Defects in Solids*, Vol. 2, Plenum Press, New York (1972).

60. H. S. Rupprecht, New advances in semiconductor implantation, *J. Vac. Sci. Technol.* **15**(5), 1669 (September–October, 1978).

61. J. Gibbons, W. S. Johnson, and S. Mylroie,*Projected Range Statistics in Semiconductors*, 2nd ed., Wiley, New York (1975); D. K. Brice, *Ion Implantation Range and Energy Deposition Distributions*, Plenum Press, New York (1975); K. B. Winterbon, *Ion Implantation Range and Energy Deposition Distributions*, Vol. 2, Plenum Press, New York (1975).

62. L. A. Christel, J. F. Gibbons, and S. Mylroie, An application of the Boltzmann transport equation to ion range and damage distributions in multi-layered targets, *J. Appl. Phys.* **51**(12), 6176 (December, 1980); D. H. Smith and J. F. Gibbons, *Ion Implantation in Semiconductors, 1976*, p. 333, Plenum Press, New York (1977); R. A. Moline, G. W. Reutlinger, and J. C. North, *Proceedings of the Fifth International Conference on Atomic Collisions in Solids*, Plenum Press, New York (1976).

63. Z. L. Liau and J. W. Mayer Limits of composition achievable by ion implantation, *J. Vac. Sci. Technol.* **15**(5), 1629 (September–October, 1978).

64. I. Krafcsik, L. Kiralyhidi, P. Riedl, J. Gyulai, and M. Fried, Implantation with ion pulses, *Phys. Status Solidi A* **94**, 855 (1986).
65. J. W. Cleland, K. Lark-Horovitz, and J. C. Pigg, Transmutation-produced germanium semiconductors, *Phys. Rev.* **78**, 814 (1950).
66. H. M. Janus, in: *Neutron Transmutation Doping in Semiconductors* (J. M. Meese, ed.), Plenum Press, New York (1978).
67. H. M. Janus and O. Malmros, Application of thermal neutron irradiation for large scale production of homogeneous phosphorus doping of float zone silicon, *IEEE Trans. Electron Devices* **ED-23**(8), 797 (August, 1976).
68. J. M. Meese, in: *Neutron Transmutation Doping in Semiconductors* (J. M. Meese, ed.), Plenum Press, New York (1978).
69. S. Prussin and J. W. Cleland, Application of neutron transmutation doping for production of homogeneous epitaxial layers, *J. Electrochem. Soc.* **125**(2), 350 (February, 1978).
70. M. I. Current, C. B. Yarling, and W. A. Keenan, in: *Ion Implantation Science and Technology*, 2nd ed. (J. F. Ziegler, ed.), p. 377, Academic Press, New York (1988); L. A. Larson, A comparison of wafer resistivity probing tools, presented at the 7th International Conference on Ion Implantation Technology, Kyoto, Japan (June, 1988).
71. Implant Evaluation, Editorial, *Semiconductor International* **81** (November, 1988).
72. A. Rosencwaig, in: SRI Report, Microelectronics—Applications, Materials and Technology, *Microscience* #6 (1985).
73. A. Rosencwaig, in: *VLSI Electronics: Microstructure Science*, Vol. 9 (N. G. Einspruch, ed.), p. 227, Academic Press, New York (1985).
74. A. Rosencwaig, in: SRI Report, Microelectronics—Photonics, Materials, Sensors, and Technology, *Microscience* #7 (1985).
75. T. Motooka and K. Watanabe, Damage profile determination of ion-implanted Si layers by ellipsometry, *J. Appl. Phys.* **51**(8), 4125 (August, 1980).
76. W. Chu, J. W. Mayer, and M. A. Nicolet, *Backscattering Spectrometry*, Academic Press, New York (1978).
77. Y. E. Strausser, in: *Proc. Fourth Natl. Vacuum Congress*, pp. 469–472 (1968).
78. R. E. Honig, Vapor pressure data for the solid and liquid elements, *RCA Rev.* **23**(4), 567–586 (1962).
79. G. M. McCracken, The behavior of surfaces under ion bombardment, *Rep. Prog. Phys.* **38**, 241–327 (1975).
80. G. K. Wehner, Controlled sputtering of metals by low-energy Hg ions, *Phys. Rev.* **102**(3), 670–704 (1956).
81. G. Carter and J. S. Colligon, *Ion Bombardment of Solids*, Elsevier, Amsterdam (1968).
82. L. B. Loeb, *Basic Processes of Gaseous Electronics*, University of California Press, Berkeley (1955).
83. H. S. Butler and G. S. Kino, Plasma sheath formation by radiofrequency fields, *Phys. Fluids* **6**(9), 1346 (September, 1963).
84. I. Brodie, C. T. Lamont, and D. O. Myers, Substrate bombardment during RF sputtering, *J. Vac. Sci. Technol.* **6**(1), 124 (1969).
85. T. A. Wade, Polyimides for use as VSCS multilevel interconnection, dielectric and passivation layer, *SRI Ser. Microsci. Microelectron.* **5**, 59–130 (1983).
86. W. J. Daughton and F. L. Givens, An investigation of the thickness-variation of spun-on on thin films commonly associated with the semiconductor industry, *J. Electrochem. Soc. Solid-State Sci. Technol.* **129**, 173–179 (January, 1985).
87. W. F. Flack, D. S. Soong, A. T. Bell, and D. W. Hess, A mathematical model for spin coating of polymers, *J. Appl. Phys.* **56**, 1199–1206 (1984).
88. L. V. Gregor, Polymer dielectric film, *IBM J. Res. Dev.* **12**(2), 140–162 (March, 1968).
89. R. M. Handy and L. C. Scala, Electrical and structural properties of Langmuir films, *J. Electrochem. Soc.* **113**, 105–115 (1966).
90. G. G. Roberts, An applied science perspective of Langmuir–Blodgett films, *Adv. Phys.* **34**, 475–512 (1985); R. H. Tredgold, The physics of Langmuir Blodgett films, *Rep. Prog. Phys.* **50**, 1609–1656 (1987).
91. H. E. Ries, Monomolecular films, *Sci. Am.* **204**, 152 (1961).
92. A. T. Bell, Fundamentals of plasma chemistry, *J. Vac. Sci. Technol.* **16**(2), 418–419 (March–April 1979).

93. H. Yasuda, *Plasma Polymerization*, Academic Press, New York (1985).
94. H. K. Yasuda, Competitive ablation and polymerization (CAP) mechanisms of glow discharge polymerization, *ACS Symp. Ser.* **108**, 37–52 (1979).
95. E. B. Priestley, P. J. Wojtowicz, and P. Sheng, *Introduction to Liquid Crystals*, Plenum Press, New York (1975).
96. F. J. Kahn, G. N. Taylor, and H. Schanhorn, Surface-Produced alignment of liquid crystals, *Proc. IEEE* **61**, 823 (1973).
97. G. J. Sprokel (ed.), *The Physics and Chemistry of Liquid Crystal Devices*, Plenum Press, New York (1980).
98. M. F. Schiekel and K. Fahrenschon, Deformation of nematic liquid crystals with vertical orientation in electrical fields, *Appl. Phys. Lett.* **19**, 391 (1971).
99. E. Kaneko, Liquid crystal matrix displays, *Advances in Image Pick Up and Displays* **4**, 1–86 (1981).
100. L. E. Tannas, Jr. (ed.), *Flat Panel Displays and CRTs*, Van Nostrand-Reinhold, Princeton, N.J. (1985).
101. D. B. Lee, Anisotropic etching of silicon, *J. Appl. Phys.* **40**, 4569–4574 (1969).
102. E. Bassons, Fabrication of novel three-dimensional microstructures by anisotropic etching of $\langle 100 \rangle$ and $\langle 110 \rangle$ silicon, *IEEE Trans. Electron Devices* **ED-25**(10), 1178–1185 (October, 1978).
103. K. E. Bean, Anisotropic etching of silicon, *IEEE Trans. Electron Devices* **ED-25**(10), 1185–1193 (October, 1978).
104. R. L. Bersin and R. F. Reichelderfer, The dryox process for etching silicon dioxide, *Solid State Technol.* **20**(41), 78–80 (April, 1977).
105. H. R. Kaufman, Technology of electron-bombardment ion thrusters, *Adv. Electron. Electron Phys.* **36**, 265–373 (1974).
106. R. G. Poulsen, Plasma etching in IC manufacture—A review, *J. Vac. Sci. Technol.* **14**(1), 266–274 (January–February, 1977).
107. R. Kumar, C. Tadas, and G. Hudson, Characterization of plasma etching for semiconductor applications, *Solid State Technol.* **19**(10), 54–59 (October, 1976).
108. J. W. Coburn and H. F. Winters, Plasma etching—A discussion of mechanisms, *J. Vac. Sci. Technol.* **16**,391 (March–April, 1979).
109. A. E. Bell, Review and analysis of laser annealing, *RCA Rec.* **40**, 295 (September, 1979).
110. R. T. Young, C. W. White, G. J. Clark, J. Narayan, W. H. Christie, M. Murakami, P. W. King, and S. D. Kramer, Laser annealing of boron-implanted silicon, *Appl. Phys. Lett.* **32**(3), 1139 (February, 1978).
111. G. K. Celler, J. M. Poate, and L. C. Kimerling, Spatially controlled crystal growth regrowth of ion-implanted silicon by laser irradiation, *Appl. Phys. Lett.* **32**(8), 464 (April, 1978).
112. P. Baeri, S. U. Campisano, G. Foti, and E. Rimini, Arsenic diffusion in silicon melted by high-power nanosecond laser pulsing, *Appl. Phys. Lett.* **33**(2), 137 (July, 1978).
113. J. C. Muller, A. Grob, J. T. Grob, R. Stuck, and P. Siffert, Laser-beam annealing of heavily damaged implanted layers on silicon, *Appl. Phys. Lett.* **33**(4), 287 (August, 1978).
114. A. Gat, J. F. Gibbons, T. J. Magee, J. Peng, P. Williams, V. Deline, and C. A. Evans, Jr., Use of a scanning cw Kr laser to obtain diffusion-free annealing of B-implanted silicon, *Appl. Phys. Lett.* **33**(5), 389 (September, 1978).
115. T. N. C. Venkatesan, J. A. Golovchenko, J. M. Poate, P. Cowen, and G. K. Celler, Dose dependence in the laser annealing of arsenic-implanted silicon, *Appl. Phys. Lett.* **33**(5), 429 (September, 1978).
116. C. W. White, W. H. Cristie, B. R. Appleton, S. R. Wilson, P. P. Pronko, and C. W. Magee, Redistribution of dopants in ion-implanted silicon by pulsed-laser annealing, *Appl. Phys. Lett.* **33**(7), 662 (October, 1978).
117. P. Baeri, S. U. Campisano, G. Foti, and E. Rimini, A melting model for pulsing-laser annealing of implanted semiconductors, *J. Appl. Phys.* **50**(2), 788 (February, 1979).
118. D. H. Auston, J. A. Golovchenko, and T. N. C. Venkatesan, Dual-wavelength laser annealing, *Appl. Phys. Lett* **34**(9), 558 (May, 1979).
119. S. S. Lau, J. W. Mayer, and W. F. Tseng, Comparison of laser and thermal annealing of implanted-amorphous silicon in laser–solid interactions and laser processing, *AIP Conf. Proc.* No. 50 (1979).
120. A. Gat and J. F. Gibbons, A laser-scanning apparatus for annealing of ion-implantation damage in semiconductors, *Appl. Phys. Lett.* **32**(3), 142 (February, 1978).
121. R. B. Fair, Modelling laser-induced diffusion of implanted arsenic in silicon, *J. Appl. Phys.* **50**(10), 6552 (October, 1979).

122. J. R. Lineback, How lasers will give chip making a big boost, *Electronics* **59**(1), 70–72 (1986).

123. L. L. Burns, Laser pantography, in: *Wafer Scale Integration* (G. Saucier and J. Trilhe, eds.), Elsevier/North-Holland, Amsterdam (1986).

124. M. W. Geis, D. C. Flanders, and H. I. Smith, Crystallographic orientation of silicon on an amorphous substrate using an artificial surface-relief grating and laser crystallization, *Appl. Phys. Lett.* **35**(1), 71–74 (July, 1979).

125. E. I. Givargizov, *Oriented Crystallization on Amorphous Substrates*, Plenum Press, New York (1991).

126. M. W. Geis, D. C. Flanders, H. I. Smith, and D. A. Antoniadis, Grapho-epitaxy of silicon on fused silica using surface micropatterns and laser crystallization, *J. Vac. Sci. Technol.* **16**(6), 1640 (November–December, 1979).

127. V. I. Klykov and N. N. Sheftal, Diataxial growth of silicon and germanium, *J. Cryst. Growth* **52**, 687 (1981). E. Kaldis (ed.), *Current Topics in Materials Science*, Vol. 10, pp. 1–53, North-Holland, Amsterdam (1982).

128. S. Rice and J. Jain, Reciprocity behavior of photoresists in excimer laser lithography, *IEEE Trans. Electron Devices* **EDO-5**(2), 32–35 (1984).

129. D. J. Ehrlich, J. Y. Tsao, D. J. Silversmith, J. H. C. Sedlacek, R. W. Mountain, and W. G. Graber, Laser micromachining techniques for reversible restructuring of gate-array prototype circuits, *IEEE Electron Device Lett.* **EDL-5**(2), 32–35 (1984).

130. D. J. Ehrlich and J. Y. Tsao, in: *VLSI Electronics: Microstructure Science*, Vol. 7, Academic Press, New York (1983).

131. Y. S. Liu, C. P. Yakymyshyn, H. R. Phillip, H. S. Cole, and L. M. Levinson, Laser-induced selective deposition of micron-sized structures on silicon, *J. Vac. Sci. Technol.* **B3**(5), 1441–1444 (1985).

132. L. Waller, Cut-and-patch lasers speed chip repairs, *Electronics* **59**(24), 19–20 (1986).

133. K. A. Jones, Laser assisted MOCVD growth, *Solid State Technol.* **28**(10), 151–156 (1985).

134. C. D. Rose, Laser writing takes a step forward, *Electronics* **59**(23), 15 (1986).

135. L. L. Burns and A. R. Elsea, in: *Wafer Scale Integration* (G. Saucier and J. Trilhe, eds.), Elsevier/North-Holland, Amsterdam (1986).

136. B. M. McWilliams, I. P. Herman, F. Mitlitsky, R. A. Hyde, and L. L. Wood, Wafer-scale laser pantography: Fabrication of n-metal-oxide semiconductor transistors and small-scale integrated circuits by direct-write laser-induced pyrolytic reactions, *Appl. Phys. Lett.* **43**(10), 946–948 (1983).

137. D. J. Ehrlich, Early applications of laser direct patterning: Direct writing and excimer projection, *Solid State Technol.* **28**(12), 81–85 (1985).

138. A. R. Kirkpatrick, J. A. Minnucci, and A. C. Greenwald, Silicon solar cells by high-speed low-temperature processing, *IEEE Trans. Electron Devices* **ED-24**(4), 429 (April, 1977).

139. R. G. Little and A. C. Greenwald, An advancement in semiconductor processing, *Semicond. Int.* **2**(1), 81 (January–February, 1979).

140. K. N. Ratnakumar, R. F. W. Pease, D. J. Bartelink, and N. M. Johnson, Scanning electron beam annealing with a modified SEM, *J. Vac. Sci. Technol.* **16**(6), 1843 (November–December, 1979).

141. A. Neukermans and W. Saperstein, Modeling of beam voltage effects in electron-beam annealing, *J. Vac. Sci. Technol.* **16**(6), 1847 (November–December, 1979).

142. L. A. Wasselle, Electron Beam Curing of Polymers, Process Economics Program, Report 116, Stanford Research Institute, Menlo Park, Calif. (1977).

143. C. K. Crawford, Electron beam machining, in: *Introduction to Electron Beam Technology* (R. Bakish, ed.), Wiley, New York (1962).

144. T. Ichihashi and S. Matsui, *In situ* observation on electron beam induced chemical vapor deposition by transmission electron microscopy, *J. Vac. Sci. Technol.* **B6**(6), 1869–1872 (November–December, 1988).

145. M. A. McCord, D. P. Kern, and T. H. P. Chang, Direct deposition of 10-nm metallic features with the scanning tunneling microscope, *J. Vac. Sci. Technol.* **B6**(6), 1877–1880 (November–December, 1988).

146. S. Matsui and K. Mori, New selective deposition technology by electron beam induced surface reaction, *J. Vac. Sci. Technol.* **B4**(1), 299 (January–February, 1986).

147. L. R. Harriott, in: *VLSI Electronics: Microstructure Science*, Vol. 21, Academic Press, New York (1989).

148. J. Melngailis, Focused ion beam technology and applications, *J. Vac. Sci. Technol.* **B5**(2), 469 (March–April, 1987).

149. A. J. Muray and J. J. Muray, Microfabrication with ion beams, *Vacuum* **35**(10–11), 467–477 (1985).
150. W. L. Brown, Recent progress in ion beam lithography, *Microelectron. Eng.* **9**, 269–276 (1989).
151. I. H. Wilson, The topography of ion bombarded surfaces, *Proc. Int. Eng. Congr. ISIAT and IPAT*, Kyoto, Japan (1983).
152. R. L. Kubena, R. L. Seliger, and E. H. Stevens, High resolution using a focused ion beam, *Thin Solid Films* **92**, 165–169 (1982).
153. H. Yamaguchi, A. Shimase, S. Haraichi, and T. Miyauchi, Characteristics of silicon removal by fine focused gallium ion beam, *J. Vac. Sci. Technol.* **B3**(1), 71–74 (January–February, 1985).
154. H. Morimoto, Y. Sasaki, Y. Watakabe, and T. Kato, Characteristics of submicron patterns fabricated by gallium focused-ion-beam, *J. Appl. Phys.* **57**(1), 159–160 (January, 1985).
155. A. Macrander, D. Barr, and A. Wagner, Resist possibilities and limitations in ion beam lithography, *SPIE Conf.* **33**, 142–151 (1982).
156. J. Melngailis, C. R. Musil, E. H. Stevens, M. Utlaut, E. M. Kellogg, R. T. Post, M. W. Geis, and R. W. Mountain, The focused ion beam as an integrated circuit restructuring tool, *J. Vac. Sci. Technol.* **B4**(1), 176–180 (January–February, 1986).
157. A. Wagner, Applications of focused ion beams, *Nuclear Instrum. Methods Phys. Res.* **218**, 355 (1983).
158. J. E. Jenson, Ion beam resists, *Solid State Technol.* **1984**, 145–150 (June, 1984).
159. M. Komuro, N. Atoda, and H. Kawakatsu, Ion beam exposure of resist materials, *J. Electrochem. Soc. Solid-State Sci. Technol.* **126**(3), 483–490 (March, 1979).
160. J. Melngailis, in: SRI Report, Microelectronics—Photonics, Materials, Sensors and Technology, Vol. 10.
161. J. S. Williams, Materials modification with ion beams, *Rep. Prog. Phys.* **49**, 491–587 (1986).
162. I. Yamada, I. Nagai, M. Horie, and T. Takagi, Preparation of doped amorphous silicon films by ionized cluster-beam deposition, *J. Appl. Phys.* **54**, 1583–1587 (1983).
163. J. A. Venables, G. D. T. Spiller, and M. Hanbruken, Nucleation and growth of thin films, *Rep. Prog. Phys.* **47**, 399–459 (1985).
164. J. P. Hirth and G. M. Pound, *Condensation and Evaporation: Nucleation and Growth Kinetics*, Macmillan Co., New York (1963).
165. D. Walton, T. N. Rhodin, and R. Rollins, Nucleation of silver on sodium chloride, *J. Chem. Phys.* **38**(11), 2698 (June, 1963).
166. G. Zinsmeister, in: *Basic Problems in Thin-Film Physics* (R. Neidermayer and H. Mayer, eds.), p. 33, Vandenhoeck & Ruprecht, Göttingen (1966).
167. H. J. Poppa, Heterogeneous nucleation of Bi and Ag on amorphous substrates, *J. Appl. Phys.* **38**(10), 3883 (September, 1967).
168. K. L. Chopra, Growth of thin metal films under applied electric field, *Appl. Phys. Lett.* **7**(5) 140 (September, 1965).
169. K. Nishiyama, M. Arai, and L. N. Watanabe, Radiation annealing of boron implanted silicon with a halogen lamp, *Jpn. J. Appl. Phys. Lett.* **19**(10), 256 (1980).
170. R. Singh, Rapid isothermal processing, *J. Appl. Phys.* **63**(8), R59–R114 (April, 1988).
171. J. Nulman and J. P. Krusius, Rapid thermal processing of thin gate dielectrics, nitridation of thermal oxides, *IEEE 1984 IDEM Technical Digest*, San Francisco (1984).
172. W. G. Pfann, Temperature gradient zone melting, *J. Met.* **1955**, 961–964 (September, 1955).
173. T. R. Anthony and H. E. Cline, Lamellar devices processed by thermomigration, *J. Appl. Phys.* **48**, 3943–3949 (1977).
174. M. J. Rand, Plasma-promoted deposition of thin inorganic films, *J. Vac. Sci. Technol.* **16**(2), 420–427 (March/April, 1979).

5

Pattern Generation

5.1. INTRODUCTION

While thin-film deposition enables one dimension of a device to be made with remarkable precision, down to, say, 10 Å, the other two dimensions are more difficult to form with the same degree of accuracy. Two-dimensional patterns are usually made on the surface by the process of lithography,[136] which derives from printing plate technology. The trend in microfabrication is toward increased circuit complexity and reduced pattern dimensions. Line widths of 0.5 μm and tolerances of 0.1 μm are now common for microcircuits. In order to meet these requirements, the need has arisen for pattern generation and lithography systems with superior performance specifications.

Patterns are formed by exposing a wafer coating with a thin film of resist material that is sensitive to a corresponding pattern of the appropriate radiation and developing the resultant latent image. Depending on the radiation source and the type of resist used, we characterize these techniques as optical, X-ray, electron beam, or ion beam. The ways in which these lithographies are utilized for mask making, image transfer, and direct writing are summarized in Fig. 5.1.

The optical lithography process usually starts with a computer-aided pattern design, followed by photoreduction of the physical pattern layout. The mask or "reticle" pattern (demagnified 2–10 times) made by photoreduction is used directly in the printing process. In the "step-and-repeat" reduction process, the sample (mask or wafer) is mechanically stepped between exposure sites to cover the entire sample so that a limited-field-diameter lens system can be used. Since many patterns may be made sequentially, the overlay accuracy of the patterns becomes important. Typically the overlay precision should be to at least one-fifth of the smallest linewidth in the patterns.

In scanning systems (electron, ion beam), the computer-stored pattern is directly converted to address and switch (blank) the beam, enabling the pattern to be exposed sequentially point by point over the wafer area.

In this chapter, reference will be made to the resolution, linewidth, edge acuity, accuracy, distortion, and precision of the pattern. The physical definition of resolution as the distance at which two points can be clearly recognized as distinct has

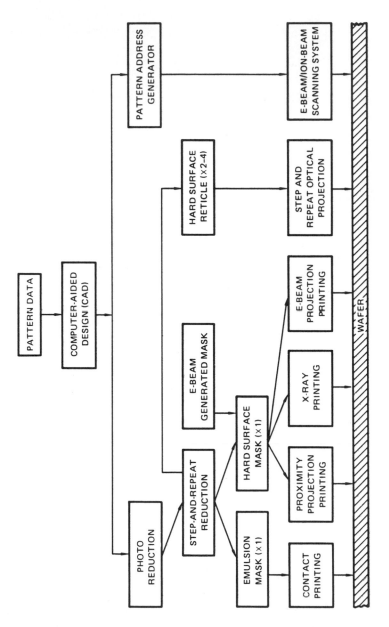

FIGURE 5.1. Use of lithography techniques for mask masking, image transfer, and direct writing.

little utility in lithography, but the engineering resolution measurement in terms of resolvable high-contrast line pairs unit distance is directly related. In lithography the term *resolution* is interchangeable with *linewidth* and indicates the smallest width of a line, or the smallest distance between two lines that is reproduced. *Edge acuity* refers to the sharpness of the lines attained. *Accuracy* is an indicator for spatial dimensions in the standard unit. *Distortion* is a measure of the relative dimensions of a pattern, and *precision* specifies how closely the patterns made by the same process match one another. Semiconductor lithography is treated comprehensively in Ref. 1.

5.2. OPTICAL LITHOGRAPHY

5.2.1. Contact Printing

An early method used for producing patterns on semiconductor wafers was contact printing. A mask transparency containing the circuit patterns is positioned on the top of the photoresist-covered wafer and exposed to light, creating exposed and unexposed areas. Development (selective etching) removes the resist according to its exposure state. Resolution is satisfactory for features as small as 2 μm, the limit being set by diffraction effects between adjacent lines (see Fig. 5.2). If partially

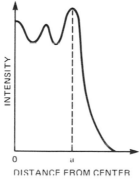

(a)   INTENSITY PROFILE

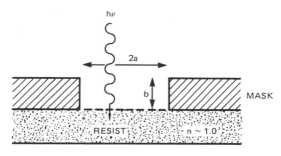

(b)   CROSS-SECTIONAL REPRESENTATION

FIGURE 5.2.   Contact printing geometry. (a) Intensity profile; (b) cross-sectional representation.

coherent light is used for illumination, we obtain increased contrast, increased ringing in the image due to coherent diffraction effects, and sharper edge gradients (intensity versus position). Uniformity of both the exposure and the photosensitive material can be maintained so that linewidth control is within acceptable tolerances.

Mask defects accumulated during successive mask uses are the main limitation of contact printing. The pressure during contact printing damages both the mask and substrate. The accrued damage and particles of resist adhering to the mask are printed on the following exposures, causing a rapid buildup of defects. Depending on the scale of integration and the durability of the mask surface, only a limit number of uses can be tolerated.

Emulsion masks are commonly used only for ten or less exposures for large silicon ICs. Chromium, iron oxide, or other hard-surface masks have become practical alternatives because they can be cleaned periodically and used for more exposures during their lifetimes. In practice, up to 100 uses may be possible with these hard-wearing materials.

Resolution in contact printing can be improved by reducing the wavelength of the exposing radiation,[2] thus decreasing the diffraction effects.

As chip areas grow larger and features grow smaller, the need for lower defect densities and higher dimensional integrity to maintain yields has encouraged the development of alternate exposure methods, particularly proximity and projection printing.

5.2.2. Proximity Printing

Spacing the mask away from the substrate minimizes contact and eliminates most of the defects that result from contact. However, diffraction of the transmitted

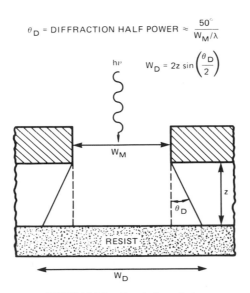

FIGURE 5.3. Proximity printing.

light, which increases with increasing spacing, causes a reduction in resolution and increases the distortion of individual photoresist features. The degree to which this occurs depends on the actual mask-to-wafer spacing, which can be variable across the wafer.

In large wafers, flatness variations, which are especially troublesome at small nominal spacings, and diffraction effects, which become troublesome at large spacings, limit the feature size. In practice, approximately 7-μm features are the smallest used in proximity exposure (see Fig. 5.3).

5.2.3. Projection Printing

Projection printing is a method[3] by which an image of the photomask is projected directly onto the photoresist-covered wafer by means of a high-resolution lens. In this system the mask life is potentially unlimited except for handling damage. As a consequence, the use of the highest-quality masks can be justified. The depth of focus must cover the ±10.0-μm flatness errors common in wafers after high-temperature processing. This limits the aperture of the lens, and therefore the resolution, to marginal values. It is also difficult to fabricate a lens system with both a uniform (diffraction-limited) image quality and a uniform light intensity over the full area of currently used wafers (10-cm diameter). The scattering of light associated with the use of glass optics has generally limited the choice of light-sensitive material to positive resists. Conventional projection systems have been used with success for features greater than 5 μm. However, as wafer diameters have increased, stationary projection systems have become less attractive.

In a scanning (or step-and-repeat) projection system, a portion of the mask is imaged onto the corresponding part of the wafer. A much smaller area (on the order of 1 cm^2) is exposed and the exposure is repeated by either scanning or stepping the image over the wafer.[3, 4]

Figure 5.4 shows the basic components of a projection alignment system. The major components are as follows:

- A light source
- A lens system
- A mask holder
- A wafer holder
- An alignment system

In conventional photolithography, light of the spectral region 3300–4000 Å is commonly adopted. Shorter wavelengths are not used because almost all of the known combinations of photoresists, intensive light sources, and illuminating optics cease to be effective at wavelengths shorter than 3300 Å. Useful light sources for projection systems are Hg-arc lamps in the spectral region 3300–4000 Å and Xe-Hg-arc lamps and deuterium spectral lamps for deep-UV lithography.

Modern projection printers employ optics that are essentially diffraction limited. This implies that the design and fabrication of the optical elements are such that their imaging characteristics are dominated by diffraction effects associated with the finite apertures in the condenser and projection optics, rather than by aberrations. Figure 5.5 shows the basic printing lens performance parameters.

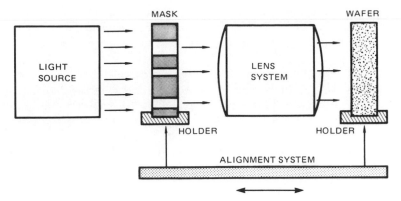

FIGURE 5.4.   Basic components of a projection system.

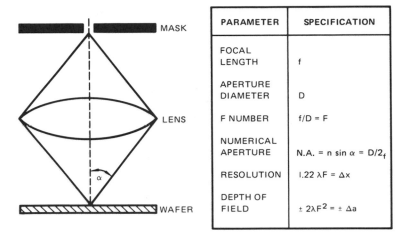

PARAMETER	SPECIFICATION
FOCAL LENGTH	f
APERTURE DIAMETER	D
F NUMBER	f/D = F
NUMERICAL APERTURE	N.A. = n sin α = D/2$_f$
RESOLUTION	1.22 λF = Δx
DEPTH OF FIELD	± 2λF^2 = ± Δa

FIGURE 5.5.   Basic printing lens performance parameters.

It is useful to recall the definitions of several parameters used to specify a multielement imaging system. The numerical aperture (NA) is given by $NA = n \sin \alpha$, where n is the refractive index (usually unity) in image space and 2α is the maximum cone angle of rays reaching an image point on the optical axis of the projection system. The effective F number of a projection system is given by $F = 1/(2NA)$, It can be shown that $F = (1 + M)f$, where f is the F number of the system with the object at infinity and M is the operating magnification.

When light from a very small point is imaged by a diffraction-limited lens, the image consists of rings of light surrounding a central bright spot, called the Airy disk. The width w at the half-intensity points of the Airy disk is given by $w = 0.5\lambda/NA = 1.0 \lambda F$ and is a rough measure of the smallest dimension that can be printed in a resist by a diffraction-limited projection system.

The defocusing aberration (or error) of

$$\Delta a = \pm n\lambda/2(NA)^2 = \pm 2n\lambda F^2 \qquad (5.1)$$

causes an optical path difference (OPD) of $\pm\lambda/4$ in the image plane. This OPD of $\pm\lambda/4$ does not seriously affect the image quality, since it causes a loss of only 20% of the light from the Airy disk, with hardly any change in its diameter. Δa is called the depth of focus. For an $F/3$ optical system, $w = 1.2\,\mu m$ and $\Delta a = \pm 7.3\,\mu m$. Although Δa is a useful guide, it is more satisfactory in photolithography to define the depth of focus of a projection system as the range over which the image plane can vary for a specified variation in image size.

The optical imaging quality of a projection printer can also be characterized in terms of the modulation transfer function (MTF) curve,[5] in which the modulation of the output image is plotted against the spatial frequency of a square-wave many-bar pattern having 100% modulation. In practice, the MTF curve is obtained by imaging sinusoidal gratings placed in the object plane (see Fig. 5.6). Each grating is characterized by a frequency (measured in line pairs per millimeter) and a modulation $M_0 = I_{max} - I_{min}/I_{max} + I_{min}$, where I_{max} and I_{min} are the local maximum and minimum light intensities emerging from the spaces and lines. The ratio I_{max}/I_{min} is called the contrast C. The corresponding modulation $M_i(v)$ in the image plane can be measured by scanning a very small photodetector across an image of a grating. The MTF at frequency v is given by $MTF(v) = M_i(v)/M_0$. For a diffraction-limited optical system, the MTF also can be calculated from[6]

$$MTF(v) = \frac{2}{\pi}\left\{\cos^{-1}\frac{v}{v_0} - \frac{v}{v_0}\left[1 - \left(\frac{v}{v_0}\right)^2\right]^{1/2}\right\} \qquad (5.2)$$

where v is the spatial frequency variable and v_0 is the optical cutoff frequency of the system, determined by its numerical aperture NA and the wavelength λ, and is given by $v_0 = 2NA/\lambda$.

Optical imaging systems for projection printers are classified as either coherent or incoherent, depending on the type of illumination employed. If a mask (or grating) is illuminated by a narrow-angle light beam originating from a point source, the imaging of the projection printer is coherent or partially coherent[5,7] (see Fig. 5.7a), since the light diffracted by the grating is coherent in amplitude at the image (wafer) plane. Information about the spatial frequency of the grating is contained only in the diffracted light beam. The direction of the first diffraction peak is given by the grating formula

$$n(a + b)\sin\theta = N\lambda \qquad (5.3)$$

and the spatial frequency v is given by

$$v = (a + b)^{-1} = \frac{n\sin\theta}{x} \qquad (5.4)$$

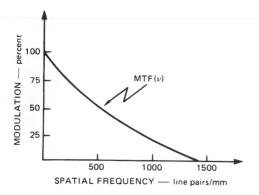

(a)   **MODULATION TRANSFER FUNCTION (MTF)**

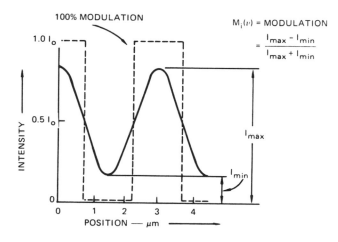

(b)   **CORRESPONDING MODULATION (M$_i$)**

FIGURE 5.6.   MTF(v) and the $M_i(v)$ for output light density for 333 line pairs/mm corresponding to 1.5-μm lines and 1.5-μm spaces between lines. (a) MTF; (b) corresponding modulation (M_i).

for $N = 1$. For imaging it is required that $\theta = \alpha$, where α is defined by the numerical aperture of the projection optics (NA $= n \sin \alpha$). Therefore, the highest grating frequency that can be imagined by a coherent illumination system is given by

$$v_{max} = \frac{n \sin \alpha}{\lambda} = \frac{NA}{\lambda} = \frac{1}{2\lambda F} = v_c \qquad (5.5)$$

where v_c is the cuttoff frequency. Figure 5.7b shows a projection system where the grating is illuminated by rays from all portions of an extended incoherent light source. In this case, each ray is diffracted by the grating and forms its own image in the wafer plane. Since there is no fixed phase relationship between the different rays

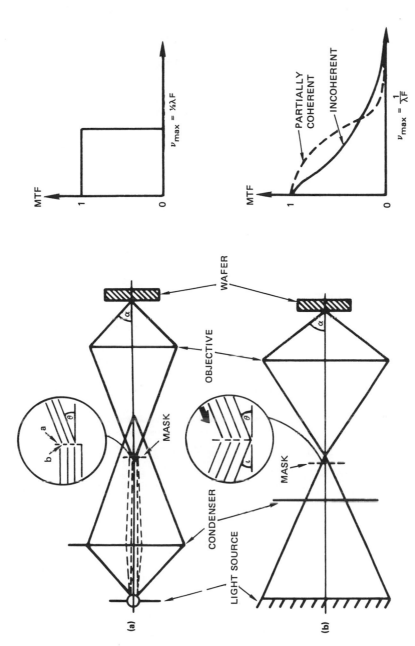

FIGURE 5.7. Coherent and incoherent projection printers. The inserts show the diffraction of coherent and incoherent light by a grating pattern. The modulation transfer functions are also shown for coherent and incoherent printers.

from the extended source, the various images at the wafer plane add in an incoherent manner.

To illuminate rays incident at an angle in the direction of the first diffraction maxima is given by

$$n(a + b)(\sin i + \sin \theta) = \lambda \tag{5.6}$$

For image formation it is required that $i = \theta = \alpha$. Therefore,

$$v_{max} = (a + b)^{-1} = \frac{2NA}{\lambda} = \frac{1}{\lambda F} = v_c \tag{5.7}$$

Hence, $v_{max}(\text{incoherent}) = 2v_{max}(\text{coherent})$.

MTF curves for coherent, partially coherent, and incoherent illuminations are shown also in Fig. 5.7.

The MTF for an incoherent source is at $v = 0$ and decreases monotonically to zero at the cutoff frequency $v_c = 1/\lambda F$. In contrast, the MTF for a coherent source is unity out to its cutoff frequency $v_c = 1/2\lambda F$, where it drops to zero. Mathematically, the incoherent MTF is the weighting function for the intensity components of the image spectrum. If the object spectrum and MTF are known, an incoherent image can be reconstructed. Coherent systems are more difficult to interpret, being sensitive to the phas of the image spectrum components as well as to their amplitude. Since the MTF does not containphase information, an image cannot be reconstructed from its MTF.

It is generally accepted that a 60% (MTF) is required for a minimum working feature in a positive resist.

If the incoherent MTF(v) is known from measurements or from calculation, the intensity distribution of the image (line pattern) can be calculated from the intensity distribution of the object (mask) by the use of the following transformation sequence.

The object spatial frequency distribution $I_v(v)$ is given by the Fourier transform F_T of the intensity distribution of the object, $I_0(x)$, as in

$$I_0(v) = F_T I_0(x) \tag{5.8}$$

In practice, a discrete Fourier transform is used with an implied periodicity. The image spatial frequency distribution $I_i(v)$ is given by

$$I_i(v) = MTF(v)[I_0(v)] \tag{5.9}$$

and the intensity description, $Ii(x)$, is given by the inverse Fourier transform of the image spatial frequency distribution,

$$I_i(x) = F_T^{-1}[I_i(v)] \tag{5.10}$$

Figure 5.8 shows the calculated intensity distribution for the image of a nominal 1-μm line object produced by a NA 0.45 lens with a 435.7-nm wavelength.[5]

The incident intensity distribution (in W/cm^2) multiplied by the exposure time (s) give the incident energy density distribution (J/cm^2) or dose across the surface of the resist film.

Most projection printers are designed so that the source only partially fills the objective. The imaging behavior of the system is then described as partially coherent.

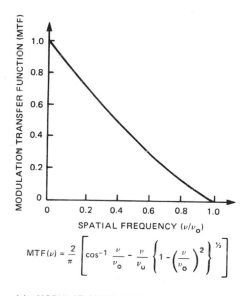

$$MTF(v) = \frac{2}{\pi}\left[\cos^{-1}\frac{v}{v_o} - \frac{v}{v_u}\left\{1 - \left(\frac{v}{v_o}\right)^2\right\}^{\frac{1}{2}}\right]$$

(a) MODULATION TRANSFER FUNCTION (MTF)

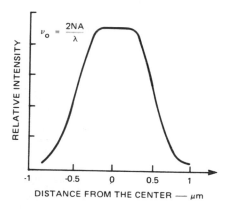

$$v_o = \frac{2NA}{\lambda}$$

(b) INTENSITY DISTRIBUTION

FIGURE 5.8. Calculated intensity distribution for the image of a normal 1-μm line object produced by a NA 0.45 lens with a 435.7-nm wavelength. (a) MTF; (b) intensity distribution. From Ref. 5.

Contact and near-contact printers make use of highly collimated coherent illumination. If incoherent illumination were employed, the image would be blurred by its penumbra in the wafer plane. In practice, the illumination system is decollimated by a few degrees in order to blur out the otherwise intense diffraction effects.

5.3. THE PHYSICS OF PHOTORESISTS

5.3.1. Positive Photoresist Exposure

Optical lithography, as applied in microelectronics, uses photoresists that can be applied as thin-film coatings on surfaces (wafers) where a pattern is to be delineated. The photoresist film is then exposed to an optical pattern, using blue or ultraviolet light, thus creating exposed and unexposed areas. Development selectively removes the resist according to its exposure state. The remaining pattern of photoresist on the surface is used to delineate etching, plating, sputtering, evaporation, or other processes commonly used in microelectronics.

Optical lithography is similar to conventional photography[8] in that a light-sensitive thin film is chemically altered by light so that an image can be produced by a subsequent development process. However, both the chemical processes and the resulting images are different. Photography involves multiphoton absorption by silver compounds, whereas optical lithography for positive photoresists involves the disruption of an organic compound by single-photon adsorption. The photographic image is one of optical density; the lithographic one is one of surface profile. Development in photography involves converting light-activated silver halide crystals to silver; the development of positive photoresists is an etching process. The differences, as shown in Table 5.1, are sufficient so that a different mathematical basis must be built for understanding optical lithography.

Photoresists are available for both positive and negative processes, i.e., development causes removal where the photoresist was exposed or unexposed. First we will discuss a quantitative basis for the lithographic process as applied to positive photoresists. Significant differences between the behavior of positive and negative resists make the mathematical treatment of exposure and development for one inapplicable to the other.

Positive photoresists are typically three-component materials consisting of a base resin, which gives the resist its film-making properties, a photoactive compound,

TABLE 5.1.   Differences between Photography and Optical Lithography

Process of characteristic	Photography	Optical lithography
Exposure	Multiphoton absorption	Single-photon absorption (breaking bonds for positive resists; polymerizing for negative resists)
Image	Optical density	Surface profile
Development	Plating silver crystals on latent image	Etching process

and volatile solvents to make the material liquid for application. In a dried film (typically 0.3–2 μm thick), the photoactive compound serves to inhibit dissolution of the photoresist in an alkaline aqueous developer solution. We refer to this compound as an inhibitor to emphasize this role. Destruction of the inhibitor by light creates by-product compounds that allow dissolution of the resist, resulting in an increased rate of removal of the resist by the developer.[9]

Lithography can be better understood it we consider the exposure and development processes separately. Exposure modifies the photoresist chemically by destruction of the inhibitor by light in a suitable wavelength range. The effect of chemical modification is highly localized around the point where the photon is absorbed, and the exposure level can vary widely within the resist film. Development is an etching process that selectively removes the photoresist at a rate related to the amount of inhibitor destroyed. The link between these two processes is the inhibitor distribution after exposure.

5.3.2. Resist Characterization

The photoresist film, after application to the substrate and drying, must be a uniform isotropic medium. It must have a uniform thickness (on both micro and macro scales) over the region used for characterization. It should be chemically isotropic so that its response to exposure and development is uniform throughout the film. Particularly, radial (starburst) thickness variations often associated with the application of photoresist by spinning must be avoided.

Since optical absorption measurements are an important part of resist characterization, several qualifications must be met. Optical scattering must be small so that the material can be described optically by its index of refraction and an absorption coefficient α.

The complex index of refraction for the photoresist can be written[6]

$$\bar{n} = n - ik \tag{5.11}$$

where n is the real part of the index ($n = 1.68$ for AZ1350J photoresist at a 435.7-nm wavelength) and k is the extinction coefficient:

$$k = \frac{\alpha\lambda}{4\pi} \qquad k = 0.02 \text{ for AZ1350J at 404.7 nm} \tag{5.12}$$

so that initially the incoming light is attenuated in the resist by a factor

$$\exp\left(-\frac{4\pi kx}{\lambda}\right) = \exp(-0.58x) = 0.56 \tag{5.13}$$

per micron of path length.

After exposure, the inhibitor concentration is much reduced and the attenuation factor is typically $\sim 0.10/\mu$m. For much of the exposure, therefore, the resist will be transparent.

The inhibitor is the principal contributor to the absorption and it is usually assumed that the absorption spectra and spectral sensitivity curves for positive resists are equivalent.

5.3.3. Exposure

The inhibitor compound typically represents about 30% of a dried photoresist film. The strong absorption associated with its photosensitivity contributes significantly to the optical absorption of the material at exposing wavelengths. As the inhibitor is destroyed, this absorption is removed as well. This can be described in terms of a relative inhibitor term $M(z, t)$, which is the fraction of the inhibitor remaining (at any position z and exposure time t) as compared to the inhibitor concentration before exposure. There is little scattering in most photoresist films, so that the absorption constant α is given by

$$\alpha = AM(z, t) + B \tag{5.14}$$

where A and B are measurable material parameters that describe the exposure-dependent and independent absorptions.

The rate of destruction of the inhibitor is dependent on the local optical intensity $I(z, t)$, the local inhibitor concentration, and a measurable optical sensitivity term C, as given by

$$\frac{\partial M}{\partial t} = I(z, t)M(z, t)C \tag{5.15}$$

A, B, and C depend upon the photoresist material and exposure wavelength. For Shipley Chemical Company's AZ1350J photoresist, one of the commonly used high-resolution resist materials, $A = 0.86\ \mu m^{-1}$, $B = 0.07\ \mu m^{-1}$, and $C = 0.018\ cm^2/mJ$ for a 404.7-nm exposure wavelength. The light intensity in an infinitely thick resist, or in a film on a matched substrate, is described by

$$\frac{\partial I(z, t)}{\partial z} = I(z, t)[AM(z, t) + B] \tag{5.16}$$

Figure 5.9 shows the normalized inhibitor concentration as a function of the distance from the resist–air interface for different total exposure energies.[10] In the matched environment the inhibitor distribution does not depend upon resist thickness, making these calculations useful for films up to 2 μm. Note that even for this relatively simple exposure environment, inhibitor concentration is never uniform once exposure has started.

Generally, Eqs. (5.15) and (5.16) can be solved by straightforward numerical integration techniques for $M(z, t)$ and $I(z, t)$ once A, B, C, and I_0 are specified. The experimental techniques for measuring the parameters (A, B, C) used in this image-forming model are described in the literature.[10]

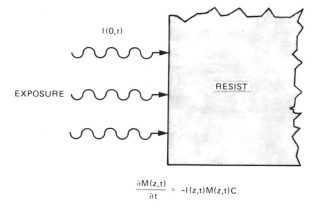

$$\frac{\partial M(z,t)}{\partial t} = -I(z,t)M(z,t)C$$

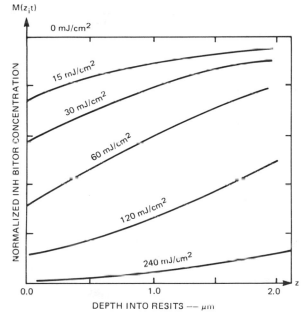

FIGURE 5.9. Inhibitor distribution in a "thick" resist. From Ref. 10.

5.3.4. Interference Effects

Optical interference can cause close-spaced intensity variations in the photoresist material when the index of refraction of the resist is not optically matched to the index of the substrate. The optical intensity of the exposing light on bare silicon has an intensity profile as shown in Fig. 5.10a.[6, 8] The absorption causes a decrease in the average intensity from the surface to the substrate. However, this intensity pattern is modulated by the interference of the standing waves reflected from the resist–substrate interface. The inhibitor concentration also shows a modulation effect due to the intensity changes in the resist film (Fig. 5.10b). These interference effects can be analyzed with some simplifying assumptions to illustrate the main points.[7]

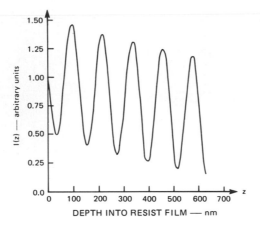

(a) PLOT OF INTENSITY OF EXPOSING LIGHT WITHIN A 630-nm AZ1350J
PHOTORESIST FILM ON BARE SILICON AT THE BEGINNING OF EXPOSURE
AT A 404.7-nm WAVELENGTH

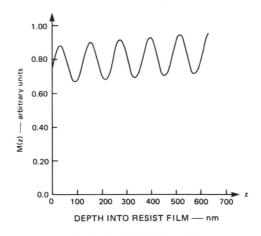

(b) INHIBITOR CONCENTRATION AS A FUNCTION OF DEPTH INTO AN
AZ1350J PHOTORESIST FILM ON BARE SILICON EXPOSED WITH
15.7 mJ/cm² AT A WAVELENGTH OF 404.7 nm

FIGURE 5.10. Interference effects of the standing waves reflected from the resist–substrate interface. (a) Plot of intensity of exposing light within a 630-nm AZ1350J photoresist film on bare silicon at the beginning of exposure at a 404.7-nm wavelength. (b) Inhibitor concentration as a function of depth into an AZ1350J photoresist film on bare silicon exposed with 15.7 mJ/cm² at a wavelength of 404.7 nm. From Ref. 1.

The incident monochromatic light is imaged normally into a weakly absorbing film of thickness d, coating a perfect reflector (substrate, $r_s = 1$) located in the plane $z = d$, as shown in Fig. 5.11a. If incident light wave 1 has unit amplitude, then light wave 2 in the film can be represented by

$$E_2(z) = E_2 \sin(\omega t - kz + \phi)\tag{5.17}$$

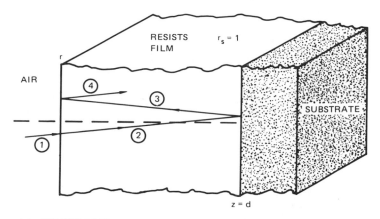

(a) **MULTIPLE REFLECTION AT THE RESIST–SUBSTRATE INTERFACE**

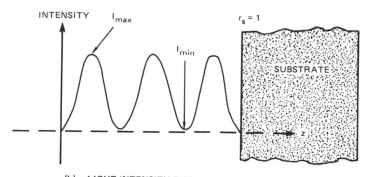

(b) **LIGHT INTENSITY DISTRIBUTION IN THE RESISTS**

FIGURE 5.11. Standing-wave effects of incident monochromatic light. (a) Multiple reflection at the resist–substrate interface; (b) light intensity distribution in the resists.

where $E_2 = (1 - r^2)^{1/2}$ and adsorption effects are neglected. Here $r = (n - 1)/(n + 2)$ is the reflection coefficient at the air–film interface, $k = 2\pi n/\lambda$, and n is the real part of the film dielectric constant, $\bar{n} = kn - ik$. The amplitude of reflected wave 3 is

$$E_3(z) = E_2 \sin[\omega t - k(2d - z) + \phi + \pi] \tag{5.18}$$

A phase change of π is assumed during reflection. Adding E_2 and E_3 gives a standing wave $E_{23}(z)$:

$$E_{23}(z) = 2E_2 \sin[k(d - z)] \cos(\omega t - kd + \phi) \tag{5.19}$$

The locations of the maxima (antinodes) and minima (nodes) are clearly independent of the arbitrary phase $\omega t + \phi$ of the incoming wave. The envelope function $(I \sim E^2)$

for the intensity of the standing wave is

$$I_{23}(z) = 4I_2 \sin^2[k(d-z)] \tag{5.20}$$

Measured from the substrate, the locations of the intensity extremes are given by the following conditions, were $N_1 = 0, 1, 2, \ldots$

$$n(d-z) = \lambda/4, 3\lambda/4, \ldots, (2N_1 + 1)\lambda/4 \tag{5.21}$$

for antinodes (maxima) and

$$n(d-z) = \lambda/2, \lambda, \ldots, N_1\lambda/2 \tag{5.22}$$

for nodes (minima). Taking into account additional waves (4, 5, . . .) does not affect the location of the maxima and minima but changes the amplitude of the standing wave.

Using "nonideal" substrates (Si, Al, Pt, and Au), we cannot assume total reflection ($r_s = 1$) at the film–substrate interface. Also, the phase change is less than π. These cause a node shift and the appearance of finite light intensity at the minima. It can be shown[7] that the locations of the intensity nodes relative to the substrate and the maxima/minima intensity conditions are unaffected by absorption in the resist. However, the modulation of the standing waves is, in general, dependent on resist absorption.

To calculate the absolute intensity distribution in the resist material, the source (lamp) and the reflective characteristics of the optical elements have to be taken into account. In optical lithography, the light source is usually a high-pressure mercury-arc lamp, which emits in the range 200–600 nm. For printers using reflective optics, the light spectrum at the wafer is essentially the source spectrum, $S(\lambda)$, times the transmission characteristic of the printer, $F(\lambda)$. That is, the light spectrum reaching the photoresisted wafer is $S(\lambda)F(\lambda)$. If $R(\lambda)$ is the spectral sensitivity of the resist, the effective light intensity at the resist surface in the wavelength range $\lambda \pm \delta\lambda/2$ is proportional to $S(\lambda) F(\lambda) R(\lambda)\delta\lambda$. For a given wavelength, standing waves in the resist films cause light intensity variations given by Eq. (5.20). Therefore, the resulting light distribution in the resist due to several wavelengths (λ_i) is proportional to

$$I(z) \approx \Sigma_i S(\lambda_i)F(\lambda_i)R(\lambda_i)\delta\lambda_i \sin^2[k_i(d-z)] \tag{5.23}$$

Since the refractive indices of common resists ($n_r \sim 1.6$) are very close to those of SiO_2 and Al_2O_3, this simplified model is also useful for resist–SiO_2/Si systems.

An important parameter used to describe the interaction of radiation $I(z)$ with the resist is the depth–dose function. This relates the rate of energy dissipation (energy loss dE/dz in the electron-beam, ion-beam resist) to the penetration distance z into the resist film.

The dose at penetration $z, D(z)$, is the photon or electron intensity at z, $I(z)$, multiplied by the exposure time τ, i.e.,

$$D(z) = I(z)\tau \tag{5.24}$$

The intensity at z for the photon case is $I(z) = I_0 F(z)$, where I_0 is the incident intensity and $F(z)$ is the depth–dose function. Hence,

$$D(z) = I_0 F(z)\tau = D_0 F(z) \tag{5.25}$$

where D_0 is the incident dose at $z = 0$. $F(z)$ varies as a function of z owing to light interference in the resist film.

The adsorbed energy per unit volume is responsible for the chemical reations (polymerization, bond breaking) in the resist. If dE/dt is the light intensity in photons $\text{cm}^{-2}\,\text{s}^{-1}$ absorbed by a slab of resist of thickness dz at a depth z, then the rate of energy absorption is

$$\left(\frac{dE}{dt}\right)_z = h\nu\left(\frac{dI}{dz}\right)_z = \alpha h\nu I(z) = \alpha h\nu I_0 F(z)\ \text{joules} \cdot \text{m}^3/\text{s} \tag{5.26}$$

where $h\nu$ is the photon energy and α the linear absorption coefficient. The energy absorbed in time τ is obtained by integration. For chemical reactions at penetration z, the energy that must be absorbed per unit volume is

$$E_\text{a}(z) = \alpha h\nu D_0 F(z)\ \text{joules} \cdot \text{m}^{-3} \tag{5.27}$$

requiring an incident dose D_0. The depth–dose function in photolithography is the result of the interference in the resist film. In electron-beam lithography, the depth–dose function $\Lambda(z)$ is determined by scattering and energy loss. The linear absorption coefficient (corresponding to α for light) is R_G^{-1} for the case of electron-beam lithography, where R_G is the Grun range. The photon energy $h\nu$ is replaced by kinetic energy of the electron, $E(z)$, at penetration z.

5.3.5. Resist Development

The development of a positive photoresist is a surface rate-limited etching reaction. The parameters that control this rate are the resist and developer chemistry and the inhibitor concentration of the resist at the surface exposed to the developer. An experimentally determined curve relating the development rate R to the inhibitor

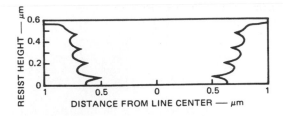

(a)   **EDGE PROFILE**

PARAMETER	SPECIFICATION
SUBSTRATE	Si
SiO$_2$	60 nm
PHOTORESIST	AZ1350J
	Lot 30000J
THICKNESS	583.6 nm
EXPOSURE CONSTANTS	A = 0.54/μm
	B = 0.03/μm
	C = 0.014 cm^2 /mJ
	n = 1.68
EXPOSURE WAVELENGTH	435.8 nm
EXPOSURE ENERGY	57 mJ/cm^2
DEVELOPMENT	1:1 Az DEVELOPER: H$_2$O
	22°C
DEVELOPMENT RATE	E$_1$ = 5.27
	E$_2$ = 8.19
	E$_3$ = – 12.5

(b)   **EXPOSURE PARAMETERS**

FIGURE 5.12.   Positive photoresist development profile. Example is for an edge profile for a normal 1-μm line in AZ1350 photoresist developed for 85 s in 1 : 1 AS developer–water. (a) Edge profile; (b) exposure parameters. From Ref. 10.

concentration M provides the link between exposure and development. From experimental data[10] the $R(M)$ relationship can be written in the form

$$R(M) = \exp(E_1 + E_2 M + E_3 M^2) \tag{5.28}$$

The values of E_1, E_2, and E_3 are shown[10] with the development profile in Fig. 5.12.

5.3.6. Negative Photoresists

A negative photoresist can be modeled as a photosensitive material with an effective threshold energy E_T.[11] If the energy E incident on the resist is less than E_T, the resist will be washed away in the developing cycle, but if $E > E_T$, the resist will be insoluble in the developer and the resulting image will be an effective etching or plating mask. This definition of E_T embraces a host of resist processing variables. For example, E_T will increase as the resist gets thicker and will decrease if the substrate has a greater than "normal" reflectance. It obviously also depends on the resist formation. E_T can be found approximately from the characteristic curve of the resist (developed thickness as a function of exposure energy).

The size of the image defined in the photoresist is determined by combining the effective threshold exposure energy of the photoresist with the energy distribution in the diffraction pattern from a mask edge, as shown in Fig. 5.13. In the vicinity of the mask edge, the intensity incident on the photoresist can be approximated by

$$I = kI_0 \exp(-mV) \tag{5.29}$$

or in terms of energy,

$$E = kE_i \exp(-mV) \tag{5.30}$$

where E_i is the energy incident on the mask, E is the energy incident on the resist, V is the dimensionless parameter defined in Fig. 5.13, k is the value of I/I_0 at $V = 0$, and m is the slope of I/I_0 near the mask edge. The constants k and m can depend on the optical system in the exposure tool and/or on the pattern that is printed. The effective edge of the image in the photoresist will occur at a value at V for which the energy incident on the resist is equal to the resist's effective exposure threshold, $E = E_T$. This value of V is found to be

$$V = \frac{1}{m} \ln\left(\frac{E_1}{E_T}\right) + \ln(k) \tag{5.31}$$

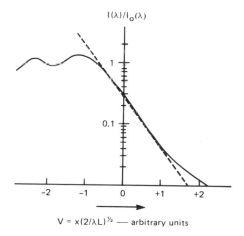

$I(\lambda)/I_0(\lambda)$

$V = x(2/\lambda L)^{1/2}$ — arbitrary units

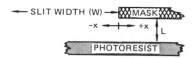

SLIT WIDTH (W) — MASK

$-x$ $+x$ L

PHOTORESIST

FIGURE 5.13. Fresnel diffraction pattern at a straight edge. From Ref. 13.

In Fig. 5.13, V is defined by

$$V = x\left(\frac{2}{\lambda L}\right)^{1/2} \tag{5.32}$$

where x is the distance from the mask edge, λ is the wavelength of the incident light, and L is the mask-to-wafer separation. The effective resist edge occurs a distance x from the mask edge and δ, the difference between the size of a clear mask feature and the size of it image in the photoresist, is $2x$ or

$$\delta = 2x = \frac{2}{m}\left(\frac{\lambda}{2}\right)^{1/2} L^{1/2}[\ln E_i - (\ln E_T - \ln k)] \tag{5.33}$$

for $\delta > 0$ the resist image of a mask slit is larger than the mask slit, and smaller if $\delta < 0$. Note that $\delta = 0$ when $E_i = E_T/k = E_0$.

5.4. PROJECTION SYSTEMS

The most widely used optical projection system[3] is shown in Fig. 5.14a. This system operates at unity magnification and utilizes a mirror-lens system. Only a crescent of the mask is illuminated at one instant, and this portion of the pattern is transferred to the sample with a parity that allows the mask and wafer to be scanned in the same direction and therefore to be mounted on a single holder. The entire wafer is exposed in a single scan. The magnification of the mirror optical system is $1 \times$, so the mask has the same dimensions as the pattern. The resolution is adequate to reproduce 3-μm linewidths or comparable to conventional UV proximity printing. Alignment accuracy is ± 1 μm and the depth of focus is ± 5 μm for 2-μm lines and ± 12.5 μm for 3-μm lines.

A broad spectrum of radiation from the mercury arc is used to expose the resist. This produces an exposure time of under 1 min, with a minimum of 6 s. The time for manual alignment must also be considered when estimating throughput. A full description of the system is given in Ref. 6.

Step-and-repeat reduction–projection systems similar in principle to that shown in Fig. 5.14b are designed to produce master masks for contact printing and to write directly on the wafer.[4] Demagnifications between $4 \times$ and $10 \times$ are used between the mask and the wafer. The sample is mechanically stepped between exposure sites to cover the entire sample. The depth of focus for the case of a 1-μm linewidth is very small (about ± 1 μm), which makes refocusing for every chip essential. Accurate sample leveling is also required. Current step-and-repeat systems have used two forms of alignment. In the first case an initial mask-to-wafer alignment is made and an accurate laser interferometric system is used to keep track of the sample during the stepping process. Systems of this type have relatively high throughput because only one alignment is required per water.

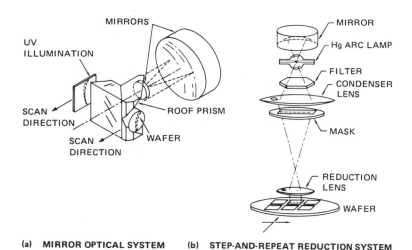

(a) **MIRROR OPTICAL SYSTEM**   (b) **STEP-AND-REPEAT REDUCTION SYSTEM**

PARAMETER	SPECIFICATION
PROJECTION OPTICAL SYSTEM	ALL REFLECTING 1:1, f/1.5
IMAGE FIELD	3-INCH DIAMETER
RESOLUTION	2.0 μm WIDE LINES AND SPACES IN THE PHOTORESIST
DEPTH OF FOCUS	± 5.5 μm FOR 2.0-μm RESOLUTION
	± 12.5 μm FOR 3.0-μm RESOLUTION
UNIFORMITY OF ILLUMINATION	± 10.0 percent
DISTORTION PLUS 1:1	
MAGNIFICATION TOLERANCE	± 1.0 μm
PHOTORESIST CAPABILITY	NEGATIVE AND POSITIVE
EXPOSURE TIME	FASTEST EXPOSURE: 6 SECONDS
MASK CAPABILITY	CHROME, EMULSION, AND IRON OXIDE
ALIGNMENT ACCURACY	± 1.0 μm
VIEWING	SPLIT-FIELD BINOCULAR MICROSCOPE; MULTICOLORED DIRECT WAFER VIEWING

FIGURE 5.14. Common projection systems. (a) Mirror optical system; (b) step-and-repeat reduction system. From Ref. 3.

The new step-and-repeat system[4,12] exposes fifty 10-cm wafers per hour (0.3 s exposure, 0.3 s table stepping), with a minimum linewidth below 2 μm and an alignment accuracy better than ±0.5 μm. In these printers the mask-to-wafer alignment is performed at each chip site. An automated system that aligns at every chip can also compensate for any wafer distortion error that might occur due to hot processing and/or wafer holding. An overall accuracy of 0.25 μm may be achievable and, in principle, the resolution of a reduction–projection system should be adequate to fabricate linewidths approaching 1 μm.

5.5. HOLOGRAPHIC LITHOGRAPHY

In holographic lithography[13] a substrate is exposed by placing it in a region where two beams from a laser interfere to produce a standing wave. This technique is useful primarily for exposing periodic and quasi-periodic patterns. Grating periods approaching 50% of the laser wavelength can be exposed. Refractive index-matching techniques can be employed to shorten even more the effective period.[14] To calculate the spacing of the interference fringes produced in the photoresist film, denote the complex amplitudes of the waves (see Fig. 5.15) as

$$E_1 = A \exp[ik(x \sin \theta - z \cos \theta)]$$
$$E_2 = A \exp[ik(-x \sin \theta - z \cos \theta) - i\phi]$$
(5.34)

respectively; the intensity on the surface is

$$|E_1 + E_2|^2_{z=0} = |A|^2|2 + 2\cos[2(kx \sin \theta - \phi)]|$$
(5.35)

Hence, the intensity is modulated in the x direction with a period

$$\Lambda = \frac{\lambda}{2n \sin \theta}$$
(5.36)

where n is the index of refraction of the resist ($n = 1.6$) and λ is the vacuum wavelength of the laser beams.

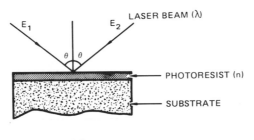

(a) EXPOSURE

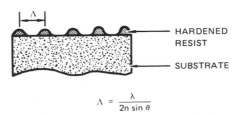

$$\Lambda = \frac{\lambda}{2n \sin \theta}$$

(b) PATTERN AFTER RESIST DEVELOPMENT

FIGURE 5.15. Holographic lithography setup for grating fabrication. (a) Exposure; (b) pattern after resist development. From Ref. 14.

Using visible light, gratings with up to 10^3 lines/mm have been produced.[15]

If two branches of a sufficiently coherent segment of a synchrotron-generated X-ray beam are used to produce the interference pattern, gratings with up to about 1.5×10^5 lines/mm could be manufactured.[16]

Such a grating could serve as a diffractive or focusing element in X-ray imaging. In addition, it may be utilized as a circuit element in nanoelectronic devices in the range 10–100 Å or as building blocks to produce variable two-dimensional model "crystals."

5.6. X-RAY LITHOGRAPHY

5.6.1. The X-ray Lithography System

X-ray lithography[17] uses contact/proximity printing with low-energy (1–10 keV) X rays rather than optical radiation. By using X rays, the diffraction problems common to photolithography are avoided, as well as is the backscattering problem encountered in electron-beam lithography. Figure 5.16 shows the basic principle of the X-ray technique.

Because of the X ray's short wavelength ($\lambda = 4$–50 Å), there are no simple mirrors or lenses that can be used to collimate the rays as in optical, UV, or electron-beam systems. Thus, an X-ray source of finite size has to be far enough away from the mask and resist for the X ray to appear to arrive with small divergence. The source size and beam divergence cause penumbral and geometric distortions, respectively. These distortions are illustrated in Fig. 5.16. Penumbral distortion blurs the definition of lines in the resist and determines the minimum attainable lithographic feature resolution Δ:

$$\Delta = s(d/D) \tag{5.37}$$

where s is the spacing between the mask and the wafer, d is the source diameter, and D is the distance from the source to the mask.

In a high-resolution system, Δ should be controlled to within 0.1 μm. The spacing s should be large enough so that large-diameter masks can be used without much risk of contacting the resist, which would greatly increase the occurrence of defects.

A noncontact projection system is desirable to maintain a low defect density. Geometric distortion arises from the fact that the image is projected from the mask to the wafer by a diverging beam. The degree of distortion z depends upon the distance of the image from the central axis of the beam:

$$z = s(w/V) \tag{5.38}$$

where w is the distance on the wafer from the image to the central axis of the beam. For whole wafer exposure, w is equal to one-half the wafer's diameter. The distortion itself is not as important as the registration problem when multiple projections must be made on a single wafer. The variations in the gap, ds, lead to variations in the

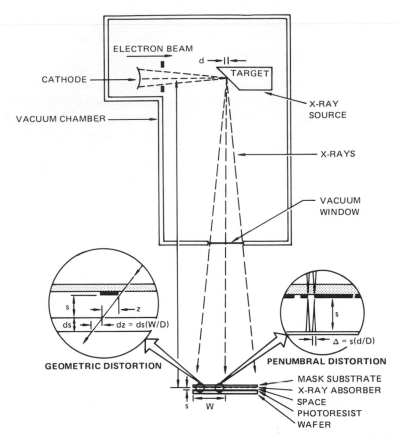

FIGURE 5.16. Arrangement for exposing resists, illustrating penumbral and geometric distortion in projection printing.

geometric distortion dz, since

$$dz = ds(w/D) \qquad (5.39)$$

Since the variation in geometric distortion also affects the minimum attainable resolution, dz must be held to 0.1 μm.

   Step-and-repeat lithography, of course, would keep the value of w small. However, in this analysis we consider a whole 3-inch-diameter wafer where $w = 38$ mm. The mask-to-wafer spacing and the degree of control of this parameter, ds, are quantities less readily defined. For purposes of estimation here we chose a spacing $s = 10$ μm that is controlled to an accuracy of $ds \cong 1$ μm. Putting these values into Eqs. (5.16) and (5.17), we get

$$D = 380 \text{ mm} \quad \text{and} \quad d = 3.8 \text{ mm}$$

Three additional parameters are required to define the needed source strength. These are the fraction of X rays transmitted through the window and mask to the resist, the desired wafer throughput rate, and the resist sensitivity.

5.6.2. X-ray Sources

X rays are produced when electrons incident on a target material are suddenly slowed down. The maximum X-ray energy is the energy E of the incident electrons. If E is greater than the excitation energy (E_c) of characteristic line radiation of the atoms of the material, then the X-ray spectrum will contain these lines.

X-ray generation by electron bombardment is a very inefficient process, most of the input power being converted into heat in the target. The X-ray intensity that can be produced is generally limited by the heat dissipation possible in the target. With electrons focused to a spot 1 mm in diameter onto an aluminum target on a water-cooled stem, 400–500 W is a typical upper limit for the input power. The X-ray power produced is still only on the order of 10 mW, and this is distributed over a hemisphere. The X-ray power is proportional to the electron current. The power in the line radiation is also proportional to $E - E_c$ for a target thin enough so that the absorption of X rays by the target itself can be neglected.[18] However, as E increases and the electrons penetrate a thick target more deeply, the characteristic X rays produced must on the average pass through more material on the way out and are thus absorbed more strongly.

A reasonable thermal model for the soft X-ray source is one where the electron-beam energy is dissipated in a circle of diameter d (cm) on a semi-infinite solid of thermal conductivity K (W/°C · cm). If the electron beam has a Gaussian profile with Gaussian diameter d, then the temperature rise ΔT at the center of the circle is $\Delta T = W/1.78 \, dK$. The maximum input power in watts is

$$W_{\max} = 1.78 \Delta T \, dK \tag{5.40}$$

Water-cooled rotating anodes provide the most intense X-ray sources for the most demanding situations.[19] One chooses the anode material on the basis of its fatigue strength, thermal capacity, and thermal conductivity, as well as on the characteristic and Bremsstrahlung X radiation it produces. A 20-cm-diameter aluminum anode operating at 8000 rpm can dissipate up to about 20 kW in a 6-mm spot. The maximum beam power scales as the 3/2 power of the spot diameter, so the brightness of the source actually decreases with increasing beam power. To reduce penumbra blurring, short turnaround time exposure systems must be built with small source diameters and short working distances between the source and the mask.

Figure 5.17 shows a rotating-target X-ray lithography system developed at Bell Laboratories with a built-in optical alignment. In this system the mask–wafer combination is aligned under an optical microscope and then shifted under the X-ray source for exposure. It is capable of 1-μm resolution printing for bubble memory production.[20]

In a new version of the Bell X-ray printer, a 4-kW beam is focused onto a stationary water-cooled palladium cone target, which in turn emits X rays at a characteristic wavelength of 4.36 Å. With this printer a resolution as small as 0.3 μm

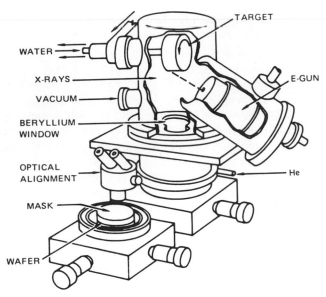

PARAMETER	SPECIFICATION
ANODE MATERIAL	Pd, ℓ = 4.37 Å
INPUT POWER	4.4 kW IN 3 mm SPOT AT 25 kV
VACUUM WINDOW	50 μm Be
MASK SUBSTRATE	25 μm KAPTON (POLYIMID)
EFFECTIVE TRANSMISSION OF WINDOW AND SUBSTRATE	0.7
SOURCE WAFER DISTANCE	50 cm
X-RAY FLUX AT WAFER	4 mJ cm^{-2} min^{-1}
X-RAY FLUX ABSORBED IN PMMA	0.5 J cm^{-3} min^{-1}
EXPOSURE TIME FOR PMMA	1000 min FOR VERTICAL WALLS
POWER CONVERSION EFFICIENCY TO X-RAY (E)	4.76 x 10^{-4}

FIGURE 5.17.   X-ray lithography system developed at Bell Telephone Laboratories. From Ref. 20.

in a 0.3-μm-thick resist containing chlorine has been achieved, using a 1.5-min expo-sure to Pd$_{La}$ X rays.[21]

5.6.3. Plasma Sources

Hot plasmas generated by high-power lasers or electric discharge produce pulses of soft X rays.[22] These sources, although as yet untapped, may play an important role in microfabrication. Because conventional X-ray tubes radiate very inefficiently at low photon energies (1 keV), much attention is currently centered on pulsed devices. In these machines the efficiency of conversion of electrical energy to soft X rays with a laser is found to be an order of magnitude higher than for electron-beam bombardment sources. The process of plasma formation from a solid surface starts with vaporization of the surface layer.[23] The energy necessary for this step is negligible compared to that required for ionization and heating the plasma to keV temperatures. The laser energy absorbed by the plasma goes almost entirely into ionization and heating of the resulting electrons, with the thermal energy of the ions being negligible for high-Z plasmas.

Three processes contribute to the radiation from the plasma: Bremsstrahlung continuum from free–free transitions, recombination continuum from free–bound transitions, and line radiation from bound–bound transitions. Each of these rates of radiation depends on the level of ionization attained within the short times available.

Following the absorption of laser energy by the target, the plasma comprises heavy ions and electrons with Maxwellian energy distributions. Figure 5.18 shows the required energy conversion processes for plasma radiation.

The limiting factor for plasma heating by laser beam is the electron density in the plasma, which has to be less than, but near, a critical value n_p for which the plasma frequency v_p equals the laser frequency v_L, namely,

$$n_p = 1.24 \times 10^{-8} \, v_L^2 \, \text{cm}^{-3} \tag{5.41}$$

At present, the most powerful lasers employ Nd glass emitting at a wavelength of 0.16 μm, corresponding to $n_p = 1.0 \times 10^{21} \, \text{cm}^{-3}$. The absorption proceeds through the process of inverse Bremsstrahlung, in which electrons accelerated by the electric fields of the focused laser light undergo momentum-transfer collisions with the ions via their Coulomb interaction.

In another approach,[24] a plasma is generated in a coaxial discharge tube, with the ions being heated primarily through the collapse of the quasi-cylindrical current sheet in a manner reminiscent of the linear pinch effect.

In addition to the electron bombardment and plasma X-ray sources, the synchrotron radiation source is currently being explored for high-intensity exposure (see Section 5.7).

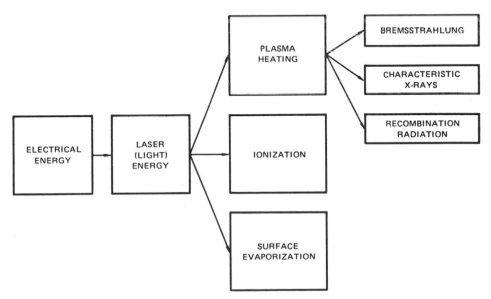

FIGURE 5.18.   Energy conversion processes in plasma X-ray generation.

5.6.4. X-ray Masks

Many different approaches have been taken in fabricating X-ray lithography masks. The primary problem is to produce a thin but strong substrate that is transparent to X rays. Two types of thin films have been tried: organic and inorganic membranes. Organic films include Mylar, Kapton and Pyrolene (Dupont trade names for polyester and polyimide films), and are held stretched to form a planar pellicle. Inorganic types include silicon, silicon oxide, silicon metals, aluminum oxide, and silicon carbide. The X-ray absorber pattern can be ion-beam etched, sputter etched, or electroplated through a resist pattern. The absorber film typically consists of two metal layers: first a thin layer of chromium for adhesion to the substrate, and then a layer of gold. The X-ray attenuation of the mask material is a function of the X-ray wavelength, as shown in Fig. 5.19.[25] Figure 5.20 shows the energy loss as a function of the photon energy for light elements usually used in the mask support material.[26]

The selection criteria for mask materials can be summarized in the following simple optimization rules.

Minimize the transmission of the mask:

$$T_M = \exp(-\mu_M z) \tag{5.42}$$

Minimize the absorption of the support materials (Si, Be, polymers):

$$A = 1 - \exp(-\mu_s z) = 1 - T_s \tag{5.43}$$

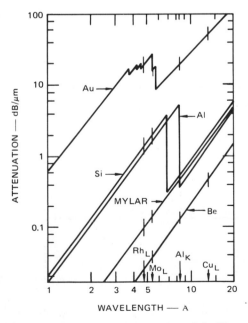

FIGURE 5.19.   X-ray attenuation as a function of the X-ray wavelength.

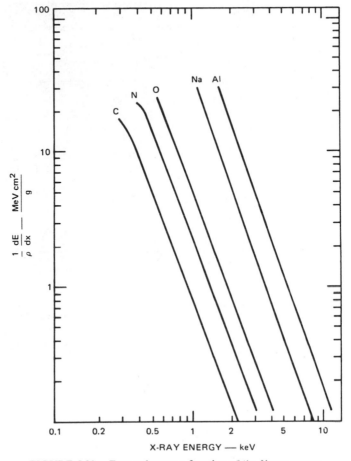

FIGURE 5.20. Energy loss as a function of the X-ray energy.

Maximize the contrast:

$$c = \frac{T_M}{T_s} \tag{5.44}$$

Figure 5.21 shows the fabrication steps for a silicon membrane with a gold absorber pattern[25] and the processing sequence of a gold mask using an intermediate titanium mask.[20]

5.6.5. X-ray Resists

X-ray resists are basically electron resists that are sensitive to the photoelectrons created when X rays are absorbed in the resist. The long exposures (hours) that are

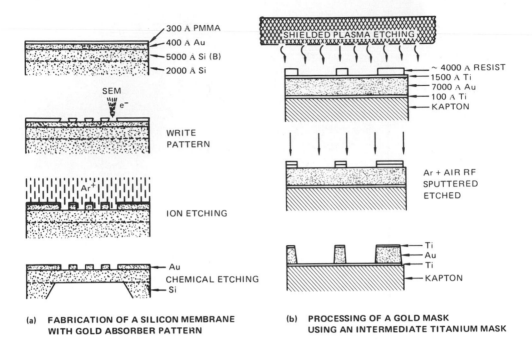

FIGURE 5.21. Processing sequence for the fabrication of a gold mask on silicon and kapton substrates. (a) Fabrication of a silicon membrane with gold absorber pattern; (b) processing of a gold mask using an intermediate titanium mask. From Refs. 20 and 25.

commonly used are necessary because of the low energy conversion of the X-ray source, the low absorption of currently available X-ray resists, and the flux loss associated with the small solid angle that the substrate must subtend at the X-ray source for collimation reasons.

All of the many resists reported absorb less than 10% of the incoming flux and are therefore relatively insensitive to X rays. Some resists have been formulated to increase absorption by resonance with certain characteristic wavelengths. For example, the palladium characteristic line at 4.37 Å is strongly absorbed by chlorine-containing resists. Although there is considerable room for improvement, current resist technology suffices for the fabrication of ICs.

Conventional electron bombardment sources generate X-ray fluxes incident on the wafer in the range of 1 to 10 mJ/cm² · min. In comparison, conventional photoresists have sensitivities of 100 mJ/cm²; electron resists COP and PBS have 175 and 94 mJ/cm², respectively, at the 4.37-Å pD_{La} wavelength.

More sensitive X-ray resists can be designed in several ways.[27] It is most important to increase the absorption coefficient for the resist by the incorporation of highweight percentages of absorbing elements for a given X-ray source. At short wavelengths, resonance absorption can result in the target. Examples are Cl at 4.36 Å, S at 5.41 Å, and F at 8.34 Å. Heavy-metal atoms (bromine) can increase the absorption in the entire energy range. However, high sensitivity is not the only requirement;

the X-ray resist must have the following qualities:

- High sensitivity to X rays
- High resolution
- Resistance to chemical, ion, and/or plasma etching

No present resist satisfies all three requirements. Sensitivity ranges from 1 mJ/cm² for experimental resists to 2 J/cm² for PMMA at a wavelength of 8.34 Å.[25]

The copolymer resist—poly(glycidal methacrylate-coethyl acrylate) or COP—represents a negative resist in which cross-linking occurs in areas exposed to X rays and prevents dissolution during developing. Although the COP resist is at least 20 times as sensitive to 8.34-Å X rays as PMMA, the control of fine geometries with COP is difficult.

Line-edge profiles can be calculated using computer simulation[28] of the exposure, development, and surface etching of the resist. Unlike in the optical case, no visible interference patterns are observable in the X-ray-exposed line-edge profiles.

5.6.6. Alignment

The attainment of the extremely high positional accuracy that X rays require for reregistration has also proved difficult. Alignment techniques using X rays themselves require the use of an active fabricated alignment structure (on the wafer or a second material) that absorbs X rays, and fluoresces, or photo-emits electrons.

Figure 5.22 shows a scheme for the registration of multiple masks in X-ray lithography.[25] This system utilizes alignment marks on the mask and wafer and a proportional counter for X-ray detection. With this registration system, 0.1-μm

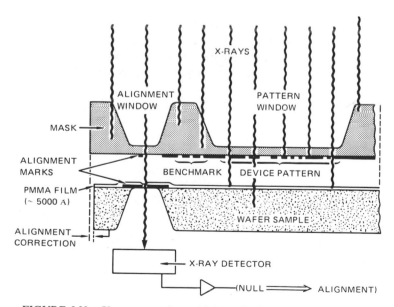

FIGURE 5.22. X-ray system for multiple mask alignment. From Ref. 25.

registration precision has been achieved. Other mask-alignment systems using fluorescent X rays or backscattered electrons have been built in laboratory X-ray printers.[29]

5.6.7. Linewidth Control

The trend toward higher packing densities and smaller features in microelectronic devices has demanded tighter linewidth tolerances at critical patterning steps. As features are packed closer together, the margin for error on feature size is obviously reduced. Furthermore, the successful performance of most devices depends upon the control of the size of critical structures, such as gate electrodes in MOS devices. Since the allowed size variation on such structures is generally a fixed fraction of the nominal feature size, reduced dimensions also imply tighter tolerances. Linewidth control of better than $\pm 0.5 \, \mu m$ from the design value is not an uncommon requirement for critical mask levels.

One of the more difficult problems with resist pattern generation is to achieve good linewidth control, high resolution, and good step coverage simultaneously. Often the requirements appear to be mutually exclusive; good step coverage requires a thick resist; high resolution, however, is more easily obtained in a thin resist. This is true for all resists, both positive and negative.

With any resist the necessary conditions to obtain high resolution and good linewidth control are a flat surface and a thin resist (3000–4000 Å). The flat surface ensures that the resist has very little variation in thickness, and as a result there will be little variation in resist linewidth. Even so, resist linewidth variations still occur when lines traverse a step. As device wafers have steps due to previously patterned layers, thick resists (7000–15,000 Å) must be applied to achieve a flat surface when covering the stepped features on the substrate.

Bell Laboratories developed a technique[30] for generating high resolutions and steep profile patterns in an organic layer a few micrometers thick that covers the steps on a wafer surface and is planar on top. A thick layer (2–3 μm) of photoresist serves as the thick organic layer. It is patterned using an intermediate masking layer (0.12 μm SiO_2) covered by a 4000-Å-thick layer of chlorine-based negative X-ray photoresist.

After X-ray exposure and development of the top X-ray resist layer, the intermediate layer of SiO_2 is etched by CHF_3 reactive plasma etching. The thick organic layer is then etched by O_2 reactive plasma etching, with the SiO_2 acting as a mask. The various steps required to define a steep resist profile are shown in Fig. 5.23.

The X-ray resist is a mixture of a host polymer and a volatile monomer that is locked into the host during exposure.[31] Pattern development is accomplished by heating and exposure to an oxygen plasma. This X-ray resist allows exposure times as short as 1.5 min when exposed to a 2.9 mJ/cm^2 min flux of palladium X ray (4.37 Å wavelength).

This new resist with the multilevel structure ensures submicron resolutions (0.3 μm) combined with good step coverage and linewidth control. This three-layer technique has various advantages for other lithographic systems as well.[21, 30]

For optical lithography, the thick underlying layer of resist and the layer of SiO_2 are sufficient to reduce the reflection from the wafer surface and, as a result, reduce

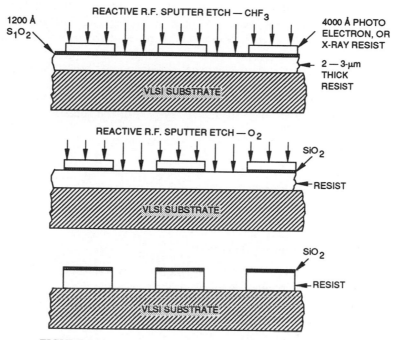

FIGURE 5.23. Process steps for the trilevel resist. From Ref. 30.

standing-wave problems. The flat surface of the thick organic layer keeps scattering down and the top resist can be made thin for high resolution.

For electron-beam lithography, backscattering from a substrate covered by 0.1 μm for SiO_2 on top of 2–3 μm of polymer should be less than that from a Si or SiO_2–Si substrate. Thus, proximity effects existing in that technology might be reduced and better linewidth control can be expected.

Although a three-layer technique [i.e., top radiation-sensitive resist, SiO_2 mask, bottom organic (resist) layer] has been discussed here, a two-level scheme without SiO_2 also works if the top patterned resist is very resistant to the reactive rf sputter etch. In addition, there are many choices for the intermediate and thick organic layers. The intermediate layer could be silicon nitride, boron nitride, or some other suitable low-defect material. For the bottom layer, many positive and negative resists familiar to photo and electron lithography may be used, as well as other polymers.

5.7. SYNCHROTRON RADIATION FOR X-RAY LITHOGRAPHY

5.7.1. Introduction

Synchrotron radiation sources are by far the brightest sources of soft X rays.[32] Electron storage rings and synchrotrons emit a much higher flux of usable collimated X rays than does any other source, thereby allowing shorter exposure times (on the order of seconds), large throughputs, less critical resist exposure conditions, and simplified geometrical conditions for applications requiring registration. Because of the high collimation of synchrotron radiation, the spatial resolution of X ray litho-graphy with synchrotron radiation is not limited by penumbral blurring, and rather large distances between the mask and the wafer can be tolerated (about 1 mm for 1-μm linewidth patterns).

The basic characteristics of synchrotron radiation are[33] the following:

- High intensity over a broad spectral range from the infrared through the visible and ultraviolet and into the X-ray part of the spectrum
- High polarization
- Extreme collimation
- Pulsed time structure
- Small source size
- High vacuum environment

5.7.2. Properties of Synchrotron Radiation

Synchrotron radiation is emitted by high-energy relativistic electrons in a synchrotron or storage ring, which are accelerated normal to the direction of motion by a magnetic field. Figure 5.24 shows a schematic of an X-ray exposure station with a synchrotron radiation source.

The flux available from a large synchrotron source is 10^4 times as powerful as from a rotating-anode-type X-ray source (10 W of X rays).

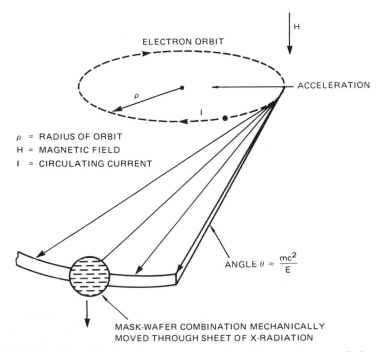

FIGURE 5.24.   Schematic of an X-ray exposure station with a synchrotron radiation source.

The energy loss per turn for the highly relativistic electrons ($\beta = v/c \cong 1$) is[33]

$$\Delta E \text{ (keV)} = \frac{88.47E^4}{\rho} \qquad (5.45)$$

where E (in GeV) $= 10^9$ eV, ρ is the radius of the orbit (m), and i is the circulating electron current in the ring. Multiplying Eq. (5.45) by the circulating current in gives the total radiated power

$$P \text{ (kW)} = \frac{88.47E^4 i}{\rho} \qquad (5.46)$$

or in terms of the magnetic field B (in gauss)

$$P \text{ (kW)} = 2.654BE^3 \qquad (5.47)$$

For machines such as SPEAR[33] (Stanford Positron–Electron Annihilation Ring) at Stanford University, the total radiated power is 105 kW, where $E = 3.5$. GeV, $\rho = 12.7$ m, $B = 9.2$ kG, and $in = 0.1$ A.

Synchrotron radiation is emitted in a cone with an opening angle θ, given by

$$\theta \approx \frac{mc^2}{E} = \frac{0.5}{E \text{ (GeV)}} \text{ (mrad)} \qquad (5.48)$$

The spectral distribution of the synchrotron radiation extends from the microwave region through the infrared, visible, ultraviolet, and into the X-ray region, decreasing in intensity below a critical energy E_c or, alternatively, the critical wavelength λ_c, where

$$E_c(\text{keV}) = \frac{2.218E^3 \text{ (GeV)}}{\rho(\text{m})} \qquad (5.49)$$

or since

$$E_c(\text{keV}) = \frac{2.218E^3 \text{ (GeV)}}{\rho(\text{m})} \qquad (5.50)$$

$$\lambda_c(\text{Å}) = \frac{5.59\rho(\text{m})}{E^3(\text{GeV})} \qquad (5.51)$$

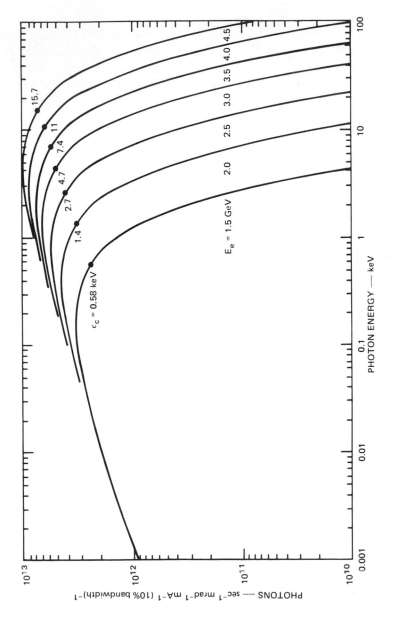

FIGURE 5.25.   Spectral distribution of synchrotron from SPEAR. From Ref. 33.

For SPEAR operating at 3.5 GeV, $E_c = 7.4 \, \text{keV}$ and $\lambda_c = 1.7 \, \text{Å}$. The spectrum of the synchrotron radiation from SPEAR is shown in Fig. 5.25. The high polarization and the pulsed time structure available from synchrotron radiation, however, are not relevant for X-ray lithography.

Using a synchrotron X-ray source for lithography, penumbral blurring can be completely eliminated and one can tolerate large distances between the mask and the wafer without loss in resolution. However, at very large distances, diffraction effects determine the resolution limit.[32]

Synchrotron radiation can form high-quality submicron images, but its main advantage is the short exposure time (a few seconds) compared to a few hours using conventional X-ray sources. Synchrotron radiation sources that provide powerful, well-collimated beams of soft X rays for research are in a number of locations worldwide.

5.8. ELECTRON-BEAM LITHOGRAPHY

5.8.1. Introduction

There are two ways in which electron beams can be used to irradiate a surface and create a pattern.[34-36] These are the parallel exposure of all pattern elements at the same time and the sequential (scanning) exposure of one pattern element (pixel) at a time.

Projection systems generally have a high throughput and are less complex than scanning systems. The pattern information is stored in masks. The scanning systems are under computer control where a finely focused beam (or beams) of electrons is used to generate the pattern, correct the distortions and proximity effects, and register the wafer position. The pattern information is stored in a digital memory.

Although an electron beam can be used in a variety of ways, direct control by a computer offers the ability to generate patterns without the need for a mask. This enables the scanning-electron-beam systems to be used for both mask making and direct wafer writing. These machines combine high spatial resolution ($<0.1 \, \mu\text{m}$) with accurate registration ($<0.1 \, \mu\text{m}$). The small electron beam (or beams) are either rastered or manipulated in a vector mode. The vector system is more efficient, since scanning is only performed over areas that are to be exposed on the resist. In the vector mode, however, the deflection system accuracy is limited, and to apply corrections a computer system is needed.

Shaped-electron-beam machines[37] that combine vector scanning and projection (hybrid machines) have been designed to further improve the efficiency. In this approach the fundamental picture elements from which a device can be built up are projected onto an appropriate wafer location by the use of a variable electron-beam shape. In character projection systems, a selected character from a mask is imaged onto the wafer, and this process may be repeated many times. It is used, for example, in bubble memory chips, random-access memories, read-only memories, or for any repetitive pattern. Here the scanning-type system is combined with a projection system in which special apertures (beam shaping) or masks form the picture elements.

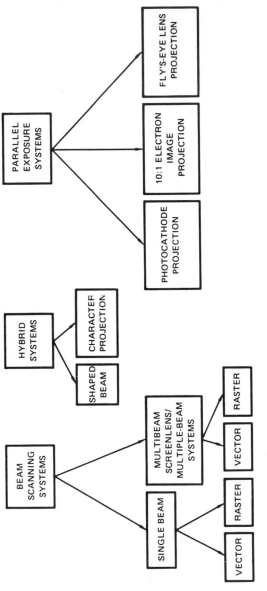

FIGURE 5.26.   Classification of electron-beam machines.

Multibeam scanning systems use a special optical element to obtain multiple beams for simultaneously generating patterns on the wafer.[38, 39]

A classification of the currently used electron-beam machines is shown in Fig. 5.26.

5.8.2. Major System Components

5.8.2.1. Projection Systems. Electron-beam projection lithography provides a means of replicating, in a single large-area exposure, submicron linewidth patterns originally made by scanning-electron-beam lithography. The technique is complementary to optical and X-ray projection lithographies. Two types of beam projection systems have been developed specifically for semiconductor device fabrication: the 1:1 and reduction projection systems.

Work on 1:1 projection systems has been under way for many years. It began at Westinghouse[40] and has been more recently pursued by Mullard (Philips) in England.[41, 42] A schematic of the Mullard system is shown in Fig. 5.27. This projection system employs a photo cathode masked with a thin metal pattern. Photoelectrons from the cathode are accelerated onto the sample by a potential of about 20 kV applied between the cathode and the sample. A uniform magnetic field focuses these photoelectrons onto the wafer (anode) with unity magnification. The mask and wafer can, in principle, be as large as desired.

In this system, image position is detected by collecting characteristic X rays from marks on the wafer with the X-ray detectors shown. (During this process, the photocathode is masked so that only the alignment marks are illuminated.) Magnetic deflection is then used to position the pattern with an accuracy of 0.1 μm. The image current density is about 10^{-5} A/cm^2 (1-s exposure for 10^{-5} C/cm^2 resist sensitivity) for cesium iodide photocathodes, which have the best lifetime and resistance to poisoning. The dominant aberration limiting resolution is chromatic aberration, and theoretical estimates of minimum linewidths vary from 0.5 to 1 μm wide.

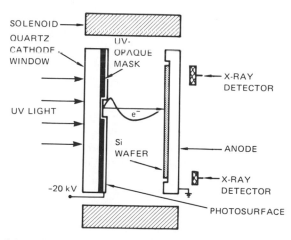

FIGURE 5.27.   Schematic of the Mullard 1:1 electron-beam projection system. From Ref. 42.

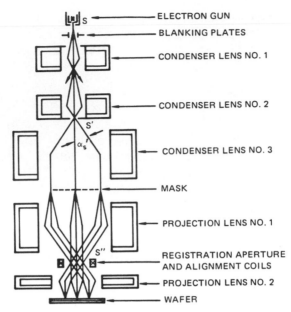

FIGURE 5.28.   Reduction-type electron-beam projection system developed at IBM. From Ref. 43.

Figure 5.28 shows the reduction-type projection system developed at IBM Research.[43] The basic concept for this system, which is the electron-optical analogy of reduction-optical projection cameras, was first described by researchers at Tubingen University, Federal Republic of Germany.[44] The mask is a freely suspended metal foil. A special electron-optical system illuminates the mask and forms a sharp demagnified image of it on the wafer. The demagnification factor is $10 \times$, and a field 3 mm in diameter and linewidths of down to 0.25 μm can be produced.

The system uses its scanning mode of operation for alignment. In this mode, the illuminating beam is focused onto the mask rather than flooding it, as in the case of image projection. The focused beam scans across the mask and an image of this focused beam scans across the sample. Scattered electrons are collected from the sample to detect sample position, and corrections are made by shifting the projected image, with deflection coils placed between the two projection lenses. This system offers very low distortion and very high resolution, factors that are problems in the 1 : 1 cathode projection system.

A parallel-image electron-beam projection system has been developed[38,45] at SRI International (see Fig. 5.29). In this system, a two-dimensional array of lenses is used for image multiplication and projection from an object mask. The screen lens, i.e., holes in a planar electrode, constitutes a particularly simple form of a multiple-imaging element. The screen lens separates two regions of uniform electrostatic fields. The image projection system is used to fabricate thin-film field-emitter electron and ion sources for various applications.[46–49]

5.8.2.2. *Beam-Scanning Systems.* In this section the basic operating principles of scanning-electron-beam systems are discussed.

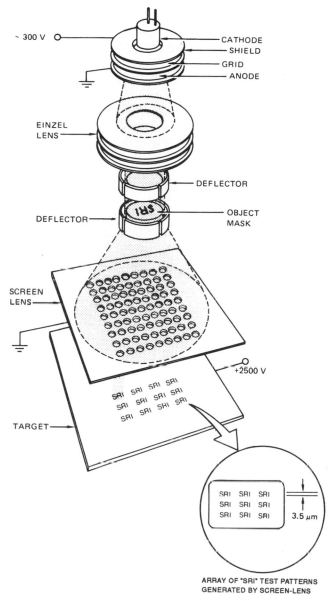

FIGURE 5.29. Screen lens parallel-image electron-beam projection system developed at SRI International.

The Electron-Optical Column. Beam-forming systems use either the Gaussian round-beam approach or the shaped-beam (square or round) approaches[36] shown in Fig. 5.30.

Gaussian systems use the conventional probe-forming concept of the scanning electron microscope, as in Fig. 5.30a. In general, two or more lenses focus the electron beam onto the surface of the wafer (or mask) by demagnifying the electron gun

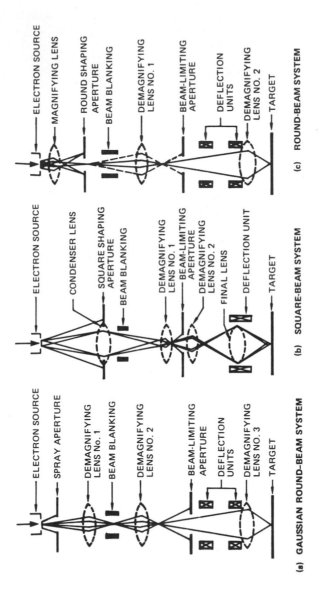

FIGURE 5.30.   Beam-forming systems. (a) Gaussian round-beam system; (b) square-beam system; (c) round-beam system. From Ref. 36.

source. High flexibility can be achieved, since the size of the final beam can be readily varied by changing the focal lengths of the electron lenses. To ensure good line definition, the beam size is generally adjusted to about a quarter of the minimum pattern linewidth.

In the square-beam approach, Fig. 5.30b, an electron source illuminates a square aperture at the center of a condenser lens placed immediately after the gun. The condenser lens images the gun crossover (1 : 1) into the entrance pupil of a second condenser lens. This lens, together with a third condenser lens, demagnifies the square aperture to form a square beam. A fourth lens images the square beam (1 : 1) onto the target plane. The size of the square beam is generally equal to the minimum pattern linewidth. To achieve an equivalent resolution, the edge slope of the intensity distribution of the square beam is matched to the intensity distribution of a Gaussian round beam between the 10 and 90% points.

In the case of the round-shaped beam shown in Fig. 5.30c, a lens focuses a magnified image of the gun crossover onto a round aperture and two condenser lenses demagnify the round aperture onto the plane of the target.

Round-beam systems (Gaussian or shaped) are generally simpler than shaped square-beam systems and have more flexibility. However, the square beam has more current in the spot (current is proportional to spot area for the same gun brightness) and therefore offers higher exposure speed in cases where speed is limited by the beam current and/or beam stepping rate. Difficulties with square-beam systems may arise when angle lines are required or when some lines have dimensions that are not integral multiples of the beam size. In the latter case, over exposure in the overlapping regions will occur.

To calculate the beam spot size on the resist, one has to know the crossover size $(d'_0 = 2r_c)$, the particular optical system for demagnification, and the lens errors

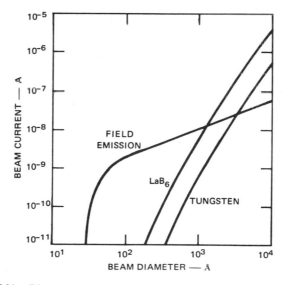

FIGURE 5.31.   Diameter and current limits of electron-beam spot. From Ref. 50.

introduced by the nonperfect optical elements. For a perfect system (no lens errors), the spot size for a three-lens system would be

$$d_0 = \frac{f_1 f_2 f_3}{L_1 L_2 L_3} d_0' \tag{5.52}$$

where the optical system demagnifies the crossover $d_0' = 2r$ to d_0, f_i denotes the focal lengths of the lenses, and L_i is the distance between lenses. For a real system, one has to take into account the enlargement of the demagnified spot resulting from the lens errors for each lens in the system (see Section 2.5.4).

Figure 5.31 shows the current limits in the electron-beam spot on target as a function of the beam spot size, using different cathodes.[50] Electron-beam machines operating in the 10^{-6}–10^{-7} A beam current mode are capable of writing 0.5 to 1.0-μm lines. The newest machine in the EBES family (EBES-4) is shown in

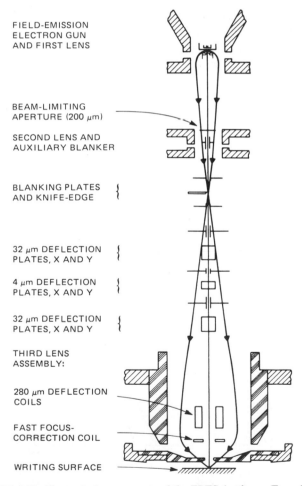

FIELD-EMISSION
ELECTRON GUN
AND FIRST LENS

BEAM-LIMITING
APERTURE (200 μm)

SECOND LENS AND
AUXILIARY BLANKER

BLANKING PLATES
AND KNIFE-EDGE

32 μm DEFLECTION
PLATES, X AND Y

4 μm DEFLECTION
PLATES, X AND Y

32 μm DEFLECTION
PLATES, X AND Y

THIRD LENS
ASSEMBLY:

280 μm DEFLECTION
COILS

FAST FOCUS-
CORRECTION COIL

WRITING SURFACE

FIGURE 5.32. The optical components of the EBES-4 column. From Ref. 51.

Fig. 5.32. This figure shows the EES-4 column[51-54] designed for 20-kV electron-beam energy and a 0.125-μm spot size with a current of 150 nA and 2-ns spot exposure time. These numbers mean that for a pixel (picture element), the number of electrons needed for exposure is given as:

$$N_e = \frac{iT}{e} \approx 2 \times 10^3 \tag{5.53}$$

and the irradiation time per circuit (die) using 10^{10} pixels is $T \times N_p = 2 \times 10^{-9} \times 10^{10} = 20$ s. If there are 100 dies per wafer, the total irradiation time per wafer is about 2000 s. If one includes the times for wafer positioning and alignment, it is easy to see that the processing time per wafer is much longer than the typical 600-s processing times on the production line. To increase the throughput of the electron-beam lithography tool, shaped beams or multibeams must be used.

Beam Size and Shape. Beam-shaping techniques[37] in electron-beam exposure systems have proved to be very effective in overcoming throughput limitations inherent to the serial exposure of scanning systems.

Figure 5.33 illustrates the spectrum of beam profiles used for pattern generation. One end of the spectrum is represented by the scanning electron microscope-type Gaussian round-beam exposure of one image point at a time. The size of the round beam, defined as the half-width of the Gaussian distribution, represents the spatial resolution and is typically four to five times smaller than the minimum pattern feature. For the shaped-beam system the spatial resolution given by the edge slope of the beam profile is decoupled from the size and shape of the beam. Consequently, a plurality of image points can be projected in parallel without loss of resolution.

In the case of the fixed-shape beam, a square aperture is projected to fit the minimum pattern feature containing 25 or more image points. The variable-shaped beam is generated by projecting two superimposed square apertures. The compound image formed by both apertures can be varied in size and shape to produce the various pattern elements containing up to 100 (or even more) image points.

Character projection represents the most efficient beam-shaping technique. Complex pattern cells containing up to 2000 image points (pixels) are addressed and projected in parallel. This technique overcomes the restriction to rectilinear geometries and is very efficient in the printing or repetitive patterns.

Character projection is an extension of the dual-aperture technique in which the second aperture is replaced by a character plate with an array of complex aperture shapes. These can be selected electronically for each exposure in full or in part to compose the image. Nonrectilinear and curved patterns can easily be included.

Pattern Generation. After the beam has been focused and shaped, it must be deflected (scanned) over a wafer by a beam-writing technique. Deflection systems for electron beams are generally electromagnetic, though electrostatic systems and even partly electromagnetic, partly electrostatic systems exist.[37,38,50]

The two basic beam-writing techniques are the raster and vector techniques. In the raster technique,[55,56] the beam is scanned over the entire chip area and is turned on and off according to the desired pattern. In the vector technique,[36] the beam addresses only the pattern areas requiring exposure, and the usual approach is to

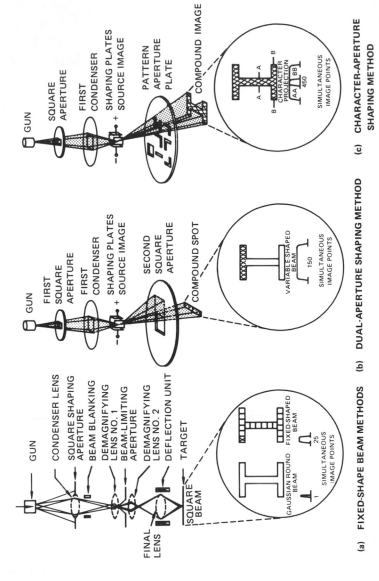

FIGURE 5.33. Beam-shaping techniques. (a) Fixed-shape beam methods; (b) dual-aperture shaping method; (c) character-aperture shaping method. From Ref. 37.

decompose the pattern into a series of simple shapes such as rectangles and parallelograms.

The raster scan places less stringent requirements on the deflection system because the scanning is repetitious and distortions caused by eddy currents and hysteresis can be readily compensated for. It can also handle both positive and negative images. The vector scan is more efficient but requires a higher-performance deflection system. In addition, it has several other advantages not readily available

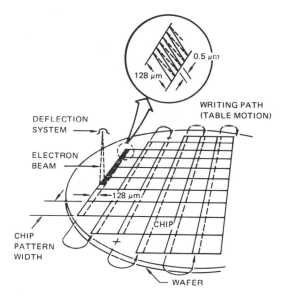

(a) EBES RASTER SCAN FEATURING ONE-DIMENSIONAL LINE SCAN (RASTER), CONTINUOUSLY MOVING TABLE, FULL WAFER ALIGNMENT, AND LASER INTERFEROMETER FEED

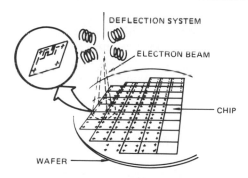

(b) IBM EL1 VECTOR SCAN FEATURING TWO-DIMENSIONAL FULL-CHIP SCAN, DISTORTION-FREE DEFLECTION, STEPPED TABLE, AND ALIGNMENT AT EVERY CHIP

FIGURE 5.34. Raster and vector scan electron-beam systems. (a) EBES raster scan featuring one-dimensional line scan (raster), continuously moving table, full wafer alignment, and laser interferometer feed; (b) IBM EL1 vector scan featuring two-dimensional full-chip scan, distortion-free deflection, stepped table, and alignment at every chip. From Refs. 36 and 55.

to the raster scan technique, e.g., ease of correction for the proximity effects of electron scattering and a significant compaction of data that can lead to a much simpler control system.

Figure 5.34a shows the operating mode of a raster scan system [electron-beam exposure system (EBES)]. In this machine a 0.50-μm-diameter writing spot exposes the area of a pixel (0.025 μm^2) in 100 ns, and thus it requires about 25 μs to expose a line consisting of 256 addresses. An interferometer-controlled table moves the substrate an address length in the x direction (32 cm/s) while the beam is sweeping out one scan line in the y direction. The alignment is performed for the full wafer.

Figure 5.34b depicts the operating mode for a vector-scan-type electron-beam system (IBM EL1). After the alignment of each chip (die) in this step-and-repeat system, the exposure is made with a two-dimensional electronic scan that covers the entire chip. Since the maximum area over which it is possible to electronically scan a submicron beam is smaller than a chip, it is necessary to move the sample mechanically. Two basic types of mechanical movement can be used: a step-and-repeat movement, where the substrate is stepped between individual exposure sites, or a continuous mechanical scanning, where the substrate moves continuously under the beam.

Stage Assembly. Two methods have been employed to control the relative position between the electron beam and the wafer, namely, field-to-field registration with bench marks and measuring the exact table location with a laser interferometer. The field-to-field registration with the bench mark technique has proved to be acceptable indirect wafer processing, but when the electron beam is used for the direct generation of master masks, a laser interferometer is invariably used to obtain sufficient step-and-repeat accuracy. The laser interferometer, operating on appropriately located mirrors on the worktable, can provide a monitoring accuracy down to several hundredths of a micron. Other important parameters such as the variation of the z position while the table is being translated in the x and y directions and the translation speed vary according to the particular writing strategy. Sophisticated servo systems, complete with multiple-feedback loops, are generally used in the control of such worktables.

Registration. In a high-resolution scanning lithography system, registration can be made on a wafer global basis (a one-time registration) relying on extreme system stability, flat and undistorted substrates, and exact stage motion during the step-and-repeat process, or on a chip-by-chip basis, making fine corrections at each site.[57] The former system can have a special mark region and considerable time can be tolerated to achieve registration. The latter system requires marks that are fully compatible with all subsequent processing steps and usually completes a registration cycle in a time that is short compared to the sum of the pattern write time and stage chip-to-chip step time.

In direct microcircuit fabrication technology, the registration mark will most likely be made of thin film steps of a high-atomic-number material or of special geometrical patterns in a material having an atomic number equal or similar to that of the substrate (see Fig. 5.35). As the electron beam is scanned across the sample surface, it passes over these differing material or topographic registration marks. Two common types are a raised pedestal or an etched hole of Si or SiO_2. The signal detected by the registration system can be thought of as having three constituents.

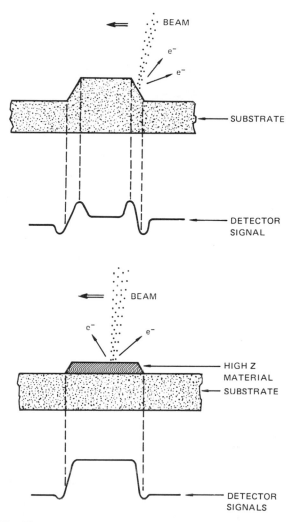

FIGURE 5.35.   Electron-beam alignment marks and detector signals for registration.

First is a steady (dc) part that represents the average number of electrons scattered and detected. The second part is a noise component that contains the noise of the primary beam (a small effect) and any other noise arising from the scattering and detection processes. The third component is the change in detected electrons resulting from change in scattering as the beam crosses a mark edge.

Backscattered electrons offer more promise in detecting registration marks over-coated with resist than do the easily absorbed secondary emission electrons. The backscattered electrons emerge in many directions and possess energies varying from the primary-beam value down to about several hundred electron volts. This broad spectrum is caused by single and/or multiple inelastic scattering of the electrons in the substrate, registration mark, and resist layer(s). To be detected, an electron emerging from the sample must be traveling in a given direction and have an energy

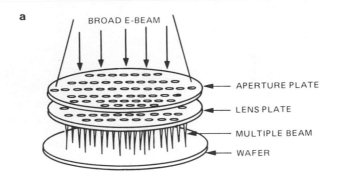

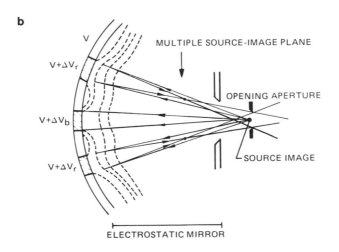

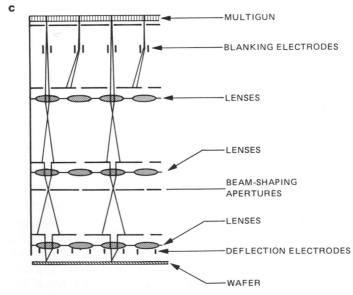

FIGURE 5.36.   Imaging principles of multiple-electron-beam machines.

above the detector threshold. Several different types of detectors (e.g., silicon diode, scintillation) have been described in the literature for detecting the backscattered electrons.[58, 59] Assuming a step-function edge as the primary beam crosses over the mark edge, the backscattered electrons must travel a much reduced distance within the base material before emerging. Hence, they have an energy nearer to the primary-beam value, and a rapid increase in backscattered electron signal is observed. The detectors would usually be used in pairs.

There is no standard set of signals or detector systems used in electron-beam lithography; each developer tends to prefer his own systems. The most commonly used are the following:

1. Signals
   - Backscattered electrons
   - Secondary electrons
   - Luminescent radiation
2. Detectors
   - Scintillation detector
   - Solid-state detector
   - Photomultiplier

The types of alignment (registration) marks also vary from system to system. There is no standard set of marks in the industry.

5.8.3. Parallel-Scanning Systems

The problems of scanning-electron-beam lithography are primarily poor throughput and, to a lesser extent, poor pattern fidelity resulting from electron scattering (proximity effect). In the past, several attempts have been made to use multiple beams to expose, simultaneously, many addresses in parallel on a wafer and increase the throughput. The proximity effect can be reduced using high-voltage (>100 keV) electron-beam lithography; in addition, the high-voltage systems reduce line width variation.

Three different kinds of multiple-beam machines have been considered, using

- Single cathodes with immersion screen lens[38,45,60]
- Single cathode with reflection (mirror) optics[61]
- Multiple cathodes with optical columns[62]

The imaging principles for these designs are shown in Fig. 5.36.

5.8.3.1. Screen Lens Tool. Figure 5.37 illustrates the principle of operation of the direct-write electron-beam system designed and built at SRI International. The electron-optical column is divided into three regions: the gun and blanking region, the drift and deflection region, and the screen lens/wafer immersion region. In the *gun region*, a high-current beam of about 1 mA is formed and accelerated to 1.0 keV. The gun is designed to illuminate an object aperture (usually square) uniformly. The beam emerges from the aperture into a field-free region with about 3° half-angle to flood the screen lens plate uniformly. The lens plate suitable for a wafer of 100-nm (4-inch) diameter is placed at a distance of about 1 m from the object aperture. The screen lens consists simply of a conductive plate with an array of beam-limiting

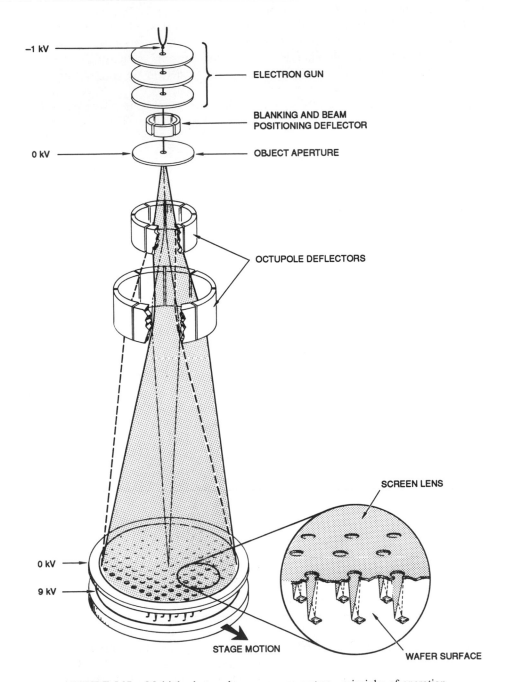

FIGURE 5.37.   Multiple-electron-beam exposure system—principles of operation.

apertures of radius r_a on the object side on an aligned array of lens apertures of radius R on the image side, one pair for each die. Electrons emerging from each lens aperture into the wafer immersion region find themselves in an accelerating field. With a voltage of about 9 kV on the wafer, which is spaced 2 cm from the screen lens, a demagnified image of the object aperture is focused on the wafer under each lens aperture. For a square object aperture 50 μm on a side, images 0.5 μm on a side can be formed under each lens.

A double electrostatic octupole deflection system in the *drift region* enables the virtual position of the object aperture to be moved while retaining the position of the flood beam over the screen lens. An object-to-screen-lens distance of 1 m was chosen so that the angle from the object aperture to the edge of the screen lens would be as small as possible without unreasonably lengthening the complete electron-optical column. The relatively low beam-drift energy (1 keV) requires that the beam be adequately shielded from external time-varying magnetic fields. This has been accomplished by using appropriate soft iron and mu-metal shielding.

When fields are applied in the *deflection region*, the images under each lens aperture of the screen move, each describing an identical pattern dictated by the time variation of the deflection field. Because deflection aberrations limit the useful amplitude to about 200 mm (or about 400 pixels) for 0- to 0.5-μm spots, a stage under laser interferometer control is used to cover a whole die (typically on the order of 5–15 mm on a side) with the required precision. Although several writing modes are possible, this machine uses the method in which the beam is deflected in one direction while the stage moves in an orthogonal direction. Thus, a series of parallel lines is drawn by the beam, forming a strip. The desired pattern is generated by sequentially turning the beam off or on at each pixel site. After a strip is exposed, the stage is moved over a distance w and a second strip is written adjacent to the first. This process is repeated until all of the pixels in each die have been addressed. In this approach a flood beam emerging from an aperture is incident on an array of holes, each of which forms an immersion lens with the substrate. This type of lens array is called a screen lens. Each lens of the array addresses a die of the wafer and focuses on it a demagnified image of the aperture. The flood beam is deflected by a double electrostatic system, enabling the beamlets to scan the dies. The amplitude of the deflection, d, is given by

$$d = \frac{l' l_D E_D}{2\sqrt{V_A}\sqrt{V_{SC}}} \tag{5.54}$$

where l' is the distance between the screen lens and the Si wafer, l_D is the effective length of the deflector electrodes, E_D is the deflecting field, V_A is the accelerating voltage (nominally 1 kV), and V_{SC} is the screen lens voltage (nominally 10 kV).

Global registration is accomplished using shaped fiduciary markers applied to the periphery of the wafer. Several outer lenslets are aligned with these markers and used to beam-address them respectively while the beam is cut off from the writing lenslets. The fiduciary marks are composed of topographic features. The registration signal is generated by elastically scattered or absorbed electrons as the fiduciary markers are interrogated.

TABLE 5.2.   Specifications for the SRI Multiple-Electron-Beam Exposure System

Quantity	Specification
Exposure rate, 100-mm-diameter wafer (wafer/h)	60
Chip size (mm)	5×5
Single beamlet deflection field (μm)	100×100
Address structure size (μm)	0.1
Deflection rate (kHz)	To 100
Blanking rate (MHz)	To 10
Beam spot size (μm)	1.0
Beam spot edge definition (μm)	<0.2
Gun cathode lifetime (h)	≥ 100
Flood beam uniformity (%)	± 2
Beamlet current into 0.1-μm spot (nA)	>20

The specification of the present prototype machine is listed in Table 5.2.

5.8.3.2. Mirror Machine. The "mirror machine" design[61] concentrates the beam splitting, beam focusing, beamlet blanking, and beamlet focusing in one electrostatic mirror. Figure 5.38 shows a multiple-beam column with mirror optics.

5.8.3.3. Multicolumn Machine. Figure 5.36c shows a multibeam machine design based on the extension of the three-cathode-ray-tube optics. This design might offer faster processing, but will require more electronics and computer control.

Another scheme using beam multiplexing and the diffraction effect in a thin single-crystal film has recently been proposed for parallel writing.[39] In this system,

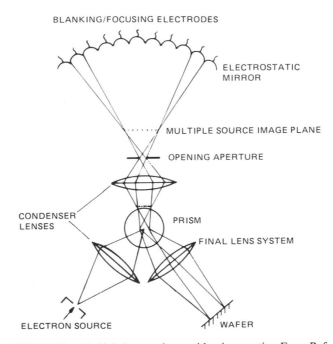

FIGURE 5.38.   Multiple-beam column with mirror optics. From Ref. 61.

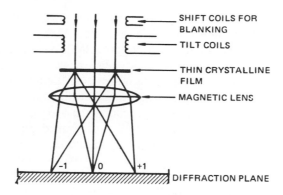

(a) TRANSMISSION-ELECTRON MICROSCOPY (TEM) MODE

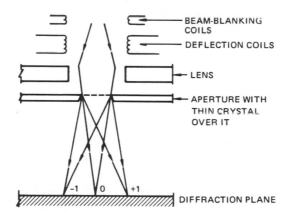

(b) SCANNING-ELECTRON MICROSCOPY (SEM) MODE

FIGURE 5.39. Beam-multiplexing method using the diffraction effect of a thin film. (a) TEM mode; (b) SEM mode. From Ref. 39.

several patterns are addressed simultaneously, using diffracted beams (see Fig. 5.39). The basic geometry for printing in the transmission electron microscope (TEM) mode is shown in Fig. 5.39a. For the scanning electron microscope (SEM) mode, the gold crystal was suspended across the final lens-limiting aperture, as shown in Fig. 5.39b. The applicability of the diffraction effect for electron-beam multiplexing has been demonstrated for both projection and scanning systems. The number of patterns that can be obtained by pattern multiplexing is an increasing function of the electron energy., At 100 kV, at least 36 patterns can be obtained.

5.8.4. Electron-Beam Resists (Ref. 41)

5.8.4.1. Resist Characterization. Electron resist must be capable of forming uniform pinhole-free films on a substrate by simple processes such as spinning, dip coating, or spraying. Usually only polymeric materials fulfill these requirements. When energetic electrons interact with organic materials, the molecules may either be broken down to smaller fragments or link together to form larger molecules. Both types of change occur simultaneously, but when any given material is irradiated, one of these processes predominates and determines the net effect upon the material. For polymers it is known that if every carbon atom in the main chain is directly linked to at least one hydrogen atom, then larger molecules are formed upon irradiation by the process of cross-linking. If a polymer is cross-linked on irradiation, it forms a three-dimensional network and becomes insoluble and infusible. This type of polymer, which is the most common, forms the basis of negative-working electron resists, as it is possible to dissolve and remove unirradiated material, while irradiated material cannot be dissolved away.

A relatively small number of polymers is known in which irradiation in moderate doses predominantly causes a break in the main polymer chain, resulting in irradiated material having a lower average molecular weight than unirradiated material. By judicious choice of a solvent or mixture of solvents, it is possible to dissolve the irradiated polymer selectively, which allows positive-working electron resists to be formulated. At higher electron exposures, however, cross-linking predominates again and the irradiated material becomes completely insoluble. Thus, there is always a limited range of electron exposures in which a positive action may be observed.

The energy available for breaking and forming chemical bonds by electron beams with accelerating voltages of 10–20 kV is very large when compared with ultraviolet radiation. Therefore, photoresists exposed by electron beams have completely different properties than do the same resists exposed to ultraviolet radiation and are normally not suitable for high-resolution delineation. Fortunately, almost all polymers and copolymers react in some manner to electron-beam irradiation, but until recently very few of these had been investigated for this application.

A positive electron resist is characterized by a reduction of molecular weight due to chain scission of the original molecules. Since polymers with different molecular weights have different solubility characteristics in a given solvent, the resist developer is required to distinguish between the fragmented molecular weight M_f and the original molecular weight M_n. The relationship between M_f and M_n and the exposure parameters is expressed as[63]

$$M_f = \frac{M_n}{1 + KQG(z)S(z)} \tag{5.55}$$

with

$$K = \frac{M_n}{100e\rho N_A} \tag{5.56}$$

where e is the electronic charge, ρ is the resist density, and N_A is Avogadro's number. Q is the incident charge per unit area, $G(z)$ is the radiation yield defined as the number of chemical events per 100 eV absorbed, and $S(z)$ is the depth–dose function for electrons expressed in terms of the normalized penetration $f = z/R_G$. The Grun range R_G (see Section 2.6.3.1) is a function of the electron energy E_0 in keV and is expressed as

$$R_G = \frac{0.046}{\rho} E_0^{1.75} \mu m \tag{5.57}$$

and ρ is the density of the resist in g/cm. Empirically, the rate of energy absorption in the resist as a function of the normalized penetration can be expressed as[64]

$$\frac{dE}{df} = -E_0 \Lambda(f) \tag{5.58}$$

where $\Lambda(f) = 0.74 + 4.7f - 8.9f + 3.5f^3$.
   The energy dissipation in terms of z is thus

$$\frac{dE}{dz} = \frac{1}{R_G} \frac{dE}{df} = -\frac{E_0}{R_G} \Lambda(f) \tag{5.59}$$

per incident electron. Backscattering is not included in this estimation. Since $(Q/e)(E/R_G)\Lambda(f)$ is the absorbed energy density (ε) in the resist, M_f can be written as

$$M_f = \frac{M_n}{1 + G_e M_n \varepsilon / 100 \rho N_A} \tag{5.60}$$

According to this, the results of electron radiation on a positive resist are expressed in terms of molecular weight. But the solubility rate of PMMA ($M_n \sim 10^5$) is expressed by the following empirical formula[63]:

$$R = R_0 + \frac{\beta}{M_f^\alpha} \tag{5.61}$$

where R_0, β, and α depend on the temperature and on the developer solvent. Hence, the development time at a constant temperature is given by

$$\tau = \int_0^{T_0} \frac{dz}{|R_0 + \beta/M_f^\alpha|} \tag{5.62}$$

where T_0 is the original thickness and the dependence of the solubility rate on z is calculated from depth–dose functions. The density of PMMA is 1.2 g/cm^3 and its initial molecular weight $M_n \cong 100{,}000$. For a typical film of 5000 Å, the sheet density is 6×10^{-5} g/cm^2. For 20-keV electrons the $G(z)$ value is ~ 1.9. The most probable molecular weight after irradiation (with a dose of 5×10^{-5} C/cm^2) is 7500. The molecular weight distribution curve is well separated from that of the scission fragments.

TABLE 5.3.   Properties of Selected Electron-Beam Resist Materials

Polymer[a]	Tone	Sensitivity (C/cm²)	Resolution (μm)
PMMA	Positive	4×10^{-5}–8×10^{-5}	0.1
P(GMA-co-EA)	Negative	3×10^{-7} (10 keV)	1.0
PBS	Positive	8×10^{-7} (10 keV)	0.5
COP	Negative	4×10^{-7}	1.0
P(GMA-co-EA)	Negative	3×10^{-7} (10 keV)	1.0
PGMA	Negative	5×10^{-7} (20 keV)	1.0
PCA	Positive	5×10^{-7} (20 keV)	0.5

[a] P(GMA-co-EA), poly(glycidyl methacrylate-co-ethyl acrylate); PBS, poly(butene-1-sulfone); PCA, copolymer of α-cyano ethyl acrylate and α-amine ethyl acrylate; PGMA, poly(glycidyl methacrylate); PMMA, polymethyl methacrylate; COP, copolymer methyl methacrylate.

PMMA resists have been routinely used in delineating microwave transistors, charge-coupled devices, surface-wave devices, bipolar memory cells, magnetic bubbles, and solar cells. For the economic fabrication of CMOS ICs by electron-beam writing, it is necessary to use an electron resist much faster than PMMA. Several high-speed electron resists have been identified during the past few years, and some of these are listed in Table 5.3.[65] Some have shown that they are capable of both high speed and high resolution.

5.8.4.2. Negative Resists. The negative electron resist is characterized by sensitivity and contrast. The dose is defined as the number of electrons (coulombs) per square centimeter at a specific voltage required to achieve a desired chemical reaction:

$$D = it/A \tag{5.63}$$

where i is the electron current (A), t is the time (s), A is the area (cm²), and D is the dose (C/cm²).

Figure 5.40 illustrates the definition of sensitivity.[66] The correct feature size is achieved with a dose that results in approximately 50% of the thickness remaining after development. This dose is denoted by $D^{0.5}$. The contrast is expressed as

$$\gamma = |\log(D^0/D_m)|^{-1} \tag{5.64}$$

where D^0 is the projected dose required to polymerize 100% of the initial film thickness and is determined by extrapolating the linear portion of the thickness–dose plot to a value of 1.0 (normalized film remaining). D_m is the minimum dose required for any detectable polymerized layer (or gel) to be formed.

The phenomenological depth–dose description (see Section 5.8.4.1) of electron energy loss can also be applied to the exposure of negative resists.[64] This model predicts a dependence of resist sensitivity on the electron energy and gives an expression for the minimum dose D_m.

The mechanism of primary-electron energy dissipation is the excitation and ionization of the resist atoms producing positive ions and electrons. The rate of

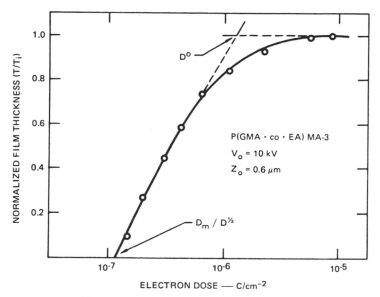

FIGURE 5.40. Fractional film thickness etched after development (normalized to final resist thickness) as a function of electron dose for P(GMA-co-EA) batch MA-3. The accelerating voltage was 10 kV and the initial resist thickness was 0.6 μm. From Ref. 66.

energy absorption by the resist film, dE/dt, is related to the incident current density J_0 by

$$\frac{dE}{dt} = \frac{J_0}{e}\left(\frac{dE}{dz}\right) = \frac{E_0}{R_G}\frac{J_0}{e}\Lambda(f)\ \text{eV cm}^{-3}\,\text{s}^{-1} \tag{5.65}$$

The rate of production of secondaries, dn_s/dt, with a mean excitation energy I_m can be written as

$$\frac{dn_s}{dt} = \frac{1}{I_m}\frac{J_0}{e}\left(\frac{dE}{dz}\right)\text{cm}^{-3}\,\text{s}^{-1} \tag{5.66}$$

or, using the G value, the number of ions or electrons produced per 100 eV lost by the primary:

$$\frac{dn_s}{dt} = \frac{G}{100}\frac{J_0}{e}\left(\frac{dE}{dz}\right) \tag{5.67}$$

and

$$G = \frac{100}{I_m} \tag{5.68}$$

If some type of chemical event (cross-linking) occurs concurrently with the production of internal secondaries, then the G value for cross-linking (G_c) is related to that for ionization as

$$G_c = \alpha G = \alpha \frac{100}{I_m} \tag{5.69}$$

With this the rate of cross-linking is found to be[64]

$$\frac{dn_c}{dt} = \left(\frac{G_c}{100}3\right)\left(\frac{J_0}{e}\frac{dE}{dz}\right)\exp(-\eta t) \text{ cm}^{-3} \text{ s}^{-1} \tag{5.70}$$

where

$$\eta = \frac{G_c}{100 S_0}\frac{J_0}{e}\left(\frac{dE}{dz}\right) \tag{5.71}$$

and S_0 is the initial concentration of available sites. After integration, the concentration of cross-links in the resist, n_c (which determines the degree of gelation, and hence the solubility of the irradiated material), is given by

$$n_c(t) = S_0[1 - \exp(-\eta t)] \text{ cm}^{-3} \tag{5.72}$$

for small ηt

$$n_c \cong \frac{G_c D_0}{100e}\left(\frac{dE}{dz}\right) \approx \eta S_0 t \text{ cm}^{-3} \tag{5.73}$$

Two parameters characteristic of a negative electron resist are the G value and the minimum number of cross-links per cubic centimeter, n_g, necessary to form an insoluble gel.

At the resist interface (i)

$$\frac{n_g}{G_c} = 10^{-2}\frac{D_g(i)}{e}\left(\frac{dE}{dz}\right) \tag{5.74}$$

or, using the energy absorption expression,

$$E_g(i) = 10^{-2} \frac{n_g}{G_c} = \frac{D_g(i)}{e} \left(\frac{dE}{dz} \right) \text{eV cm}^{-3} \tag{5.75}$$

where $E_g(i)$ is the input energy density required to produce an insoluble gel at $z = i$ and $D_g(i)$ is the minimum incident dose necessary for gel formation. Typical gel energy densities are in the range 4×10^{18}–4×10^{22} eV/cm^3.

5.8.4.3. Inorganic Resists. Inorganic resists are not widely used, although there are a variety of niches which they fill very effectively; inorganic resists generally exhibit high contrast and resolution and some are self-developing. Various exposure techniques have been demonstrated, including electron and ion beam, and excimer laser. Some examples follow.

Self-development of aluminum oxide using a 193-nm excimer laser has produced 0.1-μm features.[67] This constituted an important step in an all-day processing sequence. Al/O cermet has been used as a negative resist for ion-beam exposure.[68] The resist is developed using either chemical etch or RIE and 0.1-μm linewidths have been demonstrated. Silver-doped chalcogenide glasses have been used as high-contrast negative resists for deep-UV and electron-beam lithography.[69, 70] Following chemical development, the electron-beam lithography resulted in linewidths down to 30 nm.

Most inorganic resists require high-dose exposures; however, materials such as SrF$_2$ and BaF$_2$ can be used as electron-beam negative resists and require doses comparable to PMMA.[71] Linewidths of 0.1 μm have been attained.

Very high resolution ($\sim$1-nm features) has been demonstrated in materials such as NaCl, Al$_2$O$_3$, MgF$_2$, CaF$_2$, AlF$_3$, TiO$_2$, and M$_g$O using high-current-density ($\sim$10^7 A/cm^2 at 100 kV) electron-beam exposures (see Section 5.8.12). These very-high-resolution positive resists are self-developing. Table 5.4[72] summarizes the properties of some of these high-resolution inorganic resists. The high resolution of these

TABLE 5.4. Characteristics of High-Resolution Positive Resists

Resist	Minimum linewidth	Typical aspect ratio	Deposition	Dose at 100 keV to expose 500-nm-thick layer (C/cm^2)	Mechanism of exposure
PMMA	8–10 nm	45	Spinning	5×10^{-4}	Bond breaking
NaCl	1.5 nm	>40	Sublimation 40-Å grain	10^2–10^3	Dissociation of Cl$_2$ Diffusion of Na
LiF	1.5 nm	>40	Sublimation 50-Å grain	10^{-1}–10^{-2}	Dissociation of F$_2$ Diffusion of Li
MgF$_2$	1.5 nm	>40	Sublimation 50-Å grain	1–10^{-1}	Dissociation of F$_2$
AlF$_2$	1.5 nm	>40	Amorphous	1–10	Dissociation of F$_2$ Diffusion of Al
KCl	1.5 nm	>40	Deposition 50-Å grain	1–10	Dissociation of Cl$_2$ Diffusion of K
Metal–alumina	1.5 nm	>40	Cut thin-film slabs	1×10^7 (2000 Å thick)	

TABLE 5.5. Material Surface Modification under Charged Particle Bombardment

Target	Particle and energy	State of extreme outer surface	Method of surface analysis
AgB, LiF, $ZnBr_2$	0.5 kV e^-	Metal	Metal signal small or absent
BaF_2, CaF_2, SrF_2	$\cong 3$ kV e^-	Im part metal	Auger
CaF_2, MgF_2	3 kV e^-	Metal	Auger
CsBr, CsI, RbBr, RbCl, RbI	0.54 kV e^-	In part metal	Thermal dN/dE for metal
$CuCl_2$, FeF_3	$\cong 1$ kV hv	CuCl, FeF_2	ESCA
CuF_2	$\cong 1$ kV hv	Metal	ESCA
KCl	2 kV e^-	In part metal	Auger
KCl, LiF	1.5 kV e^-	In part metal	Auger
KCl, LiF, NaCl, NaF	0.1–1.5 kV e^-	In part metal	Partial pressure
LI	1 kV e^-	In part metal	Auger
NaCl	0.05–0.1 kV e^-	In part metal	Electron energy loss
			Lack of thermal metal signal:
AgF	6 keV Xe^+	Ag	visible metallization
CdI_2, RbBr, RbCl, RbI	6 keV Xe^+	In part metal	Thermal dN/dE for metal
$CuCl_2$	2 keV Ar^+	CuCl	ESCA
CuF_2, NiF_2	2 keV Ar^+	Metal	ESCA
			Lack of metal signal:
LiI	6 keV He^+	Li	transient halogen signal
	6 keV Ne^+, Ar^+, Kr^+, Xe^+	Li	Lack of thermal metal signal
NaCl, NaI	20 keV He^+, Ne^+, Ar^+, Kr^+, Xe^+	In part metal	Thermal dN/dE for metal
RbI	6 keV Ne^+, Ar^+, Kr^+, Xe^+	In part metal	Thermal dN/dE for metal

inorganic thin films can be explained by the small molecular size and by the fact that the threshold exposure energy is higher than the secondary electron energy.

Sufficiently energetic particle bombardment of insulators results in radiation damage.[77] Significant surface modification can be apparent. Table 5.5[72] lists surface modifications of various halides due to electron, ion, and photon bombardment. A full discussion of this subject is given in Ref. 73. These surface modifications have in some cases been attributed to radiolysis and electron-stimulated desorption. A model for electron-beam self-developing inorganic halide resists, based on this background, is found in the literature.[74-76] Figure 5.41[77] summarizes the model. More recently, a model has been proposed which is based on both ionization processes and momentum transfer events; see Section 5.8.12 for details.

5.8.5. Cross-Sectional Profiles of Single-Scan Lines

In high-resolution electron lithography, it is important to know the cross section of the developed electron resist lines.[78] The shape of the cross section often has a direct influence on the results of other processing steps in the fabrication of electronic or optical devices. The cross-section shape depends, among other things, on the exposure pattern of the resist film. The simplest exposure pattern is a single line

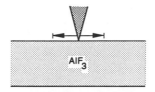

(a) BEGINNING OF EXPOSURE

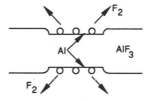

(b) AFTER SOME IRRADIATION

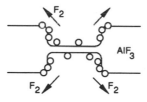

(c) MORE IRRADIATION

(d) ALL HALIDE REMOVED AND METAL FORMS ON WALL

FIGURE 5.41. Mechanism of mass loss (self-development of metal halides electron-beam bombardment). (a) Beginning of exposure; (b) after some irradiation; (c) more irradiation; (d) all halide removed and metal forms on wall.

scanned by a focused electron beam. A more complex exposure pattern may be composed of many segments of such single-scan lines.

The profile of the single-scan resist line depends on the beam energy, the current density distribution of the incident beam as well as the electron scattering effects inside the resist and the substrate. Electron scattering broadens the beam radius in the resist films. The effect of electron scattering on the resulting resist line profile depends very much on whether the resist is positive or negative working (see Fig. 5.42).

It has been shown[79] for negative resists that the line profile is parabolic when the incident radial current density and backscattered dose distributions in the resist–substrate interface plane are Gaussian. The base width W of the parabola after

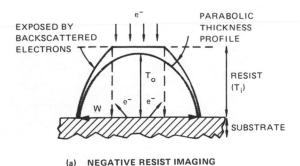

(a)   NEGATIVE RESIST IMAGING

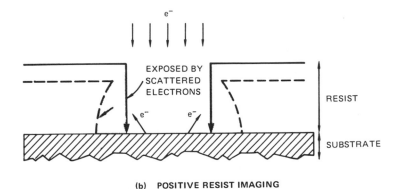

(b)   POSITIVE RESIST IMAGING

FIGURE 5.42.   Scattering effects in electron-beam imaging. (a) Negative resist imaging; (b) positive resist imaging.

development is given as

$$W = 2r' \left[\ln\left(\frac{D^0}{D_m}\right)^{1/2} \right] \tag{5.76}$$

where r' is the Gaussian radius in the resist–substrate interface and is related to the Gaussian radius of the incident beam r by Eq. (2.199) as

$$r'^2 = r^2 + \frac{4z_0^3}{3\lambda} \tag{5.77}$$

where z_0 is the resist thickness and λ is the scattering mean free path (see Section 2.6.1.4).

The height of the resist line (see Fig. 5.42) is given by

$$T_0 = T_i \frac{\gamma}{\ln 10} \ln \left(\frac{D^0}{D_m} \right)$$ (5.78)

where T_i is the initial resist thickness and γ is the contrast of the resist. Equations (5.76) and (5.78) specify completely the cross-section profile of the resist lines if the parameters $l(r)$, r, r', γ, and D_m are given.

The resolution in electron-beam lithography is limited by the electron scattering in the resist and by the backscattering from the substrate, and not by the electron-optical system. Electron beams with diameters in the range of 20 to 200 Å can be routinely produced in electron microscopes. By the use of a thin support substrate (200–500 Å) the effect of backscattered electrons can be eliminated.

The electron scattering in the resist, which increases the dimensions of the exposed spot over the size of the beam, can be drastically reduced by the use of a thin (30–1000 Å) monomolecular resist prepared by the Langmuir method (see Section 4.8.3) and a reduced accelerating voltage (5 keV). Such a resist film built from $CH_2=C-(C_2)_{20}-COOH$ molecules, superimposing monolayers, each 30 Å thick, gave patterns of 600-Å resolution after development.[80]

Lithography in the nanometer range (10–100 Å) can be achieved by the use of a thin salt resist deposited on an appropriate substrate. The salt "evaporates" under electron-beam bombardment and delineates the pattern on the substrate.[81]

Here the pattern writing and development can be carried out in one step. The selection of small molecules (alkali halides and aliphatic amino acids) for the resist material with molecular grain size yields resolutions in the range 15–20 Å.[82]

5.8.6. Development

When an organic polymer is subjected to electron-beam radiation, energy is absorbed and the polymer undergoes chemical changes involving two possible events. Polymer cross-linking results when adjacent polymer chains chemically cross-connect to form a complex, three-dimensional structure for which the average molecular weight increases. Chain scission (degradation) is the result of chemical bonds being broken by the electron radiation; this results in a lower average molecular weight. In general, both processes occur to different degrees in the same polymer. In PMMA, chain scission processes dominate cross-linking events under moderate (less than $2 \times 10^3 C/cm^2$) doses of kilovolt electron-beam radiation. In the case of chain scission, the fragment molecular weight M_f is inversely proportional to the total absorbed energy loss density ε, as given by[79] (Section 5.8.4.1)

$$M_f = \frac{1}{g\varepsilon}$$ (5.79)

where g represents an efficiency factor that is a characteristic of a given polymer, indicating its susceptibility to electron-beam degradation.

Electron-beam-sensitive resists are developed by the method of fractional solution. A simple model states that for a given polymer, organic solvents at a given temperature dissolve polymer chains of molecular weight below some critical value M_{fc}. For PMMA a solvent is used that removes the polymer chains in the irradiated volume while leaving the surrounding chains of higher molecular weight; it is therefore a positive resist.

5.8.7. Proximity Effects

As the incident electrons penetrate the resist, they undergo both elastic and inelastic scattering events; consequently, a delta-function beam will spread. The amount of spreading depends upon the cumulative scattering history of the penetrating electrons.

Proximity effects[83] are created by scattered electrons in the resist and backscattered electrons from the substrate, which partially exposure the resist up to several micrometers from the point of impact. As a result, serious variations of exposure over the pattern occur when pattern geometries fall in the micrometer and submicrometer ranges.

The proximity effect is dependent on the beam-accelerating voltage, the resist material and thickness, the substrate material, the resist contrast characteristics, and the chemical development process used. Where the beam diameter is appreciable, the current density profile of the beam must also be considered. To achieve accurate pattern delineation, it is therefore necessary to apply some method of exposure adjustment to compensate for the proximity effect. One such method is to vary the charge reaching each pattern element by controlling the scanning speed or the beam current.

5.8.8. Specifications for Electron-Beam Exposure Systems

The most important quantities that must be specified for electron-beam lithography are the following:

- τ = pixel exposure time (s)
- D = dose for large area (C/cm^2)
- fD = dose for one pixel ($f \leq 1$)
- l^2p = pixel area (cm^2)
- l^2p/τ = area (cm^2) exposed per second
- j = current density in the spot (A/cm^2)

These quantities are not independent of each other. One can see, for example, that

$$\frac{1}{\tau} = \frac{j}{fD} \qquad (5.80)$$

and

$$\frac{l_P^2}{\tau} = \frac{j l_P^2}{f D} \tag{5.81}$$

An important specification for these machines is the operating speed in pattern generation.

Both the raster scan and vector control pattern generators face basic limitations in operating speed. The rate at which a pattern is generated can be limited by two factors:

- The bandwidths of the systems that control the electron beam place limits on how fast the pattern can be written.
- Resist sensitivity can limit the writing rate of the electron beam.

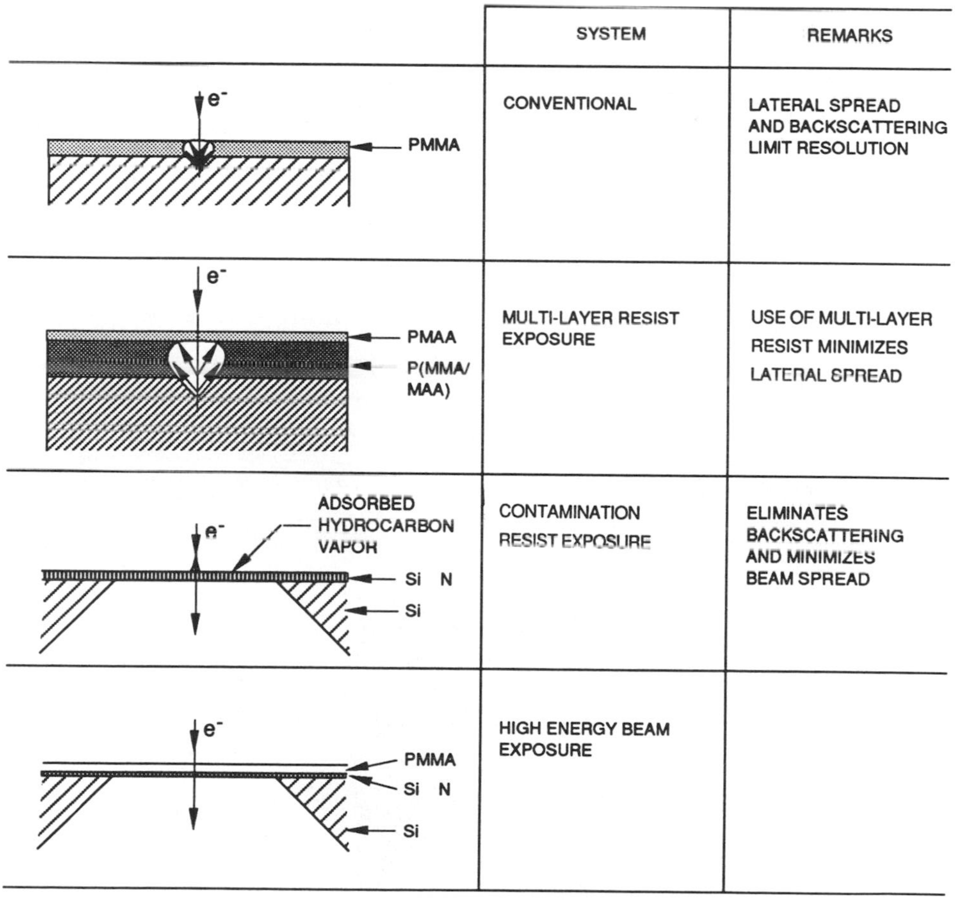

	SYSTEM	REMARKS
e⁻ PMMA	CONVENTIONAL	LATERAL SPREAD AND BACKSCATTERING LIMIT RESOLUTION
e⁻ PMMA P(MMA/ MAA)	MULTI-LAYER RESIST EXPOSURE	USE OF MULTI-LAYER RESIST MINIMIZES LATERAL SPREAD
e⁻ ADSORBED HYDROCARBON VAPOR Si N Si	CONTAMINATION RESIST EXPOSURE	ELIMINATES BACKSCATTERING AND MINIMIZES BEAM SPREAD
e⁻ PMMA Si N Si	HIGH ENERGY BEAM EXPOSURE	

FIGURE 5.43. Electron-beam exposure configurations.

5.8.9. *Special Lithography Techniques*

The minimum feature size fabricated by any electron-beam lithography process is influenced by the choice of the exposure tool, by the interaction processes in the target materials, and by development.

There are several experimental lithography techniques which try to minimize some of the adverse effects in the patterning process. In the following sections, some of these innovative approaches will be discussed. Figure 5.43 shows some of the electron-beam exposure configurations which minimize feature size.[84-91]

5.8.10. *Contamination Lithography*

The contamination lithography[92] process uses a vapor resist which condenses onto the surface of the target under electron bombardment. Contamination resist is formed when the electron beam "cracks" the thin layer of hydrocarbons at the sample surface. As this process continues, a cone of resist builds up at the point of impact of the electron beam, as shown diagrammatically in Figure 5.44. The rate of buildup can be increased by increasing the partial pressure of hydrocarbons in the immediate vicinity of the sample with a local source of vapor.

Typically, a dose of $0.1–1$ C/cm^2 is required to produce an adequate buildup of "exposed" vapor to protect 100-nm-thick metal layes (e.g., gold and niobium) from ion etching. This technique also can be used for "*in situ*" applications where the

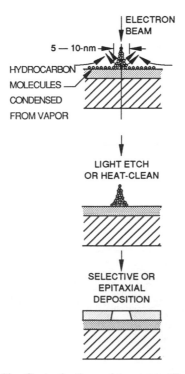

FIGURE 5.44.   Contamination resist exposure. From Ref. 94.

vapors forming the resist are selectively introduced to the chamber. A variety of nanostructures have been fabricated by this lithography[92] in the past three decades.

5.8.11. Momentum Transfer Damage Lithography

It is possible to fabricate structures smaller than the secondary electron range by using mostly energetic primaries for exposure, or causing radiation damage. For silicon a primary electron energy of 150 keV is needed to cause momentum transfer damage (see Fig. 5.45). The electron-beam-caused damage could decrease or enhance the etching rate of the material. The new method for fabricating nanometer-scale structures has been applied to make 20-nm-size relief structures using a 500-kV finely focused electron beam.[93] In this technique, the damage is located within the beam diameter which could be as small as 5 Å. It is interesting to note that probably only the primary electrons are contributing to the exposure and the secondaries emanating from a large surface (low current density) make little contribution to the exposure.

5.8.12. High-Current-Density Exposures

When working with very high electron-beam current densities ($>10^7$ A/m^2 at 100 kV) and nanometer-size probes, exposure of certain inorganic thin films results in the *direct* removal of the material—producing structures on the 1 to 30-nm scale,

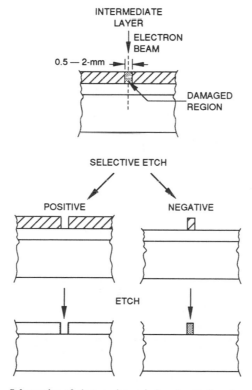

FIGURE 5.45.   Schematics of electron-beam-induced radiation-damage lithography.

which require no development.[94, 95] Suitable inorganic materials are many fluorides and oxides, and some alkali halides (e.g., amorphous AIF_3, various aluminas, NaCl). Very high doses are required for direct material removal; e.g., 25 C/cm^2 for amorphous AIF_3 at 100 kV (which can be compared with 10^{-3} C/cm^2 for PMMA, with the same beam voltage. The mechanism of structure formation is thought to vary from one material to another; it is found to depend on both ionization processes and momentum transfer events caused by the primary electron.[94,96,97] The process of nanometer structure formation depends on the primary electrons and is insensitive to secondary electrons[98]; consequently, very high packing densities of structures can be achieved; e.g., 2-nm-diameter holes at 4-nm center-to-center spacing in amorphous AIF_3.[75]

5.8.13. High-Energy Electron-Beam Lithography

High-voltage (>150 kV) lithography offers a number of advantages over conventional lithography using electron energies in the range of 15 to 50 keV. The most important effects are the reduced proximity effect, better linewidth control, and higher aspect ratios (line height to gap width) for lines using a single resist layer.

Figure 5.46 shows the result of a Monte Carlo calculation of the scattered electron trajectories at 50, 100, 250, and 500 kV.[99] These results indicate that with higher energies the scattering is increasingly confined in the forward direction. This reduction in lateral scattering makes it possible to expose nanometer-size lines with high (>10) aspect ratios. The only disadvantage of the high-energy exposure process is that the critical exposure dose increases with the beam energy.

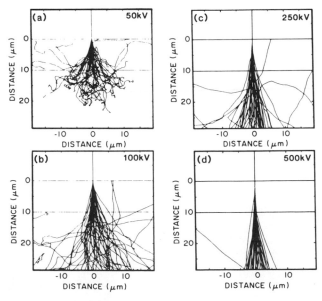

FIGURE 5.46.   Electron trajectory simulation at 50, 100, 250, and 500 kV using Monte Carlo calculations. From Ref. 99.

The high-voltage lithography tool could be very useful in building three-dimensional structures where small linewidths with nanometer dimensions and high aspect ratios (≥ 30) are required.[100]

5.8.14. Low-Energy Electron-Beam Lithography

Another way to minimize the interaction volume in the resist is to use low-energy electrons. Using low-energy (≤ 2 kV) electrons for lithography reduces the lateral scattering and the penetration into the resist. Because most of the electrons are used for exposure under the spot in thin resists, the efficiency of the exposure

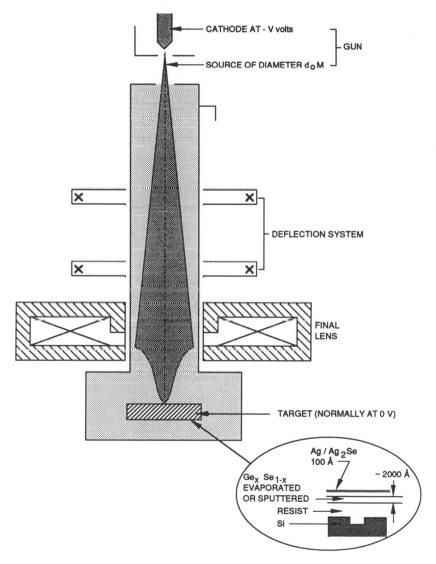

FIGURE 5.47. Low-energy electron-beam column for lithography. From Ref. 99.

process is high. The difficulties center around the generation of a small ($\leq 1\,\mu$m) spot size with high current density, because lens elements have poorer performance at low energies due to space-charge effects.

The electron-optical column of a low-energy lithography tool is shown in Fig. 5.47.[101] This optical column uses a retarding field type of optics to generate the low-energy electrons to expose a multilayer resist. The structure of the resist system is shown in Fig. 5.47. The price paid for lithographic efficiency is increased technical difficulties (e.g., positioning the wafer and preparing the multilayer resist).

Another low-energy (<100 eV) electron-beam lithography system uses the tunneling effect as a source of electrons. The sources in scanning tunneling micros-copy (STM) are capable of producing microfocused, low-energy ($\cong 10$ eV) electron beams.[102–104] Because low-energy electrons (5–50 eV) cause radiation damage in the surface layers of thin-resist films, they find applications in lithography and surface modifications. For patterning, the following thin-resist films can be used:

- Contamination resist
- Metal halides
- Langmuir–Blodgett films
- PMMA

Using Langmuir–Blodgett (LB) films at low doses (10^{-4} C/cm^2), linewidths on the order of 10 nm could be obtained.[105, 106] The LB films can be deposited pinhole-free and with uniform thickness.

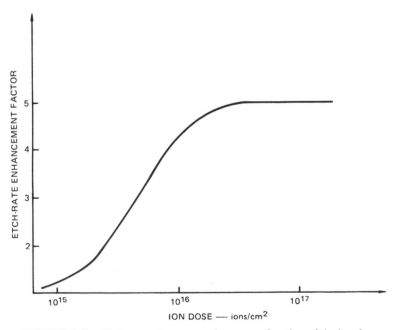

FIGURE 5.48.   Etch rate enhancement factor as a function of the ion dose.

5.9. ION-BEAM LITHOGRAPHY

5.9.1. Introduction

Ion beams will become increasingly more important for submicron lithography. They can be used not only for direct writing onto wafers[107] but also for mask fabrication.[108, 109] In these operating modes the ion-beam systems can directly

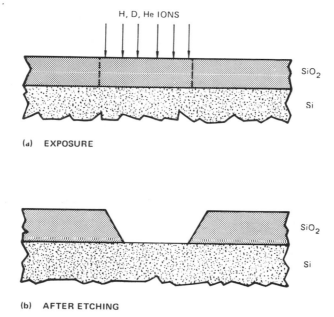

(a) EXPOSURE

(b) AFTER ETCHING

FIGURE 5.49.   Radiation damage patterning of SiO_2 with H, D, and He Ions. (a) Exposure; (b) after etching.

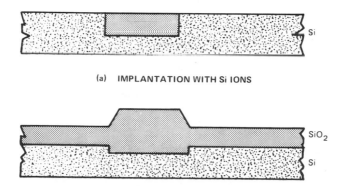

(a) IMPLANTATION WITH Si IONS

(b) THERMAL OXIDATION (~950°C) ENHANCED LINEAR OXIDATION RATE
      IN THE ION-IMPLANTED AREA

FIGURE 5.50.   Variable-thickness SiO_2 pattern generation on silicon. (a) Implantation with Si ions; (b) thermal oxidation (~950°C) enhanced linear oxidation rate in the ion-implanted area.

replace the electron-beam machines. They can also be used for ion implantation, direct ion milling, or for producing bombardment damage that will enhance a follow-up sputtering or etching step. Figure 5.48 shows the dependence of the etch rate factor in the ion proton) dose. Oxide layers can be patterned without any resist because of the enhanced etch rate initiated by radiation damage. The etch rate dependence on the ion dose shows a saturation for SiO_2 with a dose of 2×10^{16} protons/cm^2. Figure 5.49 depicts the etching behavior of SiO_2. The increase in the linear oxidation rate of implanted silicon can be used to generate oxide layers of

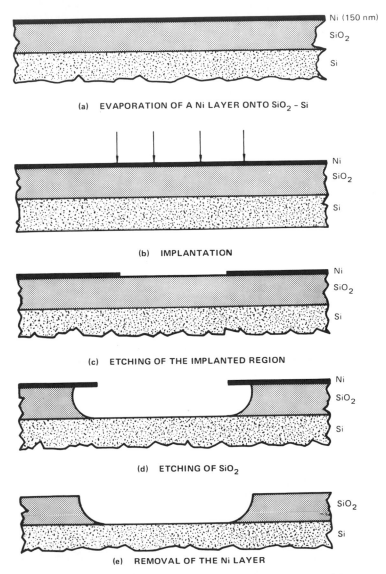

FIGURE 5.51.   Pattern generation with metallic resist. (a) Evaporation of a Ni layer onto SiO_2–Si; (b) implantation; (c) etching of the implanted region; (d) etching of SiO_2; (e) removal of the Ni layer.

variable thickness (see Fig. 5.50). After silicon-ion implantation and thermal oxidation, the SiO_2 thickness increases in the irradiated area. Nitrogen implantation into Si retards the oxidation rate of the irradiated areas and the nitrogen-ion bombardment of the thin films changes the resistivity of the film.

Ion irradiation can generate patterns directly in metal layers (Ni or Mo) in a resistless process. Thus, thin Ni or Mo layers can be used as metallic ion-sensitive resists. The pattern generation process for a metallic resist is shown in Fig. 5.51.

The resolution of ion-beam lithography is inherently good because the secondary electrons produced by an ion beam, being of low energy, have a short diffusion range and practically no backscattering occurs.

Ion-projection machines with or without mask demagnification can be used for the resistless patterning of semiconductors (Si, GaAs, poly-Si), insulating layer (SiO_2, Si_3, N_4), metals (Al, Ni, Mo, Au), and also patterning organic thin films (resists). A further possible approach is shadow printing through thin freestanding and self-supporting mask patterns. The use of such masks is limited to cases where no free-standing opaque areas are needed. Employing a thin amorphous carrier membrane sufficiently transparent for light ions can solve this problem, but the trade-off is the wide-angle scattering of the energetic ions in the membrane. This angular dispersion makes the replication of patterns by means of proximity exposure difficult. Contact exposure, on the other hand, results in a decreased yield and reliability.

A different experimental approach for masking is based on the use of a $\langle 110 \rangle$ silicon membrane 3–6 μm thick and a thin (1000 Å Au) metal pattern to be

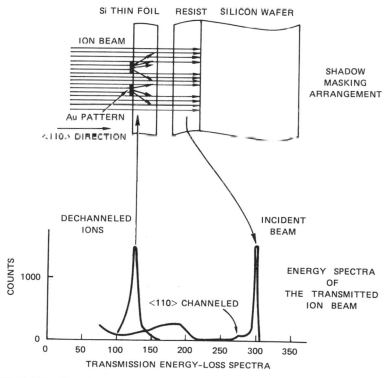

FIGURE 5.52. Experimental arrangement for ion-beam shadow printing. From Ref. 110.

duplicated.[110] The resist film is exposed through the single-crystal silicon foil in the channeling direction, using a collimated helium beam with an initial energy of 300–2000 keV. The masking technique employs the dechanneling effect of a thin metal layer as well as the difference in energy loss for random and channeling directions (see Fig. 5.52).

Other examples of different masks are shown in Fig. 5.53.[111] Stencil masks have been used for experimentation, but since these cannot support structures in open areas they are of limited use.[112–114] Extremely thin (0.1 μm) masks of materials such as aluminum oxide have been used, but there are difficulties in maintaining the flatness of such a thin mask over a large area, especially when it is heated. Channeling masks seem to show considerable promise. These have been made from both ⟨100⟩ and ⟨110⟩-oriented silicon, the former at a thickness of 0.7 μm and the latter a thickness of 6 μm, with gold absorbers or scatterers (in the case of the thick mask)

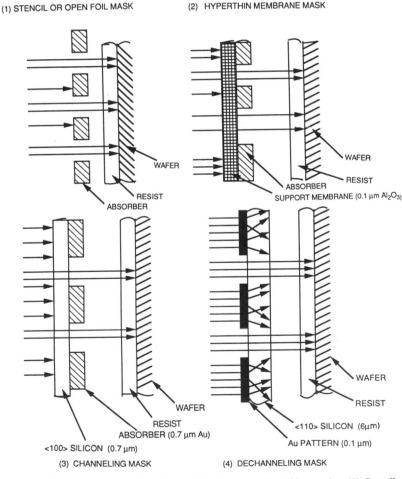

FIGURE 5.53. Different types of masks used for ion projection lithography. (1) Stencil or open foil mask; (2) hyperthin membrane mask; (3) channeling mask; (4) dechanneling mask. From Ref. 111.

to define the pattern. It is most important that the beam not diverge on passing through the mask.

Microfocused ion beams can be used to expose resists, machine materials, and change the electrical and mechanical properties of semiconductors by implantation. All of these can be accomplished directly without masks.

Until recently the limited intensity of ion-beam sources made these applications impractical, but the development of high-intensity and stable liquid-metal (LM) ion sources[115, 116] are highly encouraging for future applications.

5.9.2. Ion-Beam Resists

Chemical changes of resist materials can be induced effectively also by ion-beam exposure. The amount of energy deposited per unit path length of an ion beam is much greater than that of an electron beam of the same energy. Therefore, sensitivity to the ion beam measured in terms of the incident charge per unit area is expected to be higher. High resolutions also can be expected, since the lateral spread of ion beams in resist films is much less than in the case of electron beams. Using ion beams, the energy dissipation of ions in the resist is caused by electronic and nuclear collisions (see Sections 2.7.2.1 and 2.7.2.2). Since the effects of these two components are considered to be independent of each other, the radiation yields corresponding to both components can be estimated separately.[117] The two components are different in their contribution to chemical events, and energy deposition rates corresponding to nuclear and electronic collision losses are also different in their dependence on the penetration depths, as shown in Fig. 5.54; thus, the radiation yield is separated into two components and the number of scissions is given by the summation of both contributions. Therefore, the average molecular weight M_f after ion exposure is given as (see Section 5.8.4.1):

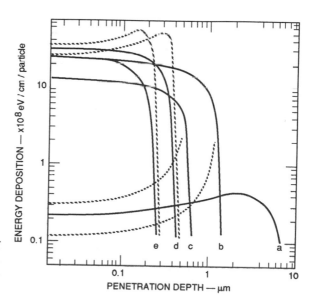

FIGURE 5.54. Calculated curves of the energy deposition rate in PMMA resist as a function of the penetration depth. From Ref. 117.

$$M_f = \frac{M_n}{1 + KQ[G_e(z)S_e(z) + G_n(z)S_n(z)]} \tag{5.82}$$

where $G_e(z)$ and $G_n(z)$ are the radiation yields for chain scission caused by electronic and nuclear collisions, respectively.

The solubility rate of the polymer can be expressed (see Section 5.8.4.1) as

$$R = R_0 + \beta M_f^{-\alpha}(z) \tag{5.83}$$

where R_0, β, and α are empirically determined constants. When the resist film is very thin, $S_e(z)$ and $S_n(z)$ can be regarded as constants, and the resist thickness removed, d, after development is given as

$$d = RT \tag{5.84}$$

where T is the development time. For thick films, the relationship between T and the dissolved thickness d becomes

$$T = \int_0^d \frac{dz}{R_0 + \beta M_f^{-\alpha_1}(z)} \tag{5.85}$$

for PMMA the best fit between the theoretical curves calculated from Eq. (5.85) and the experimental results is found when[117]

$$R_0 = 84 \text{ Å/min}$$
$$\beta = 3.9 \times 10^8 \text{ Å/min}$$
$$\alpha = 1.41 \tag{5.86}$$
$$G_e = 1.7$$
$$G_n = 0.9$$

Then the radiation yields for scission caused by the electronic collisions of ions are nearly equal to those in an electron exposure, where $G = 1.9$.

The radiation yields caused by nuclear collisions are one-half in PMMA in comparison with the radiation yields in electron-beam exposures.

5.9.3. Liquid-Metal (LM) Ion Sources

LM ion sources have been developed[115–118] for a fairly wide variety of metallic elements (Ga, Bi, Ge, Hg, Au, Woods metal, and alkali metals).

The mechanism of the LM ion source operation (see Section 2.3.5) appears to involve the formation of a field-stabilized cone of the liquid metal ("Taylor cone") from which field evaporation and ionization occur. The supply of material to the cone apex occurs via viscous flow along the emitter surface driven by the capillary forces and the electrostatic field gradients near the emitter tip.

In a scanning ion-beam machine,[107] a LM Ga source has been utilized in the optical column to focus 0.1 nA of Ga + current into a 0.1-μm beam diameter. The experimentally measured brightness value is

$$\beta = \frac{1}{\pi a^2 A_s} = 3.3 \times 10^6 \, \text{A/cm}^2 \, \text{sr} \tag{5.87}$$

at the image plane of the 57 kV Ga + beam. Here A_s is the emitting area, i.e., $A_s \cong \pi a_0^2 r_a^2$, where $a_0 \cong 0.2$ rad is the emission half-angle and $r_a \cong 600$ Å. Extracted currents of 1–10 μA can be achieved for gallium beams.

Energy distribution measurements[119] have shown that the energy spread (ΔE) increases with both current and particle mass. The energy spread for Ga, In, and Bi are 5, 14 and 21 eV, respectively, at an angular intensity of 20 μA/sr.

Thus, a small acceptance angle (~1 mrad) still allows a few nanoamperes of ion-beam current to be realized.

5.9.4. Scanning Systems

Figure 5.55 shows a high-intensity scanning-ion probe with a LM source.[107] The imaging of the source is carried out by a unit-magnification accelerating lens

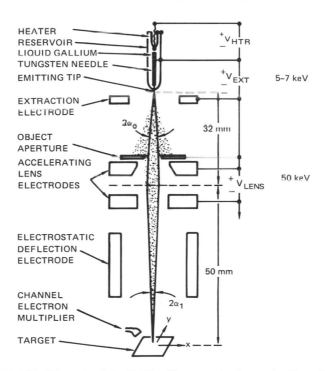

FIGURE 5.55. Schematic of the 57-kV gallium scanning-ion probe. From Ref. 107.

with a post lens deflector to form the scanned beam. The electrostatic deflector enables the beam to be scanned in the x and y directions.

The deflector under computer control can move the beam in a line or raster scan at a speed of $5 \times 10^4 \, \mu m/s$.

The spot diameter is limited by chromatic aberration and the lens acceptance angle α_0. The spot diameter is given by

$$d = C_c \alpha_0 \frac{\Delta E}{E_0} \approx 1000 \text{ Å} \qquad (5.88)$$

where

$$
\begin{aligned}
\alpha_0 &= 1.2 \text{ mrad} \\
C_c &= 33 \text{ min} \\
\Delta E &\approx 14 \text{ eV} \\
E_0 &\approx 5.7 \times 10^4 \text{ eV}
\end{aligned}
\qquad (5.89)
$$

This fine microbeam can be used for micro machining, doping, or resist exposure. It was found, however, that the exposure properties of resists are different from what would be expected on the basis of their electron exposure properties.[120]

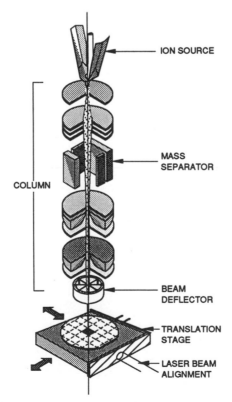

FIGURE 5.56. Schematic of a focused ion-beam system including source, column, and sample stage. From Ref. 121.

A complete focused ion-beam machine is depicted in Fig. 5.56 showing the main parts, i.e., the ion source, the ion optical column, and the table for the sample.[121]

In conclusion, one can say that the liquid ion sources have excellent properties for applications in microfabrication. Ion currents of $1-100\,\mu A$ may be obtained, corresponding to source angular intensities of up to $10^{-4}\,A/sr$, which is approximately 100 times greater than the gas-phase field-ion source.

The UV, X-ray, electron-, and ion-beam systems have resolutions below $1\,\mu m$. The ion-beam spot size could be inherently even smaller, with a resolution as low as $100-500\,Å$. From the discussions in this chapter, it is clear that there are several avenues for the realization of submicron lithographies and fabrication processes. The actual method used in a given submicron fabrication process will ultimately depend on the characteristics of the fabricated device itself.

5.9.5. Projection Systems

Techniques using focused ion beams in parallel exposure [ion projection lithography (IPL)] offer opportunities for sub-0.5 μm device fabrication, combining lithography with ion implantation and the ion beam modification of materials.

With demagnifying ion projection, high-volume-production techniques can be developed in the nanometer (500–100 nm) range.[122]

Inserting a grid reticle into the ion-optical wafer stepper, the multiple focused ion beams on the wafer can be scanned simultaneously generating a powerful production

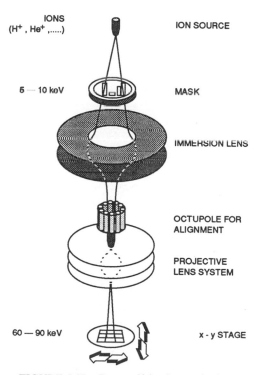

FIGURE 5.57. Demagnifying ion projection.

method for regular structures (e.g., memory elements, gratings). Figure 5.57 shows the basic principles of demagnifying ion projection: ions are extracted from a source which is connected to an ExB mass analyzer to obtain the desired ion species. The resulting divergent ion beam illuminates the mask with an ion energy of about 5 keV. An immersion lens accelerates the ions penetrating the mask openings to a final energy of 60 keV to 90 keV, or higher (up to 400 keV, depending on system design). The projection lens system generates an ion image of the mask structures with 1 : 10 (or 1 : 5) demagnification.[123–126]

Since IPL is a parallel exposure technique, the exposure times are shortened by several orders of magnitude.

Figure 5.58 summarizes the calculated exposure times for an IPL machine which uses ion sourcse of different current densities. Inserting a grid mask into an IPL machine, an ion multi-beam scanning machine can be implemented. This ion multi-beam scanning machine can be used for the production of line grids with sub-0.5-μm periodicity for integrated optics, surface acoustic wave devices, and many other applications such as graphoepitaxy.

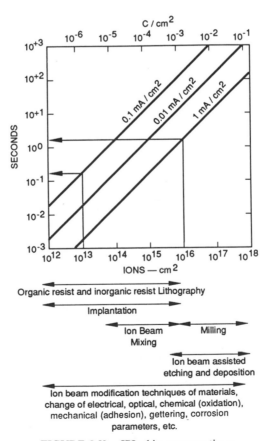

FIGURE 5.58.   IPL chip exposure times.

5.9.6. *Multibeam Systems*

Figure 5.59 shows the optical column design for a multiple-beam machine using the fly's-eye principle[127–129] (see Section 5.8.3). In this machine the multiple beams are created by a screen lens, which consists of an array of holes in a planar metal electrode. The actual screen lens consists of several metal and semiconductor layers so as to minimize sputtering damage. A similar screen lens has been developed for a multi-beam electron machine that is used for large-area display lithography and pattern transfer for micromechanical structures. The screen lens is used in conjunction with a single-object aperture, which is illuminated by an ion beam. This object

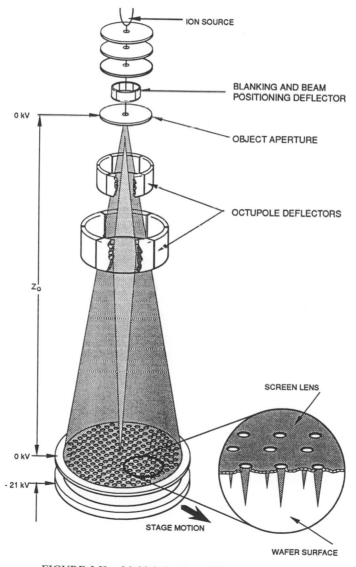

FIGURE 5.59. Multiple ion-beam lithography machine.

aperture is then demagnified by every hole in the screen lens, simultaneously forming multiple images of the object (i.e., one image for each lens in the array). Each beamlet can be simultaneously deflected by the deflector system. The deflector system consists of two octupole deflectors operating at a low (<100 V) voltage level for faster deflection. The top octupole deflector shifts the beam away from the screen lens array; the second octupole deflector returns the beam onto the screen lens. The net results is a shift of the virtual position of the object aperture in its plane and a corresponding demagnified shift in the image. This ion beam tool is designed for specific applications where only one lithography step is needed.

The most important design formulas for both the electron- and the ion-beam machines can be summarized as follows:

- The demagnification of the object aperture on the wafer is given as

$$M \cong \frac{Z_0}{2Z_i} \tag{5.90}$$

where Z_0 is the object screen lens distance and Z_i is the distance between the screen lens and the wafer.
- The focus condition in first order is given by the formula

$$V_i = 9V_0 \tag{5.91}$$

where V_i is the potential on the wafer (screen lens is ground) and eV_0 is the drift energy of the electrons or ions.
- The amplitude of the deflection for a deflection and refraction system is given as

$$d = \frac{Z_i Z_D E_D}{2\sqrt{V_0}\sqrt{V_i}} \tag{5.92}$$

where E_D is the deflection field (V/cm), Z_D is the effective length of the deflector electrodes, and the other quantities are as previously defined.

5.9.7. Resolution

Based on some experimental data,[130, 131] one can theorize that the resolution for ion-beam lithography, similar to electron-beam lithography, is related to the mean energy of the secondary electrons. Experiments indicate that the mean energy of ion-produced secondaries is less than the electron-produced secondaries, for the same primary beam energy. Now if one relates the measured secondary electron energy values to the mean free path of these electrons, one can get a mean range (~resolution) on the order of 1–10 nm. However, this ignores any beam broadening in the target. Perhaps a better estimate can be given by considering the shapes, the surface area of the interaction volumes, and the secondary electron fluxes from these surfaces.

TABLE 5.6. Lateral Resolution Limits Δ from the Secondaries

Primary particles (keV)	Secondary particles	
	Electrons (nm)	Ions
Electrons		
>50 keV	<100 nm	—
10–50 keV	≅100 nm	—
<5 keV	<100 nm	—
Ions	0.5–10 nm	Range of Secondary ions, R_s (E)

Also, since the interaction volume is usually larger and closer to the surface for electrons, one expects an inferior inherent resolution for electron-beam lithography.

One can conclude that the resolution limit for ion-beam lithography will satisfy the needs in nanoelectronics, but not in molecular electronics where 0.1 to 1-nm resolution will be required. Table 5.6 shows the estimated lateral resolution limit Δ which is due to the secondaries (electrons, ions, photons) that are created by charged primary particles (electrons, ions).

5.10. SUMMARY

5.10.1. Beam Processing

Particle beams (photons, electrons, and ions) are frequently used for micro- and nonfabrication processes. The electron beam is the leading tool for mask making, conventional (with resist) and direct-wire lithographies, and ion beams used for doping. Table 5.7 lists the fabrication processes and the types of particle beams used. From Table 5.7 it is evident that ion beams can be used for the implementation of all of the fabrication processes in microscience.

Ion-beam fabrication is also a candidate for nano structure technologies. Laser beams can be used as a general tool set for fabrication but are limited by the wavelength of the laser. X rays and UV light are especially interesting in device fabrication where large aspect ratios are needed (sensors) and in micromachining.

There is a trend in fabrication to implement all of the process steps under UHV in the same vacuum system by the usage of a few particle beams. This concept will not require "breaking the vacuum" and therefore would contribute to increased device yield and turnaround time.

5.10.2. The Road to Nanotechnologies

Presently the most familiar route to the fabrication of nanostructures is electron-beam lithography followed by dry etching and electron-beam testing.[91, 99, 132–135] There are other alternatives as shown in Table 5.8 but the research community is the most familiar with electron-beam processing. Ion-beam lithography and other assisted processes are in competition with electron-beam processing especially for very

TABLE 5.7. Microfabrication Processes with Particle Beams

Technology[a]	Lithography	Etching	Doping	Deposition	Epitaxy	Material modification	Annealing	Diagnoses
Atoms beams								
Neutral, BB		Prod		Prod	Prod			Lab
Neutron, BB			Prod					
Cluster, FB				Feas	Feas			
Cluster, BB				Prod	Feas			
Ion, FB	Prod	Prod	Prod	Prod		Prod	Feas	Lab
Ion, BB	Prod[c]	Prod	Prod	Prod		Prod	Feas	
Electron beams								
FB	Prod	Lab		Lab			Lab	Prod
BB	Prod[c]			Lab			Lab	
Proton beams								
X ray								
FB(laser)								Lab
BB	Prod						Feas	
Ultraviolet								
FB(laser)	Lab	Feas	Feas	Feas	Lab(?)		Feas	Lab
BB	Prod						Feas	
Visible								
FB(laser)	Lab						Prod	Lab
BB	Prod[c]						Prod	

[a] BB, broad beam; FB, focused beam.
[b] Feas, feasible technology; Lab, laboratory technology; Prod, production technology.
[c] Plus mask.

TABLE 5.8. Alternatives to Electron-Beam Lithography

Lithographies	Pattern range
Electron-beam vector/raster Ion beam (IPL, stencil) X ray (synchrotron)	Submicron region (0.5–1 μm)
Scanning from emission Scanning tuneling—electron beams Ion projection (100 nm) Scanning X ray	Deep submicron region (0.1–0.5 μm)

High-energy electron beam (2 nm)
    Contamination
    High current density
    Damage—electron beams
Ion beams (30 nm)
Special nanolithographies
    Shadow deposition
    Step edge
	Nanometer region
	(<100 nm)

small structures (<10 nm). Ion beams can be focused to less than 30 nm but high-energy electron beams can be focused to a spot less than 2 nm in diameter.[132] This topic is treated comprehensively in Ref. 136. The limits for fabrication will be discussed in more detail in Chapter 7.

PROBLEMS

1. a. Distinguish between (i) resolution, (ii) linewidth, (iii) edge acuity, and (iv) accuracy.

   b. Discuss the difference between contact, proximity, and projection printing lithography. In particular, describe the differences in resolution, accuracy, depth of field, and linewidth.

   c. How are resists characterized? What are the advantages and disadvantages of positive and negative resist?

2. As IC designs are pushing the limits of optical lithography, new tools are being developed for submicron work. Consider two examples, direct write electron-beam lithography and X-ray lithography. Discuss the following:

   a. The main technical differences.

   b. The choice of resist materials.

   c. How is the alignment carried out between wafer and mask and how is registration achieved between consecutive layers?

3. a. How do conventional electron-beam resists function?

   b. Give some examples of inorganic resists and describe how they operate.

   c. What are the advantages of inorganic resists over conventional electron-beam resists?

4. a. Describe how ion-beam lithography works.

   b. What kinds of resist are used and how do they operate?

   c. Describe several alternatives for the fabrication of masks suitable for ion-beam lithography.

5. A mask consists of a quartz plate coated on one side with a thin layer of chromium on which the chromium has been removed from a circular area of radius 1 μm. A silicon substrate coated with a resist layer 1 μm thick is placed in nominal contact with the chromium-coated side of the mask. This leaves an average spacing of $\frac{1}{4}$ μm between the chromium and the resist surfaces. Discuss the effects of diffraction and interference when the mask is illuminated with ultraviolet light. Sketch the cross section through the resist after exposure and development. Assume incoherent light at a wavelength of 350 bnm.

REFERENCES

1. W. M. Moreau, *Semiconductor Lithography*, Plenum, N.Y. (1988)
2. B. J. Lin, Deep UV lithography, *J. Vac. Sci. Technol.* **12**(6), 1317 (November–December, 1975).
3. M. C. King, Future development for 1 : 1 projection photolithography, *IEEE Trans. Electron Devices* **ED-26**(4), 705 (April, 1979); M. C. King and E. S. Muraski, New generation of 1 : 1 optical projection-mask aligners, *Proc. SPIE* **174**, 70 (1979); J. W. Bossung and E. S. Muraski, Optical advances in projection lithography, *Proc. SPIE* **135**, 16 (1978); J. W. Bossung, Projection printing characterization, *Proc. SPIE* **100**, 80 (1977).
4. H. Binder and M. Lacombat, Step-and-repeat projection printing for VLSI circuit fabrication, *IEEE Trans. Electron Devices* **ED-26**(4), 698 (April, 1979); G. L. Resor and A. C. Tobey, The role of direct step-on-the-wafer in microlithography strategy for the '80's, *Solid State Technol.* **22**(8), 101 (August, 1979); H. E. Mayer and E.W. Leobach, A new step-by-step aligner for very large scale integration (VLSI) production, *Proc. SPIE* **221**, 9 (1980).
5. M. C. King and M. R. Goldrick, Optical MTF evaluation techniques for microelectronic printers, *Solid State Technol.* **19**(2), 37 (February, 1977).
6. F. H. Dill, A. R. Neureuther, J.A. Tuttle, and E. J. Walker, Modeling projection printing of positive photoresists, *IEEE Trans. Electron Devices* **ED-22**(7), 456 (July, 1975).
7. J.D. Cuthbert, Optical projection printing, *Solid State Technol.* **20**(8), 59 (August, 1977).
8. F. H. Dill, Optical lithography, *IEEE Trans. Electron Devices* **ED-22**(7), 440 (July, 1975).
9. M. J. Bowden and L. F. Thompson, Resist materials for fine line lithography, *IEEE Trans. Electron Devices* **ED-22**(7), 456 (July, 1975).
10. F. W. Dill, W. P. Hornberger, P. S. Hauge, and J. M. Shaw, Characterization of positive photoresist, *IEEE Trans. Electron Devices* **ED-22**(7), 455 (July, 1975).
11. D. A. McGillis and D. L. Fehrs, Photolithographic linewidth control, *IEEE Trans. Electron Devices* **ED-22**(7), 471 (July, 1975).
12. Mann 4800, manufactured by CICA-Mann Corp. Burlington, Mass.
13. W. W. Ng, C.-S. Hong, and A. Yariv, Holographic interference lithography for integrated optics, *IEEE Trans. Electron Devices* **ED-25**(10), 1193 (October, 1978).
14. C. V. Shank and R. V. Schmidt, Optical technique for producing 0.1 μ periodic surface structures, *Appl. Phys. Lett.* **23**(3), 154 (August, 1973).
15. H. W. Schnopper, L. P. Van Speybroeck, J. P. Delvaille, A. Epstein, E. Kaline, R. Z. Bachrach, J. Dijkstra, and L. Lantward, Diffraction grating transmission efficiencies for XUV and soft X-rays, *Appl. Opt.* **16**(4), 1088 (April, 1977).
16. P. L. Csonka, Holographic X-ray gratings generated with synchrotron radiation, *J. Appl. Phys.* **52**(4), 2692 (April, 1981).
17. D. L. Spears and H. I. Smith, High-resolution pattern replication using soft X-rays, *Electron. Lett.* **8**, 102 (February, 1972).
18. M. Green and V. E. Cosslett, Measurement of K, L, and M shell X-ray production efficiencies, *Br. J. Appl Phys.* **1**, 425 (1968).
19. M. Yoshimatsu and S. Kozaki, in- *X-Ray Optics, Application to solids* (H. J. Queisser, ed.), Springer-Verlag, Berlin (1977).
20. D. Maydan, G. A. Coquin, J. R. Maldonado, S. Somekh, D. Y. Lou, and G. N. Taylor, High speed replication of submicron features on large areas by X-ray lithography, *IEEE Trans. Electron Devices* **ED-22**(7), 429 (1975).

REFERENCES

21. M. P. Lepselter, Scaling the micron barrier with X-rays, *IDEM Tech. Dig.* **1980**, 42 (December 1980).
22. D. J. Nagel, in: *Advances in X-ray Analysis* (W. L. Pickles, C. S. Barrett,. J. B. Newkirk, and C. O. Ruud, eds.), Vol. 18, p.. 1, Plenum Press, New York (1975).
23. J.J. Muray, Photoelectric effect induced by high-intensity laser light beam in quartz and borosilicate glass, *Dielectgrics* **2**, 221 (February, 1964).
24. R. A. Gutcheck, J.J. Muray, and D. C. Gates, An intense plasma X-ray source for X-ray microscopy, *Proceedings of the Brookhaven Conference on the Optics of Short Wavelengths* (November, 1981).
25. H. I. Smith, D. L. Spears, and S. E. Bernacki, X-ray lithography: A complementary technique to electron beam lithography, *J. Vac. Sci. Technol.* **10**(6), 913 (November–December, 1973).
26. J. Kirz, D. Syare, and J. Dilger, Comparative analysis of X-ray emission microscopies for biological specimens, *Ann. N.Y. Acad. Sci.* **306**, 291–305 (1978).
27. G. N. Taylor, X-ray resist materials, *Solid State Technol.* **23**(5), 73 (May, 1980).
28. A. R. Neureuther, Simulation of X-ray resist line edge profiles, *J. Vac. Sci. Technol.* **15**(3), 1004 (May–June, 1978).
29. J. H. McCoy and P. A. Sullivan, Precision mask alignment for X-ray lithography, *Proceedings of the Seventh International Conference on Electron and Ion Beam Science and Technology*, p. 536, Electrochemical Society, Princeton, N.J. (1976).
30. J. M. Moran and D. Maydan, High resolution, steep profile, resist patterns, *Bell Syst. Tech. J.* **58**(5), 1027 (May–June, 1979).
31. G. N. Taylor and T. M. Wolf, Plasma-developed X-ray resists, *J. Electrochem. Soc.* **127**(12), 2665 (December, 1980).
32. E. Spiller, D. E. Eastman, R. Feder, W. D. Grobman, W. Gudat, and J. Topalian, Application of synchrotron radiation to X-ray lithography, *J. Appl. Phys.* **47**(12), 5450 (December, 1976).
33. H. Winick and A. Bienenstock, Synchrotron radiation research, *Annu. Rev. Nucl. Part. Sci.* **28**, 33–113 (1978).
34. G. R. Brewer (ed.), *Electro-Beam Technology in Microelectronic Fabrication*, Academic Press, New York (1980); G. Owen, Electron lithography for the fabrication for microelectron devices, *Rep. Prog. Phys.* **48**, 795 (1985).
35. P. R. Thornton, Electron physics in device microfabrication, I: General background and scanning systems, *Adv. Electron. Electron Phys.* **48**, 272 (1979).
36. A. N. Broers and T. H. P. Chang, High resolution lithography for microcircuits, IBM Research Report No. 7403, IBM Thomas J. Watson Research Center, Yorktown Heights, N.Y. (November, 1978).
37. H. C. Pfeiffer, Recent advances in electron-beam lithography for high-volume production of VLSI devices, *IEEE Trans. Electron Devices* **ED-26**(4), 663 (April, 1979).
38. I. Brodie, E. R. Westerberg, D. Cone, J. J. Muray, N. Williams,and L. Gasiorek, Electron beam wafer exposure system for high throughput, direct write, sub-micron, lithography, *IEEE Trans. Electron Devices* **ED-28**(11), 1422 (November, 1981).
39. W. Krakow, L. A. Howland, and G. McKinley, Multiplexing electron beam patterns using single-crystal thin films, *J. Phys. E* **12**, 948 (1979).
40. P. R. Malmberg, T. W. O'Keeffe, M.M. Sopira, and M. W. Levi, LSI pattern generation and replication by electron beams, *J. Vac. Sci. Technol.* **10**(6), 1025 (November–December, 1973).
41. E. D. Roberts, Electron resists for the manufacture of integrated circuits, *Philips Tech. Rev.* **35**, 41 (1975).
42. M. P. Scott, Recent progress on electron image projector, *J. Vac. Sci. Technol.* **15**(3), 1016 (May–June, 1978).
43. M. B. Heritage, Electron-projection microfabrication system, *J. Vac. Sci. Technol.* **12**(6), 1135 (November–December, 1973).
44. H. Koops, On electron projection systems, *J. Vac. Sci. Technol.* **10**(6), 909 (November–December, 1973).
45. L. N. Heynick, E. R. Westerberg, C. C. Hartelius, Jr., and R. E. Lee, Projection electron lithography using aperture lenses, *IEEE Trans. Electron Devices* **ED-22**(7), 399 (July, 1975).
46. C. A. Spindt, A thin-film field-emission cathode, *J. Appl. Phys.* **39**(7), 3504 (June, 1968).
47. C. A. Spindt, I. Brodie, L. Humphrey, and E. R. Westerberg, Physical properties of thin-film field emission cathodes with molybdenum cones, *J. Appl. Phys.* **47**(12), 5248 (December, 1976).

48. W. Aberth, C. A. Spindt, and K. T. Rogers, Multipoint field ionization beam source, *Record of the 11th Symposium on Electron, Ion, and Laser Beam Technology* (R. F. M. Thornley, ed.), pp. 631–636, San Francisco Press, San Francisco (1971).

49. W. Aberth and C. A. Spindt, Characteristics of valcano field ion quadrupole mass spectrometer, *Int. J. Mass Spectrom. Ion Phys.* **25**, 183 (1977).

50. K. Amboss, Electron optics for microbeam fabrication, *Scanning Electron Microsc.* 1, 699 (1976).

51. D. S. Alles, C. J. Biddick, J. H. Bruning, J. T. Clemens, R. J. Collier, E. A. Gere, L. R. Harriott, F. Leone, R. Liu, T.J. Mulrooney, R.J. Nielsen, N. Paras, R. M. Richman, C. M. Rose, D. P. Rosenfeld, D. E. A. Smith, and M. G. R. Thomson, EBES4: A new electron-beam exposure system, *J. Vac. Sci. Technol. B* **5**(1), 47 (January–February, 1987).

52. M. G. R. Thomson, R. Liu, R. J. Collier, H. T. Carroll, E. T. Doherty, and R. G. Murray, The EBES4 electron-beam column, *J. Vac. Sci. Technol. B***5**(1), 53 (January–February, 1987).

53. R. J. Nielsen, J. H. Bruning, R. M. Richman, C. J. Biddick, J. Giacchhi, G. J. W. Kossyk, D.R. Bush, S. J. Barna, and D. S. Alles, A hydraulic X-Y stage system for application in electron beam exposure systems, *J. Vac. Sci. Technol. B***5**(1), 57 (January–February, 1987).

54. D. S. Alles and M. G. R. Thomson, in *VLSI Electronics, Microstructure Science*, Vol. 16, Academic Press, New York (1987).

55. D. R. Herriott, R. J. Collier, D. S. Alles, and J. W. Stafford, EBES: A practical electron lithographic system, *IEEE Trans. Electron Devices* **ED-22**(7), 385 (July, 1975).

56. R. F. W. Pease, J. P. Ballantyne, R. C. Henderson, A. M. Voshchenkov, and L. D. Yau, Application of the electron beam exposure system, *IEEE Trans. Electron Devices*, **ED-22**(7), 393 (July, 1975).

57. D. E. Davis, R. D. Moore, M. C. Williams, and O. C. Woodard, Automatic registration in an electron-beam lithographic system, *IBM J. Res. Dev.* **21m**(6), (November, 1977).

58. A. D. Wilson, T. HJ. P. Chang, and A. Kern, Experimental scanning electron-beam automatic registration system, *J. Vac. Sci. Technol.* **12**(6), 1240 (November–December, 1975); E. R. Westerberg, D. R. Cone, J. J. Muray, and J.C. Terry, Reregistration system for a charged particle beam exposure system. U. S. Patent 4,385,238 (1983).

59. E. D. Wolf, P. J. Coane, and F.S. Ozdemir, Composition and detection of alignment marks for electron beam lithography, *J. Vac. Sci. Technol.* **12**(6), 1266 (November–December, 1975).

60. J. J. Muray, Characteristics and applications of multiple beam machines, *Microelectron. Eng.* **9**, 305 (1989).

61. B. J. G. M. Roelofs and J.E. Barth, Feasibility of multi-beam electron lithography, *Microelectron. Eng.* **2**, 259 (1984).

62. T. Sasaki, A multibeam scheme for electron-beam lithography, *J. Vac. Sci. Technol.* **19**(4), 963 (November–December, 1981).

63. J. S. Greeneich, Development characteristics of poly-(methyl methacrylate) electron resist, *J. Electrochem. Soc.* **122**(7), 970 (July, 1975).

64. T. E. Everhart and P. H. Hoff, Determination of kilovolt electron energy dissipation vs. penetration distance in solid materials, *J. Appl. Phys.* **42**(13), 5837 (December, 1971).

65. G. W. Martel and W. B. Thompson, A Comparison of commercially available electron beam resists, *Semicond. Int.* **2**(1), 69 (January–February, 1979).

66. L. F. Thompson, J. P. Ballantyne, and E. D. Feit, Molecular parameters and lithographic performance of poly(glycidylmethacrylate-co-ethyl acrylate): A negative electron resist, *J. Vac. Sci. Technol.* **12**(6), 1280 (November–December, 1975).

67. S. W. Pang, R. R. Kunz, M. Rothschild, R. B. Goodman, and M. W. Horn, Aluminum oxides as imaging materials for 193-nm excimer laser lithography, *J. Vac. Sci. Technol. B* **7**(6), 1624–1629 (November–December, 1989).

68. J. Melngailis, D. J. Ehrlich, S. W. Pang, and J. N. Randall, Cermet as an inorganic resist for ion lithography, *J. Vac. Sci. Technol. B* **5**(1), 379–382 (January–February, 1987).

69. K. J. Polasko, C. C. Tsai, M. R. Cagan, and R. F. W. Pease, Silver diffusion in $Ag_2Se/GeSe_2$ inorganic resist system, *J. Vac. Sci. Technol. B* **4**(1), 418–422 (January–February, 1986).

70. B. Singh, S. P. Beaumont, A. Webb, P. G. Bower, and C. D. W. Wilkinson, High resolution patterning with Ag_2S/As_2S_3 inorganic electron-beam resist and reactive ion etching, *J. Vac. Sci. Technol. B* **1**(4), 1174–1178 (October–December, 1983).

71. A. Scherer and H. G. Craighead, Barium fluoride and strontium fluoride negative electron beam resists, *J. Vac. Sci. Technol. B* **5**(1), 374–379 (January–February, 1987).

72. J. M. Macaulay, R. M. Allen, L. M. Brown, and S. D. Berger, Nanofabrication using inorganic resists, *Microelectronic. Eng.* **9**, 557 (1989).

73. R. Behrisch (ed.), *Sputtering by Particle Bombardment*, Vols. I and II, Springer-Verlag, Berlin (1981, 1983).

74. M. Isaacson and A. Muray, In situ vaporization of very low molecular weight resists using 1/2 nm diameter electron beams, *J. Vac. Sci. Technol.* **19**(4), 1117 (November–December, 1981).

75. A. Muray, M. Scheinfein, I. Adesida, and M. Isaacson, Radiolysis and resolution limits of inorganic halide resists, *J. Vac. Sci. Technol.* **B3**(1), 367 (1985).

76. A. Muray, M. Isaacson, and I. Adesida, AlF_3—A new very high resolution electron beam resist, *Appl. Phys. Lett.* **45**(5), 589 (September, 1984).

77. R. Kelly, Phase changes in insulators produced by particle bombardment *Nucl. Instrum. Methods* **1832–183** (part 1) 351 (1981).

78. R. D. Heidenreich, L. F. Thompson, E. D. Feit, and C. Melliar-Smith, Fundamental aspects of electron beam lithography, I. Depth–dose response of polymeric electron beam resists, *J. Appl. Phys.* **44**(9), 4039 (September, 1973).

79. J.S. Greeneich and T. Van Duzer, An exposure model for electron-sensitive resists, *IEEE Trans. Electron Devices* **ED-21**(5), 286 (May, 1974).

80. A. Barraud, C. Rausilio, and A. Rauaudel-Tiexier, Monomolecular resists—A new approach to a high resolution electron beam microlithography, *J. Vac. Sci. Technol.* **16**(6), 2003 (November–December, 1979).

81. M. S. Isaacson and A. Muray, Nanolithography using in situ electron beam vaporization of very low molecular weight resists, Workshop on Molecular Electronic Devices, Naval Research Laboratory (March 23–24, 1981).

82. M. Isaacson and A. Muray, In situ vaporization of very low molecular weight resists using 1/2 nm diameter electron beams, *J. Vac. Sci. Technol.* **19**(4), 1117 (November–December, 1981).

83. T. H. P. Chang, Proximity effect in electron-beam lithography, *J. Vac. Sci. Technol.* **12**(6), 1271 (November–December, 1981).

84. A. N. Broers and T. H. P. Chang, High resolution lithography for microcircuits, IBM Research Report No. 7403, IBM Thomas J. Watson Research Center, Yorktown Heights, N.Y. (November, 1978).

85. D. F. Kyser and K. Murata, in: *Proceedings of the 6th International Conference on Electron and Ion Beam Science and Technology*, p. 205, Electrochemical Society, Princeton, N.J. (1974).

86. A. N. Broers, Resolution limits of PMMA resist for electron beam exposure, *J. Electrochem. Soc.* **128**, 166 (1981).

87. A. N. Broers, in: *Scanning Electron Microscopy 1969, Proceedings of the 3rd Annual Scanning Electron Microscopy Symposium*, p. 1, ITT Research Institute, Chicago (1970).

88. S.A. Rishton, S. P. Beaumont, and C. D. W. Wilkinson, in: *Proceedings of Microcircuit Engineering 82*, p. 341, Grenoble (1982).

89. J. J. Muray, Electron beam processing, *VLSI Electronics, Microstructure Science*, **21**, 113 (1988).

90. A. N. Broers, Resolutions limits for electron-beam lithography, *IBM J. Res. Dev.* **32**(4), 502 (July, 19883.

91. A. N. Broers, Limits on thin-film microfabrication, *Proc. R. Soc. London Ser. A* **416**, 1 (1988).

92. A. N. Broers, in: *Proceedings of the 1st International Conference on Electron and Ion Beam Science and Technology* (R. Bakish, ed.), p. 191, Wiley, New York (1964); A. N. Broers, W. W. Molzen, J. J. Cuomo, and N. D. Wittels, Electron beam fabrication of 80 Å metal structures, *Appl. Phys. Lett.* **29**, 596 (1976); A. N. Broers and R. B. Laibowitz, in: *Future Trends in Superconductive Electronics*, B (S. Deavor, Jr., et al., eds.), p. 289, American Institute of Physics, New York (1978).

93. G. A. C. Jones, S. Blythe, and H. Ahmed, in: *Proceedings of Microcircuit Engineering* 86, p. 263, North-Holland, Amsterdam (1986).

94. J. M. Macaulay, The production of nanometre structures in inorganic materials by electron beams of high current density, Ph.D. thesis, Cambridge University Press, London (1989).

95. J. M. Macaulay, R. M. Allen, L. M. Brown, and S. D. Berger, Nanofabrication using inorganic resists, *Microelectron. Eng.* **9**, 557 (1989).

96. T. J. Bullough, C. J. Humphreys, and R. W. Devenish, Electron beam induced hole-drilling and lithography on a nanometre scale in Al, MgO and $a\text{-}AlF_3$ in a STEM, *Proc. MRS*, Boston (1989).

97. S. D. Berger, J. M. Macaulay, L. M. Brown, and R. M. Allen, High current density electron beam induced desorption, *Mat. Res. Soc. Symp. Proc.* **129**, 515 (1989).

98. J. M. Macaulay and S. D. Berger, Characterization of the lithographic properties of inorganic resists with nanometre resolution on bulk substrates, *Microelectron. Eng.* **6**, 527 (1987).

99. A. Muray, Electron and ion beam nanolithography, Ph.D. thesis, Cornell University, Ithaca, N.Y. (1984).

100. A. N. Broers, J. J. Cuomo, and W. Krakow, Method for producing lithographic structures using high energy electron beams, *IBM Tech. Discl. Bull.* **24**, 1534 (1981).

101. K. J. Polasko, Y. W. Yau, and R. F. W. Pease, Low energy electron beam lithography, *SPIE* **333**, 76–82 (1982).

102. M. A. McCord and R. F. W. Pease, Scanning tunneling microscope as a micromechanical tool, *Appl. Phys. Lett.* **50**(10), 569 (9 March, 1987).

103. M. A. McCord and R. F. W. Pease, Exposure of calcium fluoride resist with the scanning tunneling microscope, *J. Vac. Sci. Technol. B* **5**(1), 430 (January–February, 1987).

104. M. A. McCord and R. F. W. Pease, A scanning tunneling microscope for surface modification, *J. Phys. Colloq.* C2, Suppl. to No. 3, Vol. 47 (March, 1986).

105. G. G. Roberts, An applied science perspective of Langmuir–Blodgett films, *Adv. Phys.* **34**(4), 475 (1985).

106. A. N. Broers and M. Pomerantz, Rapid writing of fine lines in Langmuir–Blodgett films using electron beams, *Thin Solid Films* **99**, 323–329 (1983).

107. R. L. Seliger, J. W. Ward, V. Wang, and R. L. Kubena, A high-intensity scanning ion probe with submicrometer spot size, *J. Appl. Phys. Lett.* **34**(5), 310 (March, 1979).

108. B. A. Free and G.A. Meadows, Projection ion lithography with aperture lenses, *J. Vac. Sci. Technol.* **15**(3), 1028 (May–June, 1978).

109. G. Stengl, R. Kaitna, H. Loschner, P. Wolf, and R. Sacher, Ion projection system for IC production, *J. Vac. Sci. Technol.* **16**(6), 1883 (November–December, 1979); G. Stengl, P. Wolf, R. Kaitna, H. Loschner, and R. Sacher, Experimental results and future prospects of damagnifying ion-projection systems, Paper presented at the Third International Conference on Ion Implantation Equipment and Techniques, Kingston, Canada (July 8–11, 1980).

110. L. Csepregi, F. Iberl, and P. Eichinger, Ion-beam shadow printing through thin silicon foils using channeling, *Appl. Phys. Lett.* **37**(7), 630 (October, 1980).

111. J. L. Bartelt, C. W. Slayman, J. E. Wood, J. Y. Cheyn, and C.M. Mckenna, Masked ion-beam lithography: A feasibility demonstration for submicrometer device fabrication, *J. Vac. Sci. Technol.* **19**, 1166 (1981).

112. F.-O. Fong, D. P. Stumbo, S. Sen, G. Damm, D. W. Engler, J. C. Wolfe, J.N. Randall, P. Mauger, and A. Shimkunas, Low stress silicon stencil masks for sub-100 nm ion beam lithography, *Microelectron. eng.* **11**, 449–452 (1990).

113. D.P. Stumbo, G. A. Damm, D. W. Engler, F.-O. Fong, S. Sen, J. C. Wolfe, J. N. Randall, P. Mauger, A. Shimkunas, and H. Loschner, Advances in mask fabrication and alignment for masked ion beam lithography, *SPIE Conf.* (1990).

114. D. P. Stumbo, S. Sen, F.-O. Fong, G. A. Damm, D. W. Engler, J. C. Wolfe, and J. N. Randall, Alignment for masked ion beam lithography using ion-induced fluorescence, *Electron, Ion and Photon Beam Conf.* (1990).

115. L. W. Swanson, G. A. Schwind, A. E. Bell, and J.E. Brady, Emission characteristics of gallium and bismuth liquid metal field ion sources, *J. Vac. Sci. Technol.* **16**(6), 1864 (November–December, 1979).

116. R. J. Culbertson, T. Sakurai, and G. H. Robertson, Ionization of liquid metals, gallium, *J. Vac.Sci. Technol.* **16**(2), 574 (March–April, 1979).

117. M. Komura, N. Atoda,and H. Kawakatsu, Ion beam exposure of resist materials, *J. Electrochem. Soc.* **126**(3), 483 (March, 1979).

118. T. M. Hall, Liquid gold ion source, *J. Vac. Sci. Technol.* **16**(6), 1871 (November–December, 1979).

119. L. W. Swanson, A. E. Bell, G. A. Schwind, and J. Orloff, A comparison of the emission characteristics of liquid ion sources of gallium, indium and bismuth, Paper presented at the Ninth International Conference on Electron Science and Ion Beam Technology, St. Louis, Missouri (May 11–16, 1980).

120. T. M. Hall, A. Wagner, and L. F. Thompson, Ion beam exposure characteristics of resists, *J. Vac. Sci. Technol.* **16**(6), 1889 (November–December, 1979).

121. T. Shiokawa, P. H. Kim, K. Toyoda, and S. Namba, 100 keV focused ion beam system with E × B mass filter for maskless ion implantation, *J. Vac. Sci. Technol. B.* **1**(4), 1117 (October–December, 1983).

122. W. Fallmann, F. Paschke, G. Stengl, L.-M. Buchamann, A. Heuberger, A. Chalupka, J. Fegerl, R. Fischer, H. Loschner, L. Malek, R. Nowak, G. Stengl, C. Traher, and P. Wolf, Ion Projection Lithography: Electronic Alignment and Dry Development of IPL Exposed Resist Materials, *AEU* (Austria), Vol. 44, Section 3 (1990).

123. G. Stengl, H. Loschner, E. Hammel, E. D. Wolf, and J. J. Muray, Ion projection lithography, in: *Emerging Technologies for in Situ Processing* (D. J. Ehrlich and V. T. Nguyen, eds.), Nijhoff, The Hague (1988).

124. G. Stengl, H. Loschner, and J. Muray, Sub-0.1-μm ion projection lithography (extended abstract). Conference on Solid State Devices and Materials, pp. 29–32, Tokyo (1986).

125. G. Stengl, H. Loschner, and J. Muray, Ion projection lithography, *Solid State Technol.* **29**(2), 119–126 (1986).

126. G. Stengl, H. Loschner, W. Maurer, and P. Wolf, Ion projection lithography for submicron modification of materials, Materials Research Soc. Fall Meeting, Boston (1984).

127. M. Ando, Preliminary experimental study of the multiple ion beam machine, *J. Vac. Sci. Technol. B* **6**(6), 2120 (November–December, 1988).

128. J. J. Muray, Characteristics and applications of multiple beam machines, *Microelectron. Eng.* **9**, 305 (1989).

129. M. Ando and J. Muray, Spatial resolution limit for focused ion-beam lithography from secondary-electron energy measurements, *J. Vac. Sci. Technol. B* **6**(3), 986 (May–June, 1988).

130. I. Adesida, A. Muray, M. Isaacson, and E. D. Wolf, in: *Proc. Microcircuit Engineering '83* (H. Ahmed, J. R. Cleaver, and G. A. C. Jones, eds., pp. 151–156, Academic Press, New York (1983).

131. M. Ando and J. Muray, Spatial resolution limit for focused ion-beam lithography from secondary-electron energy measurements, *J. Vac. Sci. Technol. B* **6**(3), 986 (May–June, 1988).

132. C. D. W. Wilkinson, Nanofabrication, *Microelectron. Eng.* **6**, 155 (1987).

133. T. H. P. Chang, D. P. Kern, E. Kratschmer, K. Y. Lee, H. E. Luhn, M. A. McCord, S.A. Rishton, and Y. Vladimirsky, Nanostructure technology, *IBM J. Res. Dev.* **32**(4), 462 (July, 1988).

134. R. F. Howard and D. E. Prober, in: *VLSI Electronics, Microstructure Science* V (N.G. Einspruch, ed.), (1982).

135. E. D. Wolf, Nanofabrication: Opportunities for interdisciplinary research, *Microelectron. Eng.* **9**, 5–11 (1989)

136. K. A. Valiev, *The Physics of Submicron Lithography*, Plenum, New York (1992).

6

Microcharacterization

6.1. INTRODUCTION

Having built a device consisting of elements made to nanometer tolerances, it is often necessary to examine it to see whether the device was built as specified and whether the components are of the required materials with the desired physical properties.

We consider the area of an object being viewed (A) as divided into picture elements or pixels of the smallest area (a) that needs to be resolved. In a microscope the intention is to display a magnified image of A that can be comfortably viewed by the human eye with the magnified pixel likewise comfortably resolved by the eye. The optimum useful magnification of any microscope (M_0) is that which will enlarge the smallest resolvable detail in the microscope (δ) to a size resolvable by the unaided human eye (0.2 mm). That is,

$$M_0 = \frac{2 \times 10^{-4}}{\delta} \tag{6.1}$$

for δ measured in meters.

The ultimate for nanofabrication is to resolve atoms requiring

$$\delta \approx 2 \, \text{Å} \quad \text{giving} \quad M_0 = 10^6 \tag{6.2}$$

In order to construct an image showing contrast (or detail), the surface of the object must be radiating detectable particles at a rate that varies from pixel to pixel. Furthermore, there must be some means for generating the particles and counting the particle emission rate or intensity from each pixel. The imaging of the pixels can be carried out in parallel, as in optical microscopes, or serially (one pixel at a time), as in scanning electron microscopes. The image, which consists essentially of a map of the number of a specific type of particles emitted from each pixel in a given time, may be stored in photographic film, magnetic disk or tape, or computer memory, and may be viewed in a photograph, displayed on a screen, or plotted with ink on paper.

Parallel imaging requires that the entire viewed area A be subjected to a uniform excitation source which generates the particles to be detected from each pixel, and the use of large-aperture lenses appropriate to the radiation being detected, to focus an enlarged image on an array of detectors. In this case the wavelength λ corresponding to the radiation being imaged must be less than the pixel size, but in principle the exciting particle wavelength could be larger.

Serial imaging requires that the area of the exciting source be confined to the pixel area. Only a single detector which subtends a large solid angle to the object area is needed so that imaging particles can be received from any pixel. In this case the illuminating particles are focused by small-aperture lenses into a small spot and must be of a wavelength smaller than the pixel size. The particles used for detection can be of larger wavelength provided they originate only within the pixel area. If the illuminating particles are monoenergetic, the energy spectrum of the detected particles may be of use in determining the chemical and physical nature of the pixel it is addressing. If diffraction of monoenergetic illuminating particles is being used, then the direction of the reflected particles contains useful information, as in low-energy electron diffraction (LEED).

Since 1960 the serial or scanning mode of operation has become of increasing importance as the pixel size to be resolved has become smaller. Because the depth of focus of the scanned beam can be made large, this makes this approach more satisfactory for imaging three-dimensional surfaces.

The smallest pixel size is limited by:

- The wavelength of the exciting or detected particles (Section 2.1)
- The damage caused by the excitation source (Sections 2.6, 2.7, and 2.9)[1]
- Aberrations in the lenses used for imaging or focusing (Section 2.5.4)[2]
- Particle counting statistics in the detector (Section 7.3.3)

While in principle the average number of particles counted per pixel can be made as large as necessary, this number is limited by practical considerations of the time it takes to do this, the number and rate of exciting particles available or can be absorbed by the pixel without substantial damage, and errors due to gross movements due to slow temporal changes in the ambient conditions (e.g., temperature, electric and magnetic fields, vibrations).

The intensity of each pixel may be characterized by dividing the number of particles detected in a given time interval from the most intense pixel into m equal steps of j particles each. Thus, the detector count for any given pixel would be in a range of $j, 2j, 3j \ldots sj \ldots mj$ depending on the intensity of the pixel. If i is the actual count for a given pixel, it is assigned a step value s if $(s-1)j \leq i \leq sj$. Due to the statistics of particle emission, usually assumed to have a Poisson distribution (see Section 7.3.3), there is always a probability that a given pixel will be incorrectly assigned. For convenience, this may be put in the form that for every R pixels assigned on average one will be in error. The experimenter can then choose what value of R is tolerable and this will fix the minimum value for j. The probability of a count of value i coming from a pixel of average count $\bar{i}$ is given for a Poisson distribution (see Section 7.3.3) by

$$P(i) = \frac{\bar{i}^i}{i!} e^{-\bar{i}}$$

(6.3)

In general, R decreases with increasing s and increases with increasing j, and the largest error occurs in assigning the last (nth) step.

If the maximum tolerable time to record an image of n_0 pixels is t_0, and only a fraction f of the particles emitted reach the recording device, then in the case of parallel imaging the particle emission rate from the most intense pixel (area a) is given by

$$n_p = \frac{jm}{f_p t_0 a} \text{ particles per unit area per second}$$

(6.4)

However, for serial imaging where the time allowed for each pixel count is t_0/n_0

$$n_s = \frac{jmn_0}{f_s t_0 a} \text{ particles per unit area per second}$$

(6.5)

Thus, parallel imaging requires an emission flux density smaller by the number of pixels per image than for serial imaging. This is partly compensated by a larger f for serial imaging.

Table 6.1[3] summarizes the means used for excitation and detection and Table 6.2 lists many of the different techniques used for image analysis and their acronyms.

As shown in Tables 6.1 and 6.2 the number of instruments that have been devised for microcharacterization is extremely large. Each instrument has special applications for which it is uniquely suited. In this chapter we attempt to give an overview of this complex field which utilizes the physics discussed in previous chapters, and discuss some of the more important instruments that are currently used for micro/nano-device research.

6.2. PARALLEL IMAGING

6.2.1. Optical Compound Microscope

The optical microscope, the archetypical parallel imaging microscope, was perfected in the 19th century. We assume that the reader is familiar with the theory of this microscope,[3] but note that we are concerned here with only a few important aspects, namely:

- Field of view (A)
- Magnification (M)
- Smallest resolvable element (area a, linear dimension δ)
- Depth of focus (D)

TABLE 6.1. Matrix of Interaction Processes (in *italics*) between Different Primary Beams and Solid Matter Emitting Different Secondary Particles and Common Methods for Thin-Film Analysis Arranged According to the Excitation Source and the Detected Particles; Principal Methods in Boxes[a,b]

	Detection			
Excitation	Photons (*hv*)	Electrons (*e*)	Ions (i)	Neutrals (n)
Photons (*hv*)	*Fluorescence, reflection*	Photo- and Auger electron emission	*Photodesorption and local heating*	
	XRFA , TRXRFA, X-diffraction, light microscopy, IR, UV, Raman scattering, LOES, ellipsometry	ECSC = XPS XAES UPS	LMP	Sampling by laser-induced emission for OES, AAS
Electrons (*e*)	*Electron-induced photon emission*	*Secondary electrons, backscattered and Auger electrons*	Electron-stimulated desorption	Electron-stimulated desorption
	EMP-X , *SEM-X, TEM-X, APS,* cathodo-luminescence	AES , SAM, TEELS, TEM, SEM, DAPS, TEM-ED, LEED		
Ions (i)	*Ion-induced photon emission* PIXE, (NRA), GDOES, BLE = SCANIIR	Ion-induced AES	*Secondary ion emission* SIMS , *IMP,* GDMS	*Sputtering* SNMS
		(Ion neutralization spectrometry)	*Ion scattering* LEIS = ISS RBS = HEIS NRA (inelastic)	
Neutrals (n)	*Neutral-particle-induced photon emission BLE = SCANIIR*	*Neutral-particle-induced electron emission*	Neutral-particle-induced ion emission FAB-MS	*Sputtering*
Heat (*kT*)	*Thermal radiation*	*Thermionic emission*	*Thermal ion desorption*	*Thermal evaporation*
Electric field (*E*)		Field electron emission FEM, STM	Field ion emission FIM	
Mechanical (*p*)	*Triboluminescence*	Exoelectron emission		*Peeling, ultrasonic desorption* Thin-film sampling
Chemical (*μ*) (from solution)	*Chemiluminescence*	(*Cathodic reduction*)	*Chemical/ electrochemical dissolution*	Chemical dissolution
			Thin-film sampling	

[a]Source: Ref. 37.
[b]The common acronyms used for the methods are listed in Table 6.2.

TABLE 6.2. Acronyms and Associated Terms of the Different Methods Used for Analysis

AAS	Atomic absorption spectrometry
ABSS	Atomic beam surface scattering
AEAPS	Auger electron appearance potential spectroscopy
AES	Auger electron spectroscopy
AFM	Atomic force microscopy
APPH	Auger peak-to-peak height
APS	Appearance potential spectroscopy
ATR-IR	Attenuated total reflection-infrared spectrometry
BLE	Bombardment-induced light emission (=SCANIIR) (=IBSCA)
CMA	Cylindrical mirror analyzer
CSOM	Confocal scanning optical microscope
DAPS	Disappearance potential spectroscopy
EELS	Electron energy-loss spectrometry (=ELS)
EEM	Electron emission microscope
ELL	Ellipsometry
EMM	Electron mirror microscope
EMP(-X)	Electron probe microanalysis, used in two modes, viz., wavelength-dispersive X-ray analysis (WDX), or energy-dispersive X-ray analysis (EDX)
ESCA	Electron spectroscopy for chemical analysis (=XPS)
EXAFS	Extended X-ray absorption fine structure
FAB-MS	Fast atom bombardment-mass spectrometry
FEM	Field electron microscopy
FIM	Field ion microscopy
FWHM	Full width at half-maximum of a spectral line
GDMS	Glow discharge mass spectrometry
GDOES	Glow discharge optical emission spectrometry (=GDOS) (=GDS) (=GDL-OES)
HEIS	High-energy ion scattering (=RBS)
IBSCA	Ion bombardment surface chemical analysis
IETS	Inelastic tunneling spectroscopy
ILEED	Inelastic low-energy electron diffraction
ILS	Ionization-loss spectroscopy
IMFP	Inelastic mean free path
IMMA	Ion microprobe mass analysis
IMP	Ion microprobe
IR	Infrared absorption spectrometry
ISS	Ion scattering spectrometry (=LEIS)
LAMMA	Laser microprobe mass analysis (=LMP)
LEED	Low-energy electron diffraction
LEIS	Low-energy ion scattering (=ISS)
LERM	Low-energy reflection microscope
LMP	Laser microprobe analysis
LOES	Laser optical emission spectrometry
MFP	Mean free path
MS	Mass spectroscopy
NAA	Neutron activation analysis
NRA	Nuclear reaction analysis
NSOM	Near-field scanning optical microscope
OES	Optical emission spectrometry
PAM	Photo-acoustic microscopy ($\neq$ AM)
PIXE	Proton-induced X-ray emission
PTFE	Polytetrafluoroethylene
RBS	Rutherford backscattering spectrometry (=HEIS)
RHEED	Reflected high-energy electron diffraction
SALI	Surface analysis by laser ionization
SAM	Scanning Auger microprobe
SCANIIR	Surface chemical analysis by neutral- and ion-induced radiation (=BLE) (=IBSCA)
SDA	Spherical deflection analyzer
SEAM	Scanning electron acoustic microscope

(continued)

TABLE 6.2.   (*Continued*)

SEM	Scanning electron microscopy
SEXAFS	Surface-extended X-ray absorption fine structure
SIMS	Secondary-ion mass spectrometry
SLEEP	Scanning low-energy electron probe
SNMS	Sputtered neutrals mass spectrometry
SOM	Scanning optical microscope
SSMS	Spark source mass spectrometry
STEM	Scanning transmission electron microscopy
STM	Scanning tunneling microscopy
SXAPS	Soft-X-ray appearance potential spectroscopy
SXDA	Soft-X-ray emission depth analysis
TEELS	Transmission electron energy-loss spectrometry
TEM	Transmission electron microscope
TEM-ED	Transmission electron microscope–electron diffraction
TRXRFA	Total reflection X-ray fluorescence analysis
UPS	Ultraviolet photoelectron spectrometry
UV	Ultraviolet absorption spectrometry
XES	X-ray emission spectrometry
XAES	X-ray-induced Auger electron spectrometry
XPD	X-ray photoelectron diffraction
XPS	X-ray photoelectron spectrometry
XRFA	X-ray fluorescence spectrometric analysis

The theory[3] gives

$$\delta = \frac{\lambda}{\text{NA}} \tag{6.6}$$

where NA is the numerical aperture of the objective lens defined by

$$\text{NA} = \frac{n \times \text{radius of objective lens}}{\text{focal length}} \tag{6.7}$$

where n is the refractive index of the material between the object and the lens. For the best modern objective lenses, NA may be as great as 0.95 in air, and by filling the space between the object and the objective lens with oil, the NA may be increased to as much as 1.5. Since the useful range of the visible spectrum is only 0.4–0.7 μm, even with the best optical microscopes we cannot observe details much smaller than 0.3 μm (or 3000 Å).

The depth of focus is the distance, measured along the optical axis, along which the resolving power is unaffected by defocusing, and it is given by

$$D = \frac{\delta}{\sin \alpha} \tag{6.8}$$

where 2α is the angle of divergence of the imaging rays from the object. Since $\sin \alpha \cong 1$ in a high-powered objective, we see that the depth of field in optical microscopes is approximately equal to the resolution.

Clearly, optical microscopes are barely adequate for viewing device elements with tolerances much less than a micron.

6.2.2. *Transmission Electron Microscope*

The transmission electron microscope (TEM) (see Fig. 6.1)[4,5] is essentially the electron analogue of the optical compound microscope. TEMs were developed in the 1930s, using the technologies gained in building CRT displays. In concept, a thin specimen is uniformly irradiated on one side by an electron beam of a given energy formed by an illuminating system. On the other side, a powerful objective lens images

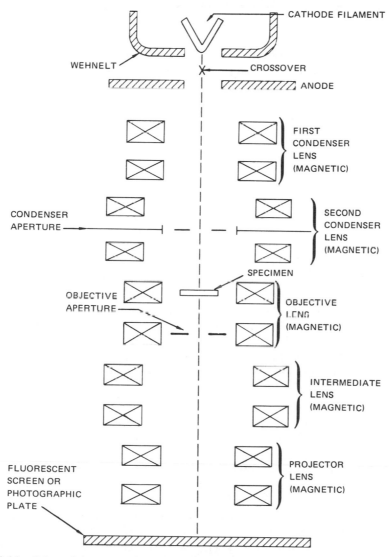

FIGURE 6.1.   Schematic representation of a transmission electron microscope (TEM). From Ref. 4.

the plane containing the specimen. The objective lens is followed by a projection system that displays the final magnified image of the specimen on a fluorescent screen to a size that can be comfortably viewed by the eye. The fluorescent screen can be replaced by a photographic film, plate, or video camera to record the image.

The specimen must be thinner than the mean free path for elastic collisions of the electrons illuminating it, and in any case it must be as thin as possible, otherwise a vast amount of overlapping detail appears in the image because of the high resolving power and great depth of focus of the electron microscope. As a rule of thumb, the thickness should not exceed ten times the resolving power required. Ultramicrotomes have been developed for biological specimens that can prepare sections from 1000 to 100 Å thick, and microetching techniques have been developed to obtain films of metals, alloys, and semiconductors in the same range, so that the full resolution capabilities of the microscope can be utilized.[6]

Contrast in the magnified image is obtained because some electrons traveling through the specimen collide with atomic particles, resulting in scattering or energy loss, which prevents their being imaged by the objective lens. To collect the scattered electrons, an aperture stop is placed in the image focal plane of the objective lens, thus preventing them from forming a background fog on the fluorescent screen. This aperture also serves the function of reducing the spherical aberration of the objective lens, and its size is optimally reduced to the point where the image blurring due to spherical aberration is equal to that due to diffraction. Further reduction would increase the diffraction blurring and would therefore not be useful. The higher the electron energy, the smaller is its wavelength and the smaller the aperture stop that can be used. At the usual working voltages of 50–100 kV, the optimum aperture angle turns out to be $1/2°$, as compared with over $80°$ for optical microscopes. For a focal length of the objective lens of 2 mm, this corresponds to an aperture diameter of 40 μm.

To view the image at the fluorescent screen or to record the image on a photographic film in the same plane in a reasonable time, the illuminating electron flux at the specimen must be as intense as possible. The illuminating system is designed with this and other features in mind. It usually consists of an electrostatic triode gun (see Section 2.4) followed by two magnetic condenser lenses. A crossover is formed between the cathode and the anode of the triode gun, with a divergence after leaving the anode aperture of about 10^{-2} rad ($1/2°$). The first condenser lens, which is always operated at full strength, forms a demagnified image of the crossover inside its lower pole piece. This image is magnified by the second condenser lens by a small factor in the range 1–2 to form an illuminating spot at the specimen. The diameter of the spot illuminating the specimen is controlled by varying the strength of the second condenser lens and need not be larger than the area to be viewed. The beam current is determined by the size of the condenser aperture, the voltages on the gun electrodes, and ultimately by the maximum emission brightness of which the cathode is capable. Typically, the first condenser lens demagnifies the crossover by a factor of 10–15. The diameter of the illuminating spot on the specimen can be reduced to 1 μm, using standard V-shaped filaments, and it can be made even smaller with point electron sources, if required.

The objective lens (usually magnetic; see Section 2.5) has a focal length that can be made as small as 1 mm, and the specimen is positioned essentially at the focal point. The division of magnification between the objective, intermediate, and projection lenses is approximately 100:20:100, respectively, for a magnification of 200,000 times. The function of the intermediate lens is to shorten the microscope column length over that required for a single projection lens, given the difficulty of making magnetic lenses with focal lengths any smaller than 1 mm.

The condensing and magnifying systems are made using magnetic lenses (see Section 2.5.2), because of the smaller aberrations obtained in these lenses in comparison with electrostatic lenses. In TEMs the axial positions of the specimen and all of the lenses are fixed. Focusing and magnification changes are accomplished by simply varying the focal lengths of the objective, intermediate, and projection lenses by changing the currents through their magnetizing coils. The stability of these currents and of the gun voltages is therefore of prime importance in obtaining the best performance. In addition, the alignment of the optical axes of the lenses and apertures must be maintained to a high degree of precision.

For viewing living biological matter, a greater range of electrons is required to produce greater penetration and less radiation damage. A 100-keV beam will not penetrate even the smallest living organism to produce a good image

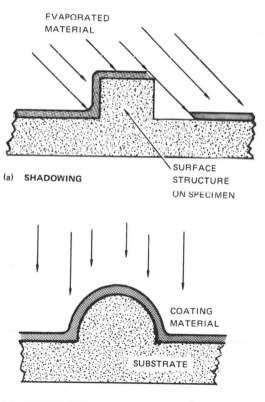

FIGURE 6.2. Methods of viewing surface structures with the transmission electron microscope. (a) Shadowing; (b) replicating.

(see Section 2.6.1). Hence, microscopes utilizing up to 3-MeV beams have been developed for this purpose. The high-voltage TEM resolves a diffraction image of samples of thin single crystals that clearly shows the arrangements of the atoms and dislocation planes.[7] The range of general usefulness of the TEM can be considered at this time to take over from optical microscopes for resolutions below 1 μm–10 Å for specimens that can be obtained in the form of very thin films.

Surface structures may be viewed with the TEM by the techniques of shadowing in the case of a thin sample, and by replicating in the case of a thick sample. In shadowing, an electron-dense material is evaporated on the surface at a shallow angle so that it condenses thicker on one side than the other, as shown in Fig. 6.2a. An enhanced contrast is then obtained in normal viewing. In replicating (Fig. 6.2b), a thin layer of coating material is deposited on the surface. The substrate is then removed, usually by dissolving it, leaving the replica behind. Coating materials may be organic, evaporated inorganic materials, electrolytically deposited metals, or oxides of the specimen itself. Replicas formed by these techniques are usually of constant thickness (in the range 100–1100 Å) and although they accurately follow the contours of the surface, replicas are usually "shadowed" before viewing in the electron microscope.

6.2.3. Electron-Reflection Microscopes

In electron-reflection microscopes the surface to be viewed is held at a potential such that all or a large fraction of the incident electrons do not physically strike the sample surface. Those that do strike the surface do so with low energies (in a range up to only a few volts) so that the surface is barely perturbed by the illuminating electron beam. Reflection principles can be utilized in both the parallel and serial modes.

The parallel imaging device[8] is called the electron-mirror microscope (EMM), and the principle of its operation is shown in Fig. 6.3. A collimated electron beam is directed normal to the sample surface. As electrons pass through the final lens aperture, they are rapidly decelerated, with a turning point determined by the potential of the sample surface with respect to the cathode and the field strength at the sample surface. After turning, the electrons are accelerated back through the lens and an enlarged image is projected onto a cathodoluminescent screen. Additional magnification may be obtained by separating the outgoing beam from the incoming beam with a weak magnetic field and inserting an additional magnifying lens in the path of the outgoing beam.

Contrast in the outgoing beam is obtained from several sources, namely,

- Surface topography
- Variations in electric potential over the surface
- Variations in magnetic fields over the surface

Any deviation from flatness of the sample surface will influence the shape of the equipotentials near the surface. Thus, lateral components of velocity will be given to reflected electrons by a bump on the surface, depending on the detailed shape of the turning equipotential. These electrons will be deflected and give rise to a decrease in the intensity of the enlarged image corresponding to the position and size of the

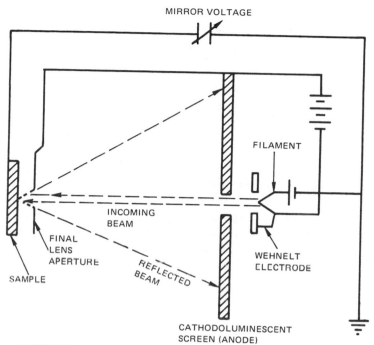

FIGURE 6.3   Principle of the electron-mirror microscope. From Ref. 8.

projection. In practice, resolution steps about 24 Å high can be comfortably observed.[8] However, lateral resolutions of only about 500 Å are observed, despite the fact that a theoretical limit of about 40 Å is predicted. This has been recently improved by using low energy illuminating electrons (below 10 eV) and resolutions to 20 Å have been observed.[9,10]

When a conductive polycrystalline material is polished, each exposed crystallite surface may exhibit a contact potential difference with its neighboring crystallites. This is because of differing exposed crystalline planes if they are of the same material, or differences in crystal structure if they are of different materials. Contact potential differences in the range of tens to thousands of millivolts are typically obtained. These cause abrupt changes in the equipotentials at the lines of contact, analogous to the case for geometric perturbations. Resolutions down to 50 mV have been observed. The voltage applied to the sample may be adjusted so that one set of crystallites reflects the incident electrons before they reach their surfaces, while another set of crystallites is at a potential so as to just collect the incident electrons. This gives rise to a very high contrast. Insulating particles on the surface may reach high negative potentials, as some energetic electrons in the beam are able to reach them. This again gives rise to high contrast. A charge pattern distributed over a flat insulating surface will similarly change the path of the reflected electrons, giving rise to high contrast in the image.

Changes in the local magnetic field over a sample surface will also give rise to perturbations in the paths of the reflected electrons. The problem of estimating the magnetic contrast is more complex, since the radial component of the electron velocity interacts with the normal magnetic field in accordance with the Lorentz equation (see Section 2.5.2). However, the effect is large enough so that information stored on magnetic tapes or disks may be read out with the EMM.

The electric field E in the space between the final lens aperture and the sample is limited by vacuum breakdown and lies typically in the range 10^5–10^6 V/cm. If the sample is held at $-V$ volts with respect to the cathode, then the electrons will turn at a distance $d = V/E$ cm from the sample. This assumes that all electrons leave the lens with the same energy. In fact, an average energy spread δV exists owing to factors discussed in Section 2.2. Thus, the turning distance d lies in a range $\delta d = \delta V/E$. For $E = 10^6$ V/cm and $\delta V = 1$ V, then $\delta d = 10^{-6}$ cm $= 100$ Å. The actual step-height resolution is somewhat better than this at the center of illumination.[9]

Reflection principles can also be used in the serial mode. The device is similar to a conventional scanning electron microscope (Section 6.3.1) except that at the spot focal point the beam is rapidly decelerated and reflected before the spot has time to grow appreciably. In one form, called SLEEP (Scanning low-energy electron probe),[11] this device is used to investigate the work-function distribution over the surface of thermionic cathodes in various stages of activation and life.

6.2.4. Electron-Emission Microscopes

Under special circumstances a sample surface can be made electronically self-luminous. That is, it emits electrons directly from its surface, such as arises from heating the sample (thermionic emission), applying a strong field to the sample surface (field emission), or by particle bombardment of the surface (secondary emission). Such electrons are usually emitted in random directions, with low energies and with a small energy spread (see Section 2.2). In the first two cases the electrons can be directly focused into an enlarged image by electron lenses, without the complications of an external source of electrons or other particles. In the third case, illumination of the sample surface can, with some ingenuity, be accomplished without interfering with the electron-optical system.

In the emission microscope shown in Fig. 6.4,[12] the sample surface is made an electrode of the system, forming an immersion lens with the Wehnelt electrode and the anode. A homogeneous electric field E is assumed to exist in the cathode-to-anode region. To improve the resolution, an aperture stop is placed in the focal plane of the lens to intercept electrons with high thermal lateral velocities. It has been shown that the resolution of this device, δ, is given by

$$\delta = \frac{\phi}{|E|} \left(\frac{r}{f}\right)^2 \tag{6.9}$$

where f is the focal length of the lens, ϕ is the potential at the focal plane, and r is the radius of the aperture stop.

With $\phi = 40$ kV, $|E| = 10^5$ V/cm, $r = 10$ μm, and $f = 5$ mm, we obtain $\delta = 160$ Å. The aperture cannot be made too small because of loss in intensity in the

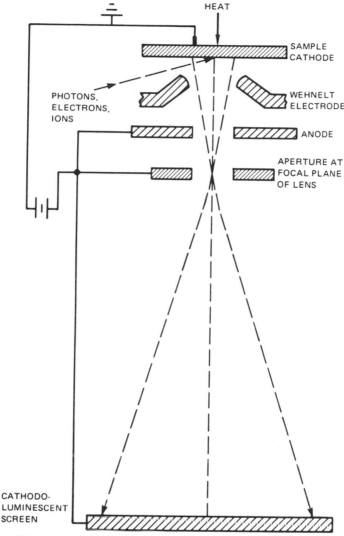

FIGURE 6.4. Principle of the emission microscope. From Ref. 12.

image, and, of course, need not be made smaller than the electron diffraction limit for the aperture (see Section 2.4.5.2).

The thermionic emission microscope is restricted to materials with a low work function and low volatility that can be heated to emit reasonable current densities. Use of the photoelectric effect vastly increases the range of materials that can be used.[13] Intense ultraviolet sources are focused over the area of interest and the photoelectronic image is projected onto the fluorescent screen. Since the energy distribution of the photoelectrons is typically much larger than that for thermionic electrons, the use of the aperture stop is imperative even for modest resolutions (fractions of a micron). Similarly, the surface can be bombarded with high-energy electrons and only the low-energy secondaries will be imaged. Secondary electrons can also be generated by ion bombardment. This tends to etch the surface by sputtering, producing beautiful patterns for metallography.

The emission microscope can be used only in the projection mode. However, in principle, scanning microscopes are emission microscopes when they utilize the secondary particles generated at the surface by the scanning beam.

The field-emission microscope (FEM), shown in Fig. 6.5,[14] must be the simplest microscope ever devised. It consists of a needle of tip radius r a distance R from a cathodoluminescent screen held at a potential V. However, it requires an ultrahigh vacuum ($p < 10^{-10}$ Torr) to operate for useful periods of time. The field at the tip is given by

$$E = \frac{V}{kr} \tag{6.10}$$

where k is a constant that depends on the shape of the needle (a value of about 7 is generally used). To obtain electrons from the tip is substantial quantities (10^4–10^8 A/cm^2) by field emission, fields in the range 10^7–10^8 V/cm must be applied. The needles are usually annealed by the application of heat and electric fields into single crystals of tip radius of about 1 μm, so voltages on the order of 10 kV have to be applied.

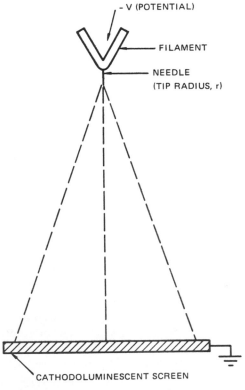

FIGURE 6.5.   Principle of the field emission microscope (FEM). From Ref. 14.

The electrons simply travel on radial paths and the magnification is given by

$$M = \frac{R}{r}$$ (6.11)

for $R \sim 10$ cm, and $r \sim 1 \mu m$, $M \sim 10^5$.

The resolution is dependent on the velocity distribution of the emitted electrons, and the momentum uncertainty and is given by[15]

$$\delta = \left(\frac{2\hbar}{mM}\right)^{1/2} (1 + 2m\tau v_0^2/\hbar M)^{1/2}$$ (6.12)

or

$$\delta = 2.72\left(\frac{r}{\sqrt{V}}\right)^{1/2}\left(1 + 0.222\frac{r}{\sqrt{V}}\right)^{1/2} \text{Å}$$ (6.13)

for r in angstroms and V in volts, where τ is the transit time from tip to screen, v_0 is the average transverse velocity, m is the mass of the electron, and the other symbols are as conventionally or previously defined.

Studies of the variation of the work function of single-crystal faces and the effects of adsorbed species have been made using this microscope. Resolutions down to 10 Å are conventionally obtained and large adsorbed organic molecules can be made visible.[14] For tip radii below 2000 Å, it is difficult to obtain a crystal structure. However, the evidence strongly suggests, in agreement with the theory, that for tip radii below 100 Å in radius[16] atomic resolution can be observed, but the conditions are such that only relatively few (1–20) atoms can be observed in this way owing to the small tip radius. Such small-radius tips exist in metallic whiskers formed by various methods, including electrolytic etching.[16]

6.2.5. Field-Ion Microscope

The field-ion microscope (FIM)[17] is unique in that until recently it was the only microscope in which arrays of atoms could be routinely observed. The principle, as shown in Fig. 6.6, is similar to that of the field electron emission device except that electrons are replaced by positive ions.

If a free atom or molecule is in an increasing electric field approaching 10^8 V/cm (or 1 V/Å), it first becomes polarized and, as a critical field is exceeded, an electron tunnels out of its shell to the vacuum region leaving a positively charged ion. Around a sharp tip, with appropriate voltages applied, there is both a high field and an increasing field gradient as a gas molecule approaches the tip. The action of the field gradient pulls the polarized particle toward the tip and, at a critical distance, an electron tunnels from the molecule forming a positively charged ion. The electric field now forces the ion away from the tip along a radial path toward the screen. Since the ionization usually takes place within 1 Å or so from the surface, perturbation of

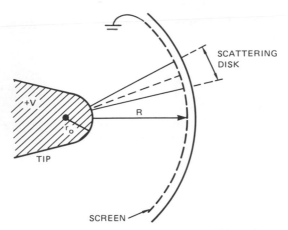

FIGURE 6.6.   Schematic diagram of the field ion microscope.

the local electric fields by the surface atoms plays an important role, since the position of a molecule when it is ionized is comparable to that of adjacent surface atoms. This effect is made stronger by cooling the tip to liquid-hydrogen temperatures. The inert gas helium is preferred, since it is monatomic, chemically inert, and has a low sputtering coefficient, causing less wear on the screen. Pressures of about 10^{-3} Torr are used. The equations governing resolution are the same as for the FEM, except that the mass of the helium ions is 7.4×10^3 greater than that of the electron and a substantially smaller resolution is obtained.

Originally, ion-luminescent phosphor screens were used to view the image; however, the ion current densities reaching the screen are quite small, and the screen tends to sputter erode. Thus, to view the image in a reasonable time and without sputter damage a "channeltron" multiplier is used. This device consists of a two-dimensional array of tiny capillaries, with an electric field applied parallel to the axis of the capillaries. When an ion strikes a capillary, it generates secondary electrons, which in their passage through the tube are amplified by a chain reaction of collisions with the capillary walls. These electrons emerging from the other side of the capillary are then accelerated to a cathodoluminescent screen, where a bright amplified image is formed.

A problem with the FIM is that the electric fields at the tip necessary for field ionization are close to values that will pull the surface atoms away from their crystals, i.e., field evaporation. Many materials field evaporate at lower fields than those at which field ionization occurs and hence cannot be viewed. However, the effect can be used to advantage to identify surface atoms. A short voltage pulse is applied to the system to exceed the evaporation field of some surface atoms. The time these atoms take to reach the screen is then measured so that an estimate of the mass of the particle can be made. This device is called an atom probe[17-19] and has proved to be of value in the study of impurities in metal crystals.

6.2.6.   X-ray Microscope

A point source of illumination will cast a magnified image of a partially transparent sample on a screen. The magnification is given by the ratio of the distance of the sample from the screen to its distance from the source. The source size should be smaller than the smallest feature to be resolved to reduce penumbral blurring and the wavelength similarly sufficiently small to avoid diffraction effects. However, with electron bombardment generation of X rays it is difficult to obtain sources of much less than 1 μm in size with sufficient intensity to expose film in an acceptable time period.[20] An advantage of point source X-ray microscopy is that even in a thick sample all of the features are in sharp focus. Thus, two exposures displaced a little from each other can be used to obtain excellent stereoscopic images. The availability of intense synchrotron radiation sources and recent developments in X-ray optical components could enable this technique to be greatly improved.

As a general rule, the X-ray energy (or wavelength) should be such that on average about half the radiation passes through the sample. Thus, for very small objects, such as a biological cell, which are on the order of 1–10 μm in size, an X-ray energy of about 2 keV is suitable. Polymeric resists (such as PMMA) are sensitive to X-ray photons in this range and their effective grain size is extremely small. Thus, they can be used for directly recording X-ray radiographs since the depth polymerized is a function of the X-ray dose. The image is recorded as a topographic profile of the resist film. The resist is then coated with a thin layer of metal and viewed in a scanning electron microscope (see 6.3.1).

6.3. SERIAL (SCANNING) IMAGING

6.3.1. Scanning Electron Microscopes

The SEM (see Fig. 6.7) requires an illumination system in which the spot at the specimen is made as small as possible and can be deflected to raster over the surface.[21,22] In its conventional mode of operation, a fraction of the secondary electrons generated by the primary beam are collected and accelerated onto a cathodoluminescent surface situated at one end of a glass rod, which acts as a "light pipe." At the other end of the light pipe, usually external to the vacuum, is placed a photomultiplier tube, which is essentially a low-noise, high-sensitivity, high-gain, high-speed amplifier. The output from the multiplier is used to modulate the intensity of the display CRT. The display is a map of the number of electrons collected by the detection system at each point of the scan. This depends not only on the variation of the secondary electron emission coefficient over the surface but also on the solid angle subtended to the collector system of the particular area on which the spot is impinging. Thus, the topography of even a surface of uniform secondary-electron coefficient is effectively displayed with an astonishing three-dimensional appearance when the specimen is tilted slightly from the plane of the normal to the optical axis of the system. Because the spot size does not change appreciably over a few microns along the axis, the depth of the field is large, making it especially useful for viewing microcircuits and devices, not to mention insects, cells, and fossils. The ability of SEMs to view surface structures, almost completely lost in optical microscopy (because of the small depth of field) and transmission microscopy (due to the thin sections used), has proved to be of great value, even at low magnifications.

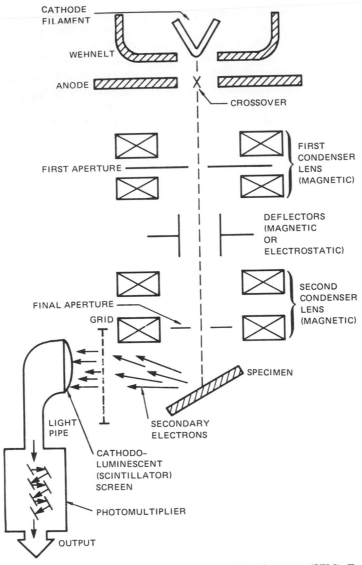

FIGURE 6.7.   Schematic representation of a scanning electron microscope (SEM). From Ref. 4.

The key electron-optical element in the SEM is the illuminating system, which differs somewhat from that of the TEM in that as small a spot as possible is required and lower beam energies in the spot are necessary (5–25 kV). This latter requirement is because high-energy secondaries generated deep in the specimen can emerge from the surface a distance away from the spot where the primary beam was incident (see Section 2.6.2), resulting in a loss of resolution. The electron gun produces a crossover close to the anode. The first weak condensing lens contains an aperture and, combined with the second strong lens, enables a demagnified image of the crossover to be focused down to the limit set by spherical aberration in the gun optics, the energy distribution of the electrons leaving the cathode, and the Boersch effect (Section 2.4.5). Using point-field-emission sources or point-thermionic sources of lanthanum hexaboride (see Section 2.2), spots down to 10 Å in diameter can be obtained with sufficient current (10^{-11} A).

The number of electrons of the primary beam striking the illuminated spot per second must be sufficiently high that measurements of the detected secondary particles are not limited by statistical fluctuations (see Section 7.3.3). This number is about 10^8 electrons/s (10^{-11} A) and is independent of the diameter of the focused spot. Thus, the current density in a spot diameter d Å

$$j_s = \frac{4 \times 10^{-11}}{\pi d^2 \times 10^{-16}} \leq j_c \left(\frac{eV}{kT}\right) \sin^2 \alpha \tag{6.14}$$

(the Langmuir limit, see Section 2.4.3) where V is the beam voltage, T the cathode temperature in K, and α the convergence angle of the beam on the spot. Thus, the minimum resolvable spot diameter is given by

$$d_{min} \approx A\left(\frac{T}{Vj_c}\right)^{1/2} \tag{6.15}$$

where A is a constant for a given electron-optical system.

In a typical SEM system operating at 2×10^4 V using a tungsten filament for which $j_c \sim 1$ A/cm^2 at $T = 2500$ K gives $d_{min} \sim 100$ Å. The same system using a lanthanum hexaboride cathode for which $j_c \sim 20$ A/cm^2 at $T = 2000$ K therefore gives $d_{min} \leq 20$ Å. Thermionic cathodes cannot be improved to give much more than 20 A/cm^2; however, single-crystal field emission tips can easily exceed 10^6 A/cm^2 and are of sufficient emitting area to deliver 10^{-11} A. This advantage enables lower voltages to be used in scanning microscopes at the highest resolution.

If very thin samples and higher beam energies are used, the microscope can be used in the transmission mode (STEM). In this case the elastically scattered electrons are collected on an annular ring scintillator formed at the end of a glass tube light pipe, and the unscattered primary beam and small-angle inelastically scattered electrons are lost down the center. In this way, resolutions of 3–5 Å have been obtained and heavy single atoms have been viewed.[23] Alternatively, an energy spectrum of the entire transmitted beam can be obtained with an electron energy analyzer (see Section 2.5.3). Electrons that have suffered inelastic collisions lose a certain amount of energy characteristic of the atomic species involved in the collision, so that an analysis of the material through which the electrons have passed can be obtained. This is called energy loss spectroscopy (ELS).[24,25]

In the conventional SEM, we utilize all of the types of secondary electrons. If, however, we first pass them through an energy analyzer, we can select only a particular energy range for image modulation. Now an electron vacancy in an atom, created by collision of the primary beam, can be filled by an orbital electron from a less tightly bound state, with the simultaneous emission of a second electron from another less tightly bound state. The emitted electron has an energy characteristic of the atom concerned. This is the Auger effect (see Section 2.6.2). Since Auger electrons have relatively low energies, they can only escape from the surface layer of atoms. Thus, by selecting an Auger peak of a given element, we can obtain a map of that element over the surface.[26] Because of the low numbers of electrons counted in the Auger peaks, it is necessary to use slow scanning times, a large scanning spot, and an efficient electron energy analyzer to obtain a signal of measurable amplitude above the noise in a reasonable time. Nevertheless, fractional micron resolutions can be obtained. Because the Auger electrons only arise from the surface atomic layer, analysis can only be carried out in ultrahigh vacuums after the surface contaminants have been removed, usually by argon sputtering. This device is known as a scanning Auger microprobe (SAM).

When the scanning spot interacts with the surface, it also generates X rays characteristic of the material being bombarded (X-ray fluorescence; see Section 2.6.3). When electronic vacancies are filled by electrons from less tightly bound states, the emission of photons characteristic of the atom concerned occurs. Since X-ray photons can escape from much deeper in the material than electrons of the same energy, they are more representative of the bulk composition, and hence poorer vacuums than necessary for Auger spectroscopy (a surface effect) can be used. The X rays are detected by solid-state counters that count individual photons and measure their energy. If only photons of a given energy characteristic of a given material are counted, a map of that material over the scanned area is obtained. Intensity considerations again generally limit the resolution to barely submicron dimensions. X-ray crystal spectrometers operating on the principle of Bragg reflections are used instead of counters when more precision is required for lower-energy photons originating from low-atomic-number materials. This device is known as the scanning X-ray microprobe.

Of course, in these latter variations of the SEM we could hold the beam at a single point of interest on the specimen, first identified by using it in the standard detection mode, to measure the composition of a particular point of interest. Such instruments are simply called microprobes.[27]

In order to examine the deep interior of a device with microprobes, an area may be sputtered away layer by layer using an ion gun, typically argon, between analyses. This technique is known as "depth profiling" and is most often combined with Auger analysis, but it is of course a destructive technique.

The higher the energy of the incident electrons, the greater is the depth from which the secondary X rays arise. This effect has been used for nondestructive depth profiling (SXDA).[28]

If the scanning beam strikes the surface of an operating semiconductor device, the currents and voltages induced alter the paths of the secondary electrons. Such effects are used to analyze the mechanism of operation and locate faults in microcircuits. In the voltage-contrast technique,[29,32] the change of secondary-electron signal gives a measure of the local potentials on the surface of an IC. In voltage-contrast SEM micrographs, positive-biased components of an IC appear dark compared to areas of lower potential. This contrast results from the existence of retarding fields above positive-biased specimen areas, which caused a decrease in the secondary-electron signal.

TABLE 6.3. Operational Modes of SEM for the Characterization of Semiconductor Devices

Mode of operation	Physical effect	Information and applications	Spatial resolution
Secondary emission (SE)	Dependence of SE on angle of electron-beam incidence and materials	Pattern structure, magnetic domain imaging	0.01 μm
Backscattered emission (BE)	Dependence of backscattered coefficient on the angle of incidence	Surface structure, measure of mean atomic number, chemical crystal orientation	0.01–1 μm
Current mapping (SE) + (BE)	Current change in the probe is due to the changes in SE	Topological contrast, material contrast, and BE on the surface	0.01–1 μm
Channeling	Dependence of SE and BE on crystal orientation	Crystal structure	1–10 μm
Voltage contrast	Influence of probe potential and probe-detector geometry on SE signal	Mapping of potential distribution in IC surface structures, location and height of barriers, resistivity changes	1–10 μm
Induced current (EBIC)	Incident electrons create excess current, which induces current in an external circuit	Detects built-in barriers in inhomogeneous semiconductors, diffusion length determination, depth and thickness of p–n junctions, defect location, resistivity	0.1 μm
Cathodoluminescence (CL)	Emission of photons (infrared to uv) from the sample after electron-beam irradiation	Characteristics of local band gaps, doping distribution, relaxation times for radiative process	0.1 μm
X-ray analysis	Emission of characteristic radiation	Qualitative and quantitative elemental analysis	1–10 μm
Auger electrons	Measurement of Auger electrons (100–1000 eV) with an energy spectrometer; displacement of Auger peaks resulting from local fields	Material analysis (light elements) element mapping, local measurement of potential	Depth 2–10 Å 0.1 μm
Transmission	Elastic scattering Phase/diffraction contrast	Atomic movements, nucleation studies, energy-loss spectroscopy, nanolithography	10–100 Å
Acoustic	Generation of an elastic wave	Thermal elastic properties	0.1–1 μm

However, this voltage-contrast measurement gives only qualitative results in determining the local surface voltage because of the following restrictions:

- The retarding fields close to the surface of ICs depend not only on the geometry and voltages of the spot, but also on the voltage distribution of the whole specimen surface.
- A nonlinearity exists between the signal and the specimen voltage, caused by the energy distribution of the secondary electrons.
- The voltage contrast in SEM micrographs is superimposed on the material and topography contrast.

Electron-beam-induced current technique (EBIC) in semiconductors is another important family of measurements for microelectronic devices.[33-35] In this technique, a high-energy (E_B) electron beam is focused on a small area of a microcircuit and can penetrate through several layers of the device structure. In the semiconducting regions, electron–hole pairs are generated by incident electrons of energy E_B of number $E_B/3E_G$ per incident electron, where E_G is the band gap of the semiconductor. With appropriate external voltages applied to the microcircuit, the currents generated by the newly created charge carriers are measured. This technique permits the observation and study of inversion and depletion layers, breakdowns, carrier multiplication effects, and the phenomena of impact ionization. It also provides a method for determining carrier diffusion lengths, evaluating doping profiles near a junction, and measuring surface recombination velocities.

The combined application of EBIC with other techniques (X-ray, Auger) permits analysis of the chemical nature of the precipitates that generate the localized breakdown. It can be used jointly with cathodoluminescence to relate light-emitting regions to microplasmas or to evaluate the doping distribution of p–n junctions. Furthermore, the EBIC technique allows the study of dislocations and other defects in semiconductors.

The SEM is the single most useful instrument for microcircuit analysis. Table 6.3 lists the different kinds of operational modes of SEMs for the characterization of semiconductor devices, showing the various physical effects, obtainable information, and expected spatial resolutions.

6.3.2. X-ray and Ion Microprobes

A number of corresponding methods for chemically analyzing surfaces and films are available that utilize beams of X-ray photons and ions rather than electrons.[36]

A beam of X rays impinging on a surface is more penetrating than an electron beam of the same particle energy and can therefore provide more information concerning the constitution of the material at greater depths, although the depth is still limited by the range of the photoelectrons generated. X rays also produce photoelectrons from the inner shells that carry information concerning the chemical bonding state of the excited atoms. Thus, analysis of the energy distribution of the electrons emitted from a surface bombarded by X rays is useful for subsurface chemical analysis. Instruments operating on these principles are known as ESCA (electron spectroscopy for chemical analysis) or XPS (X-ray photoelectron spectroscopy) instruments.[37] Soft X ray sources are used for ESCA, typically using the Al K_α (1486.6 eV) or Mg K_α (1253.6 eV), and are of quite low intensities compared to electron beams. To read out the data in a reasonable time, larger spots must be used than in electron microprobes and longer integrating times must be used in the detection electron energy analyzers. However, ESCA does not have the noise from the strong secondary background to contend with as does Auger electron spectroscopy (AES), and surface charging is not as severe a problem as with electron bombardment. These properties have made ESCA particularly useful for the analysis of organic and polymeric materials.

Scanning X-ray microscopes have been developed using microfabricated zone plates to focus the X rays to a small spot on the sample. The X rays transmitted through the sample are detected using a (lithium drifted) silicon single-crystal detector to count the incoming photons. Scanning may be accomplished by placing the sample on a movable stage using piezoelectric actuators. Recently, an 8-keV scanning X-ray microscope with 10-nm pixel sizes has been reported with a dwell time on each pixel of about 10 s.[38]

Ion beams can now be obtained with relatively high intensities (see Section 2.3) and can be focused into a small spot on the surface of interest with appropriate lenses and deflected in ways similar to those used for electron beams. At low energies (10^2–10^3 eV), ions are scattered from the surface molecules principally by billiard-ball-type elastic scattering (see Section 2.7). In ion-scattering spectroscopy (ISS),[39] the energy-loss measurements are made on the scattered beam ions at a specific scattering angle. Most of the incident ions are neutralized, particularly if they penetrate the first atomic layer. However, the few percent that are reflected exhibit energy peaks mainly characteristic of elastic scattering at the surface atoms. The incident ion beam is held to a narrow energy range and the reflected intensity is measured as a function of the ratio of the scattered energy to the incident energy. Noble gas ions are mostly used, since they give sharper peaks.

The elastic backscattering of high-energy light ions (see Section 2.7), usually called Rutherford backscattering spectrometry (RBS), has also proved to be a very useful tool in the study of implantation damage and annealing, the location of impurity atoms in crystal, and the investigation of surfaces and thin films.[40]

A well-collimated beam of singly ionized 1- to 3-MeV helium ions is used. The specimen to be analyzed is placed on a goniometer and exposed to the beam in a target chamber at a pressure of 10^{-6} Torr. Information about the specimen is contained in the energy spectrum of the ions backscattered through an angle of 150°. These ions are detected by a silicon surface barrier detector. The signal is amplified and analyzed, stored, displayed, and recorded.

A silicon sample, for example, exposed to a beam of energy E_0 gives rise to a spectrum consisting of a sharp edge at 0.59 E_0 corresponding to helium ions elastically scattered by the surface silicon atoms and deflected into the detector. The sharp edge is followed by a smoothly varying yield at lower energies. Helium ions not scattered at the surface penetrate into the target, losing energy through inelastic collisions and being scattered elastically in the bulk. They eventually leave the target, to be detected after losing more energy inelastically on the way out. If the stopping power of the target and the geometry of the system are known, one can convert the energy scale into a depth scale.

If the ion beam is incident in a low index direction of a high-quality crystal specimen, it would appear to the beam to be a very open structure consisting of hollow "tubes" or "channels," with lattice atoms forming the walls of the tubes. Channeling minimizes collisions of incident ions with target atoms and reduces the observed backscattered yield for all depths in the crystal. A small surface damage peak observed only in the channeled spectrum is due to localized surface disorder. Even high-quality crystal samples show this surface-disorder peak. If the crystal has been damaged, for example, by ion implantation, and the displaced atoms occupy sites in the channel areas, the channeled ions suffer elastic collisions with these atoms, resulting in a higher scattered yield. The ratio of the yield in the channeling and nonchanneling directions is a measure of the quality of single-crystal specimens. For good-quality silicon, this ratio, often referred to as K_{min}, is 0.03.

RBS, in conjunction with channeling, can also be used to determine the concentration and position of impurity atoms in the crystal, the technique being more sensitive for heavier elements. This technique has also been applied in a variety of surface and thin-film problems.[41]

In secondary-ion mass spectrometry (SIMS), the ions sputtered from the surface by the primary-ion beam are monitored in a mass spectrometer.[37,39] Typically, quadrupole mass spectrometers are used at resolutions of 1 amu. The imaging of surface inhomogeneities can be effected by projection images of the secondary ions or by raster scanning. Resolutions on the order of 1 μm have been reported in both cases. The primary beam is also used to remove surface layers for depth profiling, and depth resolutions to 50 Å have been reported.

Figure 6.8 shows the schematic of a SIMS in which secondary ions are extracted from a sample by an electric field and analyzed according to their mass-to-charge ratio by a mass spectrometer. The SIMS instrument can operate in three modes:

- An ion microprobe may be used, in which local analysis is performed by microfocusing of the primary-ion beam.
- A true image may be formed by geometrical extraction of the secondary ions and aperturing within this image.
- Broad-beam instruments may be used, in which analysis over areas greater than 100 μm is performed instead of microanalysis.

SIMS offers the following capabilities:

- All of the chemical elements can be observed, including hydrogen.
- Separate isotopes of an element can be measured, offering the possibility of determining self-diffusion coefficients, isotopic dating, and isotopic labeling.
- The area of analysis can be as small as 1 μm in diameter.
- The secondary ions are emitted close to the surface of the sample.
- The sensitivity of this technique exceeds that of AES or electron probe microanalysis in many but not all situations, with detection limits in the 10-ppm range under favorable conditions.
- The microanalytical part of this technique can be used to prepare ion images (ion microscope) or ion area scans (ion microprobe), which can give pictorial information about the distribution of species of interest.
- Material can be eroded away in a regular and carefully controlled fashion, and a profile in depth can be prepared.

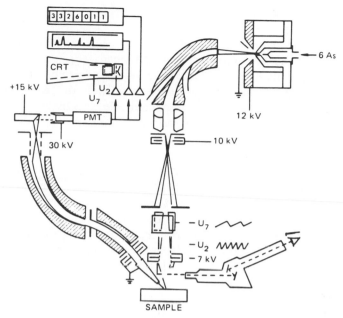

FIGURE 6.8. Schematic diagram of a secondary-ion mass spectrometry (SIMS) system. From Ref. 39.

6.3.3. Scanning Optical Microscopes

Raster scanning a focused spot over the surface of an object and detecting the reflected radiation in a photomultiplier has been made to produce an image in a manner similar to the SEM and is called the scanning optical microscope (SOM).[42] Because of the short depth of field of a very small focused spot, this can be used in high resolution only for very flat surfaces. This has been improved by using confocal imaging of the reflected light in the confocal scanning optical microscope (CSOM).[43] Here a small aperture (not diffraction limited) in the path of the reflected beam allows very little light to reach the detector from regions where the spot is not in the focal plane. Thus, an image of a three-dimensional object can be built up by moving the focal plane of the scanned spot along the optical axis. The availability of lasers with very high intensity enabled this concept to be made practical. Microscopes of the SOM and CSOM variety are limited in resolution by the smallest spot that can be focused by a lens, i.e., about the wavelength of the light being used. This limit has recently been overcome by the near-field scanning optical microscope (NSOM).[44] In this device a very small hole (about 500-Å diameter) is illuminated by an intense light source, and the evanescent wave penetrating the hole is scanned over the object. The reflected light is, as usual, detected by a photomultiplier and the image presented on a CRT display. In the NSOM, resolutions down to 10 Å have been obtained. Raster scanning in SOMs can be carried out by mechanically moving the stage on which the sample is mounted. Commercially available piezoelectric actuators are particularly useful for small controlled movements. Alternatively, the illuminating light beam can be deflected using, for example, galvanometer mirrors.

Raster scanning can also be accomplished by rotating an array of holes (Nipkow disk[42]) over the sample, as was used in the early days of television, the disk itself being illuminated by a parallel light beam.

Laser scanning is also used to investigate the inner workings of active semiconductor devices. The scanner does not damage the device and can be used to map dc and high-frequency gain variations in transistors to reveal areas of the device operating in a nonlinear manner, to map temperature within devices, to determine internal logic states in ICs, and to selectively change these states.[45] Figure 6.9 shows the light and signal paths of a dual-laser scanner.

Two lasers are used for a greater measurement flexibility. Visible or near-infrared radiation incident on silicon creates electron–hole pairs with a generation rate that decreases exponentially with distance into the material. The penetration depends on the wavelength of the incident radiation. The visible light from the 0.633-μm laser has a characteristic penetration depth of about 3 μm in silicon. Because most modern silicon devices have their active regions within a few micrometers of the surface, the 0.633-μm laser is effective in exciting active regions of such devices. The intensity at the specimen can be varied to produce junction photocurrents over a range from about 10 pA to about 0.1 mA.

Silicon at room temperature is almost transparent to the 1.15-μm infrared radiation from the second laser; the characteristic penetration depth of this radiation is about 1 cm. The infrared laser is used for three classes of measurement:

- Examination of the silicon–header interface
- Device temperature profiling
- Examination of the device through the backside of the silicon chip

Each of these application modes uses the penetrating nature of the radiation. In the first application the reflected-light circuit is used to look through the silicon wafer and observe irregularities at the silicon–header interface in the scanning mode. The second application made use of the temperature sensitivity of silicon absorption: A larger signal is produced on the display screen for areas that are warmer than others. Utilizing this sensitivity, one has an electronic technique for the thermal

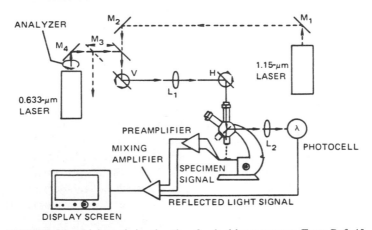

FIGURE 6.9.   Light and signal paths of a dual-laser scanner. From Ref. 45.

mapping of devices, which appears to have a number of advantages over the more traditional methods. The third application uses the penetrating infrared radiation to photogenerate carriers deep within silicon devices. This capability allows the operation of face-down bonded components, such as beam-lead devices to be examined.

6.3.4. Scanning Acoustic Microscope

The scanning acoustic microscope operates at gigahertz frequencies, has a greater resolution than the optical microscope, and can look beneath the surface of thin-layer structures such as IC chips.[46]

The structure of the acoustic microscope is shown in Fig. 6.10. Input signals of up to about 3 GHz are fed into the piezoelectric transducer consisting of a thin coating of zinc oxide on the back of a sapphire (aluminum oxide) acoustic lens. The ultrasonic waves generated by the transducer are focused by the lens, which has a radius of about 100 μm, on the object under examination. The reflected waves containing information about the object are passed back to the transducer, which converts them into electrical signals for display on a CRT. To obtain a higher resolution, very high frequencies must be used, but these high frequencies are attenuated by the material.

Focusing on layers beneath the top layer and reflections from these sublayers can provide useful information about the state of the chips, such as whether any corrosion has taken place. The points where the electrodes make contact with the semiconductor layers can also be examined and the optical and acoustic images compared. These operations can be performed without damaging the device, so that a comparison can be made of the performance of different devices that have various

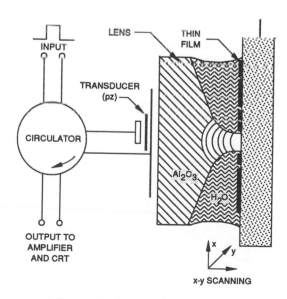

FIGURE 6.10. Arrangement for a reflection scanning acoustic microscope (SAM) system. From Ref. 47.

defects in the subsurface structure. A correlation between these defects and perform-
ances will help to identify the origin and nature of defects in semiconductors.

Another variant is the scanning electron acoustic microscope,[47] where a sound
wave is initiated by an electron beam, and detected through the sample.

6.3.5. Scanning Tunneling Microscope (STM)

The concept of the STM dates back to 1971[48] but a practical device with
atomic resolution was not reported until 1982.[49] Figure 6.11 shows the principle of
operation.[50] A metallic point with an atomically sharp tip is mounted on a stage
that can be precisely moved in three orthogonal directions, one perpendicular to the
surface of the sample. The sample is mounted on a relatively coarse positioning stage
to bring its surface within range of the point. The precision movements are obtained
using commercially available piezoelectric translators that are capable of controlled
displacements down to 10^{-4} Å. The electric field for obtaining measurable field electron emission from the atom at the tip is somewhat above 10^9 V/m; thus, with 1 V
applied, the tip will be less than 10 Å from the surface. In operation, the tip is

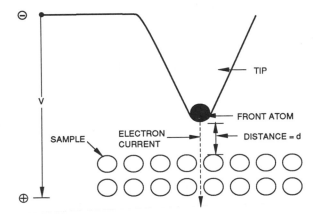

(a) **PRINCIPLE OF OPERATION DEPENDS ON CURRENT FROM ATOMIC TIP V
AND DISTANCE FROM CLOSEST SAMPLE ATOM**

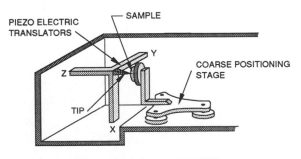

(b) SCHEMATIC OF APPARATUS

FIGURE 6.11.   Scanning tunneling microscope. (a) Principle of operation depends on current from
atomic tip V and distance from closest sample atom. (b) Schematic of apparatus. From Ref. 50.

brought close enough to the surface to obtain a measurable tunneling current (10^{-10}–10^{-9} A), and the tip is mechanically raster scanned over the surface. In one mode the current is kept constant by varying the vertical position of the tip. Since the field emission current is exponentially dependent on the field (see Section 2.21), this keeps the distance between the tip and surface essentially constant. In this way a map of tip height z versus lateral position (x, y) is obtained. An alternate mode is to scan the point in a plane parallel to the average plane of the sample surface and record the current as a function of position. In this case a map of current versus lateral position is obtained. Computer storage and processing of the map data enables an image of the surface topography to be displayed. Electrostatic microlenses for STMs are also an area of importance.[51]

An important advance in this technology has been the atomic force microscope (AFM).[52] For an STM to operate, it is required that the substrate be conducting to maintain an equipotential along its surface. The AFM avoids this requirement by measuring the force between a stylus tip and the surface and thus can be used for insulating and biological substances. The forces arise from the desire of the atom at the stylus tip to bond to the proximate atom at the surface, and range from weak van der Waals adsorption bonds to strong ionic bonds. In Ref. 49 it is estimated that these forces may range from 10^{-11} to 10^{-7} N. The principle of operation is shown in Fig. 6.12. A small diamond stylus is mounted on one end of a thin metal cantelever, and sandwiched between the sample and an STM tip. The other end of the cantilever is attached to a small piezoelectric element that can drive the cantilever at its resonant frequency. The STM tip is also mounted on a piezoelectric element so that the tunneling current gives a measure of the distance between the STM tip and the

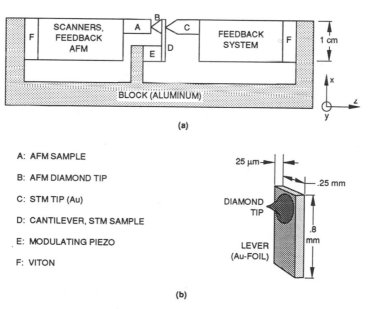

(a)

A: AFM SAMPLE

B: AFM DIAMOND TIP

C: STM TIP (Au)

D: CANTILEVER, STM SAMPLE

E: MODULATING PIEZO

F: VITON

(b)

FIGURE 6.12. Atomic force microscope: experimental setup. The lever is not to scale in (a). Its dimensions are given in (b). The STM and AFM piezoelectric drives are facing each other, sandwiching the diamond tip that is glued to the lever. From Ref. 52.

cantilever end. The sample, which is mounted on a piezoelectric x, y, z stage, is brought sufficiently close to the stylus tip that the force between them is sufficient to deflect the cantilever. In the preferred mode of operation, the AFM cantilever was vibrated at its resonant frequency (5.8 kHz) with an amplitude in the range 0.1–10 Å. As the sample is moved in the z direction, the force between the sample and the stylus deflects the cantilever, which changes the tunneling current from the STM. As the sample is scanned laterally, its z position with respect to the stylus is adjusted by a feedback circuit to maintain the STM current constant. Thus, the average z position of the cantilever with respect to the STM, and hence the average force on the stylus are maintained constant, and a map of the sample-to-stylus distance (z) as a function of lateral position (x, y) is obtained.

While the problems of shielding these microscopes from vibration due to extraneous electric, magnetic, mechanical, thermal, and Brownian fluctuations are all formidable, atomic resolution is currently attained and improvements are sure to follow. It is anticipated that major advances in the materials and biological sciences will result from the use of these microscopes. We note here that the STM has also been used for direct write patterning of structures on silicon with near single atom resolution[53] (see Section 5.8.14), and has demonstrated the ability to move atoms over the surface.[54]

TABLE 6.4. Summary of Microcharacterization Techniques

Measurement	Techniques available	Range of usage	Remarks
1. Dimensions	Optical	Dimensions >0.3 μm	Poor depth of field
	Evanescent wave		
	TEM	>3 Å	Requires thin samples
	SEM	>30 Å	Good for surface topography
	STM	>1 Å	
	EEM	>100 Å	
	FEM	>10 Å	Useful only with special preparation
	EMM	>25 Å	
	FIM	>1 Å	
	STM	>1 Å	
2. Chemical composition	ESCA	Measures constitutents to a depth of 100 μm	1 ppm
	X-ray microprobe	Measures constituents to a depth of 1 μm	100 ppm
	SIMS	Measures constituents of	1 ppm
	ISS	surface layer only	100 ppm
	Auger spectroscopy	Measures constituents through	1000 ppm
	ELS	sample thickness	
3. Device operation	SEM (EBIC)		
	SEM (voltage contrast)		See Table 6.1
	Laser scanning		
	Acoustic scanning	Subsurface layers	

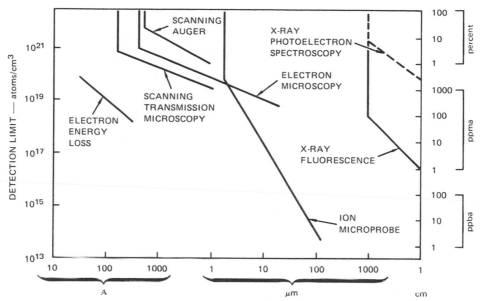

FIGURE 6.13. Effect of diameter of analyzed area on detection limit. From Ref. 56.

6.4. SUMMARY

The characterization of devices of small dimensions requires dimensional, chemical, structural, and functional measurements.[55] The general applicability of the microscopy techniques discussed in this chapter is summarized in Table 6.4.

It is apparent that to determine microscale morphology, elemental constituents, molecular species, and lateral and depth distributions, a variety of microanalytical techniques must be employed.[56] It is frequently of great interest to simultaneously determine both the surface and bulk compositions of a sample with submicron resolution. Such requirements are met by combined analytical systems. However, for nanodevice characterization, it will be necessary to develop new techniques with extremely high signal-to-noise ratios.[57] Better sensitivities at submicron geometries are also needed for chemical analysis. This can be seen from Fig. 6.13, which shows that a material must be present in a density greater than 10^{17} atoms/cm^3 for it to be detected in a pattern with dimensions of under 1000 Å.

PROBLEMS

1. a. Give the full names for the following acronyms:

   • UPS
   • AES
   • RBS

b. The energy of an Auger electron is (indicate one)

- Dependent on the energy of the incoming beam.
- Independent of the energy of the incoming beam.
- Dependent on the energy of the emitted X ray.
- The same for most elements.
- Hard to pin down.

c. Which of the following techniques is quantifiable to 0.1%?

- X-ray photoelectron spectroscopy
- Energy-dispersive X-ray spectroscopy
- Rutherford backscattering spectroscopy
- X-ray fluorescence
- X-ray photoelectron spectroscopy and energy-dispersive X-ray spectroscopy

2. a. Describe the important differences between electron reflection microscopes, scanning electron microscopes, and transmission electron microscopes.

   b. Discuss the differences between an atomic force microscope and the scanning tunneling microscope.

   c. What technique would you use to look for a void at a semiconductor/metal interface without removing the metal?

3. a. Compare electron surface analysis tools to ion-beam tools. For example, consider Auger electron spectroscopy, SPS, SIMS, and RBS.

   b. Compare X-ray analysis to electron energy analysis for XPS, energy-dispersive analysis to Auger spectroscopy and electron luminescence.

4. A MOSFET integrated circuit has been fabricated with low yield. What techniques would you use to investigate the cause of the problem and why?

REFERENCES

1. V. E. Coslett, Radiation damage and chromatic aberration produced by inelastic scattering of electrons in the electron microscope: Statement of the problem, *Ann. N.Y. Acad. Sci.* **306**, 3 (1978).
2. A. Septier, The struggle to overcome spherical aberration in electron optics, *Adv. Opt. Electron Microsc.* **1**, 204–272 (1966).
3. R. W. Ditchburn, *Light*, Blackie, Glasgow (1952).
4. B. M. Siegal (ed.), *Modern Developments in Electron Microscopy*, Academic Press, New York (1964).
5. P. W. Hawkes, *Electron Optics and Electron Microscopy*, Taylor & Francis, London (1972).
6. V. E. Coslett, *Modern Microscopy*, Bell and Sons, Glasgow (1966). D. B. Williams and D. T. Newberry, Recent advances in the electron microscopy of materials, *Adv. Electron. Electron Phys.* **62**, 162–288 (1984).
7. S. I. Ijima, High resolution electron microscopy of crystal lattice of titanium-niobium oxide, *J. Appl. Phys.* **42**, 5891–5893 (1971).
8. R. M. Oman, Electron mirror microscopy, *Adv. Electron. Electron Phys.* **26**, 217–249 (1969).
9. W. Telieps and E. Bauer, The $(7 \times 7) \leftrightarrow (1 \times 1)$ phase transition on Si(111), *Surface Sci.* **162**, 163 (1985).
10. W. Telieps and E. Bauer, Kinetics of $(7 \times 7) \leftrightarrow (1 \times 1)$ transition on silicon, *Ber. Bunsenges. Phys. Chem.* **90**, 197 (1986).
11. G. A. Haas and R. E. Thomas, Electron beam scanning technique for measuring surface work function variations, *Surface Sci.* **4**, 64 (1966).
12. G. Mollenstedt and F. Lenz, Electron emission microscopy, *Adv. Electron. Electron Phys.* **18**, 251–329 (1963).

13. L. Wegmann, The photoemission electron microscopy: Its technique and applications, *J. Microsc.* **96**, 1 (August, 1972).

14. R. H. Good and E. W. Muller, Field emission, *Handbuch der Physik* (S. Flugge, ed.), Vol. 21, pp. 176–231, Springer-Verlag, Berlin (1956).

15. D. J. Rose, On the magnification and resolution of the field-emission electron microscope, *J. Appl. Phys.* **27**, 215 (1956).

16. I. Brodie, Visibility of atomic objects in the field-electron emission microscope, *Surface Sci.* **70**, 186–196 (1978).

17. E. W. Muller and T. T. Tsong, *Field Ion Microscopy*, Elsevier, Amsterdam (1969).

18. J. A. Pantiz, The 10-cm atom probe, *Rev. Sci. Instrum.* **44**, 1034–1038 (1973).

19. D. G. Brandon, Field ion microscopy, *Adv. Electron Opt. Microsc.* **2**, 343–402 (1968).

20. V. E. Cosslett and W. C. Nixon, *X-ray Microscopy*, Cambridge University Press, London (1966).

21. C. W. Oatley, W. C. Nixon, and R. F. W. Pease, Scanning electron microscopy, *Adv. Electron. Electron Phys.* **21**, 181 (1985).

22. L. Reimer and G. Pfefferkorn, *Raster-Electronen-Mikroscopie*, Springer-Verlag, Berlin (1977).

23. A. V. Crewe, Scanning transmission-electron microscopy, *J. Microsc.* **100**, 247–259 (1974).

24. M. Isaacson, All you want to know about ELS, *Scanning Electron Microsc.* **1978**(I), 763–776 (1978).

25. R. F. Edgerton, *Electron Energy Loss Spectroscopy in the Electron Microscope*, Plenum Press, New York (1986).

26. M. Cailler, J. P. Ganachaud, and D. Roptin, Quantitative Auger spectroscopy, *Adv. Electron. Electron Phys.* **61**, 162–289 (1983).

27. M. H. Loretto, *Electron Beam Analysis of Materials*, Chapman & Hall, London (1984).

28. A. Szasz and J. Kojnok, Soft X-ray emission depth profile analysis, *Appl. Surface Sci.* **24**, 34–56 (1985).

29. L. J. Balk, H. P. Feuerbaum, E. Kubalek, and E. Menzel, Quantitative voltage contrast at high frequencies in the SEM, *Scanning Electron Microsc.* **1976**(I), 615–646 (1976).

30. J. R. Banbury and W. C. Nixon, A high-contrast directional detector for the scanning electron microscope, *J. Sci. Instrum.* **2**, 1055–1059 (1969).

31. D. L. Crosthwait and T. W. Ivy, Voltage contrast methods for semiconductor device failure analysis, *Scanning Electron Microsc.* **1974**(I), 935–940 (1974).

32. A. Gopinath and W. J. Tee, Theoretical limits on minimum voltage change detectable in the SEM, *Scanning Electron Microsc.* **1976**(I), 603–608 (1976); A. Gopinath, Voltage measurement in the scanning electron microscope, *Adv. Electron. Electron Phys.* **69**, 1–54 (1987).

33. A. J. Gonzales, On the electron beam induced current analysis of semiconductor devices, *Scanning Electron Microsc.* **1974**(I), 941–948 (1974).

34. J. F. Bresse, Electron beam induced current in silicon planar p–n junction: physical model of carrier generation, determination of some physical parameters in silicon, *Scanning Electron Microsc.* **1972**, 105–112 (1972).

35. J. F. Bresse and D. Lafeuille, SEM beam induced current in planar p–n junctions diffusion length, and generation factor measurements, *Proc. EMAG*, Institute of Physics, London (1971).

36. A. G. Michette, X-ray microscopy, *Rep. Prog. Phys.* **51**, 1525–1606 (1988).

37. H. W. Werner and R. P. H. Garten, A comparative study of methods for thin film and surface analysis, *Rep. Prog. Phys.* **47**, 221–344 (1984).

38. R. M. Bionta, E. Ables, O. Clamp, O. D. Edwards, P. C. Gabriele, K. Miller, L. L. Ott, K. M. Skulina, R. Tilly, and T. Viada, Tabletop X-ray microscope using 8 keV zone plates, *Opt. Eng.* **29**(6), 576–580 (1990).

39. A. Feurerstein, H. Grahmann, S. Kalbitzer, and H. Oetzmann, in: *Ion Beam Surface Layer Analysis* (O. Meyer, G. Linker, and F. Kappler, eds.), Plenum Press, New York (1976).

40. W. K. Chu, J. W. Mayer, and M. A. Nicolet, *Backscattering Spectrometry*, Academic Press, New York (1978).

41. J. A. Davies, in: *Material Characterization Using Ion Beams* (J. P. Thomas and A. Cachard, eds.), Plenum Press, New York (1978).

42. T. Wilson and C. F. R. Shephard, *Scanning Optical Microscopy*, Academic Press, New York (1984).

43. P. Davidovits and M. D. Egger, Scanning laser microscope, *Nature* **233**, 831 (1969).

44. R. Kopelman and M. Isaacson, Light microscopy beyond the limits of diffraction and to the limits of single molecule resolution, Proc. SPIE 1205, 60–61 (1990).

45. D. E. Sawyer, D. W. Barning, and D. C. Lewis, Laser scanning of active integrated circuits and discrete semiconductor devices, *Solid-State Technol.* p. 37 (June, 1977).

46. D. E. Yuhas and T. E. McGraw, Acoustic microscopy, SEM, and optical microscopy: Correlative investigation in ceramics, *Scanning Electron Microsc.* **1979**(I), 103 (1979).

47. L. J. Balk, Scanning electron acoustic microscopy, *Adv. Electron. Electron Phys.* **71**, 1–74 (1988).

48. R. Young, J. Ward, and F. Scire, Thermal drive apparatus for direct vacuum tunneling experiments, *Rev. Sci. Instrum.* **47**, 1303 (1976).

49. G. Binnig, H. Rohrer, C. Gerber, and E. Weibel, 7×7 reconstruction on Si (111) resolved in real space, *Phys. Rev. Lett.* **50**, 120 (1983).

50. P. K. Hansma and J. T. Tersoff, Scanning tunneling microscopy, *J. Appl. Phys.* **61**(2), R1–R23 (1987).

51. T. H. P. Chang, D. P. Kern, and M. A. McCord, Electron optical performance of a scanning tunneling microscope controlled field emission microlens system, *J. Vac. Sci. Technol. B* **7**(6), 1855 (Nov./Dec. 1989).

52. G. Binnig, C. F. Quate, and C. Gerber, Atomic force microscope, *Phys. Rev. Lett.* **56**, 930–933 (1986).

53. R. S. Becker, J. A. Golovchenko, and B. S. Schwartzentruber, Atomic scale surface modifications using a tunneling microscope, *Nature* **325**, 419–421 (1987).

54. D. M. Eigler and E. K. Schweizer, Positioning single atoms with a scanning tunnelling microscope, *Nature* **344**, 524 (1990).

55. G. B. Larrabee, in: *Microstructure Science, Engineering and Technology*, National Academy of Sciences, Washington, D.C. (1979).

56. D. E. Newbury, Microanalysis in the scanning electron microscopy: Progress and prospects, *Scanning Electron Microsc.* **1979**(II), 1–20 (1979).

57. J. Silcox, in: *Microstructure Science, Engineering and Technology*, National Academy of Sciences, Washington, D.C. (1979).

7

Limits to Nanofabrication

7.1. BACKGROUND

Devices can be made with decreasing linear dimensions until one of two limitations is reached, namely

- Limitations imposed by the physical principles by which the device operates
- Limitations imposed by our ability to fabricate the device to the required dimensions and tolerances

Essentially all of the microdevices currently built and explored are based on the bulk properties of materials. That is, the physical property concerned relies on an integrated effect over a large array of atoms: as the dimensions of the array are reduced, a critical size is reached less than which the desired physical effect follows different rules or no longer occurs. For example, if we wish to propagate an electromagnetic wave of a given wavelength down a waveguide, there are critical cross sectional dimensions that cannot be made smaller without cutting off the propagation of waves down the guide.

In Table 7.1 we list a number of such devices that are of current interest, together with the principal operational characteristic lengths that will limit their further miniaturization.

The main limitations on fabrication using planar technology are in pattern generation and etching (developing) the pattern, since thin-film technology is sufficiently advanced so that the depth dimension can be controlled to any required precision for bulk devices. At the present time, electron-beam lithography is capable of attaining linewidths of 2500 Å, with edge sharpnesses of 100 Å. Because of electron scattering in the substrate, further improvements may not be expected for routine fabrication, although techniques have been evolved to obtain smaller dimensions with electron-beam lithography in special circumstances. Ion-beam lithography, however, can take over where electron beams leave off, and we can reasonably expect that patterning down to 100 Å in linewidth may become routine. Since the ion beam can be used to sputter-etch its own path as the pattern is written, the etching problem is simultaneously brought under control.

TABLE 7.1.   Principal Limiting Factors for Several Microdevices

Device	Principal limiting characteristic length	Approximate minimum dimensions	
		Å	Atoms
MOS transistor	Channel length determined by breakdown field	2.5×10^3	1000
Bipolar transistor	Base thickness and base doping concentration	2.5×10^3	1000
Magnetic bubble memories	Bubble diameter determined by domain energetics	500	200
Surface acoustic-wave delay lines	Surface wavelength determined by attenuation in surface film	2.5×10^3	1000
Electromagnetic waveguides	Wavelength determined by the photon energy (usually in optical range for integrated optical devices)	4×10^3	1000
Charge storage devices	Number of electrons to be stored on the surface determined by breakdown field and information statistics	1000	300

Our ability to view devices with electron microscopes has already attained resolutions smaller than the device and fabrication limits, and our ability to identify the materials of which the devices are made is also capable of the resolution required for device analysis (see Chapter 6).

It seems clear from the preceding discussions that there is every reason to suppose that the limits for bulk effect devices will be closely approached over the next decade or two. We may also surmise that many new bulk devices will be invented, fabricated, and operated in the decades to come. These devices will not be limited to microelectronics but will include devices for chemical analysis, sensing, and biological functions.

We see, however, a very clear gap between the natural unit of size, namely, the atom, and dimensions up to, say, 1000 atoms, in which bulk descriptions of physical phenomena are no longer valid. This is the molecular scale. Despite the fact that nature, in the form of living cells, has built many diverse and intricate devices based on molecular phenomena, man has not yet been able to carry out such engineering on anything but a trivial level. This is primarily owing to our lack of scientific understanding of the phenomena involved at these dimensional levels. However, we believe that microdevices and instruments operating on presently known principles will enable us to gain the necessary understanding to penetrate the molecular scale in the not too distant future.

Each bulk effect device has its own set of limits as its size is reduced, depending on its detailed mechanism of operation, and each device must be treated individually. However, in accordance with the concern of this book on planar silicon devices, in Section 7.2 we discuss and quantify as a reference example the limits we are approaching in the operation, fabrication, and utilization of the MOS transistor.[1] Limits for other types of devices may be sought using a similar approach.

The limits to planar technology lie in two areas: The first is related to our ability to form layers of given thickness and with desired chemical, physical, and interface properties. Layer thicknesses can be controlled down to atomic dimensions (Chapter 4). The other properties are individual to the device requirements and cannot be discussed generally. The second area relates to our ability to generate patterns in these layers to the smallest tolerances. Such limits are more general and are discussed in Section 7.3.

7.2. LIMITS FOR MOS DEVICES

7.2.1. Scaling Laws

Large integrated systems composed of interconnected MOS field-effect transistors are implemented on a silicon substrate by metal, polysilicon, and diffusion conducting layers separated by intervening layers of insulating material (see Fig. 7.1).

In the absence of any change on the gate, the drain-to-source path is like an open switch. If sufficient positive charge is placed on the gate so that the gate-to-source voltage V_{gs} exceeds a threshold voltage V_{th}, electrons flow under the gate between the drain and the source. The basic operation performed by the MOS transistor is to use the charge on its gate to control the flow of electrons between the source and the drain.

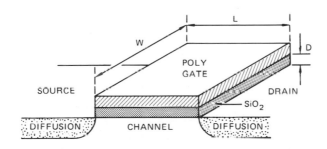

(a) MOS TRANSISTOR

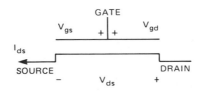

(b) SYMBOLIC REPRESENTATION OF MOS TRANSISTOR

FIGURE 7.1.   MOS transistor structure and symbolic representation. (a) MOS transistor; (b) symbolic representation of MOS transistor.

For a small voltage between the source and the drain, V_{ds}, the transit time τ required to move an elecron from the source to the drain can be written as

$$\tau = \frac{L}{v} = \frac{L}{\mu E} = \frac{L^2}{\mu V_{ds}} \tag{7.1}$$

where v is the electron velocity and μ its mobility in field $E = V_{ds}/L$. The gate, separated from the substrate by insulating material of thickness D, forms a capacitor. This gate capacitance equals

$$C_g = \frac{\varepsilon W L}{D} \tag{7.2}$$

The charge in transit and the current are given by

$$Q = -C_g(V_{gs} - V_{th}) = \frac{\varepsilon W L}{D}(V_{gs} - V_{th}) \tag{7.3}$$

and

$$I_{ds} = \frac{Q}{\tau} = \frac{\mu \varepsilon W}{LD}(V_{gs} - V_{th})(V_{ds}) \tag{7.4}$$

for low V_{ds} the device acts as a resistor ($I_{sd} = \text{const} \times V_{ds}$) and the gate voltage-controlled resistance is given as

$$R = \frac{V_{ds}}{I_{ds}} = \frac{L^2}{\mu C_g(V_{gs} - V_{th})} \tag{7.5}$$

However, when the transistor is operated near its threshold, the channel resistance is given by

$$R = R_0\, e^{-(V_{gs} - V_{th})q/kT} \tag{7.6}$$

where T is the absolute temperature, q is the charge on the electrons, and k is Boltzmann's constant. This means that below threshold, the conductance ($1/R$) has a nonzero value depending on the gate voltage and temperature, as shown in Eq. (7.6).

Now we can examine what happens when the device dimensions are made smaller by a scaling factor $n(< 1)$. The new dimensions are then

$$D' = \frac{D}{n}$$

$$L' = \frac{L}{n} \tag{7.7}$$

$$W' = \frac{W}{n}$$

The electric field cannot be increased when the dimensions are made smaller, since the devices are designed to operate close to the limiting value above which breakdown or quantum-mechanical tunneling would occur ($\sim 10^9$ V/m). Since the fields are constant, the operating voltage varies as

$$V' = \frac{V}{n} \tag{7.8}$$

From Eqs. (7.1)–(7.3), (7.7), and (7.8) the physical quantities scale down as follows (using $V = V_{gs} - V_{th}$):

$$\tau' = \frac{1}{\mu}\left[\left(\frac{L}{n}\right)^2 \Big/ \left(\frac{V}{n}\right)\right] = \frac{1}{\mu}\frac{L^2}{V_n} = \frac{\tau}{n} \tag{7.9}$$

$$C' = \varepsilon\left[\left(\frac{L}{n}\right)\left(\frac{W}{n}\right)\Big/\left(\frac{D}{n}\right)\right] = \varepsilon\frac{LW}{Dn} = \frac{C}{n} \tag{7.10}$$

$$I' = \mu\varepsilon\left(\frac{W}{n}\right)\left(\frac{n}{L}\right)\left(\frac{n}{D}\right)\left(\frac{V}{n}\right)\left(\frac{V_{ds}}{n}\right) = \frac{\mu\varepsilon V}{nLD}V_{ds} = \frac{I}{n} \tag{7.11}$$

The switching power P_{sw} and dc power scale as follows:

$$P'_{sw} = \frac{C'V'^2}{2\tau'} = \frac{1}{2}\left(\frac{C}{n}\right)\left(\frac{V}{n}\right)^2\left(\frac{n}{\tau}\right) = \frac{P_{sw}}{n^2} \tag{7.12}$$

and

$$P'_{dc} = I'V' = \left(\frac{I}{n}\right)\left(\frac{V}{n}\right)\left(\frac{n}{\tau}\right) = \frac{P_{dc}}{n^2} \tag{7.13}$$

and the switching energy

$$E'_{sw} = \frac{1}{2}C'V'^2 = \frac{1}{2}\left(\frac{C}{n}\right)\left(\frac{V}{n}\right)^2 = \frac{E_{sw}}{n^3} \tag{7.14}$$

Table 7.2 summarizes the scaling laws and the present and future operating specifications limited by the physics of the device. Thus, as the dimensions are made smaller, all but one of these variables move in a favorable direction. The operating speed increases, and the power density remains constant. However, the current density is increased by the scale-down factor, in this case 10, presenting an important obstacle to scaling.

Metal conductors have an upper current density limit imposed by electromigration.[2] Electromigration is a diffusive process in which the atoms of a solid move from one place to another under the influence of electrical forces. This effect limits the maximum current that can be carried by a conductor without its rapid destruction. For example, the current density for aluminum conductors of ICs

TABLE 7.2.   Scaling Laws for MOS Devices

Physical quantities	Scale-down factor (n)	1980	Future ($n = 10$)
Linear device dimension	$1/n$	$5\ \mu$m	$0.5\ \mu$m
Supply and logic voltage	$1/n$	5 V	0.5 V
Current	$1/n$	—	—
Capacitance	$1/n$	—	—
Gate transit (delay)	$1/n$	5×10^{-10}	5×10^{-11}
Power dissipation	$1/n^2$	—	—
Switching energy	$1/n^3$	10^{-13} J	10^{-16} J
Device density	n^2	4×10^6	4×10^8
Impedance	Constant	—	—
Power density	Constant	—	—
Current per unit surface area	n	x	$10x$

must be kept lower than 10^6 A/cm^2. Electromigration does not limit the minimum device size but, rather, limits the number of circuit functions that can be carried out by a given number of connected circuit elements per unit time. Essentially, it should be considered as a system limit. Due to electromigration, the mean time for failure for a MOS transistor scales as $1/n^4$. Recently, new results have been published analyzing device reliability in a multilevel Al interconnected systems.[2–4]

7.2.2. Fundamental Limits

Limits that are closely related to basic physical laws are called basic or fundamental limits.[5] Fundamental limits are largely concerned with energetic phenomena related to quantum and statistical physics.

7.2.2.1. Quantum Limits.   A basic concept of quantum physics is that a physical measurement performed in a time Δt must involve an energy

$$\Delta E \geq \frac{\hbar}{\alpha t} \tag{7.15}$$

where $\hbar = 1.05 \times 10^{-34}$ J $\cdot$ s/rad $= h/2\pi$ and h is Planck's constant. The energy is dissipated as heat. The power dissipated during the measurement (switching) process is

$$P = \frac{\Delta E}{\Delta t} \geq \frac{h}{(\Delta t)^2} \tag{7.16}$$

which can be considered as the lower bound of power dissipation per unit operation.

Using Eq. (7.15), we find the minimum energy dissipated in a nanosecond switching device is on the order of 10^{-25} J per operation. The actual minimum energy dissipated in a MOS transistor is on the order of 10^{-13} J, which is far removed from this quantum limit.

7.2.2.2. Tunneling.   If very thin (10–100 Å) insulators are placed between two conductors, the decay of the wave function of an electron on one side is not sufficient to give zero amplitude at the opposite side, and there is a finite probability of the electron passing through the dielectric by the process of quantum-mechanical tunneling. In this way a current can pass through a classically forbidden region (see Section 2.2.1).

In the case of a MOS transistor operation, this "tunneling current" must be much smaller than all of the circuit currents for the proper operation, setting a fundamental size limitation for the thickness of the oxide gates and depletion layers at about 10^{-3} μm.[6] The thickness of gate oxides can already be made in the range 10^{-2}–10^{-3} μm, close to this fundamental size limitation.

7.2.3. Material Limitations

Material properties (critical field for breakdown, doping concentration, dislocation density, and electromigration) also impose limits on device operation.[1,5] Dielectric breakdown limits the electric field in semiconductors and thereby limits the size and speed of the device operation. If the electric field exceeds a critical value ($E_c = 3 \times 10^5$ V/cm in Si) in the semiconductor, a rapid increase of current can occur, owing to avalanche breakdown.

Generally, the breakdown field is not a constant but depends on the doping levels and the doping profile in a junction.[5] For heavy doping, breakdown occurs by tunneling rather than via the avalanche mechanism.

To estimate the limitation set by the material properties of silicon, we consider the maximum propagation time in a cube of silicon material of length Δz. Since the potential and the linear dimensions can be written as

$$\Delta V = E \Delta z \tag{7.17}$$

and

$$\Delta V = v_{Si} \Delta t \tag{7.18}$$

for $\Delta V = kT/q$, the minimum transit time, using $E_{max} \cong 3 \times 10^5$ V/cm and $v_{max} \cong 8 \times 10^6$ cm/s, is

$$\Delta t_{min} = \frac{kT/e}{E_{max} v_{max}} = 10^{-14} \text{ s} \tag{7.19}$$

Thus, the breakdown voltage limit imposed by the material properties of silicon limits the minimum transit time to 10^{-14} s.

7.2.4. Device Limits

Device limits are limitations the depend on device operation (e.g., turn-on/turn-off voltage, gain, bandwidth, thermal limits). To illustrate an important example, consider the channel conductance ($1/R$) of a MOS transistor near the threshold voltage. The conductance near, but below, the threshold is not really zero but given from Eq. (7.6) as

$$\frac{1}{R} = \frac{1}{R_0} e^{(V_{gs} - V_{th})/kT/e} \tag{7.20}$$

At room temperature, $kT/e = 0.025$ V. If $V_{th} = 1$ and $V_{gs} = 0.5$ V, the device conductance is decreased by a factor of 2×10^8. However, if these dimensions and voltages are scaled down by a factor of 5, then the conductance decreases only by a factor of 1.8×10^2. This is a "leaky" transistor indeed. One way to cope with the problem is to operate the devices at lower than room temperature to reduce kT.

Because of thermal fluctautions, the MOS transistor may randomly switch from the conductive to the nonconductive state or vice versa. If the switching energy is E_{sw}, the thermal fluctuation leads to a failure rate of

$$\frac{1}{\tau} \exp\left(\frac{-E_{sw}}{kT}\right) \tag{7.21}$$

per second per device. For a given mean time between errors (T'), Eq. (7.21) gives the following condition for the switching energy:

$$E_{sw} \geq kT \ln\left(\frac{T'}{\tau}\right) \tag{7.22}$$

T' is typically required to be many hours and τ is $\sim 10^{-10}$ s, then $\ln(T'/\tau) \sim 30$ or $E_{sw}/e \sim V_{sw} \sim 0.75$ V for a single device. It is thus possible to reduce the error rate caused by thermal fluctuations to any desired extent by increasing E_{sw}.

It is interesting to note that in human nerve excitation the excitation energy is only a few kT, and for information processing found in biological systems $E_{sw} \sim 20 \, kT$, the same order of magnitude as in a MOS transistor.[1]

This thermal limit for the switching energy is a device and system limit for all of the electrically operated microstructures. For this reason the voltage operating the device should be many times kT/e.

Beyond the devices themselves, there are circuit and system limits. These limits deal with optimal internal and external partitioning, power dissipation, and chip architecture. These problems, which are outside the present scope, have been reviewed in a classic text on VLSIC (very-large-scale integrated circuit) design.[7]

Based on the scaling laws and fundamental device limits, we may predict future MOS devices with approximately 0.25–0.5-μm channel lengths and current densities ten times what they are today. Smaller devices might be built, but at the cost of lowering the voltage or the temperature at which the devices have to be operated. Finally, because of tunneling through the gate, the oxide thickness of the insulator is limited to a value that must be larger than 50 Å.

To give a quantitative comparison of the benefits that can be accrued from improvements in device miniaturization, the power required per bit of information

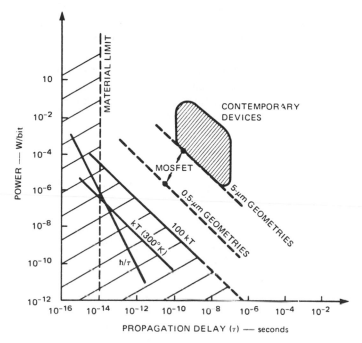

FIGURE 7.2. Power dissipation and transit time of silicon devices. From Ref. 7.

is plotted against the transit time (Fig. 7.2). The performance improvement that can be achieved by reducing the component dimensions from 5 to 0.5 μm is demonstrated through the diagonal lines. The limits imposed by fundamental $(\hbar/\tau)$ thermal and material limitations are also depicted in Fig. 7.2.

7.3. LIMITS FOR PATTERN GENERATION

The minimum size of a microdevice is determined by the material properties, the device limits, and the fabrication technology available at the given time. Lithography has received a substantial amount of attention as a limit to miniaturization.[8,9] In this section we examine some of these limiting factors.

7.3.1. Limits on Linewidths and Edge Sharpness

To transfer information about a circuit pattern from a computer memory or mask to the water, a printing tool is needed. Modern printing tools use particle beams, photons, electrons, and ions for lithography. The momentum of these particles in the beam is related to uncertainties in their location, Δl, through the Heisenberg relation

$$\Delta l \, \Delta p \geq \hbar \tag{7.23}$$

Thus, elementary particles, owing to their wave–particle duality, cannot possess both definite coordinates and definite momentum components simultaneously. Thus, their state cannot be specified by simultaneously giving these observables, as is the case with a classical particle. If in spite of this we want to specify the state of a particle by simultaneously giving these observables, we can do this only with the uncertainties (errors) in the coordinates of the particle and the uncertainties in the momentum components must be taken into account. For photons

$$\Delta l \geq \frac{\hbar c}{E} = \frac{1.23 \times 10^{-6}}{E(\text{in eV})} \text{ (in m)} \tag{7.24}$$

For electrons

$$\Delta l \geq \frac{\hbar}{(2mE)^{1/2}} = \frac{1.22 \times 10^{-9}}{[E(\text{in eV})]^{1/2}} \text{ (in m)} \tag{7.25}$$

and for ions

$$\Delta l \geq \frac{\hbar}{2ME} = \frac{2.738 \times 10^{-11}}{|(M/M_p)E|^{1/2}} \text{ (in m)} \tag{7.26}$$

where c is the velocity of light, E is the beam energy in electron volts, m is the electron mass, M is the ion mass, and M_p is the mass of the proton.

This position uncertainty (or noise) limits the sharpness of a line edge. For photon beams in the visible range, this line edge uncertainty is in the 0.5-μm range; for electron beams ($E = 10^4$ eV) it is 10^{-1} Å. However, it one includes the low-energy electron contribution to the line edge definition, $\Delta l \sim$ 1–2 nm. If one further includes the proximity effect[8] (see Section 5.8.7), the resultant edge uncertainty is on the order of 10 nm for electron beams.

From experiments on the electron microscopy of individual atoms,[9] it is also clear that the edge of a metal conductor on an insulating substrate cannot be defined sharply because of the thermal movement of the atoms at the edge.

The edge of these thin metallic lines can be considered as viscous liquid to a depth on the order of 1.0 nm. These edge atoms are in continuous thermal movement in this transition layer, where they are slowly diffusing away from the edge and thereby decreasing the thickness of the conductor. This diffusion process may be characterized by a "liquid-phase" surface diffusion coefficient similar to that involved in high-temperature liquid-phase epitaxy.

7.3.2. Resist Materials

The basic process of resist exposure is the conversion of one kind of molecule into another (to a lower molecular weight in the case of a positive resist, and to a higher molecular weight for a negative resist) by radiation damage.

Radiation damage is caused by the energy transfer from the energetic particle in the fabricating beam to the resist molecules. In inelastic collisions, the average energy loss ΔE is the range 20–30 eV (see the Bethe formula in Section 2.6.2.5) and is essentially independent of the energy of the incident particle.

If the resist has an activation energy in the range of kT ($\sim$0.025) eV, an inelastic collision can potentially damage up to 1000 molecules. In this case the grain-size radius (the radius of the interaction volume) is ten times the radius of an individual molecule or

$$R_{\text{damage}} = 10 r_{\text{mol}} \tag{7.27}$$

From target theory the number of grains that are damaged (N) depends exponentially on the number of incident particles (n) as

$$N = N_0 \left[1 - \exp\left(- \frac{nz}{\Lambda N_0} \right) \right] \tag{7.28}$$

where N_0 is the number of grains per centimeter before irradiation and z is the thickness of the resist film. Since

$$N_0 = \frac{3z}{4\pi R^3} \tag{7.29}$$

the "$1/e$" dose is given by

$$n = \frac{\Lambda N_0}{z} = \frac{3\Lambda}{4\pi R^3} \tag{7.30}$$

where Λ is the mean free path for inelastic collisions ($\Lambda = 400$ Å for PMMA at 20 kV) so that

$$n = \frac{10^{-6}}{R^3} \, \text{electrons/cm}^2 \tag{7.31}$$

or the dose

$$D = \frac{1.5 \times 10^{-25}}{R^3} \, \text{C/cm}^2 \tag{7.32}$$

From this equation we see that the "speed" of the resist is proportional to the inverse cube of the grain size. The largest tolerable grain size is set by the required edge resolution. To illustrate the role of the maximum grain-size effect for a given exposure, we estimate the electron flux for a 1-cm^2 chip with 0.1-μm pixel elements ($\sim 10^8$) and a 0.1-μm edge resolution. Here the maximum grain size is $\leq 0.1\ \mu$m. Thus,

$$n = \frac{10^{-6}}{10^{-15}} = 10^9 \text{ electrons/cm}^2 \tag{7.33}$$

and the number of electrons per pixel is

$$\frac{n}{\text{number of pixels}} = \frac{10^9}{10^8} = 10 \tag{7.34}$$

We will see in the following section that because of the particle statistics in the beam, to expose 10^8 pixels with a small probability of error (8×10^{-13}) we need at least 200 electrons per pixel, or 2×10^{10} electrons/cm^2.

High-resolution resists (PMMA) limit the edge resolution to the range of the grain size, or $R \sim 2.5$ nm (25 Å).

7.3.3. Exposure Statistics

7.3.3.1. Basic Concepts.   The particles used to expose a resist (and for detection in microscopy) generally form a uniform uncorrelated beam (see Section 6.1). "Uncorrelated" means that one cannot tell from the arrival time and position of one particle, when and where the next particle will arrive. "Uniform" means that if the number of particles arriving over a given period in a given area is averaged over a number of samplings, the average value tends to a constant as the number of samplings is increased.

Let $\bar{m}$ be the average number of particles arriving in an area, a, per second and i be the actual number arriving in an exposure time, t. The probability of a particle arriving in a time interval dt is thus $\bar{m}\, dt$. Let $P_i(t)$ be the probability of i photons arriving in time t, in area a; then

$$P_i(t + dt) = P_i(t)(1 - \bar{m}\, dt) + P_{i-1}(\bar{m}\, dt) \tag{7.35}$$

or

$$P_i(t + dt) - P_i(t) = \bar{m}\, dt[P_{i-1}(t) - P_i(t)] \tag{7.36}$$

so that

$$\frac{dP_i(t)}{dt} = \bar{m}[P_{i-1}(t) - P_i(t)] \tag{7.37}$$

The solution to this differential equation is

$$P_i(t) = \frac{(\bar{i})^i}{i!} \exp(-\bar{i}) \tag{7.38}$$

This is the well-known Poisson distribution[10] for which the standard deviation is

$$\sigma^2 \equiv \bar{i}^2 - (\bar{i})^2 \qquad (7.39)$$

For cases where

$$\frac{(i - \sigma^2)}{\sigma^4} \ll 1 \qquad (7.40)$$

it can be shown by using Stirling's approximation that $P_i(t)$ approaches a value

$$P_i(t) \approx \frac{1}{\sqrt{2\pi}} \exp\left[-\frac{(i - \bar{i})^2}{2\bar{i}} \right] \qquad (7.41)$$

To expose a given area of resist correctly requires i to lie in the range $(\bar{i} - \Delta i_1)$ to $(\bar{i} + \Delta i_2)$ where $i < (\bar{i} - \Delta i_1)$ is insufficient to expose the resist, and $i > (\bar{i} + \Delta i_2)$ overexposes it. The contrast of a resist may be defined as $C = \bar{i}(\Delta i_1 + \Delta i_2)^{-1}$. The probability that the actual value of i lies outside this range can be calculated from Eq. (7.38) or (7.41). Since the exposure per unit even of a given resist is constant, as the exposed area gets smaller, i gets smaller, and the probability of incorrect exposure increases.

When patterning a layer for a device, it is customary to divide the pattern into equal squares, called picture elements or pixels, whose size is determined by the precision with which the pattern is to be cut. Ideally, the average number of correctly exposed pixels (R) per error should be very much greater than the number of pixels in the pattern. As the pixel size gets smaller, the sensitivity of the resist use must be decreased to maintain the number of particles required for a given low error rate.

To expose a resist usually requires a transfer of energy from the impinging particle to a resist molecule of only a few electron volts. The particles being used to expose the resist are classified as

1. Low-energy photons, usually in the ultraviolet range, which can only activate a single resist molecule.
2. High-energy photons, usually in the few kiloelectron volt, X-ray range. These initially produce energetic photoelectrons in the resist, which lose their energies by various pathways eventually generating the low-energy photons required to activate the resist. Clearly many resist molecules can be activated by a single X-ray photon.
3. High-energy ions, usually in the range 10–100 keV.

As the pixel area gets smaller, insensitive resists are necessary to keep i per pixel high enough to keep the probability of incorrect exposure small. Unfortunately, low-energy photons which are the least sensitive cannot be used for pixels of below about 5000-Å linear dimensions, due to diffraction effects. Thus, high-energy photons, electrons, and ions which are intrinsically more efficient in producing chemical reactions must be used. This argues that sputter erosion of a protective layer by focused ion-beam techniques has the best potential for the smallest pixel sizes since the sputter coefficient (number of atoms removed per incident ion) is controlled by the ion energy varying from unity at about 10 keV down to less than 10^{-4} at 100 eV (Section 2.7.2.5). Pixel sizes for electrons and ions are limited by the lateral range of secondary exposure effects of the incident particles, known as proximity effects (see Section 5.8.7). To take into account the probability of exposing pixels adjacent to the one where the particle entered is very complex but has been attempted by using Monte Carlo computer simulations.[11]

   7.3.3.2. Electron-Beam Exposure.   Let D be the dose in C/cm^2 delivered to an electron-beam resist and S the minimum dose (or sensitivity) required to expose the resist for development. Information theory requires that some minimum number of electrons, N_m, must be detected in a pixel[12] in order to be able to assign to it a given probability of being exposed correctly, i.e.,

$$\frac{D}{e} \geq \left(\frac{N_m}{l_p^2}\right) = \frac{S}{e} \text{ for exposure} \tag{7.42}$$

where l_p is the pixel linear dimension and e is the electron charge. Equation (7.42) shows that D must increase as l_p decreases for the probability that each pixel will be correctly exposed to remain constant. Stated another way, based on pixel signal-to-noise considerations, the minimum total number of electrons needed to reliably expose a pattern of a given complexity (i.e., with a given number of pixels) is independent of the size of pixels. Since N_m is a constant depending only on the tolerate error rate assigning a pixel to the exposed state, Eq. (7.42) shows that highly sensitive resists are useful for larger pixels, whereas less-sensitive resists are necessary for smaller pixels.

   To ensure that each pixel is correctly exposed, a minimum number of electrons must strike each pixel. Since electron emission is an uncorrelated process, the actual number of electrons striking each pixel, n, will vary in a random manner about a mean value n. Adapting the signal-to-noise analysis to the case of binary exposure of a resist, it has been shown[13] that the probability of error for large values of the mean number of electrons per pixel, n, is

$$p(\bar{n}) = \frac{e^{-\bar{n}/8}}{[(\pi/2)\bar{n}]^{1/2}} \tag{7.43}$$

This leads to Table 7.3 for the probability of error of exposure.

   To be conservative, for the correct exposure of 10^{10} pixels the average number of electrons exposing a pixel should be at least 200, i.e., $n = 200$. Thus, a pixel of

TABLE 7.3.   Probability of Error in Exposure

$\bar{n}$	Probability of error $p(\bar{n})$	Average error rate $R(\bar{n})$
50	2.2×10^{-4}	1 in 4.5×10^3
100	3×10^{-7}	1 in 3.3×10^6
150	4.7×10^{-10}	1 in 2.1×10^9
200	7.8×10^{-13}	1 in 1.3×10^{12}

dimension l_p with the minimum number of electrons striking it ($N_m = 200$) gives a charge density of

$$D = \frac{N_m e}{l_p^2} \tag{7.44}$$

Using Eq. (2.126) for the electron-beam minimum spot size (with $d_{eff} = l_p$), the minimum charge on a pixel τi can be expressed as

$$\tau i = N_m e = \frac{3\pi^2}{16} \frac{\beta}{C_s^{2/3}} \tau l_p^{-8/3} \tag{7.45}$$

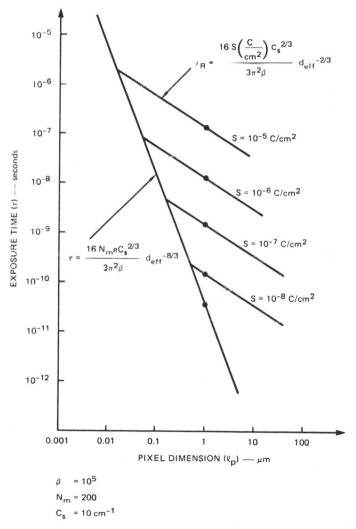

$$\beta = 10^5$$
$$N_m = 200$$
$$C_s = 10 \text{ cm}^{-1}$$

FIGURE 7.3. Exposure time as a function of pixel dimension.

TABLE 7.4. Comparison of Exposure Times at 0.5-μm Resolution for Different Sources and Different Electron-Optical Parameters[a]

Source and conditions	Exposure time τ			
	For 1 pixel	For 10^8 pixels	For 1 cm^2	Typical wafer
Electron beam	9.7 μs	970 s	107 h	—
Vector scan[b]				
Tungsten (hairpin)				
LaB$_6$ (standard)	0.26 μs	26 s	2.9 h	—
Field effect (gun and lens)	0.018 μs	1.8 s	0.2 h	—
Raster scan[c]	0.16 μs	16 s	1.7 h	—
Tungsten (hairpin)				
LaB$_6$ (standard)	0.021 μs	2.1 s	0.23 h	—
Conventional X ray: 100 mA, 10 kV, 30-cm working distance	—	—	—	1000 s
Synchrotron radiation: 100-mA, 500-MeV beam	—	—	—	1 s

[a] Source: Ref. 7.
[b] C_s = 12 cm, C_e = 1 cm, beam voltage = 25 kV.
[c] C_s = 1.8 cm, C_e = 1 cm, beam voltage = 25 kV.

where β is the source brightness, τ is the exposure time, and i is the beam curve. Hence,

$$\tau = \frac{16 N_m e}{3\pi^2} \frac{C_s^{2/3}}{\beta} l_p^{-8/3} \tag{7.46}$$

For a real resist $Q = \tau i = S l_p^2$. Then from Eq. (7.45) we get

$$\tau_R = \frac{16 S C_s^{2/3}}{3\pi^2 \beta} l_p^{-2/3} \tag{7.47}$$

Equation (7.46) expresses the fact that the time τ required to exposure a pixel increases as the pixel linear dimension l_p decreases ($N_m = 200$). This equation is plotted in Fig. 7.3. For a real resist with sensitivity S (C/cm^2), the exposure time is given by Eq. (7.47). A family of curves corresponding to resists with different sensitivities (10^{-5}–10^{-8} C/cm^2) is shown in Fig. 7.3. These curves illustrate that for electron-beam exposure, slow resists are necessary for high resolution.

From Fig. 7.3 the exposure times for a given resolution can be calculated using a different source brightness (β) and electron-optical parameters (C_s) (see Table 7.4).

7.3.3.3. X-ray Exposure.   In X-ray resist exposure, the statistical nature of photon absorption in the resist limits the exposure parameters.[10] Because photons arrive at random, the actual number of photons absorbed in different pixels will deviate from the average value. These deviations from the average number of photons (n)

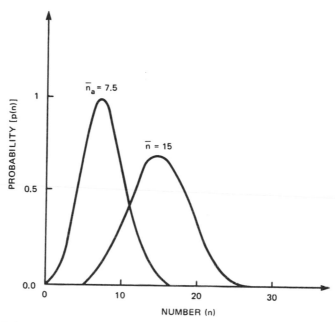

FIGURE 7.4. Poisson distributions for an exposure $\bar{n} = 15$ in the clear and $\bar{n} = 7.5$ in the opaque areas of the mask. From Ref. 10.

are responsible for the limitations in exposure. Since photon emission is uncorrelated, the probability of absorbing n photons in a pixel is given by the Poisson distribution

$$P(n, \bar{n}) = \frac{\bar{n}^n}{n!} \exp - [\bar{n}] \tag{7.48}$$

The X-ray mask consists of areas with two different transmission values. In the clear part (where only the supporting thin film absorbs) the value of the transmission coefficient levels to 1, and in the "opaque" parts we assume that the transmission coefficient is T_a. If n photons are incident on a pixel, n would be counted in a clear part of the mask and $n_a = T_a n$ in an opaque part of the mask.

The probability $p(n)$ of observing n photons in a pixel is shown in Fig. 7.4 for $n = 15$ and $n_a = 7.5$, i.e., $T_a = 0.5$. Since the two Poisson distributions overlap, one cannot tell for certain from the exposed resist whether the photons came through a transparent or an opaque region. However, one can define n_0 in such a way that if $n \leq n_0$, there is no exposure, and if $n \geq n_0$, the ideal resist will be exposed.

In this case we observe an error in assignment for a clear area when $n \leq n_0$ and in an opaque area when $n \geq n_0$. The probability for such errors (or defects) is given by

$$P_{\text{def}} = \sum_0^{n_0} p(n, \bar{n}) + \sum_{n_0 + 1}^{\infty} p(n, \bar{n}_a) \tag{7.49}$$

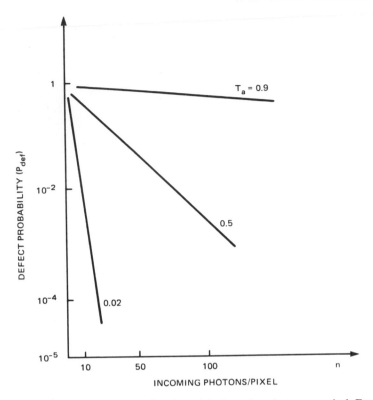

FIGURE 7.5.   Defect probability as a function of the incoming photons per pixel. From Ref. 8.

Figure 7.5 shows the probability for a defect (P_{def}) as a function of the incoming photons per pixel with different T_a values.

It is important to note that the probability for obtaining a defect becomes smaller for a larger separation of the two distributions (Fig. 7.4). This can be achieved by increasing the number of photons impinging on a pixel and/or by increasing the contrast (C) of the mask ($C = 1/T_a$). The limitation imposes strict design and selection rules for the mask materials and their thicknesses when a specific exposure time and resolution are required.

7.3.3.4. Comparison of Lithographies.   Taking into account the width to which the resist has to be developed, and the uncertainty in the development process, a more detailed statistical analysis of the lithographic process has been carried out.[14]

A grating of lines is assumed to be made up of square pixels of side ε. If the number of particles incident on a pixel, N, is above $N_d + \delta N/2$, it will, after appropriate processing, certainly develop, and if below $N_d - \delta N/2$, it certainly will not develop within $N_d \pm \delta N/2$. In exposing a grating there is a minimum pixel dose at the center of each line $\bar{N}_{max}$ and a minimum pixel dose at the center of each space $\bar{N}_{min}$ with $\bar{N}_{max} > \bar{N}_d$ and $\bar{N}_{min} < \bar{N}_d$. Along the edge of the line will be a region where $\bar{N}$ changes (assumed monatomically) from $\bar{N}_{max}$ to $\bar{N}_{min}$. This region contains m

pixels of which some may be uncertain of development. Let

$$\bar{N}_m = \frac{[\bar{N}_{max} + \bar{N}_{min}]}{2} = \text{pixel dose} \approx \bar{N}_d \tag{7.50}$$

$$K = \frac{[\bar{N}_{max} - \bar{N}_{min}]}{[\bar{N}_{max} + \bar{N}_{min}]} = \text{exposed contrast} \tag{7.51}$$

$$\Delta x = \text{line edge uncertainty} \tag{7.52}$$

To reduce the probability that a pixel in a line does not receive sufficient particles for development, it is necessary to constrain $\bar{N}_{max}$ such that n standard deviations below it produces a number of particles per pixel which is still greater than $\bar{N}_m + \delta N/2$, i.e.,

$$\bar{N}_{max} - n\sqrt{\bar{N}_{max}} > \bar{N}_m + \delta N/2 \tag{7.53}$$

leading to

$$n \le (\bar{N}_m/2)^{1/2} \tag{7.54}$$

The value of n is a matter of choice but it has a maximum value of $n = (N_m/2)^{1/2}$ for $\delta N/\bar{N}_m = 0$. Using this nomenclature it is deduced in the above reference that

$$(\Delta x/\varepsilon) = (m/2K)(\delta N/\bar{N}_m) + nm/(K\sqrt{\bar{N}_m}) \tag{7.55}$$

TABLE 7.5. Comparison of the Maximum Pixel Transfer Rate R Possible with Each of the Lithographic Processes[a,b]

Lithography technique	Footnote	N_m	S	R (Hz)
Optical projection	c	1300	0.33 mJ/cm^2	1.2×10^4–1.2×10^{17}
X ray (conventional)	d	24	23 mJ/cm^2	9.9×10^8
X ray (plasma)	e	24	15 mJ/cm^2	1.3×10^9–1.1×10^{12}
X ray (synchrotron)	f	24	15 mJ/cm^2	2.6×10^{12}
Scanning (e-beam)	g	900	$4 \text{ }\mu\text{C/cm}^2$	2.3×10^7
Scanning (ion beam)	h	29	$0.18 \text{ }\mu\text{C/cm}^2$	2.3×10^7
Masked ion beam	i	29	$0.18 \text{ }\mu\text{C/cm}^2$	2.1×10^{11}

[a] Source: Ref. 14.
[b] N_m is the minimum number of photons or charged particles per pixel, assuming infinite resist contrast, $\delta N/N_m = 0$. S is the corresponding resist sensitivity (i.e., minimum required dose). R is the corresponding maximum pixel transfer rate. Linewidth error $\Delta\omega = 2\Delta x$ is 0.1 μm except in the case of e-beam where $\Delta\omega = 0.12$ μm.
[c] Assumes $\lambda = 310$ nm, $f = 0.1$, $\varepsilon = 50$ nm, $n = 3$, $m = 6$, $K = 0.5$, and $P = 1000$ W. (S = fraction of incident photons absorbed in a single layer of pixels.)
[d] Assumes $\lambda = 0.834$ nm, $f = 0.01$, $\varepsilon = 50$ nm, $n = 3$, $m = 1$, $K = 0.82$, $P = 1000$ W, $B = 60$ W/sr, $T = 0.3$, $\gamma = 0.1$.
[e] Assumes $\lambda = 1.24$ nm, $f = 0.01$, $\varepsilon = 50$ nm, $n = 3$, $m = 1$, and $K = 0.82$. The first R value is for 1-J laser pulses at 1 Hz onto a target with 10% conversion to useful X rays emitted into 2π steradians. The second R value corresponds to a pulsed-plasma X-ray source with 600-W output into 4π steradians.
[f] Assumes $\lambda = 1.24$ nm, $f = 0.01$, $\varepsilon = 50$ nm, $n = 3$, $m = 1$, $K = 0.82$, and $P = 1$ W.
[g] Assumes $B = 10^4 \text{ A/cm}^2$ sr, $\alpha = 5 \times 10^{-3}$, $\beta_s = 60$ nm, $K = 0.5$, $\varepsilon = 60$ nm, and $m = 5$.
[h] Assumes $B = 10^4 \text{ A/cm}^2$ sr, $\alpha = 10^{-3}$, $m = 1$, $K = 1$, and $a = 10$ nm.
[i] Assumes $m = 1$, $K = 1$, $a = 10$, $\varepsilon = 50$ nm, 1 μA/cm^2 incident ion current density, and a 1 cm^2 mask area.

Using this equation, Table 7.5 was constructed with the values indicated for the various parameters and qualities, to compare the maximum pixel transfer rate R possible with each of the lithographic processes.

7.3.4. Contamination Limits

Many types of environmental contaminants can prevent the processes used for microfabrication from reaching their potential capabilities, or reduce the manufacturing yields to unacceptable levels. These include

- Impurities and particulates in reagent liquids and gases, and in the ambient air
- Variations in the temperature, humidity, background, radiation, and the presence of electrostatic and magnetic fields in the working area
- Vibrations and noise

Unwanted particulates in the environment and fluids used for processing pose the most serious threat to being able to extend feature sizes into the submicron, nanometer, and atomic scales. As a rule of thumb, yield-limiting defects are assumed to be caused by particles larger than one-tenth that of the feature size, and particles smaller than this are considered harmless. The principle sources of particles are those

- Suspended in the ambient atmosphere in which the fabrication processes are being carried out
- Associated with the personnel carrying out the processes
- Suspended in the liquids used for processing (e.g., washing and etching)
- Suspended in the gases used for processing (e.g., chemical vapor deposition)
- Generated by mechanical equipment in the workplace
- Generated by deposition equipment

Small particles are easily electrically charged by contact with other materials, the attachment of ions generated by electrical equipment, cosmic rays, local ultraviolet or radioactive sources. Because they are charged, such particles may be moved by electrostatic forces as well as bulk movement of the carrier fluids.

The processes that are carried out in the normal atmosphere require specially built clean rooms in which the airborne particulates are removed as far as possible.[15] Clean areas are classified according to the particle count per unit volume of a specified size or greater. Figure 7.6 shows the classification scheme recommended in U.S. Federal Standard 209B. Thus, for example, a class 10,000 clean room cannot contain more than 10,000 particles per cubic foot of size 0.5 μm or greater, and a class 100 cannot contain more than 100 particles in this size range. The specified particle size should, of course, be related to the smallest feature size that is to be patterned in the clean area. The number of particles with a diameter greater than a given value, a, is assumed to have a dependency as shown in Fig. 7.6 so that the number of particles above a specified size can be estimated for any given class of clean room. The air in a clean room is typically filtered through a special type of HEPA filter (High Efficiency Particle Accumulators) and the pressure maintained higher than that outside the room to prevent backstreaming of particles from doorways. Air flow in the clean room is preferred to be from a filter in the ceiling through holes in the floor to

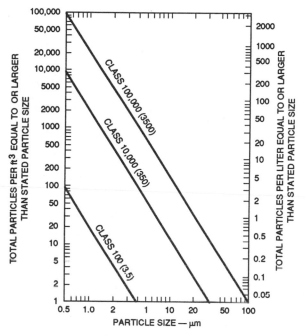

FIGURE 7.6. Particle size distribution curves from Federal Standard 209B.

minimize spreading. Work is often carried out in a chemical hood installed in the clean room in which the clean-room air is filtered for a second time, and flows in a laminar fashion from a filter in the top through holes in the working surface at a positive pressure with respect to the clean room. Thus, the ambient clean-room air cannot enter the working space except through the filter.

Airborne particles are typically monitored by optical methods. The instrument pumps the air containing the particles through a tube which traverses a small sensing zone which is illuminated by an intense light beam (see Fig. 7.7). Light scattered from a particle is detected by a photomultiplier detector, while the main beam is

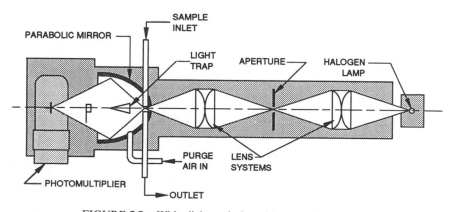

FIGURE 7.7. White light optical particle measuring system.

trapped so no light from it can directly reach the detector. A pulse of light detected by the photomultiplier indicates the presence of a particle, and the amplitude of the pulse indicates its size. A multichannel analyzer sorts and counts the pulses according to their amplitude. Calibration is accomplished by using standard spherical particles of known size. Typically, a flow rate of 1 cubic foot per minute is used. The ultimate test is to measure the particles actually setting on a water during its stay in the clean room. This is accomplished by laser surface scanners which also use light scattering by the particles to detect and size the particles. Commercially available instruments print a map of the position and size of particles over the water area.

Airborne particles can additionally be removed by deliberately charging them using gaseous ion generators, and then collecting them by passing the air through strong electrostatic fields (electrostatic precipitation).[16] Charging of insulating benches or clothing can be neutralized by ion generators provided the sign and quality of charge on these surfaces are monitored using appropriate instruments.

As the feature sizes on microdevices are reduced below, say, 0.1 μm, where particles 10 nm or greater will prevent fabrication with reasonable yields, it seems unlikely that it will be possible to improve atmospheric clean rooms to the required levels. At some point it will be necessary to carry out all of the fabrication steps at very low pressures, where the mean free path of the residual molecules is on the order of the chamber size or more. Here the generation of particles can be reduced to very small values and such as are generated will fall by gravity or if charged can be rapidly removed by electrostatic fields. This will eliminate certain processes which are now commonly used, such as resist deposition by spinning, and will require the development of new low-pressure processes to replace them.

Vibration poses a serious threat to the patterning processes as it can cause the substrate to move with respect to the projected image or writing beam. Noise from matching, air-conditioning ducts, and personnel walking about the clean room can produce difficult-to-isolate low vibrations in the range 3–10 Hz with amplitudes in the range 0.1–1 μm. Ultimately, motion of the sample with respect to the patterning image may have to be controlled by using a fast sensor to detect a change in position which in turn would control an actuator to keep the sample in its original position.

7.3.5. Process Limits

Computer simulation of the fabrication sequences enables manufacturing processing constraints to be identified without actually building a device.[17] As more process steps are required for fabrication of an integrated system, the importance of prior computer simulation to ensure high-yield manufacturing will increase. A device or integrated system consists of a number of patterned conducting and insulating layers with dimensions in the micron and submicron domains. In the computer simulation of planar technology, each process step is carefully devised to produce features with the minimum possible deviation form the ideal device behavior. For example, the Stanford University Press Engineering Models (SUPREM) program is capable of simulating most of the IC fabrication steps,[18] and other institutions have developed corresponding programs.[19]

The fabrication-step simulation is based on process models for oxidation/drive-in, predeposition (gaseous or solid), epitaxial growth, ion implantation, etching,

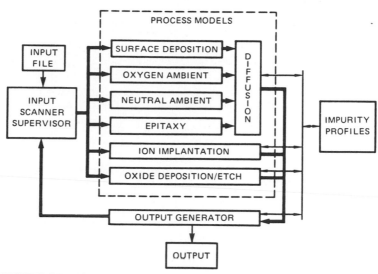

FIGURE 7.8.   Block diagram of the SUPREM process simulator. From Ref. 18.

oxide deposition, and so forth. These various process models are implemented in SUPREM as subprograms, each consisting of a number of subroutines. Figure 7.8 shows the block diagram of the SUPREM process simulator. The input specifications have been designed as process run sheet data and documentation, which includes a series of process steps. The output of the program consists of the one-dimensional profiles of all dopants present in the silicon and Si–SiO$_2$ materials.

We anticipate that the use of computer simulations to reduce process limitations will increase substantially in the future, leading to the adoption of more sophisticated manufacturing process.

7.4. NANOSTRUCTURES

Quantum-mechanical effects governing the transport properties of electrons in small structures (≤ 10 nm) become important for understanding nanodevice (structure) operation. To study these small structures, new tools and lower operating temperatures are required. Many of the new studies are fundamental in nature and carried out in liquid helium temperature ranges ($T \leq 4.2$ K).

The purpose of this section is to review the quantum effects on electron transport in nanostructures and discuss nanofabrication technologies. These experimental nanostructures can be classified as one-, two-, and three-dimensional structures (devices).

In vacuum or a small (<100 nm) metal or semiconductor structure where there is no scattering, the electron wave function is a plane wave such as

$$\psi = |\psi| e^{i\phi} = |\psi| e^{i[\vec{k}\cdot\vec{r} - (Et/\hbar)]} \tag{7.56}$$

TABLE 7.6. Experimental Values for Nanostructures

	Metal	MOSFET	GaAs
Conductor thickness, d	10 to 40 nm	≈ 3 nm	≈ 8 nm
Electron density	Fixed $\approx 10^{22}$ cm^{-3}	Controllable, $n \approx 10^{12}$ cm	$\approx$ fixed; can change with light, $n_s = 5 \times 10^{11}$ cm^{-2}
Fermi energy	Fixed ≈ 10 eV	Controllable, ≈ 10–50 meV	$\approx$ fixed
Examples	Au, Ag, Al	$\langle 100 \rangle$Si with 20-nm oxide	GaAs–Ga$_{0.7}$Al$_{0.3}$As
μ_{max} (cm^2/V-sec)	50	20,000	$\approx 2 \times 10^6$
Elastic mean-free path	≈ 10 nm	≈ 10-100 nm	≈ 10 μm
Smallest structure fabricated and studied	15 nm	≈ 20 nm	≈ 100 nm

At low temperatures, the time between inelastic (energy exchange) scattering events is relatively long ($\Delta t \geq 10^{-8}$ s), so the electron energy (μ frequency) is constant ($\tau_0 \sim 10^{-8}$ s) for relatively long times and distances (~ 1 μm). The result of this long coherence distance is the interference of electron waves in these small structures. The length of the coherence distance is given as $l = v_F \tau$ where $\tau \approx 10^{-14}$ s is the time between elastic scattering (momentum change) and v_F is the Fermi velocity of the electron.

The phase coherence time in two-dimensional structures (in thin films) is given as

$$l_\phi = (D\tau_\phi)^{1/2} \tag{7.57}$$

where l_ϕ is the coherence distance and $D = v_F^{1/2}$, the two-dimensional diffusion constant. The Fermi energy is $E_F = mv_F^2/2 = \hbar^2 k_F^2/2m$. Table 7.6 shows the experimental values used in metal semiconductor and GaAs nanostructures.[20]

The other phase-breaking scattering process in addition to the inelastic e^-–e^- scattering is the electron phonon scattering; therefore, the phase-breaking rate is given as

$$\tau^{-1} = \tau_{e^- - e^-}^{-1} + \tau_{e^- - p}^{-1} \tag{7.58}$$

A good example is the appearance of interference between electron wave functions emerging from a ring structure; the degree of interference being controlled by a magnetic field.

Structures fabricated on a 10-nm scale are much smaller than biological cells and indeed smaller than most viruses. For this reason, these nanostructures might play an important role in molecular biology.

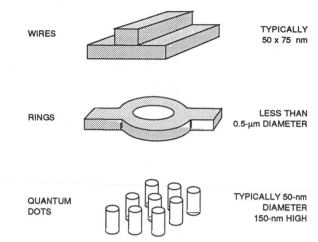

WIRES — TYPICALLY 50 x 75 nm

RINGS — LESS THAN 0.5-μm DIAMETER

QUANTUM DOTS — TYPICALLY 50-nm DIAMETER 150-nm HIGH

FIGURE 7.9.   Nanostructures for transport experiments. From Ref. 21.

Figure 7.9 shows the one- and two-dimensional nanostructures used in experiments for transport studies. These structures are smaller than the phase coherence distance ($l_\phi \approx 1\ \mu$m). The structure dimension has to be smaller than the thermal diffusion distance $I_T = (\hbar D/kT)^{1/2} \approx 0.2\ \mu$m in metal wires and the path length around the ring must be on the order of l_ϕ.

7.4.1. Nanofabrication Tools

Nanofabrication[21,22] continues to evolve from the extensive planar processing techniques used in the manufacture of semiconductor microelectronics and form new energy beam techniques such as the scanning tunneling microscope probe and focused ion beams. Typically, the sought-after attributes in nanofabrication are resolution (with linewidth, $W < 100$ nm), overlay accuracy ($\sim W/5$), large height-to-width aspect ratios (>1), and surface smoothness ($\ll W/10$). To date, these have been provided primarily by scanning electron-beam lithography, evaporation/liftoff and chemical vapor deposition (CVD) methods, and reactive-ion-etching process.

Two of the more important thin-film growth methods, in addition to thermal oxidation and various deposition techniques, are molecular-beam epitaxy (MBE) and metallorganic chemical vapor deposition (MOCVD) for compound semiconductors. These growth techniques are even more effective when combined with various energy beams, e.g., photons to promote atomic layer epitaxy (ALE) or ion beams for *in situ* processing.[23] The spatial resolution obtained in beam processing is set by the size (~ 100 Å for electron beam, ~ 10–50 Å for ion beam). The factors determining resist resolution are the secondary electron path length, and the size of the coiled-up polymer in the PMMA resist. Limits on resist resolution are summarized in Table 7.7, and some of the performance characteristics of the various exposure systems are shown in Table 7.8.

In the following section, the special nanofabrication techniques are discussed for producing experimental structures for quantum effect studies. This review will be

TABLE 7.7.   Resolution and Other Patterning Limits[a]

Mechanism	Resulting limit
Path length of low-energy electron	≈ 100 Å
Polymer structure	Not established; <100 Å
Exposure statistics	<100 Å (PMMA)
Diffraction (X-ray) or scattering in resist (electron, ion exposure)	<100 Å for thin resist layer
Electron backscattering from thick substrate	Loss of contrast
Processing temperature	<150°C, most resists

[a] Source: Ref. 33.

TABLE 7.8.   Pattern Resolution Achieved[a]

Exposure method	Exposure process	Resolution	Field of view
Optical	Microscope projection	2000 Å	200 μm
	Commercial projection ssytem	1 μm	1 cm
Electron beam	SEM, PMMA	250 Å	100 μm
	STEM, PMMA	100 Å	30 μm
	JEOL lithography system (25 kV), PMMA	≤ 200 Å	100 μm
	STEM contamination resist	50 Å	—
X-ray	C_K X-ray source, PMMA	175 Å	1 cm

[a] Source: Ref. 33.

followed by describing the various "hole-drilling" techniques used to make structures for experiments in physics and biology.

7.4.2. Nanofabrication Processes

7.4.2.1. Shadow Deposition.   The ability to fabricate structures in the ~100-Å linewidth regime makes many new devices and applications possible. In the basic technique, a smooth vertical step with precisely determined depth is produced in a substrate by reactive ion etching. This step is shadowed at a shallow angle by evaporation in order to deposit material on the vertical wall. The substrate is then etched at normal incidence with an ion beam to remove the thin layer of material outside the step. This leaves a thin rectangular line of material at the step. By alternately evaporating different materials and by adding lithography and etching steps, more complex structures can be made.

Figure 7.10 shows the fabrication sequence for producing a thin wire (<100 Å) using the shadowing technique.[24] The dimensions of the wire are precisely controlled by the pattern dimensions fabricated on the substrate.

7.4.2.2. Step-Edge Deposition.   Metal lines as narrow as 300 Å and as long as 0.5 mm have been fabricated by step-edge techniques based on substrates with surface-relief steps. Substrate steps with a square profile are formed by ion-beam etching. Metal wires of triangular cross section are produced by ion-etching a metal-coated substrate at an angle, so that the wire is formed in the shadow of the step.

Figure 7.11 shows the fabrication processes to produce fine wires by step-edge techniques.[25] These three-dimensional techniques permit nanostructure fabrication

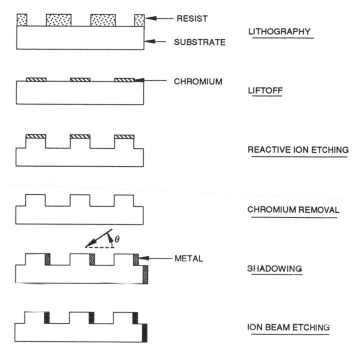

FIGURE 7.10. Fabrication of thin wires by shadowing. From Ref. 25.

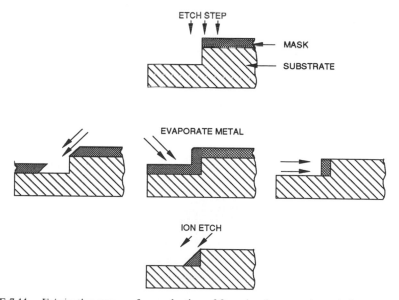

FIGURE 7.11. Fabrication process for production of fine wires by step-edge technique. From Ref. 25.

smaller than 10 nm. However, there are a variety of effects which limit their resolution. The most obvious is penumbra due to limited collimation of the ion beam or evaporant flux. This effect can be eliminated, at the cost of reduced flux, by simply using apertures or moving the source farther from the sample. (Diffraction effects are negligible.) Film residue or scattered atoms from the deposited films can cause serious problems, however since high-contrast etching processes are typically employed.

A more fundamental limit for ion-beam etching occurs because sputtered atoms are ejected from a finite volume of the substrate. This region has been estimated to be as large as 10 nm in diameter. These limits are also not yet well defined for processes using chemically active species such as reactive-ion etching or reactive-ion-beam etching. Additional limitations for etching are imposed by the microstructure of the material.

Deposition processes, such as evaporation, suffer from grain growth and surface migration problems. Adsorbed atoms tend to migrate across the surface before consensing into crystallites, thus placing a resolution limit on any shadowing process. Because of this migration, there are relatively few materials which can be deposited with grain sizes below 10 nm on room-temperature substrates (Au-Pd, Ni-Cr, W-Re).

Finally, the stability of the thin-film structure against thermal motion surface diffusion is important at this size range.

7.4.2.3. Nuclear Fragment Methods.   Ultrahigh-energy (>16 MeV) heavy ions have been used to fabricate ultrafilters with hole sizes ranging from 0.1 μm to 1 μm (see Section 2.7.3).

Fission fragments at lower energies (1–10 MeV are also used in nanofabrication for wires as small as 80 Å. The fabrication process[26] is shown in Fig. 7.12.

First, fission fragments travel through a thin sheet of mica, leaving behind damage tracks. These tracks are then etched with HF acid producing channels which are approximately cylindrical. The diameter of the channels is determined by the

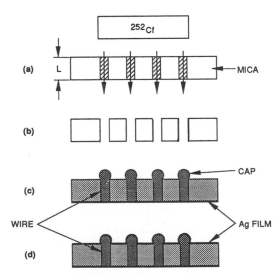

FIGURE 7.12.   Schematic of the fabrication process for 80-Å Au Wires. From Ref. 26.

etching time. After an Ag film is deposited on one side of the mica, Au is electroplated into the channels to form wires. Last, a Ag film is deposited on the other side of the mica making contact to the ends of the wires.

The minimum wire size is limited by the etched nuclear tracks. Smaller sizes are possible using more controlled etching.

7.4.3. Nanodevices

Nanometer-size holes, slots, apertures, and three-dimensional structures (spheres) are the best structures for device physics and process research. These structures are fabricated by electron- and ion-beam combinations using a scanning transmission electron microscope for lithography and ion beams for reactive-ion etching. Figure 7.13 shows the fabrication process for an ~8-nm hole for mask application.[27] Here the thin films (~200 nm) of silicon were fabricated using the technique of boron doping followed by orientation-dependent etching. The films were further thinned down to 50–60 nm using CF_4 or SF_6 reactive-ion etching. Then, the films were coated with 60 nm PMMA, and exposed in a STEM with a 0.5-nm electron beam at 100 keV. Reactive etching is carried out in SF_6.

Another interesting electron-beam technique has been reported[28] on the fabrication of 10-nm silver particles using electron-beam contamination lithography in a STEM. These particles can be made from Ag, AuPd, and used as masks on thin films for X-ray lithography for pattern replication.

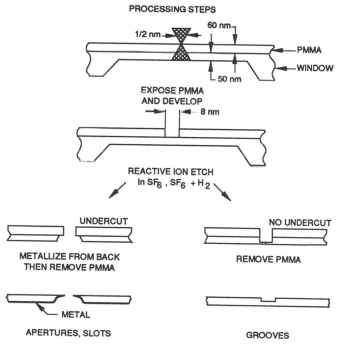

FIGURE 7.13. Processing steps for fabricating thin-film structures and masks in silicon with PMMA resist. From Ref. 27.

Some of these nanostructure filters, membranes are building blocks of living organisms.[29] These bacterial crystalline cell surfaces (S-layers) with round hexagonal and square openings (holes) can be duplicated and may be used for components in molecular electronics.

7.5. DISCUSSION

Figure 7.14[5,30,31] summarizes the improvements attained and anticipated in the most important device and fabrication sizes as a function of time and when we may expect to reach the limiting values. Based on physical limitations, the minimum channel length for MOS transistors is about 0.1 μm. Similar limits can be placed on the minimum feature size in bipolar transistors. Since the fabrication tools are available, it is expected that in the next two decades miniaturization will reach these physical limits.

Beam fabrication techniques (electron-beam, ion-beam, X-ray, dry processes) are on the horizon which would allow devices with characteristic dimensions of 100–250 Å. In this range we will encounter fundamental questions both in the form of new device physics and in the form of limitations to traditional planar approaches.

Because of the high electric fields in ultrasmall devices, the transport mechanisms of electrons and holes can change drastically.[32] The velocity of the electrons becomes very large and the time between collisions very short. In some small devices the electrons can pass through a thin region of the device ballistically without collisions.

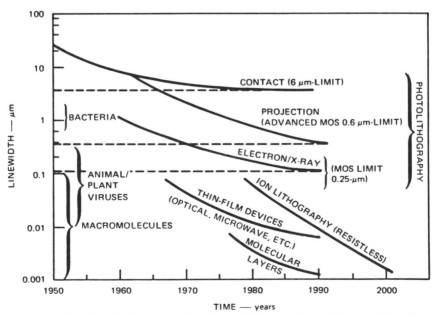

FIGURE 7.14.   Linewidth of microelectronic devices and lithographies as a function of time. Data from Refs. 5, 30, 31.

Also, at such very small dimensions the device properties will depend strongly on the environment, i.e., on other devices nearby, insulating structures, surfaces, and interfaces. The possibility exists that new cooperative phenomena will arise among the no-longer-isolated devices.

Of course, all of these new phenomena could lead to new kinds of devices and structures in the microworld. Clearly, the evolution of fabrication technology will have widespread scientific consequences in device development and opportunities for new physics.

Several materials-related phenomena will limit the advance of microelectronics. The most important effects are the following:

- Electromigration
- Breakdown
- Resistivity
- Power dissipation and heat conduction

In the range 100–250 Å, a broad range of new physical and chemical problems will be encountered. New processing (deposition, etching, doping) has to be developed to fabricate structures and new materials such as intercalated compounds and relief structures.[33]

Surfaces with specific site properties will be an important area for research.[34] Using artificially created nanometer patterns, surface sites for attachment to specific parts of organic molecules could be fabricated. These site-specific surfaces could play an important role in chemical analysis, superconductive studies, and biological research.[20,21]

PROBLEMS

1. Discuss the fundamental limits in scaling MOS devices.

2. What are the scaling limitations of lithography considering ion beam, X-ray, UV light, and electron beam? Discuss how statistics plays a role in defining this limit.

3. What type of contamination limits the fabrication of very small devices?

REFERENCES

1. B. Hoeneisen and C. A. Mead, Fundamental limitations in microelectronics. I. MOS technology, *Solid State Electron.* **15**, No. 7, 819–829 (July, 1972).
2. P. S. Howard and T. Kwok, Electromigration in metals, *Rep. Prog. Phys.* **52**, 301–348 (1989).
3. L. P. Muray, L. C. Rathbun, and E. D. Wolf, New techniques and analysis of accelerated electromagnetic life testing in multilevel metallizations, *Appl. Phys. Lett.* **53**(15), 1414 (1988).
4. L. P. Muray, Electromigration at via contacts in multilevel interconnect systems, Ph.D. dissertation, Cornell University (January, 1990).
5. R. W. Keyes, Physical limits in digital electronics, *Proc. IEEE* **63**, No. 5, 740 (May, 1970).
6. R. M. Hill, Single-carrier transport in thin dielectric films, *Thin Solid Films* **1**, 39 (1967).

7. C. Mead and L. Conway, *Introduction to VLSIC Systems*, Addison Wesley, Reading, Mass. (1979).
8. J. T. Wallmark, A statistical model for determining the minimum size in integrated circuits, *IEEE Trans. Electron Devices* **ED-26**, No. 2, 135 (February, 1979).
9. A. V. Crewe, Some limitations on electron beam lithography, *J. Vac. Sci. Technol.* **16**, No. 2, 255 (March–April, 1979). A. N. Broers, Limits of thin-film microfabrication, *Proc. R. Soc. London Ser. A* **46**, 1 (1988).
10. T. C. Fry, *Probability and its Engineering Use*, 2nd ed., Van Nostrand, Princeton, N.J. (1965).
11. K. Murata and D. F. Kyser, Monte Carlo methods and microlithography simulation for electron and X-ray beams, *Adv. Electron. Electron Phys.* **69**, 176–261 (1987).
12. E. Spiller and R. Feder, in: *X-ray Optics: Applications to Solids* (H. J. Queisser, ed.), Springer-Verlag, Berlin (1975).
13. I. E. Sutherland, C. A. Mead, and T. E. Everhart, Basic limitations in microcircuit fabrication technology, Report No. R-1956-ARPA, RAND Corporation, Santa Monica, Calif. (November, 1976).
14. H. I. Smith, A statistical analysis of UV, x-ray, and charged particle lithographies, *J. Vac. Sci. Technol.* **B4**(1), 148–153 (1986).
15. T. Ohmi, N. Mikoshiba, and K. Tsubouchi, Super clean room system–ultra clean technology for submicron LSI fabrication, *Proceedings of the First International Symposium on Ultra Large Scale Integration (VLSI)* (1987).
16. K. Dillenbeck, Characteristics of air ionization in the clean room, *Microcontamination* (June, 1978).
17. P. D. Scovell, C. N. Duckworth, and P. J. Raser, Modelling of VLSI semiconductor manufacturing processes, *Rep. Prog. Phys.* **52**, 349–388 (1989).
18. D. A. Antoniadis, S. E. Hansen, R. W., Dutton, and G. Gonzalez, SUPREM IA program for IC process modelling and simulation, Technical Report No. 5019-1, Integrated Circuit Laboratory, Stanford University (May, 1977). J D. Plummer, R. W. Dutton, J. F. Gibbons, J. D. Meindl, W. A. Tiller, L. A. Chrestel, C. P. Ho, L. Mei, K. C. Saraswat, B. E. Deal, and T. I. Kamins, Computer Aided Design of Integrated Circuit Fabrication Processes for VLSI Devices, Technical Report, Stanford Electronics Laboratories, Stanford University, California (1980).
19. W. G. Oldham, S. N. Nandgaonkar, A. R. Neureuther, and M. O'Toole, A general simulator for VLSI lithography and etching processes: Part I—Application to projection lithography, *IEEE Trans. Electron Devices* **ED-26**(4), 717–724 (1979). W. G. Oldham, A. R. Neureuther, C. Sung, J. L. Reynolds, and S. N. Nandgaonkar, A general simulator for VLSI lithography and etching processes: Part II—Application to deposition and etching, *IEEE Trans. Electron Devices* **ED-27**(8), 1455–1462 (1980).
20. D. E. Prober, Quantum transport in microstructures, *Microelectron. Eng.* **5**, 203–216 (1986).
21. C. D. Wilkinson, Nanofabrication, *Microelectron. Eng.* **6**, 155–162 (1987).
22. E. D. Wolf, Nanofabrication opportunities for interdisciplinary research, *Microelectron. Eng.* **9**, 5–11 (1989).
23. T. H. P. Chang, D. P. Kern, E. Kratschmer, K. Y. Lee, H. E. Luhn, M. A. McCord, S. A. Rishton, and Y. Vladimirsky, Nanostructure technology, *IBM J. Res. Dev.* **32**, No. 4, 462 (July, 1988).
24. D. C. Flanders and A. E. White, Application of ≈ 100 Å linewidth structures fabricated by shadowing techniques, *J. Vac. Sci. Technol.* **19**(4), 892 (November–December, 1981).
25. M. D. Feuer and D. E. Prober, Step-edge fabrication of ultrasmall Josephson microbridges, *Appl. Phys. Lett.* **36**(3), 226 (February, 1980). D. E. Prober, M. D. Feuer, and N. Giordano, Fabrication of 300-Å metal lines with substrate-step techniques, *Appl. Phys. Lett.* **37**(1), 94 (July, 1980).
26. W. D. Williams and N. Giordano, Fabrication of 80 Å metal wires, *Rev. Sci. Instrum.* **55**(3), 410–412 (March, 1984).
27. I. Adesida, A. Muray, M. Isaacson, and E. D. Wolf, Very high resolution ion beam lithography, *Microcircuit Eng.* **83**, 151–156 (1983).
28. H. G. Craighead and P. M. Mankiewich, Ultra-small metal particle arrays produced by high resolution electron-beam lithography, *J. Appl. Phys.* **53**(11), 7186–7188 (November, 1982).
29. U. B. Sleytr, M. Sara, and D. Pum, Application potentials of two dimensional protein crystals, *Microcircuit Eng.* **9**, 13–20 (1989).
30. G. E. Moore, in: *Tech. Dig. 1975 Int. Electron Devices Meet.*, pp. 11–13, IEEE, New York (1975).
31. R. N. Noyce, Large-scale integration: What is yet to come? *Science* **195**, 1102–1107 (1977).

32. D. K. Ferry, J. R. Barker, and C. Jacoboni (eds.), *Physics of Nonlinear Transport in Semi-conductors,* Ser. B, Vol. 52, Plenum Press, New York (1980).

33. D. E. Prober, in: *Percolation, Localization, and Superconductivity* (A. M. Goldman and S. Wolf, eds.), Plenum Press, Les Arcs, France (1983).

34. C. Harvey, C. Hoch, R. C. Staples, B. Whitehead, J. Comeau, and E. D. Wolf, Signaling for growth orientation and cell differentiation by surface topography in Uromyces, *Science* **235**, 1659–1662 (March, 1987).

Appendixes

A

The Error Function and Some of Its Properties

The error function is expressed as

$$\operatorname{erf} z \equiv \frac{2}{\sqrt{\pi}} \int_0^z e^{-a^2} \, da$$

The definition of the complementary error function is

$$\operatorname{erfc} z \equiv 1 - \operatorname{erf} z$$

Some of their most common properties are the following:

$$\operatorname{erf} 0 = 0$$

$$\operatorname{erf} \infty = 1$$

$$\operatorname{erf} z \approx \frac{2}{\sqrt{\pi}} z \qquad \text{for } z \ll 1$$

$$\operatorname{erfc} z \approx \frac{e^{-z^2}}{\sqrt{\pi}} z \qquad \text{for } z \gg 1$$

$$\frac{d \operatorname{erf} z}{dz} = \frac{2}{\sqrt{\pi z}} e^{-z^2}$$

$$\int_0^z \operatorname{erfc} z \, dz = z \operatorname{erfc} z + \frac{1}{\sqrt{\pi}} (1 - e^{z^2})$$

$$\int_0^\infty \operatorname{erfc} z \, ez = \frac{1}{\sqrt{\pi}}$$

C

Useful Physical Constants in Microscience

Charge on the electron	$e \cong 1.6 \times 10^{-19}$ coulomb
Electron volt	$eV \cong 1.6 \times 10^{19}$ joules
Planck's constant	$h \cong 6.6 \times 10^{-34}$ joule $\cdot$ second/cycle
	$\cong 4.1 \times 10^{-15}$ eV $\cdot$ second/cycle
h-bar	$\hbar = h/2\pi \cong 1.05 \times 10^{-34}$ joule $\cdot$ second/radian
	$\cong 6.6 \times 10^{-16}$ eV $\cdot$ second/radian
1 cycle per second	1 cps $= 2\pi$ radians/second $= 1$ hertz
Flux quantum	$\Phi_0 = h/2q \cong 2.1 \times 10^{-15}$ volt $\cdot$ second [or weber]
Boltzmann's constant	$k \cong 1.4 \times 10^{-23}$ joule/K
	$\cong 8.6 \times 10^{-5}$ eV/K
Permeability of vacuum	$\mu_0 = 4\pi \times 10^{-7}$ henry/meter
Permittivity of vacuum	$\varepsilon_0 \cong 8.85 \times 10^{-12} \cong 10^{-9}/36\pi$ farad/meter
Velocity of light in vacuum	$c = (\varepsilon_0\mu_0)^{-1/2} \cong 3.0 \times 10^8$ meters/second
Wavelength of visible light in vacuum	0.4–0.7 μm (4000–7000 Å)
Avogadro's number	$A_0, N_A = 6.022 \times 10^{23}$ molecules/mole
Thermal voltage	$V_t = kT/e$
At 80.6°F (300 K)	0.025860 volt
At 68°F (293 K)	0.025256 volt
Free electron mass	$m_0 = 9.11 \times 10^{-31}$ kilogram

CONVERSIONS

1 angstrom	$= 10^{-1}$ nm $= 10^{-4}$ μm $= 10^{-8}$ cm $= 10^{-10}$ m
1 mil	$= 10^{-3}$ inch $= 25.4$ μm
1 electron volt	$= 1.602 \times 10^{-19}$ joule
1 joule	$= 10^7$ ergs $= 6.242 \times 10^{18}$ eV
	$= 2.389 \times 10^{-1}$ calorie

1 degree $= 60$ minutes $= 0.01745$ radian

λ (nm) $= 1240/E$ (ev)

Electron rest mass, mc^2 $= 0.511$ MeV

Bohr radius, a_0 $= \hbar^2/me^2 = 0.529$ Å

Bohr velocity, v_0 $= e^2/\hbar = 2.188 \times 10^8$ cm/s

Fine structure constant, α $= e^2/\hbar = 7.297 \times 10^{-13} \cong 1/137$

Hydrogen binding energy, $e^2/2a^0 = 13.606$ eV

Classical electron radius, r_e $= e^2/mc^2 = 2.818 \times 10^{-13}$ cm

Electron Compton wavelength, $\lambda = \hbar/mc = 3.861 \times 10^{-11}$ cm

Photon energy wavelength, hc $= 12{,}398.5$ eV Å

Author Index

The author listing below gives the chapter number in **bold** type followed by the reference in that chapter where the author can be found.

Subject Index